# Lecture Notes in Artificial Intelligence     10984

Subseries of Lecture Notes in Computer Science

More information about this series at http://www.springer.com/series/1244

Zhiyong Chen · Alexandre Mendes
Yamin Yan · Shifeng Chen (Eds.)

# Intelligent Robotics and Applications

11th International Conference, ICIRA 2018
Newcastle, NSW, Australia, August 9–11, 2018
Proceedings, Part I

 Springer

*Editors*
Zhiyong Chen (iD)
University of Newcastle
Callaghan, NSW
Australia

Yamin Yan (iD)
University of Newcastle
Callaghan, NSW
Australia

Alexandre Mendes (iD)
University of Newcastle
Callaghan, NSW
Australia

Shifeng Chen (iD)
Shenzhen Institutes of Advanced
    Technology
Shenzhen
China

ISSN 0302-9743          ISSN 1611-3349  (electronic)
Lecture Notes in Artificial Intelligence
ISBN 978-3-319-97585-6        ISBN 978-3-319-97586-3   (eBook)
https://doi.org/10.1007/978-3-319-97586-3

Library of Congress Control Number: 2018950091

LNCS Sublibrary: SL7 – Artificial Intelligence

This Springer imprint is published by the registered company Springer Nature Switzerland AG
The registered company address is: Gewerbestrasse 11, 6330 Cham, Switzerland

# Preface

The 11th International Conference on Intelligent Robotics and Applications (ICIRA 2018) was hosted by the School of Electrical Engineering and Computing, Faculty of Engineering and Built Environment, The University of Newcastle, Australia, during August 9–11, 2018. The conference aimed at bringing researchers from several countries and organizations together to encourage new collaborations in the areas of robotics, automation, and mechatronics.

In this edition of the event, we had some outstanding contributions in the areas of multi-agent systems and distributed control, mobile robotics and path planning, robotic vision, human–machine interaction, and robot intelligence and learning. In addition, there were many contributions in application areas, including rehabilitation robotics and industrial robots.

The conference attracted a total of 129 submissions by researchers from 13 countries, and 81 manuscripts were accepted for presentation and inclusion in the conference proceedings. All manuscripts accepted were peer-reviewed by at least two reviewers, and represent valuable contributions to the state of the art in robotics research.

We acknowledge the contribution of the Program Committee as well as all the reviewers in supporting this event. We also thank the four plenary speakers who participated in the conference: Emeritus Laureate Professor Graham Goodwin from The University of Newcastle, Australia, Professor Toshio Fukuda from Nagoya University, Japan, Professor Eduardo Nebot from The University of Sydney, Australia, and Professor Richard Middleton from The University of Newcastle, Australia. Special thanks go to Laureate Professor Graham Goodwin for his contributed paper, included in these proceedings. We also thank all authors who chose this event to submit their work and hope it met their expectations. Finally, we thank Springer for the support and for providing access to the conference organization system, which streamlined the entire process.

June 2018

Zhiyong Chen
Alexandre Mendes
Yamin Yan
Shifeng Chen

# Organization

## General Chair

Zhiyong Chen             The University of Newcastle, Australia

## Program Chair

Alexandre Mendes       The University of Newcastle, Australia

## Program Co-chair

Stephan Chalup          The University of Newcastle, Australia

## Publicity Chairs

Haitao Zhang            Huazhong University of Science and Technology, China
Hongyu Zhang           The University of Newcastle, Australia

## Publication Chairs

David Cornforth         The University of Newcastle, Australia
Shifeng Chen            Chinese Academy of Sciences, China
Yamin Yan              The University of Newcastle, Australia

## Award Chair

Minyue Fu               The University of Newcastle, Australia

## Financial Chair

Raymond Chiong        The University of Newcastle, Australia

## Invited Session Chairs

Lijun Zhu                The University of Hong Kong, SAR China
Aurelio Tergolina Salton    PUCRS, Brazil
Yuenkuan Yong          The University of Newcastle, Australia

## Local Arrangements Chairs

| | |
|---|---|
| Mohsen Zamani | The University of Newcastle, Australia |
| Min Xu | University of Technology Sydney, Australia |
| Weihua Li | University of Wollongong, Australia |

## Conference Secretariat

| | |
|---|---|
| Jayne Disney | The University of Newcastle, Australia |

## Advisory Committee Chairs

| | |
|---|---|
| Jorge Angeles | McGill University, Canada |
| Tamio Arai | University of Tokyo, Japan |
| Hegao Cai | Harbin Institute of Technology, China |
| Xiang Chen | Windsor University, Canada |
| Gamini Dissanayake | University of Technology Sydney, Australia |
| Toshio Fukuda | Nagoya University, Japan |
| Huosheng Hu | University of Essex, UK |
| Sabina Jeschke | RWTH Aachen University, Germany |
| Yinan Lai | National Natural Science Foundation of China, China |
| Jangmyung Lee | Pusan National University, Korea |
| Ming Li | National Natural Science Foundation of China, China |
| Peter Luh | University of Connecticut, USA |
| Zhongqin Lin | Shanghai Jiao Tong University, China |
| Brett Ninness | The University of Newcastle, Australia |
| Xinyu Shao | Huazhong University of Science and Technology, China |
| Xiaobo Tan | Michigan State University, USA |
| Dacheng Tao | University of Sydney, Australia |
| Michael Wang | Hong Kong University of Science and Technology, China |
| Yang Wang | Georgia Institute of Technology, USA |
| Youlun Xiong | Huazhong University of Science and Technology, China |
| Huayong Yang | Zhejiang University, China |
| Haibin Yu | Chinese Academy of Science, China |

# Contents – Part I

**Human-Machine Interaction**

**Rehabilitation Robotics**

## Industrial Robot and Robot Manufacturing

## Sensors and Actuators

# Contents – Part II

## Mobile Robotics and Path Planning

## Robotic Vision, Recognition and Reconstruction

## Robot Intelligence and Learning

# A Critique of Observers Used in the Context of Feedback Control

Graham C. Goodwin$^{(\boxtimes)}$ (iD)

University of Newcastle, Callaghan, NSW 2308, Australia
graham.goodwin@newcastle.edu.au

**Abstract.** One of the core tenets of feedback control is that a system's state contains all of the information necessary to predict a system's future response given future inputs. If the state is not directly measured then it can be estimated using a suitably designed observer. This is a powerful idea with widespread consequences. This paper will present a critique of the use of observers in feedback control. Benefits and drawbacks will be highlighted including fundamental design limitations. The analysis will be illustrated by several real world examples including robots executing a repetitive task, relay autotuning in the presence of broadband disturbances, power line signalling in AC microgrid power systems, Type 1 diabetes management and harmonic suppression in power electronics.

**Keywords:** Fundamental limitations · Harmonic suppression
Observers · Periodic disturbances · Type 1 diabetes management

## 1 Introduction

Observers play a central role in many fields [9–11,13,14,87,91]. They allow the current state of a system to be estimated by combining prior information (including the system model) with real-time measurements. The literature on this subject is immense. For example [86] lists over 350 core references with a history dating back to the 1700's. Our goal in the current paper is not to attempt a comprehensive survey. Instead, we focus on inherent limitations associated with observers especially when used in the context of feedback control. In the context of feedback control, the state provides the link between past behaviour and future behaviour.

Many different design tools are available, including but not limited to:

- Lumberger Observer [15,42]
- Kalman Filter [12,16,17]
- $H_\infty$ design [80–84,86]
- Extended Kalman Filter [17,19,20,25,26,28,29,35]
- Unscented Kalman Filter [18]
- Nonlinear observer related to Luenberger observer [8]
- High Gain Observers [21–24,27]

© Springer Nature Switzerland AG 2018
Z. Chen et al. (Eds.): ICIRA 2018, LNAI 10984, pp. 1–24, 2018.
https://doi.org/10.1007/978-3-319-97586-3_1

- Nonlinear Observers based on output injection [6, 7]
- Sliding Mode Observers [1–3]
- Multiple Model Observers [4, 5]
- Moving Horizon Estimators [32, 34]
- Particle Filtering [30, 31, 33, 34, 36–38]
- Dual Control Methods [39]

This plethora of techniques may give the impression that the final word has already been written on this subject. However, certain issues are not yet fully resolved.

The current paper will focus on the benefits, and drawbacks, of observers with special emphasis on their use in feedback control. Observers are a key component of a feedback system designer's tool kit. Both controllable and uncontrollable states are of importance. Specifically, a simplified view of control is that the overall aim is to steer the controllable part of a system so as to cancel (as far as is feasible) two components, namely (i) the future output response arising from unanticipated changes in the initial state of the controlled part of the system and (ii) the future output response arising from the uncontrolled part of the system. In many applications, the latter aspect eclipses the former. Hence some emphasis will be placed in the current paper on estimating uncontrollable modes including sinusoidal disturbances.

Alas, observers of all forms come with a set of fundamental limitations [40] which inhibit their performance. In some scenarios, these limitations can be so severe that high performance control is simply impossible. This suggests that one has no option but to invest time and effort into developing additional physical sensors rather then relying upon a "soft sensor" or an observer.

For pedagogical reasons we will restrict our analysis to simple cases but will point to generalizations and occasional open problems.

## 2   A Class of Linear Observers

As a starting point, consider a linear time-invariant multi-input multi-output system described in transfer function form by

$$y^o = T^o_{yu} \cdot u + T^o_{yw} \cdot w \tag{1}$$

where $y^o \in \mathbb{R}^p$, $u \in \mathbb{R}^p$, $w \in \mathbb{R}^w$ denote output, input and process noise, respectively. Here, and in the sequel, $T^o_{ba}$ denotes the direct linear mapping from signal $a$ to signal $b$, in either continuous or discrete time, with the appropriate operator and dimensions.

Consider an auxiliary variable, $\eta \in \mathbb{R}^n$, given by

$$\eta = T^o_{\eta u} \cdot u + T^o_{\eta w} \cdot w, \tag{2}$$

having the property that $y^o = T^o_{y\eta} \cdot \eta$. The quantity $\eta$ can be a performance variable, the system state, some suitable combination of the states, or some other variable of interest. We have $T^o_{yu} = T^o_{y\eta} \cdot T^o_{\eta u}$ and $T^o_{yw} = T^o_{y\eta} \cdot T^o_{\eta w}$.

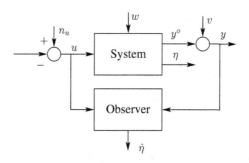

**Fig. 1.** Observer architecture

We restrict attention to a class of linear observers of the following generic form:

$$\hat{\eta} = F_y \cdot y + F_u \cdot u \tag{3}$$

where $\hat{\eta}$ is an estimate of $\eta$, $F_y$, $F_u$ are linear $(n \times p)$ stable transfer functions and where $y$ is the measured output:

$$y = y^o + v \tag{4}$$

for some measurement noise process $v \in \mathbb{R}^p$. The general set-up is shown schematically in Fig. 1.

## 3 Unbiased Linear Observers

A desirable property for an observer is that it should "on average" yield the correct result, i.e., is unbiased [40].

### 3.1 The Standard Linear Model

For the standard linear model, (1), an unbiased observer has the property that the error between $\hat{\eta}$ and $\eta$ does not depend upon the input [41]. Unbiased observers include Luenberger observers and the steady state Kalman filter as special cases. However, unbiased observers are more general since, for example, the degree of the observer may differ form that of the plant. Motivations for this could include providing robustness against modeling errors in certain frequency bands or discriminate against non-white noise. (The latter property is also an integral part of the standard Kalman filter formulation [12,16,17].)

The following result is immediate.

**Lemma 1.** *A necessary and sufficient condition for* (3) *to be unbiased, i.e.* $T^o_{\hat{\eta}u} = T^o_{\eta u}$, *is that the following identity holds:*

$$F_y \cdot T^o_{yu} + F_u = T^o_{\eta u} \tag{5}$$

*Proof* (See [40], Lemma 7.3.1). The transfer function from $u$ to $\hat{\eta}$ is

$$T_{\hat{\eta}u}^o = F_u + F_y[T_{yu}^o] \tag{6}$$

For unbiasedness it is required that

$$T_{\hat{\eta}u}^o = T_{\eta u}^o \tag{7}$$

Equating (6) and (7) yields (5).                     □□□

We define the observer error as $\tilde{\eta} = \hat{\eta} - \eta$. Using (3), (4), and (1) for $\hat{\eta}$, and using (2) and (5) for $\eta$, it follows that

$$\tilde{\eta} = \underbrace{F_y}_{T_{\tilde{\eta}v}} \cdot v + \underbrace{\left(F_y T_{yw}^o - T_{\eta w}^o\right)}_{T_{\tilde{\eta}w}} \cdot w \tag{8}$$

## 3.2    Impact of Errors-in-Variables

Some caution is necessary when demanding that an observer be unbiased. As an illustration, consider the following, seemingly simple, state estimation problem:

$$\dot{x}_t = A[u_t]x_t + Bu_t + \omega_t \tag{9}$$
$$y_t^m = C[u_t]x_t + v_t \tag{10}$$

where $\{\omega_t\}, \{v_t\}$ are noise processes and where $A[u_t], C[u_t]$ denote functions of the input, $u_t$. If the input $\{u_t\}$ were exactly known, then the above problem would fall within the standard time-varying Kalman filter framework [85]. However, say that $\{u_t\}$ is measured in the presence of noise, i.e., the available measured input satisfies

$$u_t^m = u_t + \tilde{u}_t \tag{11}$$

where $\tilde{u}_t$ denotes a measurement error.

In this case, the above state estimation problem is of the Errors-in-Variables class [88,89] and the standard Kalman filter may be biased.

**Example:** To illustrate the above ideas, consider the following simple discrete-time state estimation problem:

$$x_{t+1} = x_t + \omega_t; E\left\{\omega_t^2\right\} = Q \tag{12}$$
$$y_t^m = x_t u_t + v_t; \ E\left\{v_t^2\right\} = R \tag{13}$$
$$u_t^m = u_t + \tilde{u}_t; \ E\left\{\tilde{u}_t^2\right\} = V \tag{14}$$

where $\{\omega_t\}, \{v_t\}, \{\tilde{u}_t\}$ are white noise sequences. If we simply replace $\{u_t\}$ by $\{u_t^m\}$, then the standard time-varying Kalman filter would be

$$\hat{x}_{t+1} = \hat{x}_t + J_t \left\{y_t^m - \hat{x}_t u_t^m\right\} \tag{15}$$

where the gain $J_t$ satisfies:

$$J_t = P_t u_t^m \left( R + P_t \left( u_t^m \right)^2 \right)^{-1} \tag{16}$$

$$P_{t+1} = P_t - P_t \left[ R + P_t \left( u_t^m \right)^2 \right]^{-1} \left( u_t^m \right)^2 P_t + Q \tag{17}$$

Note that this estimate is, in general, biased when $V \neq 0$. Indeed, the error $\tilde{x}_t = \hat{x}_t - x_t$ satisfies

$$\tilde{x}_{t+1} = J_t \left[ u_t + \tilde{u}_t \right] \left\{ u_t \tilde{x}_t - \hat{x}_t \tilde{u}_t \right\} \tag{18}$$

It can be seen that the error $\tilde{u}_t$ appears via a product which is the source of the bias issues. In the absence of additional information, then the best one can say is that the state lies in a range of possibilities [90]. Additional side information can, in some cases, be used to yield a unique estimate [88]. This problem has been extensively studied in the context of parameter estimation but, seemingly, less so in the context of state estimation. One possible remedy is to treat $u_t^m$ and $y_t^m$ both as measurements. If the noise characteristics are known, then it would be possible to utilize a nonlinear maximum likelihood estimator. This will be, at least, asymptotically unbiased.

## 4    Linear Observer Sensitivity Functions

We return to the standard linear model of Sect. 1.

It would be tempting to use $T_{\tilde{\eta}v}$ and $T_{\tilde{\eta}w}$, as in (8), to define observer sensitivity functions. However, the variables $v$ and $\omega$ have different dimensions and units. We thus follow the ideas of [40] and introduce "normalized" observer sensitivity functions.

**Definition 1.** *The observer complementary sensitivity is defined as $M = \left[ T_{\eta y}^o \right]^{-1} \cdot T_{\tilde{\eta}v}$, and the observer sensitivity function as $P = I - M$.*

The observer complementary sensitivity, $M$, captures the relative effect of the measurement noise, $v$, on the estimation error $\tilde{\eta}$. The observer sensitivity function captures the relative effect of process noise $\omega$ on the estimation error. These relationships can be readily seen, at least in the scalar case, by noting that

$$M = T_{\tilde{\eta}v} \left[ T_{\eta y}^o \right]^{-1} \tag{19}$$

and

$$T_{\tilde{\eta}\omega} = F_y T_{y\omega}^o - T_{\eta\omega}^o \tag{20}$$

$$= \left[ F_y T_{y\eta}^o - I \right] T_{\eta\omega}^o \tag{21}$$

$$= \left[ M - I \right] T_{\eta\omega}^o \tag{22}$$

Hence

$$T_{\tilde{\eta}\omega} \left[ T_{\eta\omega}^o \right]^{-1} = -P \tag{23}$$

# 5   Sensitivity Trade-Offs for Observers

## 5.1   Sum of Sensitivity and Complementary Sensitivity

Since $P + M = I$, it is evident that the observer error, $\tilde{\eta}$, cannot simultaneously have low sensitivity to both measurement noise and process noise but, instead, the sum of the (normalized) sensitivities must always be $I$ (1 in the scalar case). Thus observers, no matter how designed, must always make a trade-off between sensitivity to measurement noise and sensitivity to process noise.

In practice, the situation can be worse than might be, at first sight, apparent from the trade-off mentioned above. In particular, when considered as a function of frequency, $P$ and $M$ are complex numbers. Hence it is entirely possible for $P$ and $M$ to add to 1 but have a magnitude, at any given frequency, which is much greater then 1.

This is illustrated in Fig. 2 for the scalar case.

## 5.2   Relative Degree Issues

A particular difficulty arises due to relative degree. Say, for example, that one wishes to design an observer when the input is unknown, is measured in the presence of large noise, or there exists large unmeasured input disturbances. In these cases, the dependence of the observer on the input can be reduced by letting $F_u \rightarrow 0$. However, it then follows from (5) that $F_y \rightarrow [T_{\eta u}^o][T_{yu}^o]^{-1}$.

Hence, if the relative degree, $\rho(T_{\eta u}^o)$, is less than that of $T_{yu}^o$, then the observer will approach an $\bar{r}$ fold differentiator, where $\bar{r} = \rho[T_{yu}^o] - \rho[T_{\eta u}^o]$. In practice, one must use bandlimited differentiations to yield a causal observer. However, the degree of bandlimiting will necessarily impact performance. These ideas are summarized in Table 1:

**Table 1.** Observer extremes

|  |  |
|---|---|
| M = 1: | Full sensitivity to measurement noise |
| P = 0: | No sensitivity to input errors |
|  | Requires output differentiation |
| P = 1: | No sensitivity to measurement noise |
| M = 0: | Full sensitivity to Model and input |
| Intermediate values: | Somewhere between the above extremes |

The fact that observers lie somewhere between a filter that differentiates the output and one that use an open loop model driven by known inputs is well known in practice, see for example [92]. Thus, "optimality" of an observer may not necessarily mean that it performs well but only that it makes the best from a bad situation.

A real world application of the impact of this trade-off will be given in Sect. 8.9.

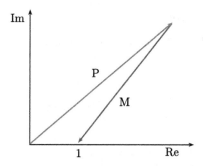

**Fig. 2.** Sum of observer sensitivity - complementary sensitivity

### 5.3   Bode Sensitivity Integrals and Impact of Non-minimum Phase Zeros

The conceptual idea illustrated in Fig. 2 leads to an associated question, namely, when might one expect that the magnitudes $|P|$ and $|M|$ are much greater than 1 yet $P + M = 1$. This is a well studied problem in the context of control [40]. Interestingly, there exist dual results which apply to the sensitivity functions of observers. For example, $P$ and $M$ satisfy Bode type integral constraints analogous to those that hold for the control sensitivity functions – see [40].

For example, when $T^o_{\eta u}$ is minimum phase but $T^o_{yu}$ is non-minimum, then the following integral constraint holds for the complementary observer sensitivity function:

$$\int_0^\infty log\left|\frac{M(j\omega)}{M(0)}\right|\frac{d\omega}{\omega^2} = \frac{\pi}{2}\frac{1}{M(0)}\lim_{s\to 0}\frac{dM(s)}{ds} + \pi\sum_{i=1}^{n_q}\frac{1}{q_i} + \frac{\pi}{2}\tau \qquad (24)$$

where $\{q_i : i = 1, \ldots, n_q\}$ is the set of open right half plane zeros of $T^o_{yu}$ and $\tau$ is a pure delay. (The above result is established in [40] on p.57 save for the delay term which holds analogously to the corresponding control result.)

An immediate implication of (24) is that all observers satisfy a conservation of "sensitivity dirt" principle similar to that which applies in feedback control [43]. Thus reducing sensitivity to either measurement noise or process noise in one frequency band inevitably leads to increased sensitivity in another band.

It is also clear from (24) that "small" non-minimum phase zeros or delays in $T^o_{yu}$ will make the sensitivity trade-off more difficult.

More will be said in Sect. 7 regarding the relationship between observer sensitivity functions and performance of output feedback controllers.

## 6   Estimating Sinusoidals in Noise

An interesting, and widely applicable, use of observers arises in the context of the estimation of sinusoidal components from signals having broad band noise.

Associated design issues are discussed below.

## 6.1  A Suitable Observer

The core idea is to model a sinusoidal signal (of frequency $\omega_r$) by a second order system of the form:

$$\dot{x}_1(t) = x_2(t) + \omega_1(t) \tag{25}$$

$$\dot{x}_2(t) = \omega_r^2 x_1(t) + \omega_2(t) \tag{26}$$

$$y(t) = x_1(t) + v(t) \tag{27}$$

where $\omega_1(t)$, $\omega_2(t)$, $v(t)$ denote noise sources.

A suitable observer then takes the following generic form:

$$\dot{\hat{x}}_1(t) = \hat{x}_2(t) + K_1 \left[ y(t) - \hat{x}_1(t) \right] \tag{28}$$

$$\dot{\hat{x}}_2(t) = -\omega_r^2 \hat{x}_1(t) + K_2 \left[ y(t) - \hat{x}_1(t) \right] \tag{29}$$

$$\hat{y}(t) = \hat{x}_1(t) \tag{30}$$

The gain $[K_1, K_2]$ can be defined by any suitable method. A simple choice is

$$[K_1, K_2] = [2\xi\omega_r, 0]; \ 0 < \xi < 1 \tag{31}$$

The corresponding observer transfer function is easily seen to be

$$F(s) = \frac{2\xi\omega_r s}{s^2 + 2\xi\omega_r s + \omega_r^2} \tag{32}$$

where $0 < \xi \ll 1$.

The frequency response of $F(s)$ is shown in Fig. 3 for different values of the damping ratio, $\xi$.

The observer (28)–(32) recursively generates the Fourier Transform at $\omega_r$.

## 6.2  Sampled Data Implementation

Equation (32) assumes "continuous" implementation of the filter. In practice, one will need to use a sampled data implementation. For small $\xi$ care is needed with numerical precision. In particular, incremental implementation is advisable, see [52–54].

Let the sampling period be $\Delta$ and the sinusoidal frequency of interest be $\omega_r$. Then, a possible state space realisation, in incremental (or $\delta$) form, is ( [52–54])

$$x_{k+1} = x_k + \Delta \left[ \begin{matrix} \frac{C-1}{\Delta} & \frac{S}{\omega_r \Delta} \\ \frac{-\omega_r S}{\Delta} & \frac{C-1}{\Delta} \end{matrix} \right] x_k + \left[ \begin{matrix} \ell_1 \\ \ell_2 \end{matrix} \right] (y_k - \left[ 1 \ 0 \right] x_k),$$

$$y_k^F = \left[ 1 \ 0 \right] x_k, \tag{33}$$

where

$$C \doteq \cos(\omega_r \Delta), \ S \doteq \sin(\omega_r \Delta), \ \ell_1 = 2\xi\omega_r \Delta C, \ \ell_2 = \xi\omega_r^2 \Delta\omega_r (2C^2 - 2 + \xi\omega_r \Delta)/S.$$

Note that the filter (33) has the property that for $y_k = A\cos(\omega_r k\Delta)$, then $y_k^F = A\cos(\omega_r k\Delta)$ as required.

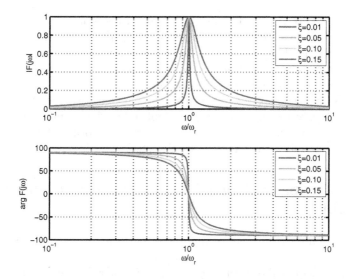

**Fig. 3.** Observer frequency response for estimating a sinusoid.

Also, if multiple sinusoids at different frequencies need to be estimated, then the observer should be implemented as a set of parallel second order sections.

*Remark 1*

1. Equations (33) correspond to a discrete steady-state Kalman Filter for the problem of estimating a sine wave buried in noise, see e.g., Example 10.7.3 in [54]. Two important caveats apply to this observation. Firstly, if one assumes a perfect sine wave in noise then the time-varying Kalman filter gain will converge to zero [88]. This effect can be avoided by assuming the presence of a small amount of process noise in the model for the sine wave. Secondly, if there is high uncertainty associated with the initial state, then the Kalman filter gain will be time varying. This could be achieved by implementing the appropriate Ricatti equation of optimal filtering [85,88].
2. Note that, as $\Delta \to 0$, the filter (33) converges to the continuous time filter (32).

### 6.3 Design Considerations and Performance Limitations

Key properties of the above observer are that it has unity gain at $\omega_r$ and that the frequency response is focused on a narrow frequency band. Several consequences arise from this property:

(i) Transient Time – When $\xi$ is reduced then so is the "bandwidth" of the observer (see Fig. 3). Hence making $\xi$ smaller gives greater frequency selectivity. However, narrow frequency selectivity comes at the cost of a greater settling time. This represents an unavoidable trade-off.

(ii) Sensitivity Dirt Trade-off – Since the observer has a narrow frequency band, then minimal reallocation of "frequency dirt" is needed. A caveat on this observation is that, when one uses a large number of such observers in parallel for multiple sinusoids, then the amount of "dirt" to be shifted can accumulate to the point where it becomes non-negligible.

(iii) Phase Shift – The observer provides significant positive and negative phase shift relatively to small changes in amplitude. The property is evident in Fig. 3. This can be exploited to develop a relay autotuner for systems have large broadband disturbances – see Sect. 8.4.

# 7    State Feedback Control Versus Output Feedback Control

Here we investigate the link between observer performance and output feedback control.

For simplicity we will restrict attention to the linear SISO case. Multivariable extensions are addressed in [93].

## 7.1    State Feedback

We begin by considering the case when the full state is measured.

Consider a linear system having state space description

$$\dot{x} = Ax + Bu \tag{34}$$

$$y = Cx \tag{35}$$

Say we have access to the full state vector, then we can use a feedback control law of the form:

$$u = -\eta + r; \ \eta = Kx \tag{36}$$

Define

$$T^o_{yu} = C(sI - A)^{-1}B = \frac{b(s)}{a(s)} = \frac{b_{n-1}s^{n-1} + \ldots + b_0}{s^n + a_{n-1}s^{n-1} + \cdots + a_0} \tag{37}$$

$$T^o_{\eta u} = K(sI - A)^{-1}B = \frac{h(s)}{a(s)} = \frac{h_{n-1}s^{n-1} + \ldots + h_0}{s^n + a_{n-1}s^{n-1} + \cdots + a_0} \tag{38}$$

The closed loop transfer function resulting from the use of the law (36) is

$$T^o_{yr} = \frac{T^o_{yu}}{[1 + T^o_{\eta u}]^{-1}} = \frac{b(s)}{[a(s) + h(s)]} \tag{39}$$

## 7.2   State Variable Feedback Sensitivity Functions

We define state variable feedback sensitivity function as follows:

**Definition 2.** *The state variable feedback complementary sensitivity function,* $T_{SF}^o$, *and sensitivity function,* $S_{SF}^o$, *are defined as:*

$$T_{SF}^o = (I + T_{\eta u}^o)^{-1} T_{\eta u}^o \tag{40}$$

$$S_{SF}^o = (I + T_{\eta u}^o)^{-1} = I - T_{AF}^o \tag{41}$$

Elementary control design considerations suggest that one should avoid non-minimum phase zeros in $T_{\eta u}^o$ if one wishes to have good closed loop performance. Further justification for this statement is obtained by noting that for high gain feedback, the closed loop poles tend to the open loop zeros of $T_{\eta u}^o$ plus other poles that tend to $-\infty$. Hence one should avoid non-minimum phase zeros in $T_{\eta u}^o$.

## 7.3   Output Feedback Sensitivity Functions

In the case of output feedback, Eq. (37) should be replaced by

$$u = -\hat{\eta} + r \tag{42}$$

where $\hat{\eta}$ is an estimate of $\eta$ provided by an observer as in Eq. (3). This yields $u = -F_y y - F_u u + r$ or equivalently, $u = (1 + F_u)^{-1}(-F_y y + r)$.

Hence, the closed loop output feedback control complementary sensitivity and sensitivity functions are respectively

$$T_{OF} = \frac{T_{yu}^o F_y}{(1 + F_y T_{yu}^o + F_u)}; \quad S_{OF} = 1 - T_{OF} \tag{43}$$

## 7.4   Relating Control and Observer Sensitivity Functions

An insightful connection can now be established as follows:

**Lemma 2** *(see [93]). Consider* $T_{OF}, T_{SF}^o$ *and* $M$ *as in* (43), (40) *and* (19) *respectively. Then,*

$$T_{OF} = [T_{SF}^o][M] \tag{44}$$

*Proof.* From (43)

$$T_{OF} = \frac{T_{yu}^o F_y}{(1 + F_y T_{yu}^o + F_u)} \tag{45}$$

Assuming that an unbiased observer is utilized, then substituting (5) into (45) yields

$$T_{OF} = \frac{T_{yu}^o F_y}{(1 + T_{\eta u}^o)} \tag{46}$$

$$= (T_{\eta u}^o)^{-1} T_{SF}^o T_{yu}^o T_{\eta y}^o M \tag{47}$$

$$= T_{SF} M \tag{48}$$

□□□

As argued above, one would normally choose $\eta$ so that $T_{SF}^o$ is devoid of sensitivity peaks when considering state feedback control. This lead to the inevitable conclusion from (44) that the observer complementary sensitivity function is the prime source of control difficulties in output feedback control.

Thus, any limitations inherent in the observer will manifest themselves in output feedback control performance degradation. In some cases, it may be impossible to achieve suitable observer performance. In such a case, good control performance requires that new physical measurements be developed.

# 8    Selected Applications of Observers

To illustrate the design trade-offs inherent in observers, a summary of a number of applications carried out by the author and his colleagues will be given below:

## 8.1    Estimating Rotor Blade Disturbances in Helicopters

This application is a straightforward application of estimating sinusoidal components in noise. Details are given in [44]. Note that, in this application, time is measured relative to engine speed so as to give "constant frequency" sinusoids when the engine speed changes.

## 8.2    Eccentricity Compensation in Rolling Mills

This application is very similar to that discussed in Sect. 8.1 save that here time is measured relative to the rolling speed. This is again aimed at achieving constant frequency in the face of roll speed changes. Details are given in [45].

## 8.3    Control of a Robot Executing Repetitive Tasks

A repetitive (or periodic) task can (via Fourier decomposition) be thought of as being composed of a sum of harmonically related sinusoids. Hence, one can extract the error associated with executing a repetitive task by using a bank of sinusoidal observers. The output of the observers can then be used to reduce the tracking error to zero. An early publication describing this idea is [46]. Many follow up embellishments of this idea exist in the literature.

## 8.4    Relay Autotuning in the Presence of Large Broadband Disturbances

Relay autotuning was first introduced in 1984 by Åström and Hägglund [55]. Since that time the method has been widely adopted in industry. Many associated commercial products are available. There has been substantial follow up research and embellishments of the original idea. For example, [56] discusses practical features aimed at industrial application of the method.

In its basic form, the relay autotuner is sensitivity to broadband disturbances. One idea for dealing with this issue is discussed in [47] where narrow band sinusoidal observers are added to the basic relay circuit so as to extract the relay induced oscillation in the presence of large broadband disturbances. The basic set-up is shown schematically in Fig. 4 where $F(s)$ is an observer of the type given in Eq. (32), $G_p(s)$ is the plant transfer function and where $u_{ext}, u, d_u, d_m, \hat{y}, y$ denote, respectively, an external input, the plant input, an unmeasured disturbance, a measured disturbance, a set-point and the plant output.

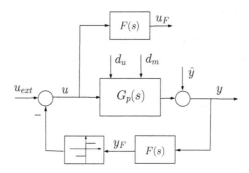

**Fig. 4.** Modified autotuner

The phase shift provided by $F(s)$ can be utilized to cause an oscillation to occur near the center frequency of the filter. The plant frequency response at the oscillation frequency can then be obtained by dividing the narrow-band signal $y_F$ by the narrow band signal $u_F$. A full description is given in [47].

Several examples of the use of this modified relay autotuner are described in [47] including application to insulin sensitivity estimation in a Type 1 diabetes patient.

## 8.5  Harmonic Suppression in Power Electronic Inverters

Inverters play a central role in modern society including integration of wind and solar energy into the grid, high voltage DC power transmission and electronic motor speed control. An inverter uses high frequency switching to produce (nominally) sinusoidal voltage waveforms from a DC source. An inevitable consequence of the use of switching is that harmonics are generated. These can lead to mechanical load resonances and power supply difficulties. Thus there has been on-going research aimed at minimizing harmonics at specific frequencies. A valuable strategy, in this context, is to utilize a set of observers to extract the harmonic content at particular frequencies of interest. A suitable observer is a parallel combination of observers of the form of (33). As remarked in Sect. 6.3, since the observers have narrow bandwidth, their impact on "sensitivity dirt" is relatively small unless many frequencies are considered simultaneously. The estimated harmonic components can then be fed-back (with loop gain 1) so as to suppress these components from the waveform. Various practical embellishments are needed to ensure that this scheme operates satisfactorily in practice including attention to numerical issues and non-linear delay compensation in the feedback path – see discussion in [50].

## 8.6   Power Line Communication

The capability of easily estimating sinusoidal components from a signal leads to an interesting possibility for Power Line Communication, see [51]. The basic idea is that a required message can be coded into a binary representation and then used to turn-on or -off harmonic components via switching electronics (e.g., three frequencies would allow one to code eight alternatives). The presence, or otherwise, of the harmonics can then be detected at some remote point using appropriate sinewave observers. The message can then be decoded. Other harmonics can similarly be used for onward communication. An application of this idea to an AC Microgrid is described in detail in [51].

## 8.7   Control of Highly Resonant Systems

Another application of narrowly focused observers is for the control of highly resonant systems. A sinusoidal observer can be used to focus a controller on a particular frequency band. Two such applications are:

– Micro gyroscope manufacture [48]
– Control of atomic force microscopes [49]

## 8.8   Inverted Pendulum State Estimation

An inverted pendulum is often used to illustrate state feedback and output feedback control, see e.g., [75]. Our interest here focuses on the estimation of the pendulum cart angle when one measures only the pendulum cart position. A source of difficulty in this problem is that the transfer function from input to cart position contains a non-minimum phase zero. Hence, the Bode sensitivity constraint given in (24), implies that observer sensitivity issues are inevitable. Reference [40] on p.194, considers this problem in detail. It is shown that estimating the angle of the pendulum from the cart position is necessarily associated with large sensitivity peaks. For example, consider a pendulum where the ratio of the mass at the end of pendulum to the mass of the cart is 0.1. Sensitivity peaks, at the order of 50:1 are a consequence of the plant non-minimum phase zero. On the other hand, angle information is critical to achieving satisfactory control. Hence, in practice, it is essentially mandatory to provide a direct physical measurement of the angle rather than attempting to estimate it via an observer.

## 8.9   Type 1 Diabetes Management

Type 1 diabetes is an autoimmune disease in which the pancreas is unable to produce the insulin necessary to control Blood Glucose Levels (BGL) [57]. Excessively high BGL has long term health consequences, including blindness and limb amputation. Excessively low BGL has short term health consequences including dizziness, coma and even death in extreme cases. Type 1 diabetes can manifest itself at any age but is particularly devastating for young people. Its causes are unknown. Dramatic improvements in survival have occurred since the discovery of the beneficial impact of injected insulin.

Because of its importance, there has been an enormous amount of work devoted to improving diabetes management. Of relevance to the current paper is work based on using some form of feedback (i.e., a closed loop) to link measured blood glucose levels (obtained from a continuous glucose sensor) to insulin infusion (via a pump) [57–69].

One of the continuing debates in this area is whether or not one can avoid "food announcements" i.e., whether pure feedback control is adequate – see e.g., [76–78]. We argue below that, given the current limitations of insulin delivery and blood glucose measurements, it is necessary to use food announcements to achieve satisfactory blood glucose regulation. Indeed, this is consistent with the recent development of, so called, hybrid closed loop algorithms [70, 74]. Many different models have been proposed for use in this area [71–73]. We will use a simplified model shown in Fig. 5, where $T(s)$ and $F(s)$ denote linear transfer functions, $f$ denotes food consumption rate (g/min), $u$ denotes injected insulin flow (units/min), $b_e$ denotes endogenous glucose production (assumed constant) $y$ denotes BGL (mmol/L). Note that $f$ is often modeled as an impulse to approximately describe a meal consumed over a short period of time.

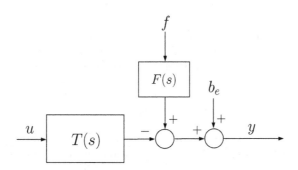

**Fig. 5.** Blood glucose model

Ignoring modelling errors, the one sided nature of the control and other relevant constraints (all of which have the potential to make the following conclusions worse), then all stabilizing feedback/feedforward control laws [75] are as in Fig. 6 where $y^*$ is the BGL set point. In this figure, $T(s)$ and $T^m(s)$ denote the true "plant" and the model respectively. In the sequel, we ignore model errors, i.e., we take $T^m(s) = T(s)$.

From Fig. 6 it is easily seen that the BGL error, $e$, satisfies

$$e = [1 - H(s)T(s)][F(s)f - D(s)T(s)f + b_e - y^*] \qquad (49)$$

We note that having $H(s)T(s) = 1$ at zero frequency removes any steady state error. This automatically yields integral action [75].

Two scenarios are considered below, namely (i) when meal announcements are available and (ii) when only the BGL response is measured. Further details are given in [94–96].

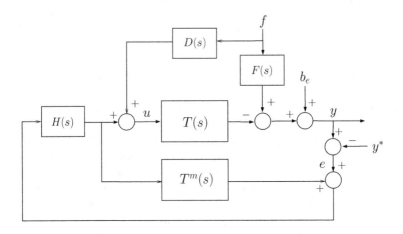

**Fig. 6.** Feedback and feedforward loops

**With Meal Announcement.** When a meal announcements are available, then this is equivalent to being able to measure the variable $f$ in Fig. 6. In this case, the BGL error, $e$, can be made small by choosing the feedforward transfer function, $D(s)$, as

$$D(s) = F(s)T(s)^{-1} \qquad (50)$$

This is relatively easy to achieve since $F(s)$ and $T(s)$ typically have the same relative degree. (In practice, $T(s)$ often contains pure delays of the order of 15 min to 20 min. This can be accounted for provided the meal announcement occurs prior to meal ingestion. Indeed many diabetes educators suggest that insulin be given 15 min prior to consuming a meal thus negating the delay.)

Figure 7 shows real data collected from a patient. The upper trace shows the BGL response (mmol/L) to a high fat/high protein meal (20 g CHo, 40 g fat, 50 g protein) whilst the lower trace shows the BGL response to a pulse of insulin. The smooth curve shows a model fit using finite dimensional transfer functions of the form shown in Fig. 5.

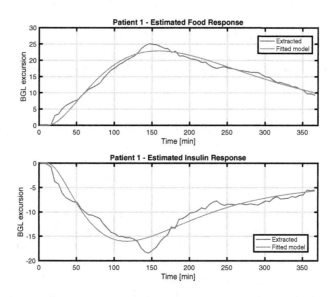

**Fig. 7.** BGL response to a high fat/high protein meal and to a bolus of insulin

The optimal insulin delivery profile was evaluated. The resulting strategy required 1.66 times the amount of insulin normally associated with the patient's Insulin to Carbohydrate Ratio, (ICR). Of this amount, 50% was delivered as a pulse (called a bolus) via the feedforward term i.e., $D(s)$ in Fig. 6. The remaining 50% was delivered over a subsequent period of two hours. The peak BGL (above set-point) predicted by the model was evaluated to be 0.55 mmol/L.

**Without Meal Announcement.** In this case, it is assumed that the food signal, $f$, is not available. We must then rely upon the term $[1 - H(s)T(s)]$ in Eq. (49) to reduce the BGL excursion. Note that $H(s)$, in part, plays the role of an observer to estimate the states of the disturbance model, namely $F(s)$. This allows the controllable part of the system, namely $T(s)$, to be steered so as to cancel the future disturbance response as foreshadowed in Sect. 1 of the paper. It follows from Eq. (49) that, the best one can do is to choose $H(s) = T(s)^{-1}$. However, two issues mitigate against this solution, namely

(i) The delay in $T(s)$ cannot be pre-compensated in a feedback solution since meal anticipation is impossible under these conditions.
(ii) The high relative degree of $F(s)$ means that any attempt to estimate the states of the disturbance model from the measured BGL response implicity involves multiple differentiation – see discussion in Sect. 5.2.

In practice, $H(s)$ can only be chosen as a bandlimited differentiator. The achievable BGL response is then a direct function of this bandwidth. To illustrate we choose

$$T(s)H(s) = \frac{e^{-s\tau}}{(bs+1)^3} \tag{51}$$

where $\tau = 15$ min, $b = 10, 50, 100$ min. The predicted BGL peaks (mmol/L) are as follows:

| | |
|---|---|
| $b = 10$ | Peak BGL deviation $= 10.67$ |
| $b = 50$ | Peak BGL deviation $= 19.73$ |
| $b = 100$ | Peak BGL deviation $= 21.95$ |

Of these, $b = 10$ is impractical since it depends upon an approximate three fold differentiation of the BGL signal. This would result in large magnification of any measurement noise by the observer. This leaves the $b = 50$ or $b = 100$ results. The BGL deviation is then predicted to be 19.73 or 21.95. (Recall that meal announcement gave a predicted peak of 0.55 mmol/L.) These results have been confirmed on the real patient where a BGL rise of 2 mmol/L was obtained with a dual policy including a feedforward bolus delivered 15 min prior to meal.

The predicted responses for the hybrid solution (including feedforward) and for the feedback solution with $b = 10$ (feedback(ideal)) and for $b = 100$ (feedback(realistic)) are shown in Fig. 8.

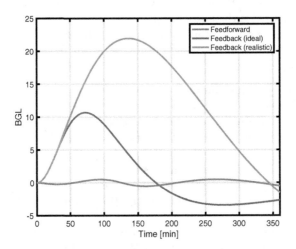

**Fig. 8.** Predicted BGL response with feedback and feedforward control

The results described above lead to the following conclusions, (i) it appears to be impractical to use feedback alone due to limitations arising from the delay and relative degree of the disturbance transfer function, $F(s)$. (ii) it is thus highly desirable to use both a feedforward bolus driven by a measurement of the meal characteristics (denoted $f$ in Figs. 5 and 6) together with some continuous insulin flow (possibly delivered by feedback). The above results support the recent emphasis on hybrid closed loop algorithms for blood glucose regulation [70, 73, 79].

# 9    Conclusions and Recommendation for Future Work

This paper has presented a critique of the use of observers in output feedback control. It has been argued that observers play a pivotal role in control. On the other hand, observers have fundamental limitations which can, in some scenarios, unduly limit the achievable performance. Indeed, in some cases, the limitations can be so severe that acceptable performance can only be achieved if new direct measurements are made. For pedagogical reasons, the discussion in the paper has been limited to simple cases. Many open problems remain, including:

– Investigating performance limitations (as in Sect. 5.3) arising from the use of observers for nonlinear systems.
– Further investigation of Errors-in-Variables problems (as in Sect. 3.2) for observers in state-space models.
– A more thorough understanding of the limitations (as in Sect. 5) associated with the use of observers and the tracking of uncontrolled disturbances in optimization based control laws such as Model Predictive Control.
– Extension of the ideas (as in Sect. 6) to the multivariable case.
– Extension of the analysis of Sect. 8.9 to include the fact the insulin has a one-sided action ($u \geq 0$) and to include nonlinear dynamics. (It is hypothesized that these constraints will actually worsen the trade-offs discussed herein.)

**Acknowledgements.** The author gratefully acknowledges significant input into the development of this paper from Maria Seron. Input into specific sections has been provided by Diego Carrasco, Adrian Medioli, Richard Middleton, Mario Salgado, Bruce King, Carmel Smart, Tenele Smith, Galina Mirzaeva and Christopher Townsend.

# References

1. Utkin, V., Guldner, J., Shi, J.: Sliding Mode Control in Electromechanical Systems. Taylor and Francis Inc., Philadelphia (1999). ISBN 0-7484-0116-4
2. Drakunov, S.V.: An adaptive quasioptimal filter with discontinuous parameters. Autom. Remote Control 44(9), 1167–1175 (1983)
3. Drakunov, S.V.: Sliding-mode observers based on equivalent control method. In: Proceedings of the 31st IEEE Conference on Decision and Control (CDC), pp. 2368–2370 (1992). https://doi.org/10.1109/CDC.1992.371368, ISBN 0-7803-0872-7
4. Narendra, K.S.: A new approach to adaptive control using multiple models. Int. J. Adap. Control Signal Process. 26(8), 778–799 (2012)
5. Bernat, J., Stepien, S.: Multi modelling as new estimation schema for high gain observers. Int. J. Control 88(6), 1209–1222 (2015). https://doi.org/10.1080/179. 2014.1000380

6. Krener, A.J., Isidori, A.: Linearization by output injection and nonlinear observers. Syst. Control Lett. **3**, 47–52 (1983). https://doi.org/10.1016/0167-6911(83)90037-3

7. Hammouri, H., Kinnaert, M.: A new procedure for time-varying linearization up to output injection. Syst. Control Lett. **28**(3), 151–157 (1996). https://doi.org/10.1016/0167-6911(96)00022-9

8. Ciccarella, G., Dalla Mora, M., Germani, A.: A Luenberger-like observer for nonlinear systems. Int. J. Control **57**(3), 537–556 (1993). https://doi.org/10.1080/00207179308934406

9. Friedland, B.: The Control Handbook. CRC Press, IEEE Press (1999). Ch. Observers, pp. 607–618

10. Chen, C.-T.: Linear Systems Theory and Design (Oxford Series in Electrical and Computer Engineering), 3rd edn. Oxford University Press, Oxford (1998)

11. Ellis, G.: Observers in Control Systems: A Practical Guide. Academic Press, Boston (2002)

12. Kalman, R.E.: A new approach to linear filtering and prediction problems. Trans. ASME J. Basic Eng. **82**(Series D), 35–45 (1960)

13. Joesph, I., Profeta, A., Vogt, W.G., Mickle, M.H.: Disturbance estimation and compensation in linear systems. IEEE Trans. Aerosp. Electron. Syst. **26**(2), 225–231 (1990)

14. Wang, W., Gao, Z.: A comparison study of advanced state observer design techniques. In: American Control Conference (2003)

15. Luenberger, D.: Observers for multivariable systems. IEEE Trans. Autom. Control **11**(2), 190–197 (1966)

16. Kalman, R.E., Bucy, R.S.: New results in linear filtering and prediction theory. Trans. ASME J. Basic Eng. **83**, 93–107 (1961)

17. Sorenson, H. (ed.): Kalman Filtering Theory and Applications. IEEE Press, New York (1983)

18. Julier, S.J., Uhlmann, J.K., Durrant-Whyte, H.: A new approach for filtering nonlinear systems. In: American Control Conference, pp. 1628–1632 (1995)

19. Ahrens, J.H., Khalil, H.K.: Closed-loop behaviour of a class of nonlinear systems under EKF-based control. IEEE Trans. Autom. Control **52**(9), 536–540 (2007)

20. Boutayeb, M., Aubry, D.: A strong tracking extended Kalman observer for nonlinear discrete-time systems. IEEE Trans. Autom. Control **44**(8), 1550–1556 (1999)

21. Deza, F., Busvelle, E., Gauthier, J.P., Rakotopara, D.: High gain estimation for nonlinear systems. Syst. Control Lett. **18**(4), 295–299 (1992)

22. Farza, M., M'Saad, M., Triki, M., Maatoug, T.: High gain observer for a class of non-triangular systems. Syst. Control Lett. **60**(1), 27–35 (2011)

23. Freidovich, L.B., Khaili, H.K.: Lyapunov-based switching control of nonlinear systems using high-gain observers. Automatica **43**(1), 150–157 (2007)

24. Khalil, H.K., Praly, L.: High-gain observers in nonlinear feedback control. Int. J. Robust Nonlinear Control (2013). https://doi.org/10.1002/rnc.3051

25. Krener, A.J.: The convergence of the extended Kalman filter. In: Rantzer, A., Byrnes, C.I. (eds.) Directions in Mathematical Systems Theory and Optimization. LNCIS, vol. 286, pp. 173–182. Springer, Heidelberg (2003). https://doi.org/10.1007/3-540-36106-5_12

26. Memon, A.Y., Khalil, H.K.: Full-order high-gain observers for minimum phase nonlinear systems. In: Proceedings of the 48th IEEE Conference on Decision and Control, 2009 Held Jointly With the 2009 28th Chinese Control Conference, (CDC/CCC 2009), pp. 6538–6543. IEEE (2009)

27. Nazrulla, S., Khalil, H.K.: Robust stabilization of non-minimum phase nonlinear systems using extended high-gain observers. IEEE Trans. Autom. Control **56**(4), 802–813 (2011)

28. Reif, K., Sonnemann, F., Unbehauen, R.: An EFK-based nonlinear observer with a prescribed degree of stability. Automatica **34**(9), 1119–1123 (1998)

29. Song, Y., Grizzle, J.W.: The extended Kalman filter as a local asymptotic observer for discrete-time nonlinear systems. J. Math. Syst. Estim. Control **5**(1), 59–78 (1995)

30. Arulampalam, M.S., Maskell, S., Gordon, N., Clapp, T.: A tutorial on particle filters for online nonlinear/non-Gaussian Bayesian tracking. IEEE Trans. Signal Process. **50**(2), 174–188 (2002)

31. Doucet, A., Godsill, S., Andrieu, C.: On sequential Monte Carlo sampling methods for Bayesian filtering. Stat. Comput. **10**, 197–208 (2000)

32. Rao, C.V., Rawlings, J.B., Mayne, D.Q.: Constrained state estimation for nonlinear discrete-time systems: stability and moving horizon approximations. IEEE Trans. Autom. Control **48**(2), 246–258 (2003)

33. Smith, A.F.M., Gelfand, A.E.: Bayesian statistics without tears: a sampling-resampling perspective. Am. Stat. **46**(2), 84–88 (1992)

34. Rawlings, J.B., Bakshi, B.R.: Particle filtering and moving horizon estimation. Comput. Chem. Eng. **30**, 1529–1541 (2006)

35. Daum, F.: Nonlinear filters: beyond the Kalman filter. IEEE Aerosp. Electron. Syst. Mag. **20**(8), 57–69 (2005). Part 2: Tutorials

36. Ho, Y.C., Lee, R.C.K.: A Bayesian approach to problem in stochastic estimation and control. IEEE Trans. Autom. Control **9**(5), 333–339 (1964)

37. Handschin, J.E., Mayne, D.Q.: Monte Carlo techniques to estimate the conditional expectation in multistage nonlinear filtering. Int. J. Control **9**(5), 547–559 (1969)

38. Gordon, N., Salmond, D., Smith, A.: Novel approach to nonlinear/non-Gaussian Bayesian state estimation. IEE Proc. F-Radar Signal Process. **140**(2), 107–113 (1993)

39. Goodwin, G.C., De Dona, J.A., Seron, M.M., Zhuo, X.W.: Lagrangian duality between constrained estimation and control. Automatica **41**, 935–944 (2005)

40. Seron, M.M., Braslavsky, J.H., Goodwin, G.C.: Fundamental Limitations in Filtering and Control. Springer, London (1997). https://doi.org/10.1007/978-1-4471-0965-5

41. Middleton, R.H., Goodwin, G.C.: Digital Control and Estimation: A Unified Approach. Prentice Hall, Englewood Cliffs (1990)

42. Luenberger, D.: Observing the state of a linear system. IEEE Trans. Mil. Electron. **8**(2), 74–80 (1964)

43. Stein, G.: Respect the unstable. IEEE Control Syst. **23**(4), 12–25 (2003)

44. Goodwin, G.C., Evans, R.J., Lozano-Leal, R., Feick, R.: Sinusoidal disturbance rejection with application to helicopter flight data estimation. IEEE Trans. Acoust. Speech Signal Process. **34**(3), 479–484 (1986)

45. Edwards, W.J., Thomas, P., Goodwin, G.C.: Roll eccentricity control for strip rolling mills. IFAC World Congr. **20**(5), 187–198 (1987)

46. Middleton, R.H., Goodwin, G.C., Longman, R.W.: A method for improving the dynamic accuracy of robot performing a repetitious task. Int. J. Robot. Res. **8**(5), 67–74 (1989)

47. Goodwin, G.C., Seron, M.M., Townsend, C.: A modified relay autotuner for systems having large broadband disturbances. Automatica, March 2018. Accepted for publication

48. Lau, K., Goodwin, G.C., M'Closkey, R.T.: Properties of modulated and demodulated, systems with implications in feedback limitations. Automatica **41**, 2123–2129 (2005)
49. Lau, K., Quevedo, D.E., Vautier, B.J.G., Goodwin, G.C., Moheimani, S.O.R.: Design of modulated and demodulated controllers for flexible structures. Control Eng. Pract. **15**(3), 377–388 (2007)
50. Mirzaeva, G., Goodwin, G.C.: Harmonic suppression and delay compensation for inverters via variable horizon nonlinear model predictive control. Int. J. Control **88**(7), 1400–1409 (2015)
51. Townsend, C.D., Mirzaeva, G., Semenov, D., Goodwin, G.C.: Use of harmonic power line communication to enhance a decentralized control method of parallel inverters in an AC microgrid. In: Proceedings of the 3rd Annual Southern Power Electronics Conference (SPEC), pp. 1–6, December 2017
52. Goodwin, G.C., Middleton, R.H., Poor, V.H.: High speed digital signal processing and control. Proc. IEEE **80**(2), 240–259 (1992)
53. Goodwin, G.C., Agüero, J.C., Cea, M.E., Salgado, M.E., Yuz, J.I.: Sampling and sampled-data models: the interface between the continuous world and digital algorithms. IEEE Control Syst. **33**(5), 34–53 (2013)
54. Middleton, R.H., Goodwin, G.C.: Digital Estimation and Control: A Unified Approach. Prentice Hall, Englewood Cliffs (1990)
55. Åström, K.J., Hägglund, T.: Automatic tuning of simple regulators with specifications on phase and amplitude margins. Automatica **20**(5), 645–651 (1984)
56. Berner, J., Hägglund, T., Åström, K.J.: Asymmetric relay autotuning - practical features for industrial use. Control Eng. Pract. **54**, 231–245 (2016)
57. Atkinson, M.A., Eisenbarth, G.S., Michels, A.W.: Type 1 diabetes. Lancet **383**(9911), 69–82 (2014)
58. Chee, F., Fernando, T.: Closed-Loop Control of Blood Glucose, vol. 368. Springer, Heidelberg (2007). https://doi.org/10.1007/978-3-540-74031-5
59. Aronoff, S.L., Berkowitz, K., Shreiner, B., Want, L.: Glucose Metabolism and Regulation: beyond Insulin and glucagon. Diabetes Spectr. **17**, 183–190 (2004)
60. Doyle III, F.J., Huyett, L.M., Lee, J.B., Zisser, H.C., Dassau, E.: Closed-loop artificial pancreas systems: engineering the algorithms. Diabetes Care **37**(5), 1191–1197 (2014)
61. Kovatchev, B., Cobelli, C., Renard, E., Anderson, S., Breton, M., Patek, S., Clarke, W., Bruttomesso, D., Maran, A., Costa, S., et al.: Multinational study of subcutaneous model-predictive closed-loop control in type 1 diabetes mellitus: summary of the results. J. Diabetes Sci. Technol. **4**(6), 1374–1381 (2010)
62. Gondhaledar, R., Dassau, E., Doyle III, F.J.: Periodic zone-MPC with asymmetric costs for outpatient-ready safety of an artificial pancreas to treat type 1 diabetes. Automatica **71**, 237–246 (2016)
63. Kumareswaran, K.: Closed-loop insulin delivery in adults with type 1 diabetes. Ph.D. thesis, University of Cambridge (2012)
64. Bequette, B.: A critical assessment of algorithms and challenges in the development of the closed-loop artificial pancreas. Diabetes Technol. Ther. **7**(1), 28–47 (2005)
65. Klonoff, D.C., Cobelli, C., Kovatchev, B., Zisser, H.C.: Progress in development of an artificial pancreas. J. Diabetes Sci. Technol. **3**, 1002–1004 (2009)
66. Harvey, R.A., Wang, Y., Grosman, B., Percival, M.W., Bevier, W., Finan, D.A., Zisser, H., Seborg, D.S., Jovanovic, L., Doyle III, F.J., Dassau, E.: Quest for the artificial pancreas: combining technology with treatment. IEEE Eng. Med. Biol. Mag. **29**(2), 53–62 (2010)

67. Bequette, B.W.: Challenges and recent progress in the development of a closed-loop artificial pancreas. Annu. Rev. Control **36**(2), 255–266 (2012)
68. Cefalu, W.T., Tamborlane, M.V.: The artificial pancreas: are we there yet? Diabetes Care **37**(5), 1182–1183 (2014)
69. Kovatchev, B., Tamborlane, W.V., Cefalu, W.T., Cobelli, C.: The artificial pancreas in 2016: a digital treatment ecosystem for diabetes. Diabetes Care **39**(7), 1123–1126 (2016)
70. Weinzimer, S.A., Steil, G.M., Swan, K.L., Dziura, J., Kurtz, N., Tamborlane, W.V.: Fully automated closed-loop insulin delivery versus semiautomated hybrid control in paediatric patients with type 1 diabetes using an artificial pancreas. Diabetes Care **31**(5), 934–939 (2008)
71. Bergman, R.N.: Minimal model: perspective from 2005. Horm. Res. **64**(3), 8–15 (2005)
72. Kanderian, S.S., Weinzimer, S., Voskanyan, G., Steil, G.M.: Identification of intraday metabolic profiles during closed-loop glucose control in individuals with type 1 diabetes. J. Diabetes Sci. Technol. **3**, 1047–1057 (2009)
73. Oviedo, S., Vehi, J., Calm, R., Armengol, J.: A review of personalized blood glucose prediction strategies for T1DM patients. Int. J. Numer. Methods Biomed. Eng. **33**(6) (2017)
74. Bondia, J., Dassau, E., Zisser, H., Calm, R., Vehi, J., Jovanovic, L., Doyle III, F.J.: Coordinated basal bolus infusion for tighter postprandial glucose control in insulin pump therapy. J. Diabetes Sci. Technol. **3**(1), 89–97 (2009)
75. Goodwin, G.C., Graebe, S.F., Salgado, M.E.: Control System Design. Prentice Hall, Upper Saddle River (2001)
76. Cameron, F.M., et al.: Closed-loop control without meal announcement in Type 1 diabetes. Diabetes Technol. Ther. **19**(9), 527–532 (2017)
77. Hovorka, R.: The future of continuous glucose monitoring closed loop. Curr. Diabetes Rev. **4**(3), 269–279 (2008)
78. Ramkissoon, C.M., et al.: Unannounced meals in the artificial pancreas: detection using continuous glucose monitoring. Sensors **18**, 884 (2008)
79. Messer, L.H., et al.: Optimizing hybrid closed-loop theory in adolescents and emerging adults using the MiniMed 670G system. Diabetes Care **41**(4), 789–796 (2018)
80. Doyle, J.C., Stein, G.: Multivariable feedback design: concepts for a classical/modern synthesis. IEEE Trans. Autom. Control **26**(1), 4–16 (1981)
81. Doyle, J.C., Glover, K., Khargonekar, P.P., Francis, B.A.: State space solutions to standard $H_2$ and $H_\infty$ control problems. IEEE Trans. Autom. Control **34**(8), 831–847 (1989)
82. Limebeer, D.J., Green, M., Walker, D.: Discrete time $H_\infty$ control. In: 28th CDC, pp. 392–396 (1989)
83. Stoorvogel, A.A., Saberi, A., Chen, B.M.: The discrete time $H_\infty$ control with measurement feedback. Int. J. Robust Nonlinear Control **4**, 457–479 (1994)
84. Zames, G.: Feedback and optimal sensitivity: model reference transformations, multiplicative seminorms and approximate inverses. IEEE Trans. Autom. Control **26**, 301–320 (1981)
85. Anderson, B.D.O., Moore, J.B.: Optimal Filtering. Dover, New York (2005)
86. Simon, D.: Optimal State Estimation Kalman, $H_\infty$ and Nonlinear Approaches. Wiley, Hoboken (2006)
87. Jazwinski, A.H.: Stochastic Processes and Filtering. Dover, New York (2007)
88. Goodwin, G.C., Sin, K.S.: Adaptive Filtering Prediction and Control. Dover, New York (2009)

89. Söderström, T.: Errors-in-Variables Methods in System Identification. Springer, Heidelberg (2018). https://doi.org/10.1007/978-3-319-75001-9

90. Agüero, J.C., Goodwin, G.C.: Identifiability of errors-in-variables dynamic systems. Automatica **44**, 371–382 (2008)

91. Stengel, R.F.: Optimal Control and Estimation. Dover, New York (1994)

92. Radke, A., Gao, Z.: A survey of state and disturbance observers for practitioners. In: Annual Control Conference, pp. 5183–5188 (2006)

93. Carrasco, D.S., Goodwin, G.C.: Connecting filtering and control sensitivity functions. Automatica **50**(12), 3319–3322 (2014)

94. Goodwin, G.C., Seron, M.M.: A gold standard for optimal insulin infusion for Type 1 diabetes ingesting a meal with slow postprandial response (2018). Submitted for publication

95. Goodwin, G.C., Medioli, A.M., Carrasco, D.S., King, B.R., Fu, Y.: A fundamental control limitation for linear positive systems with application to Type 1 diabetes treatment. Automatica **55**, 73–77 (2015)

96. Goodwin, G.C., Carrasco, D.S., Seron, M.M., Medioli, A.M.: A performance limit for a class of positive nonlinear systems. Automatica **95**, 14–22 (2018)

# Multi-agent Systems and Distributed Control

# Semi-global Leaderless Consensus of Circular Motion with Input Saturation

Bowen Xu[1,2], Haofei Meng[1,2], Duxin Chen[1,2], and Hai-Tao Zhang[1,2(✉)]

[1] Guangdong HUST Industrial Technology Research Institute,
Dongguan 523000, China
[2] Key Laboratory of Image Processing and Intelligent Control,
School of Automation, State Key Laboratory of Digital Manufacturing
Equipments and Technology, Huazhong University of Science and Technology,
Wuhan 430074, China
{xbwen94,hfmeng,chenduxin,zht}@hust.edu.cn

**Abstract.** In this paper, a control algorithm is proposed for a group of agents moving in a circular pattern. Each individual in the system is subjected to the input saturation. Then based on the low gain feedback technique, the leaderless semi-global consensus can be achieved if the dynamics of each agent is asymptotically null controllable with bounded controls, and the topology satisfies the connectivity conditions. The effectiveness of proposed control algorithm is verified through both theoretical analysis and numerical simulation results.

**Keywords:** Semi-global · Circular motion · Leaderless
Low gain feedback

## 1 Introduction

As a special issue of complex network, these years have witnessed a rapid development in the research on multi-agent systems (MASs), which has received increasing interest in a wide range of disciplines including biology, computer science, artificial intelligence, control theory, etc. Especially, the research achievements of control of MASs have been widely utilized in different fields, such as society, industry and national defence [1–4]. The aim is to achieve a control objective in the network environment based on distributed sensing, interacting communication, and intelligent computing.

In the research on MASs, consensus is a hot topic which aims at making all the agents reach a common value or synchronize with a common trajectory, only relying on the local information of neighbors. In 2004, Olfati-Saber *et al.* proposed the consensus protocol for first-order MASs [5]. Considering the

Supported by the Guangdong Innovative and Entrepreneurial Research Team Program under Grant No. 2014ZT05G304, and the National Natural Science Foundation of China under Grant No. U1713203 and No. 61673189.

ⓒ Springer Nature Switzerland AG 2018
Z. Chen et al. (Eds.): ICIRA 2018, LNAI 10984, pp. 27–38, 2018.
https://doi.org/10.1007/978-3-319-97586-3_2

time-varying directed interaction topologies, Ren and Beard [6] proved that consensus can be achieved asymptotically if the interaction graphs have a spanning tree. In [7,8], the consensus problem of multi-agent system has been extended for second-order MASs. For linear systems, Li *et al.* [9] proposed an observer-type control strategy under a time-invariant communication topology based on relative output measurements. Then, in [10], Ajorlou *et al.* studied the consensus problem of a class of continuous-time nonlinear systems. In 2013, Li *et al.* [11] proposed fully distributed adaptive protocol, which can adjust the coupling weights between neighboring agents without any global information for both linear and nonlinear systems. The MPC method for achieving stable consensus with directed switching networks and input constraints is studied in [12]. The research achievements in consensus problem have laid a solid foundation for coordination control of MASs.

In physics and practical engineering systems, it is ubiquitous that all the agents are always subjected to input saturation. Besides, it is impractical to implement unlimited control force in MASs, and the input saturation may cause the system to be uncontrollable and unstable. Hence, the research on input saturation of MASs is theoretically and practically meaningful. Following the research line, based on the static linear feedback and dynamic output feedback, Lin *et al.* solved the semi-global stability problem of linear systems [13,14]. Furthermore, Lin investigated global stability of MASs with external disturbance and uncertainty in [15]. Then Meng [16] studied leader-following consensus and global stability of second-order MASs. By using the low gain feedback method, Su *et al.* studied semi-global containment control of MASs with input saturation [17] and leader-following consensus of saturated networks [18]. In [19], Zhang *et al.* achieved leader-following consensus of second-order MASs, where the dynamics of each individual is essentially nonlinear.

In this paper, continuing the research work in [18], we will address the leaderless consensus problem of MASs performing circular motion with each agent being subjected to input saturation. In the leaderless situation, the individuals only has the local interaction with their neighbours, none of them have knowledge of the group reference. Therefore the method in [18] for leader-following case is ineffective here by defining a error system between the followers and the leader. At first, we present a general linear expression generating circular motion. Then based on the non-singular transformation, a new error system is obtained such that the consensus problem is transformed into the stabilization problem of the error system. By utilizing the low gain feedback technique, we design a control algorithm to achieve semi-global leaderless consensus under a switching topology where each topology is connected. Then, due to limitations on communication or sensing, we extend our control algorithm to a more general connectivity condition, i.e. joint connectivity. Therein, each agent with general linear dynamics moves in a circular pattern in a two-dimensional plane. Especially, there is no specified leader or global reference in the MASs.

The rest of the paper is organized as follows. Section 2 gives some basic preliminaries and the formulation of the semi-global leaderless consensus problem.

Section 3 shows the control strategy and theoretical analysis. The numerical simulation results are shown in Sect. 4 to verify the effectiveness of control method. Finally, the conclusion is drawn in Sect. 5.

## 2  Preliminaries and Problem Formulation

Generally, an undirected graph can be represented by $G = \{V, E\}$, where $V = \{v_i | i \in \{1, 2, ..., N\}\}$ is the set of $N$ vertices ($N$ agents), $E = \{(v_i, v_j) | (i, j) \in \{1, 2, ..., N\}\}$ is the set of edges that means there exists a path between $v_i$ and $v_j$, i.e., $v_j$ and $v_i$ can obtain the information from each other. Denote the adjacency matrix of graph $G$ as $A = a_{ij}$, where $a_{ij} = 1$ if $(v_i, v_j) \in E$, and $a_{ij} = 0$ if $(v_i, v_j) \notin E$. The Laplacian of graph $G$ is given by $L$, where $L_{ii} = \sum_{j=1, j \neq i}^{N} a_{ij}, L_{ij} = -a_{ij}$, and the eigenvalues of $L$ is $\{\lambda_1, \lambda_2, ..., \lambda_N\}$ with $0 = \lambda_1 \leq \lambda_2 \leq \cdots \leq \lambda_N$.

**Lemma 1.** [5] *The Laplacian of $G$ has a unique eigenvalue 0 with its corresponding eigenvector* $\mathbf{1} = [1\ 1\ \cdots\ 1]^T$, *i.e.,* $L\mathbf{1} = 0$

Now we consider a group of $N$ agents (labeled as $\{1, 2, ..., N\}$) with linear dynamics subjected to the input saturation. The dynamics of each agent can be described as:

$$\dot{x}_i = Ax_i + B\sigma(u_i), \tag{1}$$

where $A = \begin{bmatrix} 0 & -\omega \\ \omega & 0 \end{bmatrix}$, $x_i \in \mathbf{R}^n$ is the state of agent $i$, $u_i \in \mathbf{R}^m$ is the control input, $\sigma(u_i) = [sat(u_{i1})\ sat(u_{i2})\ \cdots\ sat(u_{im})]^T$, $sat(u_{ij}) = sign(u_{ij}) \min\{|u_{ij}|, \Theta\}$, with $\Theta > 0$ and $sign(\cdot)$ being the symbolic function. Without input, the systems becomes

$$\dot{p} = Ap,$$

we can obtain a single agent moving in a circular pattern, its position can be represented as:

$$\begin{cases} a(t) = r_0 \cos \omega t \\ b(t) = r_0 \sin \omega t, \end{cases}$$

where $r_0$ is the radius, $\omega$ is the angle velocity and $p = [a(t)\ b(t)]^T$.

The problem of semi-global leaderless consensus of circular motion with input saturation is given as follows: For any a prior given bounded set $\Omega \in \mathbf{R}^n$ where $x_0 \in \Omega$, the objective is to design a control law $u_i$ such that

$$\lim_{t \to \infty} \|x_i(t) - x_j(t)\| = 0, i, j = 1, 2, ..., N.$$

**Assumption 1.** *The pair $(A, B)$ is asymptotically null controllable with bounded controls(ANCBC), i.e., $(A, B)$ is stabilizable and all the eigenvalues of $A$ in the closed left-half s-plane.*

## 3   Main Results

### 3.1   Leaderless Consensus Under a Connected Switching Topology

In the problem of leaderless consensus of circular motion with input saturation, we consider that the network topology is connected between every two contiguous switching signals. Designing the control input $u_i$ as $u_i = K \sum_{j=i}^{N} a_{ij}(x_i - x_j)$, the system can be rewritten in the compact form:

$$\dot{x} = (I_N \otimes A + L(t) \otimes BK)x$$

with $x = [x_1^T, x_2^T, ..., x_N^T]^T$, and $L$ is the Laplacian. Then we make a nonsingular transformation on $x$, i.e., let $x = T_0\bar{x}$, $T_0 = T \otimes I_n$ with $T$ representing an orthogonal matrix, of which the elements in the first column are identical, $n$ is the dimension of each agent. Thus, it follows that:

$$T_0\dot{\bar{x}} = (I_N \otimes A)T_0\bar{x} + (L(t) \otimes BK)T_0\bar{x},$$

$$\begin{aligned}
\dot{\bar{x}} &= T_0{}^T(I_N \otimes A)T_0\bar{x} + T_0{}^T(L(t) \otimes BK)T_0\bar{x} \\
&= (T^T \otimes I_n)(I_N \otimes A)(T \otimes I_n)\bar{x} + (T^T \otimes I_n)(L(t) \otimes BK)(T \otimes I_n)\bar{x} \\
&= (T^TT \otimes A)\bar{x} + (T^TLT \otimes BK)\bar{x}.
\end{aligned}$$

It is easy to see that $T^TL(t)T$ equals to $\begin{bmatrix} 0 & 0 \\ 0 & H(t) \end{bmatrix}$, where all the eigenvalues of $H(t)$ have a positive real part., let $\bar{x} = \begin{bmatrix} \bar{x}_1 \\ \bar{x}_2 \end{bmatrix}$, thus

$$\begin{cases} \dot{\bar{x}}_1 = A\bar{x}_1, \\ \dot{\bar{x}}_2 = (I_{N-1} \otimes A + H(t) \otimes BK)\bar{x}_2. \end{cases} \tag{2}$$

It is easy to see that if the state of $\bar{x}_2$ is asymptotically stable, then

$$\lim_{t\to\infty} x = \lim_{t\to\infty} T_0\bar{x} = \lim_{t\to\infty} (T_{01}\bar{x}_1 + T_{02}\bar{x}_2) = 1\bar{x}_1.$$

So the formation pattern of the system depends on the state $\bar{x}_1$. Thus the system with dynamics (1) can generate circular formation.

**Assumption 2.** *The switching graph associated with system (1) is connected at any time instant.*

**Lemma 2.** *Under Assumption 1, for each $\varepsilon \in (0, 1]$, there exists a matrix $P(\varepsilon) > 0$ solving the algebraic Riccati equation (ARE)*

$$A^TP(\varepsilon) + P(\varepsilon)A - P(\varepsilon)BB^TP(\varepsilon) + \varepsilon I = 0,$$

*with $\lim_{t\to\infty} P(\varepsilon) = 0$.*

*Remark 1.* Assumption 1 is a necessary and sufficient condition for semi-global stability of a linear system with input saturation, which ensures that the state feedback is available and $P(\varepsilon)$ exists.

Based on the low gain feedback method, let $P(\varepsilon) > 0$ be the solution of ARE as follows:

$$A^{\mathrm{T}}P(\varepsilon) + P(\varepsilon)A - \beta P(\varepsilon)BB^{\mathrm{T}}P(\varepsilon) = -\varepsilon I, \varepsilon \in (0, 1], \tag{3}$$

where $\beta \le \min\{\lambda(H^{\mathrm{T}} + H)\}$. Thus in the design of controller, we choose $K = -B^{\mathrm{T}}P(\varepsilon)$ and the control input is:

$$u_i = -B^{\mathrm{T}}P(\varepsilon)\sum_{j=1}^{N} a_{ij}(t)(x_i - x_j), \ i = 1, 2, ..., N. \tag{4}$$

**Theorem 1.** *Consider a multi-agent system consisted of $N$ agents with dynamics (1). Suppose Assumptions 1 and 2 hold, there exists a constant $\varepsilon^* > 0$ such that for any $\varepsilon \in (0, \varepsilon^*]$, the control input given by (4) can guarantee the system to achieve semi-global leaderless consensus:*

$$\lim_{t \to \infty} \|x_i(t) - x_j(t)\| = 0, \ i, j = 1, 2, ..., N,$$

*provided $x_i(0) \in \Omega$ for all $i = 1, 2, ....N$.*

*Proof.* Consider the Lyapunov function

$$V(\bar{x}_2) = \bar{x}_2^{\mathrm{T}}P\bar{x}_2, \tag{5}$$

with $P = I_{N-1} \otimes P(\varepsilon)$. Let $c > 0$ be a constant that

$$c \ge \sup_{\varepsilon \in (0,1], \bar{x}_2 \in \tilde{\Omega}} \bar{x}_2^{T}(0)P\bar{x}_2(0),$$

where $\tilde{\Omega} := \{T_2 x(0) | x(0) \in \bar{\Omega}\}$ with $T = \begin{bmatrix} T_1 \\ T_2 \end{bmatrix}$, $\bar{\Omega} := \{x(0) | x_i(0) \in \Omega\}$, because of the Lemma 2, we know that $\lim_{\varepsilon \to 0} P(\varepsilon) = 0$. Then we define a level set $L_V(c) := \{\bar{x}_2 \in \mathbf{R}^{(N-1)n} : V(\bar{x}_2) \le c\}$, and let $\varepsilon^* \in (0, 1]$ such that for any $\varepsilon \in (0, \varepsilon^*]$, $\bar{x}_2 \in \mathbf{R}^{(N-1)n}$, and $\lim_{\varepsilon \to 0} P(\varepsilon) = 0$, we can obtain that

$$\|u_i\| = \left\| B^{T}P(\varepsilon)\sum_{j=1}^{N} a_{ij}(t)(x_i - x_j) \right\| \le \Theta, \ i = 1, 2, ..., N. \tag{6}$$

Then we calculate the derivative of $V$ along the trajectories of the system as

$$\dot{V}(\bar{x}_2) = \dot{\bar{x}}_2^{\mathrm{T}}P\bar{x}_2 + \bar{x}_2^{\mathrm{T}}P\dot{\bar{x}}_2 = \bar{x}_2^{\mathrm{T}}\bar{A}^{\mathrm{T}}P\bar{x}_2 + \bar{x}_2^{\mathrm{T}}P\bar{A}\bar{x}_2$$
$$= \bar{x}_2^{\mathrm{T}}(\bar{A}^{\mathrm{T}}P + P\bar{A})\bar{x}_2, \tag{7}$$

with $\bar{A} = (I_{N-1} \otimes A + H(t) \otimes BK)$ and $K = -B^{\mathsf{T}}P(\varepsilon)$. Therein:

$$
\begin{aligned}
\bar{A}^{\mathsf{T}}P + P\bar{A} =& (I_{N-1} \otimes A^{\mathsf{T}} - H(t)^{\mathsf{T}} \otimes P(\varepsilon)BB^{\mathsf{T}})P + P(I_{N-1} \otimes A - H(t) \otimes BB^{\mathsf{T}}P(\varepsilon)) \\
=& (I_{N-1} \otimes A^{\mathsf{T}} - H(t)^{\mathsf{T}} \otimes P(\varepsilon)BB^{\mathsf{T}})(I_{N-1} \otimes P(\varepsilon)) \\
& + (I_{N-1} \otimes P(\varepsilon))(I_{N-1} \otimes A - H(t) \otimes BB^{\mathsf{T}}P(\varepsilon)) \\
=& (I_{N-1} \otimes (A^{\mathsf{T}}P(\varepsilon) + P(\varepsilon))) - (H(t)^{\mathsf{T}} + H(t)) \otimes P(\varepsilon)BB^{\mathsf{T}}P(\varepsilon)). \quad (8)
\end{aligned}
$$

Due to the symmetry of $H(t)^{\mathsf{T}} + H(t)$ and all the eigenvalues of $H(t)$ are greater than 0 at any time instant $t$, there always exists some orthogonal matrix $T_H(t)$ that

$$
\begin{aligned}
& H(t)^{\mathsf{T}} + H(t) = \\
& T_H^{\mathsf{T}}(t)diag\{\lambda_1(H(t)^{\mathsf{T}} + H(t)) \; \lambda_2(H(t)^{\mathsf{T}} + H(t)) \; \cdots \; \lambda_{N-1}(H(t)^{\mathsf{T}} + H(t))\}T_H(t),
\end{aligned}
$$

then we let $\tilde{x} = (T_H(t) \otimes I_n)\bar{x}$, it follows that

$$
\begin{aligned}
\dot{V}(\bar{x}_2) =& \bar{x}_2^{\mathsf{T}}(I_{N-1} \otimes (A^{\mathsf{T}}P(\varepsilon) + P(\varepsilon)A) - (H(t)^{\mathsf{T}} + H(t)) \otimes P(\varepsilon)BB^{\mathsf{T}}P(\varepsilon))\bar{x}_2 \\
=& \bar{x}_2^{\mathsf{T}}((T_H^{\mathsf{T}}(t)T_H(t)) \otimes (A^{\mathsf{T}}P(\varepsilon) + P(\varepsilon)A)\bar{x}_2 - \bar{x}_2^{\mathsf{T}}(T_H^{\mathsf{T}}(t)diag\{\lambda_1(H(t)^{\mathsf{T}} + H(t)) \\
& \lambda_2(H(t)^{\mathsf{T}} + H(t)) \; \cdots \; \lambda_{N-1}(H(t)^{\mathsf{T}} + H(t))\}T_H(t)) \otimes P(\varepsilon)BB^{\mathsf{T}}P(\varepsilon)))\bar{x}_2 \\
=& \bar{x}_2^{\mathsf{T}}(T_H^{\mathsf{T}}(t) \otimes I_n)(I_{N-1} \otimes (A^{\mathsf{T}}P(\varepsilon) + P(\varepsilon)A)(T_H(t) \otimes I_n)\bar{x}_2 \\
& - \bar{x}_2^{\mathsf{T}}(T_H^{\mathsf{T}}(t) \otimes I_n)(diag\{\lambda_1(H(t)^{\mathsf{T}} + H(t)) \; \lambda_2(H(t)^{\mathsf{T}} + H(t)) \\
& \cdots \; \lambda_{N-1}(H(t)^{\mathsf{T}} + H(t))\} \otimes P(\varepsilon)BB^{\mathsf{T}}P(\varepsilon))(T_H(t) \otimes I_n)\bar{x}_2 \\
\leq& \sum_{i=1}^{N} \tilde{x}_{2i}^{\mathsf{T}}(A^{\mathsf{T}}P(\varepsilon) + P(\varepsilon)A - \beta P(\varepsilon)BB^{\mathsf{T}}P(\varepsilon))\tilde{x}_{2i} \\
=& -\varepsilon \sum_{i=1}^{N} \bar{x}_{2i}^{\mathsf{T}}(T_H^{\mathsf{T}}(t) \otimes I_n)(T_H(t) \otimes I_n)\bar{x}_{2i} < 0, \forall \bar{x}_2 \in L_V(c)\backslash\{0\}, \quad (9)
\end{aligned}
$$

with $\beta \leq \min\{\lambda(H(t)^{\mathsf{T}} + H(t))\}$. (9) implies the deviation between agents starting from the level set $L_V(c)$ will converge to zero asymptotically as time goes infinite, i.e.,

$$
\lim_{t \to \infty} \|x_i(t) - x_j(t)\| = 0, i = 1, 2, ..., N.
$$

So far the proof is completed.

## 3.2  Leaderless Consensus on Jointly Connected Switching Topology

In many practical scenarios, because of the communication limitation or the sensing distance, the agents may be unable to communicate with other agents continuously, i.e., the graph $G$ associated with the system will not be always connected. Thus, it would be meaningful to investigate the consensus in jointly connected condition.

**Assumption 3.** *There exists a time sequence in the finite time interval* $[t_i, t_{i+1})$, $t_{i0}, t_{i1}, ..., t_{ik}$ *with* $t_i = t_{i0}, t_{i+1} = t_{ik}$. *The length of each sub-interval*

$[t_{ij}, t_{i(j+1)})$ is greater than or equal to $\kappa$, the system topology remains fixed in each sub-interval. If the union of all proximity graph at all switching points $\bar{G} := (\bigcup_{\tau \in [t_i, t_{i+1})} G(\tau))$ is connected, then we call the system is jointly connected over $[t_i, t_{i+1})$.

**Assumption 4.** The system matrix $A$ is marginally stable, and there exists a $P(\varepsilon)$ in ARE that satisfies $A^T P(\varepsilon) + P(\varepsilon)A \leq 0$ with $P(\varepsilon) > 0$.

**Theorem 2.** Consider a multi-agent system that consisted of $N$ agents with dynamics (1). Suppose Assumptions 1, 3, 4 and Lemma 2 hold, there exists some $\varepsilon^* > 0$ such that for any $\varepsilon \in (0, \varepsilon^*]$, the control input given by (4) can make the system (2) to achieve semi-global leaderless consensus:

$$\lim_{t \to \infty} \|x_i(t) - x_j(t)\| = 0, i, j = 1, 2, ..., N.$$

*Proof.* Consider the Lyapunov function

$$V(\bar{x}_2) = \bar{x}_2^T P \bar{x}_2,$$

with $P = I_{N-1} \otimes P(\varepsilon)$. From what we deducted before, so $V(\bar{x}_2(t_i)) \geq V(\bar{x}_2(t_{i+1}))$. Using Cauchy limitation existence criteria, denote

$$\alpha > V(\bar{x}_2(t_i)) - V(\bar{x}_2(t_{i+1})) = -\int_{t_i}^{t_{i+1}} \dot{V}(\bar{x}_2(s))ds \ (\alpha > 0),$$

thus we can obtain

$$\alpha > -\int_{t_i}^{t_{i+1}} \dot{V}(\bar{x}_2(s))ds$$

$$= -\int_{t_{i0}}^{t_{i1}} \bar{x}_2^T(s)(I_{N-1} \otimes (A^T P(\varepsilon) + P(\varepsilon)A))\bar{x}_2(s)ds$$

$$+ \int_{t_{i0}}^{t_{i1}} \bar{x}_2^T(s)((H_{t_{i0}}^T + H_{t_{i0}}) \otimes P(\varepsilon)BB^T P(\varepsilon))\bar{x}_2(s)ds$$

$$- \cdots - \int_{t_{i(k-1)}}^{t_{ik}} \bar{x}_2^T(s)(I_{N-1} \otimes (A^T P(\varepsilon) + P(\varepsilon)A))\bar{x}_2(s)ds$$

$$+ \int_{t_{i(k-1)}}^{t_{ik}} \bar{x}_2^T(s)((H_{t_{i(k-1)}}^T + H_{t_{i(k-1)}}) \otimes P(\varepsilon)BB^T P(\varepsilon))\bar{x}_2(s)ds$$

$$\geq -\int_{t_{i0}}^{t_{i0}+\kappa} \bar{x}_2^T(s)(I_{N-1} \otimes (A^T P(\varepsilon) + P(\varepsilon)A))\bar{x}_2(s)ds$$

$$+ \int_{t_{i0}}^{t_{i0}+\kappa} \bar{x}_2^T(s)((H_{t_{i0}}^T + H_{t_{i0}}) \otimes P(\varepsilon)BB^T P(\varepsilon))\bar{x}_2(s)ds$$

$$- \cdots - \int_{t_{i(k-1)}}^{t_{i(k-1)}+\kappa} \bar{x}_2^T(s)(I_{N-1} \otimes (A^T P(\varepsilon) + P(\varepsilon)A))\bar{x}_2(s)ds$$

$$+ \int_{t_{i(k-1)}}^{t_{i(k-1)}+\kappa} \bar{x}_2^T(s)((H_{t_{i(k-1)}}^T + H_{t_{i(k-1)}}) \otimes P(\varepsilon)BB^T P(\varepsilon))\bar{x}_2(s)ds.$$

For any time points $j$ between $t_i$ and $t_{i+1}$,

$$\alpha > -\int_{t_{ij}}^{t_{ij}+\kappa} \bar{x}_2^{\mathrm{T}}(s)(I_{N-1} \otimes (A^{\mathrm{T}}P(\varepsilon) + P(\varepsilon)A))\bar{x}_2(s)ds$$

$$+ \int_{t_{ij}}^{t_{ij}+\kappa} \bar{x}_2^{\mathrm{T}}(s)((H_{t_{ij}}^{\mathrm{T}} + H_{t_{ij}}) \otimes P(\varepsilon)BB^{\mathrm{T}}P(\varepsilon))\bar{x}_2(s)ds. \tag{10}$$

Then

$$k \lim_{t\to\infty} \int_t^{t+\kappa} \bar{x}_2^{\mathrm{T}}(s)(I_{N-1} \otimes (A^{\mathrm{T}}P(\varepsilon) + P(\varepsilon)A))\bar{x}_2(s)ds$$

$$- \lim_{t\to\infty} \int_t^{t+\kappa} \bar{x}_2^{\mathrm{T}}(s)((H_{t_{i0}}^{\mathrm{T}} + H_{t_{i0}}) \otimes P(\varepsilon)BB^{\mathrm{T}}P(\varepsilon))\bar{x}_2(s)ds$$

$$- \cdots - \lim_{t\to\infty} \int_t^{t+\kappa} \bar{x}_2^{\mathrm{T}}(s)((H_{t_{i(k-1)}}^{\mathrm{T}} + H_{t_{i(k-1)}}) \otimes P(\varepsilon)BB^{\mathrm{T}}P(\varepsilon))\bar{x}_2(s)ds$$

$$=0.$$

Denote

$$(H_{t_{i0}}^{\mathrm{T}} + H_{t_{i0}}) + (H_{t_{i1}}^{\mathrm{T}} + H_{t_{i1}}) + \cdots + (H_{t_{i(k+1)}}^{\mathrm{T}} + H_{t_{i(k+1)}}) = \Xi,$$

due to $\Xi$ is symmetric, so it can be rewritten in the following form:

$$\Xi = T_\Xi^{\mathrm{T}} diag\{\lambda_1(\Xi)\ \lambda_2(\Xi)\ \cdots\ \lambda_{N-1}(\Xi)\}T_\Xi,$$

with $T_\Xi(t)$ is a orthogonal matrix, thus

$$k \lim_{t\to\infty} \int_t^{t+\kappa} \bar{x}_2^{\mathrm{T}}(s)(I_{N-1} \otimes (A^{\mathrm{T}}P(\varepsilon) + P(\varepsilon)A))\bar{x}_2(s)ds$$

$$- \lim_{t\to\infty} \int_t^{t+\kappa} \bar{x}_2^{\mathrm{T}}(s)((H_{t_{i0}}^{\mathrm{T}} + H_{t_{i0}}) \otimes P(\varepsilon)BB^{\mathrm{T}}P(\varepsilon))\bar{x}_2(s)ds$$

$$- \cdots - \lim_{t\to\infty} \int_t^{t+\kappa} \bar{x}_2^{\mathrm{T}}(s)((H_{t_{i(k-1)}}^{\mathrm{T}} + H_{t_{i(k-1)}}) \otimes P(\varepsilon)BB^{\mathrm{T}}P(\varepsilon))\bar{x}_2(s)ds$$

$$=k \lim_{t\to\infty} \int_t^{t+\kappa} \bar{x}_2^{\mathrm{T}}(s)(I_{N-1} \otimes (A^{\mathrm{T}}P(\varepsilon) + P(\varepsilon)A))\bar{x}_2(s)ds$$

$$- \lim_{t\to\infty} \int_t^{t+\kappa} \bar{x}_2^{\mathrm{T}}(s)(T_\Xi^{\mathrm{T}}(t)diag\{\lambda_1(\Xi)\ \lambda_2(\Xi)\ \cdots\ \lambda_{N-1}(\Xi)\}T_\Xi(t))$$

$$\otimes P(\varepsilon)BB^{\mathrm{T}}P(\varepsilon))\bar{x}_2(s)ds, \tag{11}$$

same as before, we let $\tilde{x}_2 = (T_\Xi(t) \otimes I_n)\bar{x}_2$,

$$0 = (11) \leq k \lim_{t\to\infty} \int_t^{t+\kappa} \sum_{i=1}^N \tilde{x}_{2i}^{\mathrm{T}}(s)(A^{\mathrm{T}}P(\varepsilon) + P(\varepsilon)A - \beta'(P(\varepsilon)BB^{\mathrm{T}}P(\varepsilon)))\tilde{x}_{2i}(s)ds$$

$$= -k\varepsilon \lim_{t\to\infty} \int_t^{t+\kappa} \sum_{i=1}^N \bar{x}_{2i}(T_\Xi^{\mathrm{T}}(t) \otimes I_n)(T_\Xi(t) \otimes I_n)\bar{x}_{2i}ds$$

$$\leq 0,$$

where $\beta' \leq \min\{\lambda(\Xi)\}$. Thus the deviation of system converges to zero as time goes infinity, so far the proof is completed.

*Remark 2.* For directed networks, if all the eigenvalues of $(H(t)^{\mathrm{T}} + H(t))$ associated with the switching topology are greater than zero, the control method in this paper can be extended to the directed topology case directly.

## 4  Simulation Results

The numerical simulation is performed with four agents. Thy system matrices is:

$$A = \begin{bmatrix} 0 & -1 \\ 1 & 0 \end{bmatrix}, B = \begin{bmatrix} 0 \\ 1 \end{bmatrix}.$$

It is easy to find that the pair $(A, B)$ is asymptotically null controllable with bounded controls. The initial states of each agent is chosen as $\begin{bmatrix} 5\cos\theta_i \\ 5\sin\theta_i \end{bmatrix}$ which $\theta_i$ is picked from $[0, 2\pi]$ arbitrarily. Using the mathematical function '*care*' in Matlab R2014b, the matrix $P(\varepsilon)$ in the algebraic Riccati equation is calculated as

$$P(\varepsilon) = \begin{bmatrix} 0.3230 & -0.0247 \\ -0.0247 & 0.3153 \end{bmatrix},$$

with $\varepsilon = 0.05$.

  Figure 1(a) and (b) show the four undirected connected graphs and the jointly-connected situation, respectively. The switching pattern is shown in Fig. 2.

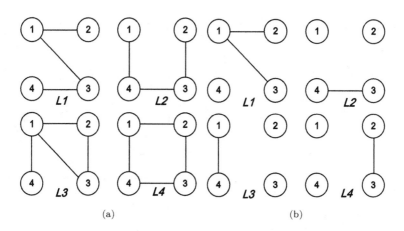

(a)                                (b)

**Fig. 1.** The switching undirected graph for connected condition and jointly-connected condition.

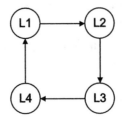

**Fig. 2.** The switching pattern of network topology.

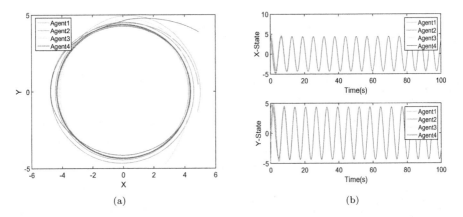

**Fig. 3.** (a) The trajectory evolution curves denoted by different colors under connected topology; (b) The state convergence of each agent under connected topology.

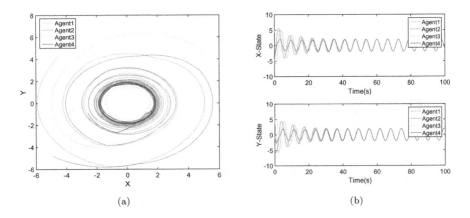

**Fig. 4.** (a) The trajectory evolution curves denoted by different colors under jointly connected topology; (b) The state convergence of each agent under jointly connected topology.

We can obtain $min\{\lambda(H(t)^{\mathrm{T}} + H(t))\} = 1.1716$, so we choose $\beta = 1 \leq min\{\lambda(H(t)^{\mathrm{T}} + H(t))\}$. Figure 3(a) presents the evolution of each agent's trajectory under switching connected topology. We can see that each agent moves in a circular pattern finally, i.e., the semi-global leaderless consensus is achieved. Figure 3(b) presents the state convergence of each agent.

When the network topology becomes jointly connected rather than connected at each time point, we can obtain $min\{\lambda(\varXi)\} = 4$, so we choose $\beta = 2.25 \leq min\{\lambda(\varXi)\}$. Figure 4 exhibits the evolution of trajectory and state convergence under jointly connected topology. From the above simulation results, it is obvious that our control algorithm can achieve semi-global leaderless consensus.

## 5  Conclusion

In this paper, we investigate the semi-global leaderless consensus of MASs with input saturation under different topology connectivity condition, With the assumption that the dynamics of each agent is asymptotically null controllable with bounded controls, a control algorithm is proposed based on the low gain feedback method, which does not need a specified leader and only need local information of neighbors. If the initial states start from a prior given bounded set, the semi-global leaderless consensus can be achieved. At last, the effectiveness of control algorithm is verified through numerical simulation.

## References

1. Zhang, H., Feng, T., Yang, G.H., Liang, H.: Distributed cooperative optimal control for multiagent systems on directed graphs: an inverse optimal approach. IEEE Trans. Cybern. **45**(7), 1315–1326 (2015)
2. Seyboth, G.S., Ren, W., Allögwer, F.: Cooperative control of linear multi-agent systems via distributed output regulation and transient synchronization. Automatica **68**, 132–139 (2016)
3. Guo, M., Xu, D., Liu, L.: Cooperative output regulation of heterogeneous nonlinear multi-agent systems with unknown control directions. IEEE Trans. Autom. Control **62**(6), 3039–3045 (2017)
4. Zhang, H., Jiang, H., Luo, Y., Xiao, G.: Data-driven optimal consensus control for discrete-time multi-agent systems with unknown dynamics using reinforcement learning method. IEEE Trans. Ind. Electron. **64**(5), 4091–4100 (2017)
5. Olfati-Saber, R., Murray, R.M.: Consensus problems in networks of agents with switching topology and time-delays. IEEE Trans. Autom. Control **49**(9), 1520–1533 (2004)
6. Ren, W., Beard, R.W.: Consensus seeking in multiagent systems under dynamically changing interaction topologies. IEEE Trans. Autom. Control **50**(5), 655–661 (2005)
7. Hong, Y., Hu, J., Gao, L.: Tracking control for multi-agent consensus with an active leader and variable topology. Automatica **42**(7), 1177–1182 (2006)
8. Ren, W.: On consensus algorithms for double-integrator dynamics. IEEE Trans. Autom. Control **53**(6), 1503–1509 (2008)

9. Li, Z., Duan, Z., Cheng, G., Huang, L.: Consensus of multiagent systems and synchronization of complex networks: a unified viewpoint. IEEE Trans. Circuits Syst. I-Regul. Pap. **57**(1), 213–224 (2010)
10. Ajorlou, A., Momeni, A., Aghdam, A.G.: Sufficient conditions for the convergence of a class of nonlinear distributed consensus algorithms. Automatica **47**(3), 625–629 (2011)
11. Li, Z., Ren, W., Liu, X., Fum, M.: Consensus of multi-agent systems with general linear and Lipschitz nonlinear dynamics using distributed adaptive protocols. IEEE Trans. Autom. Control **58**(7), 1786–1791 (2013)
12. Cheng, Z., Zhang, H.T., Fan, M.C., Cheng, G.: Distributed consensus of multi-agent systems with input constraints: a model predictive control approach. IEEE Trans. Circuits Syst. I Regul. Pap. **62**(3), 825–834 (2015)
13. Lin, Z., Saberi, A.: Semi-global exponential stabilization of linear systems subjece to input saturation via linear feedbacks. Syst. Control Lett. **21**(3), 225–239 (1993)
14. Saberi, A., Lin, Z.: Control of linear systems with saturating actuators. IEEE Trans. Autom. Control **41**(3), 368–378 (1996)
15. Lin, Z.: Global control of linear systems with saturating acturators. Automatica **34**(7), 897–905 (1998)
16. Meng, Z., Zhao, Z., Lin, Z.: On global leader-following consensus of idential linear dynamic systems subject to actuator saturation. Syst. Control Lett. **62**(2), 132–142 (2013)
17. Su, H., Jia, G., Chen, M.Z.Q.: Semi-global containment control of multi-agent systems with input saturation. IET Control Theory Appl. **8**(19), 2229–2237 (2014)
18. Su, H., Chen, M.Z.Q., Lam, J., Lin, Z.: Semi-global leader-following consensus of linear multi-agent systems with input saturation via low gain feedback. IEEE Trans. Circuits Syst. I-Regul. Pap. **60**(7), 1881–1889 (2013)
19. Fan, M.-C., Chen, Z., Zhang, H.-T.: Semi-global consensus of nonlinear second-order multi-agent systems with measurement output feedback. IEEE Trans. Autom. Control **59**(8), 2222–2227 (2014)

# Some Necessary and Sufficient Conditions for Consensus of Fractional-Order Multi-agent Systems with Input Delay and Sampled Data

Yanyan Ye[(✉)], Housheng Su[iD], Tao Geng, Xudong Wang, and Zuopeng Chen

Guangdong HUST Industrial Technology Research Institute, Guangdong Province
Key Lab of Digital Manufacturing Equipment, Image Processing and Intelligent
Control Key Laboratory of Education Ministry of China, School of Automation,
Huazhong University of Science and Technology,
Wuhan 430074, People's Republic of China
yanyanye@hust.edu.cn, houshengsu@gmail.com

**Abstract.** In this paper, the consensus of fractional-order multi-agent systems subject to input delay is investigated by sampled data method on directed graph. By applying the Laplace transform and the technique of inequality, some necessary and sufficient conditions for achieving consensus of the delay systems are obtained. It is shown that the consensus of the delay systems has relationships with the order of the derivative, the sampling period, delay, coupling strength, and communication topology. Lastly, a numerical simulation is given to illustrate the theoretical results.

**Keywords:** Consensus · Fractional-order · Multi-agent systems
Delay · Sampled data

## 1 Introduction

Consensus as one of the most essential collective behaviors in multi-agent systems has been widely studied due to its extensive applications [1–3], such as biology systems, multi-robots systems, sensor networks, to name just a few. The consensus problem is to make a group of autonomous agents converge to an agreement by only using some local information.

However, most of the researches about the consensus of multi-agent systems are focused on the integer-order dynamics, such as single-integral systems, double-integral systems. In practice, some systems cannot be exactly described by integer-order dynamics [4–6], such as porous media and electromagnetic waves, while some characteristics of these systems can be explained naturally by fractional-order dynamics. On the other hand, integer-order systems can be deemed to a special case of fractional-order systems.

© Springer Nature Switzerland AG 2018
Z. Chen et al. (Eds.): ICIRA 2018, LNAI 10984, pp. 39–47, 2018.
https://doi.org/10.1007/978-3-319-97586-3_3

Cao *et al.* discussed the consensus of fractional-order multi-agent systems firstly [6,7], and they obtained that the stability of the fractional-order multi-agent systems had something to do with the fractional order and the networked topology. Then, many works about the consensus of fractional-order multi-agent systems were done [9–14]. Nevertheless, because of the complex circumstances and the finite reaction times, the input delay always occurs in real systems. In [11], the heterogeneous input delays and communication delay were studied by the frequency-domain method and the generalized Nyquist criterion, respectively, and the consensus condition made a big difference for the different fractional order interval. In [12], a necessary and sufficient condition for the consensus of fractional-order multi-agent systems subject to input delay was obtained and a necessity condition was given for the systems with diverse input delays. In [14], consensus of fractional-order multi-agent systems with general linear and nonlinear dynamics and input delay was researched.

In practice, due to the finite transmission bandwidth and the limited resource, continuous information transmission is unreliable. Therefore, it is more practical to use the sampled data method which means that a continuous system adopts a discrete control protocol. On the other hand, it can reduce the energy and cut down the cost. As far as we know, there is only one paper which has studied the consensus of the fractional-order multi-agent systems via sampled data method [15]. In [15], the leaderless and leader consensus of fractional-order multi-agent systems with sampled data method were investigated, and some necessary and sufficient conditions were obtained.

In this paper, based on the sampled data scheme, the delay fractional-order multi-agent system was considered. Then main contributions of this paper are presented as follows. First, a sampled data method is used for the fractional-order multi-agent systems with input delay on directed network. Compared with the existing works, the system is more general and the method reduces the energy. Second, some necessary and sufficient conditions are presented, which give the relationship between the consensus of the systems and the order of the derivative, the sampling period, delay, coupling strength, and communication topology.

Section 2 introduces some preliminaries and states the problem. Section 3 presents the main results. Section 4 gives a numerical example and some conclusions are presented in Sect. 5.

## 2   Preliminaries and Problem Statement

### 2.1   Graph Theory

Let a networked graph $\mathcal{G}(\mathcal{V}, \mathcal{E}, \mathcal{W})$ be the communication topology among $N$ agents, where $\mathcal{V} = \{0, 1, \ldots, N\}$, and $\mathcal{E} \subseteq \mathcal{V} \times \mathcal{V}$ denote the set of nodes and the set of edges, respectively. $\mathcal{W} = [w_{ij}] \in \mathbb{R}^{N \times N}$ is the weighted adjacency matrix, where $w_{ij} > 0$ if $(j, i) \in \mathcal{E}$, else $w_{ij} = 0$, and $w_{ii} = 0$. If the communication topology is an undirected graph, $w_{ij} = w_{ji}$, which means the $j$-th agent can receive the information from the $i$-th agent, and vice versa. If the communication topology is a directed graph, $w_{ij} > 0$ means the $j$-th agent can receive the information from the $i$-th agent, but not the opposite. $N_i = \{v_i \in V : (v_j, v_i) \in \mathcal{E}\}$ denotes the neighbors set for the $i$-th agent. The Laplacian matrix $L$ associated with the communication topology $\mathcal{G}$ is defined as $L = [l_{ij}] \in \mathbb{R}^{N \times N}$, in which $l_{ii} = \sum_{j \neq i}^{N} w_{ij}$ and $l_{ij} = -w_{ij}, i \neq j$.

**Lemma 1.** [16] *For a directed graph, a Laplacian matrix has a simple zero eigenvalue and all of the other eigenvalues have positive real parts if and only if the directed graph has a directed spanning tree. For an undirected graph, a Laplacian matrix has a simple zero eigenvalue and all of the other eigenvalues are positive if and only if the undirected graph is connected.*

## 2.2   Caputo Fractional Operator

**Definition 1.** [17] *The definition of Caputo fractional-order integral for a function $f(t)$ can be written as*

$$_{t_0}^{C}D_t^{-\alpha}f(t) = \frac{1}{\Gamma(\alpha)} \int_{t_0}^{t} (t - \tau)^{\alpha-1} f(\tau)d\tau,$$

*where $0 < \alpha \leq 1$ denotes the order of integral, and $\Gamma(\cdot)$ is the Gamma function which is defined as*

$$\Gamma(z) = \int_0^{\infty} e^{-t} t^{z-1} dt.$$

**Definition 2.** [17] *The definition of Caputo fractional-order derivative for a function $f(t)$ can be expressed as*

$$_{t_0}^{C}D_t^{\alpha}f(t) = \frac{1}{\Gamma(1-\alpha)} \int_{t_0}^{t} \frac{f'(\tau)}{(t - \tau)^{\alpha}} d\tau.$$

*where $0 < \alpha \leq 1$ denotes the order of derivative.*

Because just the Caputo fractional operator is studied throughout this paper, we use $_0^{C}D_t^{-\alpha}f(t)$ to replace $D^{-\alpha}f(t)$ and $_0^{C}D_t^{\alpha}f(t)$ to replace $D^{\alpha}f(t)$ for convenience.

**Lemma 2.** [17] *For a constant $c$, and a function $f(t) \in \mathbb{C}^n[a, b]$,*

*(1) $D^{\alpha}c = 0$;*
*(2) $D^{-\alpha}D^{\alpha}f(t) = f(t) - f(a), 0 < \alpha \leq 1$.*

In the next, the Laplace transform of the Caputo derivative is introduction.
Let $\mathcal{L}\{\cdot\}$ denote the Laplace transform of a function. The formal definition of the Laplace transform $\mathcal{L}\{f(t)\} = \int_0^{\infty} e^{-st} f(t)dt$ is written as

$$\mathcal{L}\{D^{\alpha}f(t)\} = s^{\alpha}F(s) - s^{\alpha-1}f(0), \quad \alpha \in (0, 1].$$

## 2.3   System Description

Consider the fractional-order multi-agent system consisting of $N$ agents, the dynamics of the $i$-th agent is presented as follows:

$$D^\alpha x_i(t) = u_i(t), \quad i = 1, 2, \ldots, N, \tag{1}$$

with $0 < \alpha \le 1$. $x_i(t) \in \mathbb{R}^n$ and $u_i(t) \in \mathbb{R}^n$ represent the state and the control input for the $i$-th agent, respectively.

In practice, the input delay always cannot be negligible. Therefore, the input delay is considered in this paper. By utilizing the period sampled data method, the delay control input for the $i$-th agent is given by

$$u_i(t) = -\mu \sum_{j \in N_i} a_{ij}\big(x_i(t_k - \tau) - x_j(t_k - \tau)\big), \; t \in [t_k, t_{k+1}), \; i = 1, 2, \ldots, N, \tag{2}$$

where $\mu$ is the coupling strength. $t_k$ $(k = 0, 1, 2, \ldots)$, denotes the sampling time instants such that $0 = t_0 < t_1 < \cdots < t_k < t_{k+1} < \cdots$ and $t_{k+1} - t_k = T$, in which $T > 0$ denotes the sampling period, and $0 < \tau < T$ is the input time delay.

**Definition 3.** *The fractional-order multi-agent system (1) under the control protocol (2) is said to reach consensus, if for any initial conditions,*

$$\lim_{t \to \infty} \| x_i(t) - x_j(t) \| = 0, \quad \forall i = 1, 2, \ldots, N.$$

**Lemma 3.** [18] *For the following polynomial with $a_i \in \mathbb{R}$, $b_i \in \mathbb{R}$, $i=1,2$,*

$$f(s) = s^2 + (a_1 + ib_1)s + (a_2 + ib_2).$$

*$f(s)$ is stable if and only if $a_1 > 0$ and $a_1 b_1 b_2 + a_1^2 a_2 - b_2^2 > 0$.*

## 3   Main Results

Combining (1) and (2), the fractional-order multi-agent system can be written as

$$D^\alpha x_i(t) = -\mu \sum_{j \in N_i} l_{ij} x_j(t_k - \tau), \quad t \in [t_k, t_{k+1}), \quad i = 1, 2, \ldots, N, \tag{3}$$

described the systems in matrix form,

$$D^\alpha x(t) = -\mu(L \otimes I_n)x(t_k - \tau), \quad t \in [t_k, t_{k+1}), \tag{4}$$

where $x(t) = (x_1^T(t), x_2^T(t), \ldots, x_N^T(t))^T$.

For the Laplacian matrix $L$, there exists a nonsingular matrix $U$ satisfying $U^{-1}LU = J$, where $J$ is the Jordan form associated with $L$, and $J = diag(\Lambda_1, \Lambda_2, \ldots, \Lambda_r)$. For directed graph, the eigenvalues of $L$ may be complex,

$$\Lambda_l = \begin{bmatrix} \lambda_l & 1 & \cdots & 0 \\ 0 & 0 & \cdots & 0 \\ \vdots & \vdots & \ddots & \vdots \\ 0 & 0 & \cdots & \lambda_l \end{bmatrix}_{N_l \times N_l} \tag{5}$$

where $\lambda_i$ is the eigenvalue of $L$ with algebraic multiplicity $N_l$, $l = 1, 2, \ldots, r$ and $N_1 + N_2 + \cdots + N_r = N$.

For undirected graph, the Laplacian matrix $L$ is symmetric, so the Jordon form $J$ is a diagonal matrix with real eigenvalues of $L$.

Let $y(t) = (P^{-1} \otimes I_n)x(t)$, we obtain,

$$D^\alpha y(t) = -\mu(J \otimes I_n)x(t_k - \tau), \ t \in [t_k, t_{k+1}), \tag{6}$$

According to Lemma 1, if the networked graph has a directed spanning tree, 0 is a simple eigenvalue of $L$, thus,

$$D^\alpha y_1(t) = \mathbf{0}_n, \quad t \in [t_k, t_{k+1}), \tag{7}$$

where $\mathbf{0}_n$ denotes a n-dimensional column vector with all entries being zero.

**Lemma 4.** *Assume that the network is directed and has a directed spanning tree, the consensus of the system (4) can be reached if and only if, in (6)*

$$\lim_{t \to \infty} \|y_i(t)\| = 0, \ i = 2, 3, \ldots, N.$$

*Proof (Sufficiency):* Based on Lemma 2, integrating both sides of (7) from $t_k$ to $t$, we have $y_1(t) - y_1(t_k) = 0$, $t \in [t_k, t_{k+1})$. Therefore, $y_1(t) = y_1(t_k) = y_1(t_{k-1}) = \cdots = y_1(0)$. Because $\lim_{t \to \infty} \|y_i(t)\| = 0$, $i = 2, 3, \ldots, N$, so we can get $\lim_{t \to \infty} y(t) = [y_1(0)^T, \mathbf{0}^T, \ldots, \mathbf{0}^T]^T$. We have $\lim_{t \to \infty} x(t) = \lim_{t \to \infty}(P \otimes I_n)y(t) = (P \otimes I_n)My(0) = (P \otimes I_n)M(P^{-1} \otimes I_n)x(0)$, where the matrix $M = [m_{ij}] \in \mathbb{R}^{Nn \times Nn}$ satisfies $m_{ij} = 0$, $i = j = 1, 2, \ldots, n$, else $m_{ij} = 0$. Since the network has a directed spanning tree, so we can choose $P = [\mathbf{1}_N, p_2, \ldots, p_N]$ and $P^{-1} = [q^T, q_2^T, \ldots, q_N^T]^T$, therefore, $\mathbf{1}_N$ and $q^T$ are the right and left eigenvector of the Laplacian matrix $L$ associated with $\lambda_1 = 0$ and $q\mathbf{1}_N = 1$. So, $\lim_{t \to \infty} x(t) = (P \otimes I_n)M(P^{-1} \otimes I_n)x(0) = (\mathbf{1}_N q^T \otimes I_n)x(0)$, that is, $\lim_{t \to \infty} x_i(t) = (q^T \otimes I_n)x(0)$. The consensus is reached.

*(Necessity):* If the consensus can be reached, then there must exist a vector $x^*(t)$ such that $\lim_{t \to \infty} x_i(t) = x^*(t)$ and $\lim_{t \to \infty} x(t) = \mathbf{1}_N \otimes x^*(t)$. Because $\mathbf{0}_N = P^{-1}L\mathbf{1}_N = JP^{-1}\mathbf{1}_N = J[q_1\mathbf{1}_N, q_2\mathbf{1}_N, \ldots, q_N\mathbf{1}_N]^T$. Based on Lemma 1, if the network has a directed spanning tree, the eigenvalues of $L$ satisfy that $0 = \lambda_1 < Re(\lambda_2) \le Re(\lambda_3) \le \cdots \le Re(\lambda_N)$. Combining with the Jordan form $J$, we have $q_i\mathbf{1}_N = 0$, $i = 2, 3, \ldots, N$. Hence, $\lim_{t \to \infty} \|y_i(t)\| = \lim_{t \to \infty} \|(P^{-1} \otimes I_n)x_i(t)\| = \lim_{t \to \infty} \|(q_i \otimes I_n)x(t)\| = \lim_{t \to \infty} \|(q_i\mathbf{1}_N) \otimes x^*(t)\| = 0$, $i = 2, 3, \ldots, N$.

**Corollary 1.** *Assume that the network is directed and has a directed spanning tree, the system (4) can reach consensus if and only if the following $r - 1$ systems are asymptotically stable*

$$D^\alpha z_i(t) = -\mu\lambda_i z_i(t_k - \tau), \quad t \in [t_k, t_{k+1}), \quad i = 2, 3, \ldots, r, \tag{8}$$

*where $0 < \alpha \le 1$.*

*Proof.* The proof is similar to Corollary 1 in [15].

**Theorem 1.** *If the network has a directed spanning tree and $0 < \tau < T$, then, the fractional-order multi-agent system (1) under the control protocol (2) can reach consensus if and only if*

$$T^\alpha - (T - \tau)^\alpha < a, \tag{9}$$

*and*

$$(3T^\alpha - 2(T - \tau)^\alpha - 2a)b^2 - (T^\alpha - (T - \tau)^\alpha - a)^2(T^\alpha - 2(T - \tau)^\alpha + 2a) < 0, \tag{10}$$

*where $a = \frac{\Gamma(1+\alpha)}{\mu\|\lambda_i\|^2}Re(\lambda_i)$, $b = \frac{\Gamma(1+\alpha)}{\mu\|\lambda_i\|^2}Im(\lambda_i)$. $\alpha$ is the order of derivative. $\lambda_i$ is the nonzero eigenvalues of the Laplacian matrix $L$, $Re(\lambda_i)$ and $Im(\lambda_i)$ denote the real part and imaginary part of $\lambda_i$, respectively.*

*Proof.* For $t \in [t_k, t_{k+1})$, taking the Laplace transform of (8), we have

$$s^\alpha z_i(s) - s^{\alpha-1}z_i(t_k) = -\frac{\mu\lambda_i}{s}z_i(t_k - \tau), \tag{11}$$

so,

$$z_i(s) = \frac{1}{s}z_i(t_k) - \frac{\mu\lambda_i}{s^{\alpha+1}}z_i(t_k - \tau). \tag{12}$$

Taking the inverse Laplace transform of (12), we can obtain,

$$z_i(t) = z_i(t_k) - \frac{\mu\lambda_i}{\Gamma(1+\alpha)}(t - t_k)^\alpha z_i(t_k - \tau), \quad t \in [t_k, t_{k+1}), \tag{13}$$

and

$$z_i(t - \tau) = z_i(t_k) - \frac{\mu\lambda_i}{\Gamma(1+\alpha)}(t - t_k - \tau)^\alpha z_i(t_k - \tau). \tag{14}$$

Since the sampling time instants $0 = t_0 < t_1 < \cdots < t_k < t_{k+1} < \cdots$, $t_{k+1}-t_k = T$ and $\tau < T$, we can get

$$z_i(t_k) = z_i(t_{k-1}) - \frac{\mu\lambda_i}{\Gamma(1+\alpha)}T^\alpha z_i(t_{k-1} - \tau), \tag{15}$$

and

$$z_i(t_k - \tau) = z_i(t_{k-1}) - \frac{\mu\lambda_i}{\Gamma(1+\alpha)}(T - \tau)^\alpha z_i(t_{k-1} - \tau), \tag{16}$$

Let $w_i(t) = [z_i(t), z_i(t - \tau)]^T$, we have

$$w_i(t_k) = E(T)w_i(t_{k-1}) = E^k(T)w_i(t_0), \ w_i(t) = E(t-t_k)w_i(t_k) = E(t-t_k)E^k(T)w_i(t_0),$$

where $E(t) = \begin{bmatrix} 1 & -\frac{\mu\lambda_i}{\Gamma(1+\alpha)}t^\alpha \\ 1 & -\frac{\mu\lambda_i}{\Gamma(1+\alpha)}(t - \tau)^\alpha \end{bmatrix}$.

Therefore, $\lim t \to \infty z_i(t) \to 0$ is equivalent to $w_i(t) \to 0$ as $t \to \infty$. Because it is obvious that $E(t - t_k)$ is bounded on $t \in [t_k, t_{k+1})$, so $\lim t \to \infty z_i(t) \to 0$ is equivalent to all the eigenvalues $\gamma$ of the matrix $E(T)$ satisfy $\|\gamma\| < 1$.

Let $|\gamma I_2 - E(T)| = 0$, we can get

$$\gamma^2 + \left(\frac{\mu\lambda_i}{\Gamma(1+\alpha)}(T-\tau)^\alpha - 1\right)\gamma + \frac{\mu\lambda_i}{\Gamma(1+\alpha)}(T^\alpha - (T-\tau)^\alpha) = 0. \qquad (17)$$

Let $\gamma = \frac{s+1}{s-1}$, hence, $\|\gamma\| < 1$ if and only if $Re(s) < 0$. (17) is equivalent to

$$s^2 + 2(a_1 + ib_1)s + (a_2 + ib_2) = 0, \qquad (18)$$

where $a_1 = -1+(1-\frac{\tau}{T})^\alpha+\frac{\Gamma(1+\alpha)}{\mu T^\alpha\|\lambda_i\|^2}Re(\lambda_i)$, $a_2 = 1-2(1-\frac{\tau}{T})^\alpha+2\frac{\Gamma(1+\alpha)}{\mu T^\alpha\|\lambda_i\|^2}Re(\lambda_i)$, $b_1 = b_2 = -2\frac{\Gamma(1+\alpha)}{\mu T^\alpha\|\lambda_i\|^2}Im(\lambda_i)$. Based on Lemma 3, (19) is stable if and only if $a_1 > 0$ and $a_1b_1b_2 + a_1^2a_2 - b_2^2 > 0$. By solving the above inequation, we can get the conditions (9) and(10). Therefore, $\lim_{t\to\infty} z_i(t) = 0$ if and only if (9) and(10) are satisfied.

*Remark 1.* In Theorem 1, a necessary and sufficient condition for the fractional-order multi-agent systems subject to input delay is proposed. For a fixed directed network, we can see that the consensus of the fractional-order multi-agent systems depends on the sampling period, delay, coupling strength, communication topology, and depends on the order of the derivative.

*Remark 2.* When $\tau = 0$, the system (1) degrades as the system without input delay, and the conditions (9) and(10) simplify as

$$T^\alpha < 2a = \frac{2\Gamma(1+\alpha)}{\mu\|\lambda_i\|^2}Re(\lambda_i),$$

which is the same as the Theorem 1 in [15].

**Corollary 2.** *If the network is undirected and connected, and $0 < \tau < T$. Then the fractional-order multi-agent system (1) under the control protocol (2), can reach consensus if and only if*

$$T^\alpha - (T-\tau)^\alpha - a_1 < 0, \qquad (19)$$

*and*

$$T^\alpha - 2(T-\tau)^\alpha + 2a_1 > 0. \qquad (20)$$

*where $a_1 = \frac{\Gamma(1+\alpha)}{\mu\lambda_N}$, $\alpha$ is the order of derivative, $\mu$ is the coupling strength, and $\lambda_N$ is the largest eigenvalue of $L$.*

*Proof.* If the network is undirected and connect, we have all the eigenvalues of the Laplacian matrix $L$ are positive real number, that is $Re(\lambda_i) = \lambda_i > 0$ and $Im(\lambda_i) = 0$. Then, $a = \frac{\Gamma(1+\alpha)}{\mu\lambda_i}$ and $b = 0$. Then, the conditions in Theorem 1 degrade as (19) and (20).

## 4    Simulation Example

Consider that the fractional-order multi-agent system (4) contains four agents, and the elements of weighted adjacency matrix are $w_{12} = 0.5$, $w_{14} = 0.7$, $w_{24} = 1$, $w_{31} = 0.3$, $w_{43} = 0.8$, and $w_{ij} = 0$, otherwise. We can get that the eigenvalues of Laplacian matrix are $\lambda_1 = 0$, $\lambda_2 = 0.95 + 0.5454\mathrm{i}$, $\lambda_3 = 0.95 - 0.5454\mathrm{i}$, and $\lambda_4 = 1.4$. Choosing the order of derivative $\alpha = 0.9$, the coupling strength $\mu = 0.2$, the sampling period $T = 1s$, and the input delay $\tau = 0.5s$, which can make sure that the conditions in Theorem 1 are satisfied. The initial condition are $x_1 = -1$, $x_2 = -4$, $x_3 = -8$, and $x_4 = 2$. The state of $x_i$ ($i = 1, 2, 3, 4$) is presented in Fig. 1, which illustrates that the consensus of the fractional-order multi-agent system (4) is achieved.

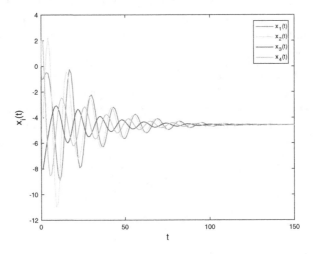

**Fig. 1.** The state of $x_i(t)$, $i = 1, 2, 3, 4$.

## 5    Conclusion

In this paper, under the period sampled data method, the consensus of fractional-order multi-agent systems with the order $0 < \alpha < 1$ subject to input delay over directed communication topology is studied. Some necessary and sufficient conditions are established which give the relationship of the achieving consensus and the systems parameters (sampling period, delay, coupling strength, communication topology and the order of the derivative). Research of the aperiodic sampled data and diverse delays for the fractional-order multi-agent systems will be done in the future work.

# References

1. Vicsek, T., Cziok, A., Jacob, E., Cohen, I., Shochet, O.: Novel type of phase transition in a system of self-driven particles. Phys. Rev. Lett. **75**(6), 1226–1229 (1995)
2. Olfati-Saber, R., Murray, R.M.: Consensus problems in networks of agents with switching topology and time-delays. IEEE Trans. Autom. Control **49**(9), 1520–1533 (2004)
3. Ren, W., Bread, R.W., Atkins, E.M.: Information consensus in multivehicle cooperative control: collective group behavior through local interaction. IEEE Control Syst. Mag. **27**(2), 71–82 (2007)
4. Arena, P., Caponetto, R., Fortuna, L., Porto, D.: Nonlinear Noninteger Order Circuits and Systems – An Introduction. World Scientific, Singapore (2000)
5. Kilbas, A., Srivastave, H., Trujillo, J.: Theory and Applications of Fractional Differential Equation. Springer, Oxford (2007)
6. Marcos, M.D.G., Duarte, F.B.M., Machado, J.A.T.: Fractional dynamics in the trajectory control of redundant manipulators. Commun. Nonlinear Sci. Numer. Simul. **13**(9), 1836–1844 (2008)
7. Cao, Y., Ren, W.: Distributed formation control for fractionalorder systems: dynamic interaction and absolute/relative damping. Syst. Control Lett. **59**(3), 233–240 (2010)
8. Cao, Y., Li, Y., Ren, W., Chen, Y.: Distributed coordinative of networked fractional-order systems. IEEE Trans. Syst. Man Cybern. Part B **40**(2), 362–370 (2010)
9. Yu, W., Li, Y., Wen, G., Yu, X., Cao, J.: Observer design for tracking consensus in second-order multi-agent systems: fractional order less than two. IEEE Trans. Autom. Control **62**(2), 894–900 (2017)
10. Zhu, W., Li, W., Zhou, P., Yang, C.: Consensus of fractional-order multi-agent systems with linear models via observer-type protocol. Neurocomputing **230**(22), 60–65 (2017)
11. Shen, J., Cao, J., Lu, J.: Consensus of fractional-order systems with non-uniform input and communication delays. Proc. Inst. Mech. Eng. Part I J. Syst. Control Eng. **226**(2), 271–283 (2012)
12. Shen, J., Cao, J.: Necessary and sufficient conditions for consensus of delayed fractional-order systems. Asian J. Control **14**(6), 1690–1697 (2012)
13. Yang, H., Zhu, X., Cao, K.: Distributed coordination of fractional order multiagent systems with communication delays. Fract. Calc. Appl. Anal. **17**(1), 23–37 (2014)
14. Zhu, W., Chen, B., Yang, J.: Consensus of fractional-order multi-agent systems with input time delay. Fract. Calc. Appl. Anal. **20**(1), 52–70 (2017)
15. Yu, Y., Jiang, J., Hu, C., Yu, J.: Necessary and sufficient conditions for consensus of fractional-order multi-agent system via sampled-data control. IEEE Trans. Cybern. **47**(8), 1892–1901 (2017)
16. Ren, W., Beard, R.W.: Consensus seeking in multiagent systems under dynamically changing interaction topologies. IEEE Trans. Autom. Control **50**(5), 655–661 (2005)
17. Podlubny, I.: Fractional Differential Equations. Academic Press, San Diego (1999)
18. Parks, P.C., Hahn, V.: Stability Theory. Prentice Hall, Upper Saddle River (1993)

# A Brief Overview of Flocking Control for Multi-agent Systems

Yaping Sun, Zhaojing Wang, Housheng Su$^{(\boxtimes)}$ ⏺, and Tao Geng

Guangdong HUST Industrial Technology Research Institute, Guangdong Province
Key Lab of Digital Manufacturing Equipment, Image Processing and Intelligent
Control Key Laboratory of Education Ministry of China, School of Automation,
Huazhong University of Science and Technology,
Wuhan 430074, People's Republic of China
SHS@hust.edu.cn

**Abstract.** In this paper, we firstly introduce two main models based on Boid Model: Vicsek Model and Couzin Model. Then, the more authoritative and representative flocking control algorithms by Olfati-Saber and Tanner are proposed. Moreover, more extensive researches of flocking algorithm are carried out. Finally, a short discussion is included to summarize the existing research and to propose several problem.

**Keywords:** Flocking control · Multi-agent system · Vicsek model
Couzin model

## 1 Introduction

Some biological groups are composed of a large number of individuals in nature. Although the individual is very simple, and individual accessing to information and exchanging information between individuals are localized, but they only use the interactions of these local information to produce complex group behavior, and help them achieve the goal of avoiding predators, hunt prey and long-distance migration efficiently, such as fish, birds quickly changing the team to escape predators hunting, the fireflies rhythmically consistent flashing in one tree [1–8]. These phenomenons are called flocking. Flocking is a form of collective behavior. The most important characteristic of flocking is that it emerges from simple local rules into a coordinated global behavior. From the view of systematics, flocking behavior is adaptive, robust, decentralized and self-organized [9].

The research of flocking behavior has gone through three stages: ($i$) The discover stage of biological flocking behaviors, many biological researchers found the unique behaviors of biological group [1]; ($ii$) The simulation stage of flocking behaviors. The researchers used the computers to simulate and test the biological flocking behaviors [3]; ($iii$) Rigorous modeling and analysis of flocking behavior, which is in progress [10,11]. The research of flocking control is generated by the rules of Reylnolds Boid model, which are Separation Cohesion and Alignment. The flocking control algorithms of multi-agent system are classified into

© Springer Nature Switzerland AG 2018
Z. Chen et al. (Eds.): ICIRA 2018, LNAI 10984, pp. 48–58, 2018.
https://doi.org/10.1007/978-3-319-97586-3_4

three kinds broadly: the algorithms which do not consider collision, the aim of these algorithms is to make agents achieve velocity consensus, and the distance between any two agents remains constant [12,13]; The algorithms which consider collision, these algorithms is to avoid the collision in addition to the above aims [14,15]; Besides, there are many algorithms which are added some new rules to satisfy the need of different tasks, such as avoiding obstacles [16], tracking a target and so on [14,17].

The remainder of this paper is organized as follows. Section 2 presented two main models based on Boid Model: Vicsek Model and Couzin Model. Section 3 introduce the control algorithms. Finally, a short discussion is included to summarize the existing research and to propose several problem in Sects. 4.

## 2   Model

### 2.1   Vicsek Model

In 1995, Vicsek proposed a model with a self-ordered particles which are driven with constant absolute velocity and each step only change theirs directions. The next step direction of a particle is the average direction of motion of particles in its neighborhood with random perturbation added [18]. In the simulations, Vicsek obtained three folds conclusions: for small densities and noise, the particles tend to disordered with random directions. When there are higher densities and noise, the particles tend to form different clusters with different directions. When the density is large and the noise is small, the particles become ordered, and all the particles tend to move in the same spontaneously selected direction. Vicsek model is the simplest model of flocking, but it exists many shortcomings. For further improving the model and accelerating the convergence rate, many researchers had made up a lot of new strategies [19,20].

In 2006, Yang et al. considered a heterogeneous influence network. The influencing ability of each agent is different each other and behaved by influencing radius, which is randomly chosen according to a power-law distribution. The conclusions were that the capacity to obtain direction convergence is enhanced as the heterogeneity of the influencing radius increases, and global consensus can be achieved only a small part of leading agents can be controlled to move in the desired direction [21]. In 2007, Li and Wang proposed an adaptive velocity model, in which each agent not only adjusts its direction but also adjusts its speed based on the degree of direction convergence among its local neighbors, and the absolute velocities agents are changeable and tended to the same maximum constant speed after a short transient process. The adaptive strategies can increase the possibility of convergence [22]. In 2009, Gao et al. presented a new weighted model by utilizing the degree in complex network and topological structure of dynamical systems, and the model convergence efficiency was more faster than Vicsek model [23]. Tian et al. studied a system of self-propelled agents with the restricted vision [24]. The field of vision of agents is a part of disk bounded by radii and the included arc, in other words, the view angle is limited. They find an optional view angle, leading to the fastest convergence.

And the optional angle is related to the density, the interaction radius, the absolute velocity and the noise. In 2014, Yang et al. proposed a dynamic model for a system consisting of self-propelled agents in which the influence of an agent on another agent is weighted by distance. They find there exists an optional value of parameter leading to the highest degree of direction convergence [25].

## 2.2 Couzin Model

In 2002, Couzin et al. proposed a new model by using three rules of Reynolds [26]. The model is a self-organizing model in three-dimensional space. There are two new rules of the new model: ($i$) Particles attempt to maintain a minimum distance among neighbors all the time. ($ii$) If particles are not behaving an avoidance manoeuvre, they will be attracted towards other agents (to avoid being secluded) and to align themselves with neighbors. The influencing area of one particle is divided into three parts: zone of repulsion ($zor$), zone of orientation ($zoo$) and zone of attraction ($zoa$). If there exist neighbors in the $zor$, the particle will move away from neighbors. In the $zoo$, a particle will tend to align with its neighbors within the zone. $zoa$ represents the tendency of particles to join groups and to avoid being on the periphery, and the particle tends to the position within the zone of attraction. The zone of orientation and attraction are spherical except for a volume behind the particle within which neighbors are undetectable. This blind volume is defined as a cone with interior angle. In the Couzin model, each particle has the same zone of repulsion orientation and attraction, but the number of neighbors is different for each particles. There are some improvements of Couzin model. In [27], a velocity adaptive strategy is presented to provide an alternative to the constant-speed setting of the Couzin model. In 2016, Zhao came up with a high-speed and fast-convergent self-organizing model with improved adaptive-velocity and weighted strategies [28]. In practice, a small number of particles effect the system greatly more than other particles, such as a leaders debate in the meeting is easily accepted. Removing some hub nodes may lead to a breakdown of the whole network although when it comes to the others, the influence is slight.

## 3   Control Algorithms

### 3.1   Algorithms by Olfati-Saber

Olfati-Saber presented a theoretical framework for the design and analysis of distributed flocking algorithms. Olfati-Saber et al. proposed three flocking algorithms: two for free-flocking and one for constrained flocking [16]. They considered a group of particles with the dynamics: $\dot{q}_i(t) = p_i(t)$, $\dot{p}_i(t) = u_i(t)$, $i = 1, ..., N$. $q_i(t)$, $p_i(t)$ denote the position and velocity of $i$th agent. In the Algorithms 1 and Algorithms 2, each -agent has a control input $u_i = s_i^g + s_i^d + s_i^\gamma$, in which $s_i^g$ is a gradient-based term, $s_i^d$ is a velocity consensus term that acts as a damping force, $s_i^\gamma$ is a navigational feedback due to a group objective. Algorithm 1 is drastically different than a group of agents applying Algorithm 2.

The algorithm 1 leads to flocking behavior only with a restricted initial states, and fragmentation. In contrast, the Algorithm 2 never leads to fragmentation. But the importance of the Algorithm 1 is fundamental structures of Algorithm 2. Algorithm 1 and 2 are for flocking in free-space. In addition, they present a distributed flocking algorithm with multiple obstacle avoidance. The main thoughts of the algorithm is regarding barriers to be agents. In the Algorithms 3, $u_i = u_i^\alpha + u_i^\beta + u_i\gamma$, in which $u_i^\alpha$ represents the $(\alpha, \alpha)$ interaction terms, $u_i^\beta$ represents the $(\alpha, \beta)$ interaction terms, $u_i\gamma$ represents navigational feedback.

These three distributed flocking algorithms were presented to result in self-organizing flocking behavior. Algorithm 1 is responsible for the creation of space order, this algorithm satisfied three rules of Reynolds but leads to regular fragmentation. Algorithm 2 and 3 are found on the basis of Algorithm 1, and these two algorithms add appropriate terms that account for system objective and obstacle avoidance, respectively. In addition, the paper presented another protocol. They assume the structural energy is finite and interaction range satisfies. In this case, under some conditions, the system is asymptotically self-assembled. On the basis of Olfati-Sabers research, more and more factors are taken into account, such as a virtual leader and multiple leaders, connectivity preserving, directed network, nonlinear factors, time delay and so on.

### 3.2   Algorithms by Tanner

In 2007, Tanner proposed flocking algorithms in fixed and switching networks. On the one hand, the paper summarized previous two papers [15,25] which analyzed the stability properties of mobile agents, and they introduced discontinuities in the agent control laws, they employed non-smooth analysis to adapt to switching topology. The main result is on the conditions that regardless of switching topology, the network remains connected at all times. The paper considered a group of mobile agents moving on the plane, and the dynamics $\dot{q}_i = p_i$, $\dot{p}_i = u_i = \chi_i + \zeta_i$, $i = 1, ..., N$. $q_i$, $p_i$, $u_i$ denote the position, velocity and control input of $i$th agent. $q_{ij} = q_i - q_j$ denotes relative position vector between agent $i$ and $j$. The input consists two parts: $\chi_i$ and $\zeta_i$. $\chi_i$ aims to align the velocity vectors of all the agents, $\zeta_i$ is a vector in the direction of the negated gradient of an artificial potential function, and it is used for collision avoidance and cohesion in the group. The sensing radius is $r$, and if agents beyond the range are assumed not to affect the direction. The algorithm of this article will make the multi-agent system to achieve the desired behavior, and the behavior has robustness.

### 3.3   Algorithms with a Virtual Leader

In 2009, Su et al. proposed a flocking algorithm which revisited the problem in the paper of olfati-saber, the difference is even when only a fraction of agents are informed, the algorithm is still able to make the informed agents move in the desired velocity, the uniformed agents also move in the desired velocity only when they have interaction with informed agents [29]. In another paper published in

2009, Gu et al. investigated a leader-follower flocking system. There are a few members of system informed, and majority members of group do not have global knowledge [30]. In this paper, as only leaders know the desired trajectory, the members have possibility to move away from the group and bring out the group splitting. But, the system is able to attain consensus led by leaders owing to all members estimating the position of flocking center via local interactions. In 2010, Yu et al. proposed a new distributed leader-follower flocking control algorithm for multi-agent dynamical systems with time-varying velocities, to avoid the assumption that the informed agents have global information [31]. Besides, in 2013, Su et al. investigated a adaptive flocking algorithm with a virtual leader [32]. And the authors introduced a distributed flocking algorithm for agents modeled by double integrators to track a virtual leader in the case that only a fraction of them are considered informed ones in [33]. [34] presented a method for robust adaptive flocking of nonholonomic multi-agent systems with an active leader. Recently the authors focused on the distributed leader-follower flocking problem with a moving leader for networked Lagrange systems with unknown parameters under a proximity graph defined according to the relative distance between each pair of agents [35].

### 3.4   Algorithm with Multiple Leaders

In reality, there are many missions in the system which include multiple leaders to obtain different goals. In 2008, Su et al. put forward a problem of tracking multiple virtual leaders with varying velocities [36]. Different agents with different leaders attain different velocity and track different corresponding leader. Luo et al. presented a flocking algorithm with multi-target tracking for multi-agent systems in 2010 [37]. Multiple targets is equal to multiple leaders. The principle of choosing target is determined by the distances from the agent to the targets and the number of agents accepted by targets. A collective potential function and repulsive potential function have been designed to deal with the problem. On the basis of previous studies, Su et al. investigated a novel flocking algorithm in Multi-agent systems with multiple virtual leaders based only on position measurements, every agent attained the desired velocity, in the process, collision can be avoided, and the final formation minimize the potential energy [38].

### 3.5   Algorithm with Connectivity Preserving

In Multi-agent system, each agent just make use of information among neighbors, not global information, the neighbors of one agent at last time may move away from the agents sensing range, and other agents neighbors may move into the sensing range of the agent. So single agent is easy to move away from group, even it leads to the fragmentation of the group [39]. In addition, interference is common in real life. Connectivity-preserving flocking algorithm can avoid these problem. Some papers flocking algorithms are based on always maintaining network connectivity [12], but in practical, we can only guarantee that the initial state is connected. Initial connectivity can not guarantee the network connected all the time, eventually may lead to splitting. Therefore, it will be of great practical significance to find a flocking control algorithm which has the ability to maintain the connectivity of the network. In 2008, Su et al. proposed a new algorithm that make all agents move with a common velocity while keeping a desired group shape under the assumption that initial network is connected [40]. In 2009, Su et al. proposed another new connectivity-preserving flocking algorithm for multiagent systems based only on position measurements [41], they proposed two rules that a potential function should obey in order to preserve the connectivity of the network as long as the initial network is connected. By means of combining the ideas of collective potential functions and velocity consensus, Wen et al. proposed a connectivity preserving flocking algorithm with bounded potential function in 2012 [42]. Inspired by the study of [41], but the communication link is totally determined by the distance between the agents. They designed a decentralized input control for connectivity-preserving flocking, each agent can not access to the velocity information, and some attractive and repulsive forces are combined to design the controller. In 2015, Dong et al. presented a new flocking algorithm with connectivity preservation problem of multiple double integrator systems, this approach is combination of potential function technique and observer design [43].

## 3.6   Algorithm with Obstacles Avoidance

Obstacles is inevitable in the application of flocking control, for each individual with autonomous multi-agent system, the ability to avoid obstacles in the process of movement is the most basic security requirements. Reynolds firstly proposed three rules including flocking centering, obstacle avoidance, and velocity matching, and they have used computer to make simulations. The obstacle avoidance control strategy is in the direction of the advance of the intelligence to construct a virtual cylinder, the diameter of the cylinder is the diameter of the agent, the height of the cylinder and the model of the individual speed is proportional to the ratio of the model [26]. If there are several obstacles, the agent will avoid the obstacles near firstly. But the strategy not considered the size of obstacles and the influence of the distance between the obstacle and the agent on the steering control force of the agent; And only the obstacle avoidance control of an obstacle is considered. Olfati-Saber investigated a new algorithm of obstacles avoidance. The obstacle is assumed to be a moving agent, and system merges this velocity and the velocity of that agent entering the region of obstacle and then regards the merged velocity as next-time velocity for the agent [16]. The system combines the velocity of the agent and the obstacles in agents influence range, and then the combined speed is used as the speed of the next time. The magnitude of velocity, but not its direction, is considered while combining velocity. When the agent is moving away from the obstacle, the obstacle is still regard as an agent. Therefore, the phenomenon will effect the behavior of the agent even the process of obstacle avoidance for these agents. In 2014, Wang et al. proposed an improved fast flocking algorithm with obstacle avoidance for multi-agent dynamic systems based on Olfati-Sabers algorithm [44]. The major contribution of this new strategy is proposing an improved algorithm for obstacle avoidance and fast form flocking towards the direction of destination point. The algorithm is illustrated based on the state of whether the flocking has formed. If flocking has not formed, agents should avoid the obstacles toward the direction of target. Otherwise, these agents have reached the state of lattice and then these agents only need to avoid the obstacles and ignore the direction of target. The experimental results show that the proposed the new algorithm has better performance in terms of flocking path length. Yang et al. also investigated the situation that the obstacles are moving, but the algorithm has no pre-judgment for whether appear obstacle collision or not, only when the agent and the obstacle distance is less than the safety value, the obstacle avoidance algorithm start up, this is a waste of energy. In 2013, Lou proposed another algorithm in his doctors dissertation. The theory proves that the flocking control strategy based on collision prediction method can effectively avoid the collision, and gives the control laws. And a more complicated obstacle avoidance control strategy for moving obstacle is designed. By adding the obstacle in the pre-collision function, the speed of the obstacle is increased and the accuracy of the judgment is improved. In the control input, the control input obstacle repulsion by means of adding the obstacles velocity vector to improve obstacle avoidance ability. In [45], the authors proposed a flocking algorithm in the presence of arbitrary shape obstacles avoidance both in 2D space and 3D space. And they presented an approach to determine the position and the velocity of beta-agent which denote the repulsive effects of obstacles.

### 3.7   More General Algorithms

(**With Time Delay**) Due to the limited transmission speed, as well as the traffic congestion and diffusion. In the reality, there are spread and propagation delay. In 2012, Yang et al. investigated a new strategy about flocking of multi-agents with time delay firstly [46]. In the paper, they consider a delay-independent flocking law, and the collision avoidance between the agents is ensured. Both leader free and virtual leader available are considered. In 2015, Zhang et al. proposed another situation of flocking control with time delay [47], which can combine the theory and practice.

(**With Nonlinear**) There are nonlinear systems in practical, the nonlinear dynamics is more general, reasonable and stronger, many nonlinear systems do not satisfy the global Lipschitz condition, only to satisfy the local Lipschitz condition. So it has practical significance to the local Lipschitz nonlinear problem of flocking control has important research. In 2010, Yu et al. proposed robust adaptive flocking control of nonlinear multi-agent systems [31]. In 2013, Su et al. introduced flocking control with a virtual leader of multiple agents governed by locally Lipschitz nonlinearity, all the agents and the virtual leader share the same intrinsic nonlinear dynamics [32]. In 2014, Peng et al. investigated neural adaptive control for leaderfollower flocking of networked non-holonomic agents with unknown nonlinear dynamics [34].

(**With Input Saturation**) Another issue inevitable in real control systems is actuator saturation, since there exists physical constraints. All real-world applications of feedback control involve actuators with amplitude and rate limitations. In particular, any physical mechanical and electrical equipment can only provide limited force, torque, flow or linear/angular rate. In 2013, the authors investigated adaptive flocking control of nonlinear multi-agent systems with directed switching topologies and saturation constraints [48]. A novel distributed adaptive flocking protocol with bounded control input. Lately, in [49], develops a model predictive flocking control scheme for second-order multi-agent systems with input constraints.

(**With Pinning Control**) Pinning control strategy is that by means of adding control inputs to a fraction of agents which are selected from the network, to control the system. This is an effective approach for the control of networked systems. In 2010, Wang et al. proposed pinning-based flocking algorithms in which navigational feedback terms are added to just a fraction of agents [50]. In 2013, Su and Wang has completed a book, which investigated synchronization, consensus and flocking of networked systems via pinning [51]. They summarized distributed pinningcontrolled flocking with a virtual leader and with preserved network connectivity.

(**With Event-Triggered Strategy**) Under the situation of resource limit, common algorithms usually cannot achieve good results. To solve this problem, The methods based event-triggered attract more and more researchers. The author in [52] proposed event-triggered flocking.

## 4  Discussions

The paper has made a brief overview on flocking control for multi-agent system. Flocking control has attracted a lot of researchers interests. The new algorithms are emerging and these algorithms are applied to many fields. Although the researches for flocking control of multi-agent systems is very sufficient, there are some unsolved problems. (*i*) (**Fractional Dynamics Model**) Most papers regard second order dynamics model as the research object, for the existing complex dynamic environment, the system with integer order dynamic model is not fit to practice situations. So the flocking behaviors with fractional characteristics is still a question. (*ii*) (**Nonlinear flocking control**) The existing algorithms for nonlinear flocking must satisfy the local Lipschitz condition, the constraint is very strong. It is a problem that how to broaden the nonlinear dynamic constraints. (*iii*) (**Flicking robust**) Since noise and external disturbances exist in the actual network, control often cannot achieve the desired results. It is significant to investigate the robust of flocking control. (*iv*) (**Multi-USV systems**) In the existing research results of flocking control, also found only little literature about multi-USV systems. The difficult point is that how to solve the control problem of multi-USV systems, when the complicated external environment are adequately considered.

# References

1. Shaw, E.: Fish in shools. Nat. Hist. **84**(8), 40–45 (1975)
2. Potts, W.K.: The chorus-line hypothesis of manoeuvre coordination in avian flocks. Nature **309**(5966), 344–345 (1984)
3. Okubo, A.: Dynamical aspects of animal grouping: swarms, schools, flocks and herds. Adv. Biophys. **22**(22), 1–94 (1986)
4. Grunbaum, D., Okubo, A.: Modeling social animal aggregations. Front. Theor. Biol. **100**, 296–325 (1994)
5. Helbing, D., Farkas, I., Vicsek, T.: Simulating dynamical features of escape panic. Nature **407**(6803), 487–490 (2000)
6. Low, D.J.: Following the crowd. Nature **407**(6803), 465–466 (2000)
7. Vicsek, T.: A question of scale. Nature **411**(6836), 421–421 (2001)
8. Parrish, K., Viscido, S.V., Grunbaum, D.: Self-organized fish schools: a examination of emergent properties. Biol. Bull. **202**(3), 296–305 (2002)
9. Beard, R.W., Lain, T.W., Nelson, D., Kingston, D., Johanson, D.: Decentralized cooperative aerial surveillance using fixed-wing miniature UAVs. Proc. IEEE **94**(7), 1306–1324 (2006)
10. Zhang, H.T., Zhai, C., Chen, Z.: A general alignment repulsion algorithm for focking of multi-agent systems. IEEE Trans. Autom. Control **56**(2), 430–435 (2011)
11. Zavlanos, M.M., Tanner, H.G., Jadbabaie, A.: Hybrid control for connectivity preserving focking. IEEE Trans. Autom. Control **54**(12), 2869–2875 (2009)
12. Lee, D., Spong, M.W.: Stable flocking of multiple inertial agents on balanced graphs. IEEE Trans. Autom. Control **52**(8), 14691475 (2007)
13. Moshtagh, N., Jadbabaie, A.: Distributed geodesic control laws for flocking of nonholonomic agents. IEEE Trans. Autom. Control **52**(4), 681–686 (2007)
14. Leonard, N.E., Fiorelli, E.: Virtual leaders, artificial potentials and coordinated control of groups. In: Proceedings of the IEEE Conference on Decision and Control, vol. 3, pp. 2968–2973 (2001)
15. Tanner, H.G., Jadbabaie, A., Pappas, G.J.: Stable flocking of mobile agents, Part I: fixed topology. In: Proceedings of the IEEE Conference on Decision and Control, vol. 2, pp. 2016–2021 (2003)
16. Olfati-Saber, R.: Folocking for multi-agnet dynamics systems: algorthms and theory. IEEE Trans. Autom. Control **51**(3), 401–420 (2006)
17. Shi, H., Wang, L., Chu, T.G.: Virtual leader approach to coordinated control of multiple mobile agents with asymmetric interactions. Phys. D **213**(1), 51–65 (2006)
18. Vicsek, T., Czirok, A., Ben-Jacob, E., Cohen, I., Shochet, O.: Novel type of phase transition in a system of self-driven particles. Phys. Rev. Lett. **75**(6), 1226–1229 (1995)
19. Chate, H., Ginlli, F., Gregoire, G., Peruani, F., Raynaud, F.: Modeling collective motion: variations on the Vicsek model. Eur. Phys. J. **64**(3), 451–456 (2008)
20. Baglietto, G., Albano, E.: Natuare of the oder-disorder transition in the Vicsek model for the collective motion of self-propelled particles. Phys. Rev. **80**, 050103 (2009)
21. Yang, W., Cao, L., Wang, X., Li, X.: Consensus in a heterogeneous influence network. Phys. Rev. **74**(2), 037101 (2006)
22. Li, W., Wang, X.: Adaptive velocity strategy for swarm aggregation. Phys. Rev. **75**, 021917 (2007)
23. Gao, J., Chen, Z., Cai, Y., Xu, X.: Approach to enhance convergence efficiency of Vicsek model. Control Decis. **24**(8), 1269–1272 (2009)

24. Tian, B.M., Wang, B.H.: Optirnal view angle in collective dynamics of self-propelled agents. Phys. Rev. **79**, 052102 (2009)
25. Yang, H., Huang, L.: Promoting collective motion of selfpropelled agents by distance-based influence. Phys. Rev. **89**(3), 032813 (2014)
26. Couzin, I., Krause, J., James, R., Ruxton, G., Franks, N.: Collective memory and spatial sorting in animal groups. J. Theor. Biol. **218**(1), 1–11 (2002)
27. Dong, H., Zhao, Y., Wu, J., Gao, S.: A velocity-adaptive couzin model and its performance. Phys. A **391**(5), 2145–2153 (2012)
28. Zhao, M., Su, H., Wang, M., Michael, Z.Q.: A weighted adaptive-velocity self-organizing model and its highspeed performance. Neurocomputing **216**(C), 402–408 (2016)
29. Su, H., Wang, X.: Flocking of multi-agents with a virtual leader. IEEE Trans. Autom. Control **54**(2), 293–307 (2009)
30. Gu, D., Wang, Z.: Leader-follower flocking: algorithms and experiments. IEEE Trans. Control Syst. Technol. **17**(5), 1211–1219 (2009)
31. Yu, W., Chen, G.: Robust adaptive focking control of nonlinear multi-agent systems. In: IEEE Multi-Conference on Systems and Control, pp. 363–367 (2010)
32. Su, H., Zhang, N., Chen, M., Wang, H., Wang, X.: Adaptive flocking with a virtual leader of multiple agents governed by locally Lipschitz nonlinearity. Nonlinear Anal. Real World Appl. **23**(9), 978–990 (2013)
33. Atrianfar, H., Haeri, M.: Flocking of multi-agent dynamic systems with virtual leader having the reduced number of informed agents. Trans. Inst. Meas. Control **35**(8), 1104–1115 (2013)
34. Peng, Z., Wang, D., Liu, H., Sun, G.: Neural adaptive control for leader-follower flocking of networked nonholonomic agents with unknown nonlinear dynamics. Int. J. Adapt. Control Signal Process. **28**(6), 479–495 (2014)
35. Ghapani, S., Mei, J., Ren, W., Song, Y.: Fully distributed flocking with a moving leader for lagrange networks with parametric uncertainties. Automatica **67**, 67–76 (2016)
36. Su, H., Wang, X., Yang, W.: Flocking in multi-agent systems with multiple virtual leaders. Asian J. Control **10**(2), 238–245 (2008)
37. Luo, X., Li, S., Guan, X.: Flocking algorithm with multitarget tracking for multi-agent systems. Pattern Recognit. Lett. **31**(9), 800–805 (2010)
38. Su, H.: Flocking in multi-agent systems with multiple virtual leaders based only on position measrements. Commun. Theor. Phys. **57**, 801–807 (2012)
39. Ji, M., Egerstedt, M.: Distributed coordination control of multiagent systems while preserving connectedness. IEEE Trans. Robot. **23**(4), 693–703 (2007)
40. Su, H., Wang, X.: Coordinated control of multiple mobile agents with connectivity preserving. In: Proceedings of the 17th World Congress (2008)
41. Su, H., Wang, X., Chen, G.: A connectivity-preserving flocking algoritnm for multi-agent systems based only on position measurements. Int. J. Control **82**, 1334–1343 (2009)
42. Wen, G., Duan, Z., Su, H., Chen, G., Yu, W.: A connectivity preserving flocking algorithm for multi-agent dynamical systems with bounded potential function. IET Control Theory Appl. **6**(6), 813–821 (2012)
43. Dong, Y., Huang, J.: Flocking with cnnectivity preservation of multiple double ingrator systems subject to external disturbances by a distributed control law. Automatica **55**, 197–203 (2015)
44. Wang, J., Zhao, H., Bi, Y., Shao, S., Liu, Q., Chen, X., Zeng, R., Wang, Y., Ha, L.: An improved fast flocking algorithm with obstacle avoidance for multiagent dynamic systems. J. Appl. Math. **2014**(4), 1–13 (2014)

45. Dai, B., Li, W.: Flocking of multi-agents with arbitrary shape obstacle. In: 2014 33rd Chinese Control Conference, pp. 1311–1316 (2014)
46. Yang, Z., Zhang, Q., Jiang, Z., Chen, Z.: Flocking of multi-agents with time delay. Int. J. Syst. Sci. **43**(11), 2125–2134 (2012)
47. Zhang, Q., Li, P., Yang, Z., Chen, Z.: Distance constrained based adaptive flocking control for multiagent networks with time delay. Math. Probl. Eng. **2015**(8), 1–8 (2015)
48. Hajar, A., Mohammad, H.: Adaptive flocking control of nonlinear multi-agent systems with directed switching to pologies and saturation constraints. J. Franklin Inst. **350**(6), 1545–1561 (2013)
49. Zhang, H., Cheng, Z., Chen, G., Li, C.: Model predictive flocking control for second-order multi-agent systems with input constraints. IEEE Trans. Circuits Syst. **62**(6), 1599–1606 (2015)
50. Wang, X., Li, X., Lu, J.: Control and flocking of networked systems via pinning. IEEE Circuits Syst. Mag. **10**(3), 83–91 (2010)
51. Su, H., Wang, X.: Distributed pinning-controlled consensus in a heterogeneous influence network. In: Su, H., Wang, X. (eds.) Pinning Control of Complex Networked Systems: Synchronization, Consensus and Flocking of Networked Systems via Pinning, pp. 103–110. Springer, Heidelberg (2013). https://doi.org/10.1007/978-3-642-34578-4_5
52. Steffen, L.: Event-triggered control of multi-agent systems with double-integrator dynamics: application to vehicle platooning and flocking algorithms. Automa. Control, 59–65 (2014)

# Optimizing Pinning Control of Directed Networks Using Spectral Graph Theory

Xuanhong Xu[1], Hui Liu[1(✉)], Jun-An Lu[2], and Jiangqiao Xu[1]

[1] School of Automation, Huazhong University of Science and Technology,
Wuhan 430074, People's Republic of China
{xhxu,hliu,xujiangqiao}@hust.edu.cn
[2] School of Mathematics and Statistics, Wuhan University, Wuhan 430072, China
jalu@whu.edu.cn

**Abstract.** Pinning control of a complex network aims at aligning the states of all the nodes to an external forcing signal by controlling a small number of nodes in the network. An algebraic graph-theoretic condition has been proposed to optimize pinning control for both undirected and directed network. The problem we are trying to solve in this paper is the optimization of pinning control in directed networks, where the effectiveness of pinning control can be measured by the smallest eigenvalue of the submatrix obtained by deleting the rows and columns corresponding to the pinned nodes from the symmetrized Laplacian matrix of the network when coupling strength and individual node dynamics are given. By analysing the spectral properties using the topology information of the directed network, a necessary condition that ensure the smallest eigenvalue greater than 0 is obtained. Upper bounds and lower bounds of the smallest eigenvalue are proved, and then an algorithm that optimize pinning scheme to improve the smallest eigenvalue is proposed. Illustrative examples are shown to demonstrate our theoretical results in the paper.

**Keywords:** Complex directed network · Pinning control
Spectral properties · Symmetrized Laplacian matrix

## 1  Introduction

Large graphs of real life can be modeled as complex networks of dynamical elements, which ranging from cell biology to epidemiology to the Internet [1–3]. Controlling a complex network to achieve synchronization has seized a rapidly growing interest in various large-scale networks [4–6]. In practical application, however, controlling all elements is very difficult and expensive, especially in large-scale networks. Hence, Wang, Chen and Li [7,8] put forward the concept

This work was supported by the National Natural Science Foundation of China under Grant Nos. 61773175 61403154, and the Natural Science Foundation of Hubei Province under Grant No. 2017CFB426.

© Springer Nature Switzerland AG 2018
Z. Chen et al. (Eds.): ICIRA 2018, LNAI 10984, pp. 59–70, 2018.
https://doi.org/10.1007/978-3-319-97586-3_5

of pinning control, which aims at aligning the states of all the nodes to an external forcing signal by only controlling a small fraction of network nodes.

Pinning control in directed networks has been extensively investigated over the past two decades [9–21], to name just a few. Reference [10] pointed out that the nodes with out-degrees bigger than in-degrees should be chosen as pinned candidates. Reference [11] proved that a complex network can be synchronized by pinning one node using linear controller with symmetric or asymmetric coupling matrix. Reference [13] discussed the pinning synchronization of strongly connected networks, networks containing at least a directed spanning tree, weakly connected networks, and directed forests.

For both undirected and directed networks, Ref. [22] has investigated the criterion for pinning synchronization and proposed that the effectiveness of pinning scheme is determined by the smallest eigenvalue ($\lambda_1$) of the submatrix obtained by deleting the rows and columns corresponding to the pinned nodes from the symmetrized Laplacian matrix of the network. The two problem we are going to solve are: *(i)* What kind of pinning scheme should we choose to ensure $\lambda_1$ greater than 0? Note that the corresponding value of $\lambda_1$ in a connected undirected network is absolutely greater than 0 with at least one node pinned [22]; *(ii)* What kind of pinning scheme does one need to choose to get relatively large $\lambda_1$ that can guarantee effective pinning control. It is an NP-hard problem to calculate all the possible eigenvalues obtained by the pinned-node combinations even when the number of pinned nodes is given, especially in a large network. Accordingly, it is meaningful to analyse the spectral properties of $\lambda_1$. Although the study of Laplacian matrices and grounded Laplacian matrices for undirected networks is fruitful [23–25], the counterpart for directed networks is limited [26–28]. By using tools from graph theory and matrix analysis [29,30], in this paper, upper bounds and lower bounds are proposed in order to provide guidance for choosing a proper pinning control scheme. Illustrative examples are shown to support our theoretical analysis.

The rest of the paper is organized as follows. In Sect. 2, we review a pinning synchronization criterion that is applicable for both undirected and directed dynamical networks. In Sect. 3, we investigate several spectral properties of $\lambda_1$ which provide effective guidance for the selection of pinned nodes. In Sect. 4, numerical simulations on specific graphs and real-world networks are given to show the applications of the proposed algorithms. Finally, conclusions are drawn in Sect. 5.

## 2    Network Model and Synchronization Criterion

Consider a directed controlled network consisting of $N$ nodes with diffusive couplings. The evolution of the state of node $i$ is given by

$$\dot{x}_i = f(x_i) - c \sum_{j=1}^{N} l_{ij} P x_j + u_i(x_1, \ldots, x_N), \quad i = 1, 2, \ldots, N, \qquad (1)$$

where each node satisfies the same dynamical function $f(\cdot)$, positive constant $c$ is the coupling strength of the network, positive semi-definite matrix $P : \mathbb{R}^n \to \mathbb{R}^n$ shows the inner coupling of two nodes, and controller applied at node $i$ is represented by $u_i$, which is to be designed. Graph $\mathbb{G} = (\mathcal{V}, \mathcal{E})$ describes the network topology, where $\mathcal{V} = \{1, \ldots, N\}$ is the node set and $\mathcal{E} \subseteq \mathcal{V} \times \mathcal{V}$ is the edge set. Laplacian matrix $L_N = [l_{ij}]_{N \times N}$ of graph $\mathbb{G}$ is defined as follows: if there is an edge from node $j$ to node $i$ ($i \neq j$), then $l_{ij} = -1$; $l_{ii} = -\sum_{j \neq i} l_{ij}$.

By assuming the target state of the network as $s(t)$, which satisfies

$$\dot{s}(t) = f(s(t)), \qquad s(0) = s_0. \tag{2}$$

Pinning control aims at aligning the states of all the nodes to $s(t)$.

**Assumption 1.** *There exists a sufficiently large positive constant $\alpha$, such that*

$$(y - z)^\top [(f(y) - f(z)) - \alpha P(y - z)] \leq -\mu \|y - z\|^2$$

*holds for some positive constant $\mu$ and for all vectors $y, z \in \mathbb{R}^n$.*

The set of controlled nodes is denoted as $\mathcal{S} = \{s_1, s_2, \ldots, s_l\}$, and $|\mathcal{S}|$ represents the number of nodes in $\mathcal{S}$, namely, $|\mathcal{S}| = l$. $\hat{L}_N = (L_N + L_N^\top)/2$ is the symmetrized matrix of $L_N$. Adopting the notations used in [29], $\hat{L}(\mathcal{S}|\mathcal{S})$ represents the $(N - l) \times (N - l)$ principal submatrix of $\hat{L}_N$ obtained by deleting the rows and columns corresponding to the nodes in $\mathcal{S}$ from $\hat{L}_N$.

Suppose that $M \in \mathbb{R}^{N \times N}$ is a real and symmetric matrix whose eigenvalues are arranged in a non-increasing order: $\lambda_N(M) \geq \lambda_{N-1}(M) \geq \cdots \geq \lambda_1(M)$, where $\lambda_1(M)$ is the smallest eigenvalue of $M$ and $\lambda_N(M)$ is the biggest one.

**Theorem 1** *(Theorem 1 from [22]). Suppose that Assumption 1 holds. Let $\mathcal{S}$ be the set of controlled nodes. The states of all the nodes in network (1) can synchronize to the target state $s(t)$ described by (2) using the adaptive pinning controllers and linear feedback controllers, if $\lambda_1(\hat{L}(\mathcal{S}|\mathcal{S})) > \alpha/c$. And, if the linear feedback controllers are used, the feedback gain must satisfies (10) in [22].*

In this paper, we focus on investigating the spectral properties of the symmetrized Laplacian matrix which give guidances for the selection of pinned nodes in directed networks.

## 3    Spectral Properties of $\lambda_1$ and Pinning Control Guidance

By using tools from spectral graph theory and matrix analysis [29,30], in this section, we mainly study the spectral properties of the symmetrized Laplacian matrix, and thus provide insights on the selection of pinning scheme.

**Lemma 1** *(A result in Chap. 1 from [29]). Let $A$ be a real symmetric $N \times N$ matrix with eigenvalues $\lambda_N(A) \geq \cdots \geq \lambda_1(A)$ arranged in a non-increasing order. Let $\|x\| = (\sum_{i=1}^N x_i^2)^{1/2}$. Then, $\lambda_N(A) = \max_{\|x\|=1}[x^\top A x]$, $\lambda_1(A) = \min_{\|x\|=1}[x^\top A x]$.*

From [22], for undirected networks, it is known that the smallest eigenvalue of $L_N$ is 0. But this is no longer valid for directed networks.

**Lemma 2.** $\lambda_1(\hat{L}_N)$ *is the smallest eigenvalue of the symmetrized matrix* $\hat{L}_N$, *then* $\lambda_1(\hat{L}_N) \leq 0$.

*Proof.* $L_N = [l_{ij}]$ is the Laplacian matrix of the directed network, and $\hat{L}_N = [\hat{l}_{ij}]$ is the corresponding symmetrized matrix. Let $x_0 = \frac{1}{\sqrt{N}}(1, \cdots, 1)^\top$ be a $N \times 1$ column vector, from Lemma 1, we have

$$\lambda_1(\hat{L}_N) = \min_{\|x\|=1} [x^\top \hat{L}_N x] \leq x_0^\top \hat{L}_N x_0$$

$$= \frac{1}{N}\sum_{i=1}^{N}\sum_{j=1}^{N} \hat{l}_{ij} = \frac{1}{N}\sum_{i=1}^{N}\sum_{j=1}^{N} \frac{l_{ij} + l_{ji}}{2} = \frac{1}{N}\sum_{i=1}^{N}\sum_{j=1}^{N} l_{ij} = 0.$$

**Definition 1** *[31]. For graph* $\mathbb{G} = (\mathcal{V}, \mathcal{E})$, *if* $\mathcal{V}' \subseteq \mathcal{V}$ *and* $\mathcal{E}' \subseteq \mathcal{E}$, *then* $\mathbb{G}' = (\mathcal{V}', \mathcal{E}')$ *is a subgraph of* $\mathbb{G}$. *Furthermore, if* $\mathbb{G}'$ *contains all the edges* $(i, j) \in \mathcal{E}$ *with* $i, j \in \mathcal{V}'$, *then* $\mathbb{G}'$ *is an induced subgraph of* $\mathbb{G}$.

**Definition 2.** *For a weakly connected directed graph* $\mathbb{G} = (\mathcal{V}, \mathcal{E})$, *if there is an induced subgraph* $\mathbb{T}$ *of* $\mathbb{G}$ *with node set* $\mathcal{T}$, *where subgraph* $\mathbb{T}$ *is strongly connected and all the edges between* $\mathcal{T}$ *and* $\mathcal{V} \setminus \mathcal{T}$ *are unidirectional with start nodes in* $\mathcal{T}$ *and end nodes in* $\mathcal{V} \setminus \mathcal{T}$. *Then, subgraph* $\mathbb{T}$ *is a* root-subgraph *of* $\mathbb{G}$.

The root-subgraph of a strongly connected graph is itself. From Definition 2, one can see that there may exist more than one root-subgraphs in a weakly connected graph $\mathbb{G}$. And root-subgraphs are obtained from every weakly connected component in the graph $\mathbb{G}$. Note that we consider an isolated node as one strongly connected component. Then, the *leaf-subgraph* is defined as the induced subgraph of graph $\mathbb{G}$ with nodes not in the root-subgraphs. Take Fig. 1 as an example, the root-subgraphs are denoted by $\mathbb{T}_i$ where $i = 1, 2, 3$, and the leaf-subgraph by $\mathbb{Q}$.

**Theorem 2.** *Suppose that* $\mathcal{S}$ *is the set of controlled nodes for a directed network* $\mathbb{G}$. *If there exists at least one root-subgraph without any pinned node, then* $\lambda_1(\hat{L}(\mathcal{S}|\mathcal{S})) \leq 0$.

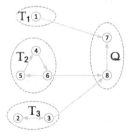

**Fig. 1.** A directed network with 8 nodes.

*Proof.* For convenience, firstly, we consider the case that the directed network only have one root-subgraph $\mathbb{T}$ with $h \leq N$ nodes, and renumber the nodes in $\mathcal{V}$ such that the nodes in $\mathbb{T}$ are the first $h$ nodes in $\mathcal{V}$, i.e., $\mathbb{T} = \{1, \cdots, h\}$. Notice that the diagonal elements of $\hat{L}_N$ are the in-degrees of the corresponding nodes and there is no edge pointing to a node in the root-subgraphs from a node in the leaf-subgraph, then for the Laplacian matrix $L_N = [l_{ij}]_{N \times N}$, we have

$$l_{ij} = 0 \text{ for } 1 \leq i \leq h \text{ and } j > h, \text{ i.e., } L_N = \begin{bmatrix} L(\mathbb{T})_{h \times h} & \mathbf{0} \\ L_{21} & L_{22} \end{bmatrix}_{N \times N}, \text{ where } L(\mathbb{T}) =$$

$[t_{ij}]_{h \times h}$ is the Laplacian matrix of the root-subgraph $\mathbb{T}$. It follows that $\hat{L}_N = \begin{bmatrix} \hat{L}(\mathbb{T})_{h \times h} & L'_{12} \\ L'_{21} & L'_{22} \end{bmatrix}_{N \times N}$, where $\hat{L}(\mathbb{T}) = [\hat{t}_{ij}]_{h \times h}$ is the symmetrized form of $L(\mathbb{T})$.

Suppose that $\mathcal{S} \subset \mathcal{V} \backslash \{1, \cdots, h\}$ for $|\mathcal{S}| = l \leq N - h$, namely, none of the nodes in root-subgraph $\mathbb{T}$ is pinned. It follows that $\hat{L}(\mathcal{S}|\mathcal{S}) = \begin{bmatrix} \hat{L}(\mathbb{T})_{h \times h} & L''_{12} \\ L''_{21} & L''_{22} \end{bmatrix}_{(N-l) \times (N-l)}$.

Let $x_0 = \frac{1}{\sqrt{h}}(1, \cdots, 1, 0, \cdots, 0)^\top$ be a $(N - l) \times 1$ column vector, where the first $h$ elements of $x_0$ are $\frac{1}{\sqrt{h}}$ and the other are 0. From Lemma 1, we have

$$\lambda_1(\hat{L}(\mathcal{S}|\mathcal{S})) = \min_{\|x\|=1} [x^\top \hat{L}(\mathcal{S}|\mathcal{S})x] \leq x_0^\top \hat{L}(\mathcal{S}|\mathcal{S})x_0 = \frac{1}{h}\sum_{i=1}^N \sum_{j=1}^N \hat{t}_{ij} = 0,$$

namely, $\lambda_1(\hat{L}(\mathcal{S}|\mathcal{S})) \leq 0$.

A similar proof also applies to the case when $\mathbb{G}$ is composed of multiple root-subgraphs.

*Remark 1.* A necessary condition for $\lambda_1(\hat{L}(\mathcal{S}|\mathcal{S})) > 0$ is obtained in Theorem 2, i.e., there is at least one node controlled in each root-subgraph. Actually, if the control signal is considered as the state of a virtual dynamical system, Ref. [16] proposed that pinning control can be achieved in a complex network of dynamical systems if and only if forcing is applied to roots of trees in a spanning directed forest of the interaction graph of $\mathbb{G}$ by analysing the Laplacian matrix. From this paper, pinning control can be achieved if there exists at least one controlled node in each root-subgraph.

In order to prove a upper bound for $\lambda_1(\hat{L}(\mathcal{S}|\mathcal{S}))$, we first introduce some notations. Let $k_i^{(in)}$ and $k_i^{(out)}$ denote the in-degree and out-degree of node $i$, respectively. For node set $\mathcal{V}$ and controlled node set $\mathcal{S}$, $k_{\mathcal{S}}^{(in)}$ and $k_{\mathcal{S}}^{(out)}$ denote the number of one-way edges from nodes in $\mathcal{V} \backslash \mathcal{S}$ to nodes $\mathcal{S}$ and the number of one-way edges from nodes in $\mathcal{S}$ to nodes in $\mathcal{V} \backslash \mathcal{S}$, respectively. Note that the two-way edge is equal to two one-way edges with different direction. Obviously, $k_{\mathcal{S}}^{(in)} = k_{\mathcal{V} \backslash \mathcal{S}}^{(out)}$ and $k_{\mathcal{S}}^{(out)} = k_{\mathcal{V} \backslash \mathcal{S}}^{(in)}$.

**Theorem 3.** *For $\mathcal{S} = \{s_1, \ldots, s_l\}$ and $\mathcal{V} \backslash \mathcal{S} = \{p_1, \ldots, p_{N-l}\}$. Let $w_{p_j}^{(in)}$ be the number of one-way edges from nodes in $\mathcal{S}$ to $p_j \in \mathcal{V} \backslash \mathcal{S}$ and $w_{p_j}^{(out)}$ be the number of one-way edges from $p_j$ to nodes in $\mathcal{S}$. Similarly, let $w_{s_j}^{(in)}$ and $w_{s_j}^{(out)}$ be the* *number of one-way edges from nodes in $\mathcal{V}\setminus\mathcal{S}$ to $s_j$ and from $s_j$ to nodes in $\mathcal{V}\setminus\mathcal{S}$, respectively. It holds that $\lambda_1\big(\hat{L}(\mathcal{S}|\mathcal{S})\big) \leq \frac{k^{(in)}_{\mathcal{V}\setminus\mathcal{S}}}{N-l} \leq \max_{p_i\in\mathcal{V}\setminus\mathcal{S}}[w^{(in)}_{p_1},\ldots,w^{(in)}_{p_{N-l}}] \leq l.$ which is equivalent to*

$$\lambda_1\big(\hat{L}(\mathcal{S}|\mathcal{S})\big) \leq \frac{k^{(out)}_{\mathcal{S}}}{N-l} \leq \frac{l\max_{s_i\in\mathcal{S}}[w^{(out)}_{s_1},\ldots,w^{(out)}_{s_l}]}{N-l} \leq l \qquad (3)$$

*since $k^{(in)}_{\mathcal{S}} = k^{(out)}_{\mathcal{V}\setminus\mathcal{S}}$.*

*Proof.* Write $\hat{L}_N = [\hat{l}_{ij}]$ as $\hat{L}_N = H + \Lambda$, where $H$ is a zero row sum matrix and $\Lambda$ is a diagonal matrix. Consider that $\hat{l}_{ii}$ is the in-degree of node $i$ and the sum of $i$th row in $\hat{L}_N$ is equal to $k^{(in)}_i - \frac{k^{(in)}_i+k^{(out)}_i}{2} = \frac{k^{(in)}_i-k^{(out)}_i}{2}$, it follows that $\Lambda = diag\{\frac{k^{(in)}_1-k^{(out)}_1}{2}, \frac{k^{(in)}_2-k^{(out)}_2}{2}, \ldots, \frac{k^{(in)}_N-k^{(out)}_N}{2}\}$. Similarly, write $\hat{L}(\mathcal{S}|\mathcal{S})$ as the sum of one zero row sum matrix and other two diagonal matrices, namely, $\hat{L}(\mathcal{S}|\mathcal{S}) = H' + \Gamma + \Lambda(\mathcal{S}|\mathcal{S})$, where $H'$ is a zero row sum matrix and $\Lambda(\mathcal{S}|\mathcal{S})$ is the corresponding deleting matrix of $\Lambda$. It is not difficult to work out that $\Gamma = diag\{\frac{w^{(in)}_{p_1}+w^{(out)}_{p_1}}{2}, \frac{w^{(in)}_{p_2}+w^{(out)}_{p_2}}{2}, \ldots, \frac{w^{(in)}_{p_{N-l}}+w^{(out)}_{p_{N-l}}}{2}\}$.
Let $x_0 = \frac{1}{\sqrt{N-l}}(1,1,\ldots,1)^\top$ be an $(N-l)\times 1$ column vector. From Lemma 1, one has

$$
\begin{aligned}
&\lambda_1\big(\hat{L}(\mathcal{S}|\mathcal{S})\big)\\
&= \min_{\|x\|=1}[x^\top(H'+\Gamma+\Lambda(\mathcal{S}|\mathcal{S}))x] = \min_{\|x\|=1}[x^\top H'x + x^\top\Gamma x + x^\top\Lambda(\mathcal{S}|\mathcal{S})x]\\
&\leq x_0^\top H'x_0 + x_0^\top\Gamma x_0 + x_0^\top\Lambda(\mathcal{S}|\mathcal{S})x_0\\
&= \frac{(w^{(in)}_{p_1}+w^{(out)}_{p_1})+\cdots+(w^{(in)}_{p_{N-l}}+w^{(out)}_{p_{N-l}})}{2(N-l)}\\
&\quad + \frac{(k^{(in)}_{p_1}-k^{(out)}_{p_1})+\cdots+(k^{(in)}_{p_{N-l}}-k^{(out)}_{p_{N-l}})}{2(N-l)}\\
&= \frac{(w^{(in)}_{p_1}+\cdots+w^{(in)}_{p_{N-l}})+(w^{(out)}_{p_1}+\cdots+w^{(out)}_{p_{N-l}})}{2(N-l)}\\
&\quad + \frac{(k^{(in)}_{p_1}+\cdots+k^{(in)}_{p_{N-l}})-(k^{(out)}_{p_1}+\cdots+k^{(out)}_{p_{N-l}})}{2(N-l)}\\
&= \frac{k^{(in)}_{\mathcal{V}\setminus\mathcal{S}}+k^{(out)}_{\mathcal{V}\setminus\mathcal{S}}}{2(N-l)} + \frac{k^{(in)}_{\mathcal{V}\setminus\mathcal{S}}-k^{(out)}_{\mathcal{V}\setminus\mathcal{S}}}{2(N-l)} = \frac{k^{(in)}_{\mathcal{V}\setminus\mathcal{S}}}{N-l}(\frac{k^{(out)}_{\mathcal{S}}}{N-l})\\
&= \frac{w^{(in)}_{p_1}+\cdots+w^{(in)}_{p_{N-l}}}{N-l}(\frac{w^{(out)}_{s_1}+\cdots+w^{(out)}_{s_l}}{N-l})\\
&\leq \frac{(N-l)\max_{p_i\in\mathcal{V}\setminus\mathcal{S}}[w^{(in)}_{p_1},\ldots,w^{(in)}_{p_{N-l}}]}{N-l}(\frac{l\max_{s_i\in\mathcal{S}}[w^{(out)}_{s_1},\ldots,w^{(out)}_{s_l}]}{N-l}) \leq l.
\end{aligned}
$$

This completes the proof.

*Remark 2.* From Theorem 3, when the number of pinned nodes is given, the upper bound of $\lambda_1(\hat{L}(\mathcal{S}|\mathcal{S}))$ can be improved by increasing the number of the edges from set $\mathcal{S}$ to set $\mathcal{V} \setminus \mathcal{S}$. This is significant for optimizing pinning control of directed networks.

**Corollary 1.** *Note that the diagonal elements of $\hat{L}(\mathcal{S}|\mathcal{S})$ are equal to the in-degrees of the nodes in $\mathcal{V} \setminus \mathcal{S}$ correspondingly. According to Theorem 6 in [22], it holds that $\lambda_1(\hat{L}(\mathcal{S}|\mathcal{S})) \leq k_{\min}^{(in)}$, where $k_{\min}^{(in)}$ is the minimal in-degree of the uncontrolled nodes.*

According to (3), the denominator of the upper bound $\frac{k_{\mathcal{S}}^{(out)}}{N-l}$ of $\lambda_1(\hat{L}(\mathcal{S}|\mathcal{S}))$ is large when the number of controlled nodes $l$ is small. In this case with $l$ is small, one should control the nodes with large out-degree to maximize $k_{\mathcal{S}}^{(out)}$. As $l$ increases, $\frac{k_{\mathcal{S}}^{(out)}}{N-l}$ also increases. At this time, $\lambda_1(\hat{L}(\mathcal{S}|\mathcal{S}))$ is still restricted by the minimal in-degree of uncontrolled nodes according to Corollary 1. Hence, one should control the nodes with smallest in-degree. Inspired by the above analysis, as the number of controlled nodes $l$ is given, we define an attribute for node $i$: $krd_i = (1-r)k_i^{(out)} - rk_i^{(in)}$, where $r = \frac{l}{N}$. Then, for a directed network with $N$ nodes, an algorithm to improve $\lambda_1(\hat{L}(\mathcal{S}|\mathcal{S}))$ is proposed as follows. where, firstly, the nodes with zero in-degree should be controlled according to Remark 1.

**Algorithm 1.** *For a given number of controlled nodes $l$:*

*(1) Calculate krd for every nodes;*
*(2) Sort the network nodes: the nodes with zero in-degree and big out-degree, followed by other nodes in non-increasing order based on their krd. For the nodes with same krd, sort them in a non-increasing order according to their out-degrees when $\frac{l}{N} \leq 50\%$, otherwise, sort them in a non-decreasing order according to their in-degrees. If the sorting value of multiple nodes are equivalent under all conditions, sort such nodes randomly;*
*(3) Choose the first $l$ nodes to be controlled.*

Note that, from Algorithm 1, the order of nodes may be different under different number of controlled nodes.

## 4    Applications

In this section, a specific graph and a real-world network are taken as examples to illustrate the effectiveness of pinning scheme according to Algorithm 1. Also comparison with the pinning scheme introduced by Remark 5 in [10] are made.

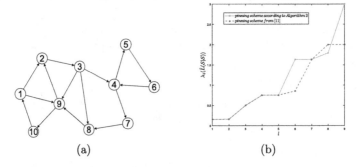

**Fig. 2.** (a) A directed network with 10 nodes. (b) The trends of $\lambda_1(\hat{L}(\mathcal{S}|\mathcal{S}))$ under different pinning scheme, as the number of controlled nodes $l$ increases from 1 to 9. (Color figure online)

**Example 1.** *Figure 2(a) is taken from the simulation part of paper [10]. According to Remark 5 of [10], it gives an order of all the nodes as follows: $k^{(out)} - k^{(in)}$:* $3, 1, 4, 5, 6, 7, 10, 9, 2, 8$. *Then, choose the first $l$ nodes to be controlled. According to Algorithm 1, the controlled node set $\mathcal{S}_l$ with $l$ nodes are shown as follows:*

$$\mathcal{S}_1 = \{3\}; \mathcal{S}_2 = \{3, 1\}; \mathcal{S}_3 = \{3, 1, 4\}; \mathcal{S}_4 = \{3, 1, 4, 5\}; \mathcal{S}_5 = \{3, 1, 4, 5, 6\};$$
$$\mathcal{S}_6 = \{3, 1, 5, 6, 7, 10\}; \mathcal{S}_7 = \{3, 1, 5, 6, 7, 10, 4\};$$
$$\mathcal{S}_8 = \{3, 1, 5, 6, 7, 10, 4, 2\}; \mathcal{S}_9 = \{3, 1, 5, 6, 7, 10, 4, 2, 8\}.$$

*The trends of $\lambda_1(\hat{L}(\mathcal{S}|\mathcal{S}))$ under different pinning schemes are shown in Fig. 2(b). The blue line is drew using our proposed algorithm, and the red one using the algorithm in [10]. One can see from Fig. 2(b) that our proposed method works no worse than the other one, only except when $l = 8$.*

**Example 2.** *Figure 3 captures innovation spread among 241 physicians in towns from Illinois, Peoria, Bloomington, Quincy and Galesburg. A node represents a physician and an edge between two physicians shows that the left physician told that the right physician is his friend or that he turns to the right physician if he needs advice or is interested in a discussion. There always only exists one edge between two nodes even if more than one of the listed conditions are true [32, 33]. Figure 4 shows the trends of $\lambda_1(\hat{L}(\mathcal{S}|\mathcal{S}))$ under the pinning schemes according to Algorithm 1 and [10], respectively.*

From Figs. 2(b) and 4, $\lambda_1(\hat{L}(\mathcal{S}|\mathcal{S}))$ under pinning scheme according to Algorithm 1 is generally larger than that under pinning scheme according to [10] when $\frac{l}{N} \geq 50\%$, especially when $l > 134$, i.e., $\lambda_1(\hat{L}(\mathcal{S}|\mathcal{S})) \geq 0.7929$, in Fig. 4.

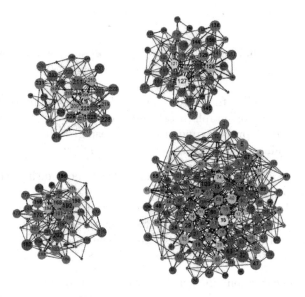

**Fig. 3.** Network captures innovation spread among 241 physicians, where the size of nodes indicates the out-degree of the nodes, the larger the node, the greater the out-degree of the node. The color depth of the nodes indicates the in-degree of the nodes, the deeper the color of the node, the smaller the in-degree of the node. (Color figure online)

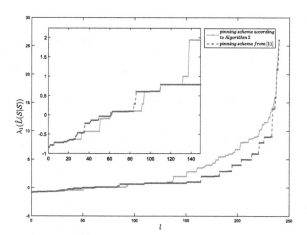

**Fig. 4.** The trends of $\lambda_1(\hat{L}(\mathcal{S}|\mathcal{S}))$ under different pinning scheme, as the number of controlled nodes $l$ increases from 1 to 240, where the small graph is the magnified graph when $l$ increases from 1 to 150.

## 5   Conclusion

Pinning synchronization criterion is reviewed for a complex directed network in this paper, which shows that the effectiveness of pinning control is measured by the smallest eigenvalue ($\lambda_1$) of the submatrix obtained by deleting the rows and columns corresponding to the pinned nodes from the symmetrized Laplacian matrix. By using tools from graph theory and matrix analysis, spectral properties of $\lambda_1$ are discussed to provide guidance for choosing a proper node set to be controlled. The main results obtained are listed as follows:

(i) A necessary condition for $\lambda_1 > 0$ is that at least one node in each root-graph of the network should be controlled;

(ii) The upper bound of $\lambda_1$ can be improved by increasing the number of the edges from the controlled nodes to the uncontrolled nodes;

(iii) The nodes with the smallest in-degree should be controlled when $\lambda_1$ is restricted by the value of the smallest in-degree.

An algorithm to achieve pinning synchronization and to improve $\lambda_1$ is also proposed, which has been tested by specific examples and real-world networks. As our future work, we are interested in investigating further spectral properties of the symmetrized Laplacian matrices and exploring more applications of the spectral properties to more typical and large-scale networks.

## References

1. Strogatz, S.H.: Exploring complex networks. Nature **410**(7), 268–276 (2001)
2. Albert, R., Barabási, A.-L.: Statistical mechanics of complex networks. Rev. Mod. Phys. **74**(1), 47–97 (2002)
3. Dorogovtsev, S.N., Mendes, J.F.F.: Evolution of networks. Adv. Phys. **51**(4), 1079–1187 (2002)
4. Jadbabaie, A., Lin, J., Morse, A.S.: Coordination of groups of mobile autonomous agents using nearest neighbor rules. IEEE Trans. Autom. Control **48**(6), 988–1001 (2003)
5. Dörfler, F., Chertkov, M., Bullo, F.: Synchronization in complex oscillator networks and smart grids. Proc. Natl. Acad. Sci. USA **110**(6), 2005–2010 (2013)
6. Zanette, D.H.: Dynamics of rumor propagation on small-world networks. Phys. Rev. E **65**(4), 041908 (2002)
7. Wang, X., Chen, G.: Pinning control of scale-free dynamical networks. Phys. A. Stat. Mech. Its Appl. **310**(3), 521–531 (2002)
8. Li, X., Wang, X., Chen, G.: Pinning a complex dynamical network to its equilibrium. IEEE Trans. Circ. Syst. I Reg. Pap. **51**(10), 2074–2087 (2004)
9. Zhou, J., Lu, J.-A., Lü, J.: Pinning adaptive synchronization of a general complex dynamical network. Automatica **44**(4), 996–1003 (2008)
10. Song, Q., Cao, J.: On pinning synchronization of directed and undirected complex dynamical networks. IEEE Trans. Circ. Syst. I Reg. Pap. **57**(3), 672–680 (2010)
11. Chen, T., Liu, X., Lu, W.: Pinning complex networks by a single controller. IEEE Trans. Circ. Syst. I Reg. Pap. **54**(6), 1317–1326 (2007)

12. Chen, G.: Pinning control and synchronization on complex dynamical networks. Int. J. Control Autom. Syst. **12**(2), 221–230 (2014)
13. Yu, W., Chen, G., Lü, J., Kurths, J.: Synchronization via pinning control on general complex networks. SIAM J. Control Optim. **51**(2), 1395–1416 (2013)
14. Wang, J.-L., Wu, H.-N., Huang, T., Ren, S.-Y., Wu, J.: Pinning control for synchronization of coupled reaction-diffusion neural networks with directed topologies. IEEE Trans. Syst. Man Cybern. Syst. **46**(8), 1109–1120 (2016)
15. Chen, L., Yu, X., Sun, C.: Characteristic modeling approach for complex network systems. IEEE Trans. Syst. Man Cybern. Syst. (2017). https://doi.org/10.1109/TSMC.2017.2737438
16. Wu, C.W.: Localization of effective pinning control in complex networks of dynamical systems. In: Proceedings of the IEEE International Symposium on Circuits and Systems, pp. 2530–2533 (2008)
17. Xu, C., Zheng, Y., Su, H., Chen, M.Z.Q., Zhang, C.: Cluster consensus for second-order mobile multi-agent systems via distributed adaptive pinning control under directed topology. Nonlinear Dyn. **83**(4), 1975–1985 (2016)
18. Liu, X., Li, P., Chen, T.: Cluster synchronization for delayed complex networks via periodically intermittent pinning control. Neurocomputing **162**, 191–200 (2015)
19. Cai, S., Jia, Q., Liu, Z.: Cluster synchronization for directed heterogeneous dynamical networks via decentralized adaptive intermittent pinning control. Nonlinear Dyn. **82**(1–2), 689–702 (2015)
20. Li, H., Chen, G., Xiao, L.: Event-triggered sampling scheme for pinning control in multi-agent networks with general nonlinear dynamics. Neural Comput. Appl. **27**(8), 2587–2599 (2016)
21. Delellis, P., Garofalo, F., Iudice, F.L.: The partial pinning control strategy for large complex networks. Autom. A J. IFAC Int. Fed. Autom. Control **89**, 111–116 (2018)
22. Liu, H., Xu, X., Chen, G., Lu, J.-A.: Optimizing pinning control of complex dynamical networks based on spectral properties of grounded Laplacian matrices. arxiv.org/abs/1804.10818v1 [math.OC] (2018)
23. Pirani, M., Sundaram, S.: On the smallest eigenvalue of grounded Laplacian matrices. IEEE Trans. Autom. Control **61**(2), 509–514 (2016)
24. Pirani, M., Sundaram, S.: Spectral properties of the grounded Laplacian matrix with applications to consensus in the presence of stubborn agents. In: Proceedings of the 2014 American Control Conference, pp. 2160–2165 (2014)
25. Pirani, M., Shahrivar, E. M., Sundaram, S.: Coherence and convergence rate in networked dynamical systems. In: Proceedings of the 54th IEEE Conference on Decision and Control, pp. 968–973 (2015)
26. Agaev, R., Chebotarev, P.: On the spectra of nonsymmetric Laplacian matrices. Linear Algebra Its Appl. **399**(1), 157–168 (2005)
27. Hao, H., Barooah, P.: Asymmetric control achieves size-independentstability margin in 1-D flocks. In: Proceedings of the 50th IEEE Conference on Decision and Control and European Control Conference, pp. 3458–3463 (2011)
28. Xia, W., Cao, M.: Analysis and applications of spectral properties of grounded Laplacian matrices for directed networks. Automatica **80**, 10–16 (2017)
29. Bapat, R.B.: Graphs and Matrices. Springer, New York (2010). https://doi.org/10.1007/978-1-84882-981-7
30. Horn, R.A., Johnson, C.R.: Matrix Analysis, 2nd edn. Cambridge University Press, New York (1994)
31. Diestel, R.: Graph Theory. Graduate Texts in Mathematics, 2nd edn. Springer, Heidelberg (2017). https://doi.org/10.1007/978-3-662-53622-3

32. Physicians network dataset - KONECT, April 2017. http://konect.uni-koblenz.de/networks/moreno_innovation
33. Coleman, J., Katz, E., Menzel, H.: The diffusion of an innovation among physicians. Sociometry **20**, 253–270 (1957)

# A Novel Variable-Gain Rectilinear or Circular Formation Algorithm for Unicycle Type Vehicles

Shuai Wang[1], Jiahu Qin[1(✉)], Qingchen Liu[2], and Yu Kang[1]

[1] Department of Automation, University of Science and Technology of China,
Hefei 230027, Anhui, China
`wsustcid@mail.ustc.edu.cn`, {`jhqin,kangduyu`}`@ustc.edu.cn`
[2] Research School of Engineering, Australian National University,
Canberra, ACT 0200, Australia
`liuqingchen1989@gmail.com`

**Abstract.** In this paper, we propose a novel variable-gain formation algorithm to steer a group of unicycle type vehicles moving in straight lines or circular orbits with three types of phase configurations (synchronized, balanced and stabilization of the average linear momentum). The algorithm design is carried out from the viewpoint of optimization theory to guarantee that control gains are variable. Specifically, a step length search algorithm used in optimization methods is employed to update the control gain at each iteration. The implementation details of the rectilinear/circular formation algorithm are given to show that the three types of phase configurations can be reached by utilizing corresponding well-designed objective functions. Furthermore, global convergence properties of the formation algorithm are analyzed. Both the results of simulations and experiments show good performance of the proposed formation algorithm.

**Keywords:** Formation control · Phase configuration
Unicycle type vehicles · Optimization methods

## 1 Introduction

A team of mobile robots moving in rectilinear and circular formation have been widely used in both military and civilian scenarios such as ground cleaning, battlefield surveillance and source seeking [1–3]. In recent years, those formation control problems have been investigated with various robot models, among which unicycle models are frequently used in describing Unmanned Aerial Vehicles (UAVs) and Wheeled Mobile Robots (WMRs) [4].

A brief review of formation control of unicycle type robots is provided as below. In [5], the authors assume that both the orientation and the speed of each unicycle can be controlled, which increases the complexity of controlling

© Springer Nature Switzerland AG 2018
Z. Chen et al. (Eds.): ICIRA 2018, LNAI 10984, pp. 71–80, 2018.
https://doi.org/10.1007/978-3-319-97586-3_6

robots. In [6], oscillatory control is applied to control constant-speed unicycle robots. However, the speeds of all the robots are required to be identical. Ref. [7] focuses on the patterns of formation. Several final formation patterns are generated by a centralized *Hungarian* algorithm for optimizing the robots' goal positions. The disadvantage of it is that all the robots can not keep moving with the final formation patterns. A more complex formation control problem is discussed in [8–12], where all the robots not only move in desired circular orbits, but also maintain a particular phase configuration. [8,9] focus on achieving synchronized and balanced phase configuration for a group of robots with all-to-all and limited communication, respectively. *In synchronized phase configuration, all the robots move in a common direction, while in balanced phase configuration, the sum of the phasors of all the robots is zero.* In [10,11], the authors propose suitable feedback control laws for achieving particular type of circular formation with synchronized or balanced phase configuration. It should be noted that all the methods proposed in refs. [6,8–11] are based on the assumption that the speeds of all the robots are identical. Under this assumption, many practical formation patterns can not be achieved for a group of robots with identical angular frequencies such as concentric circular formation with different radii. This assumption is removed in [12] and another phase configuration is defined as *the stabilization of the average linear momentum, which means that the average position of all the robots stabilize on a fixed point.* However, the control gains designed in all of the above-mentioned control schemes are fixed, which decreases the usability of these methods since selecting appropriate control gains manually is very time-consuming.

From the above discussion, we know that both the identical speeds and fixed control gains have many limitations in practice. Motivated by these facts, it is more realistic to propose a variable-gain formation algorithm for multiple unicycle type mobile robots with nonidentical speeds. In this paper, we design a variable-gain Rectilinear/Circular Formation Algorithm (RCFA) based on *Broyden-Fletcher-Goldfarb-Shanno (BFGS)* method [13], in which all the robots are required to move in straight lines or circles with three types of phase configurations (synchronized, balanced and stabilization of the average linear momentum). Those formation patterns have several potential applications such as crops harvesting and spying work in adversarial areas. The main contributions of this paper are listed as follows:

1. Unlike these existing control methods, we solve the rectilinear and circular formation control problems from the viewpoint of optimization theory. In this case, three types of phase configurations can be reached by solving corresponding optimization problems and the rectilinear/circular motion of robots can be achieved by utilizing particular control laws. In particular, we introduce a step length search algorithm into RCFA to update the control gain at each iteration. Thus, the formation algorithm proposed in this paper has variable control gains.

2. The control input designed in this paper contains only the angular frequency and there is no requirements for the speeds of robots. Thus, RCFA

acquires more flexibility than other methods. Furthermore, global convergence of RCFA can be ensured while subject to some mild assumptions.

The remainder of this paper is organized as follows. Section 2 describes the system model used in this paper. RCFA and global convergence of it are presented in Sect. 3. In Sect. 4, the algorithms are illustrated by simulations and experiments. Finally, Sect. 5 concludes the paper.

## 2    Preliminaries

### 2.1    System Model

In this section, we represent the kinematic model of the unicycle type mobile robots in the form of discrete-time integration [14]. Consider a group of $N$ unicycle robots, the kinematic model of the robot $m$ is described by

$$x_{m,k+1} = x_{m,k} + v_m \cos \theta_{m,k} \Delta t \tag{1a}$$

$$y_{m,k+1} = y_{m,k} + v_m \sin \theta_{m,k} \Delta t \tag{1b}$$

$$\theta_{m,k+1} = \theta_{m,k} + u_{m,k} \Delta t \tag{1c}$$

where $k$ denotes the $k$th iteration and $\theta_m$ is the orientation of robot $m$ and $x_m, y_m \in \mathcal{R}$ are the coordinates in the plane, for $m = 1, \ldots, N$. The robot $m$ has a positive constant speed $v_m > 0$ and $u_m \in \mathcal{R}$ is the control input (angular frequency). Note that $\Delta t$ is the sampling interval and it will be neglected in the following analysis without loss of generality.

### 2.2    Problem Formulation

In this paper, we investigate the rectilinear and circular formation control problems from the viewpoint of optimization theory. In particular, a novel variable-gain formation algorithm are proposed to make a group of robots keep moving in rectilinear/circular formation with three types of phase configurations. There are two major issues to be tackled in the design of the formation algorithm. The first issue is that how to make the phase angles of robots reach a desired phase configuration, and the other is that how to make all the robots keep moving in straight lines or circles with one of the three types of phase configurations.

## 3    Main Results

### 3.1    Achieving Rectilinear/Circular Formation with Synchronized and Balanced Phase Configuration

For the synchronized and balanced phase configuration, there is a so-called order parameter defined as

$$p_\theta = \frac{1}{N} \sum_{m=1}^{N} e^{i\theta_m}. \tag{2}$$

to measure the synchrony in the networks of $N$ coupled oscillators [15]. When all phase angles of robots are synchronized, all the robots have a common phase angle. We have $|p_\theta| = 1$. The balanced phase configuration means that the sum of the phasors of $N$ coupled robots is zero, i.e., $|p_\theta| = 0$.

We solve the first issue by defining the corresponding objective functions. In order to make a group of $N$ robots reach the synchronized or balanced phase configuration, the objective function $f_{p_\theta}(\boldsymbol{\theta}) : \mathcal{R}^N \to \mathcal{R}$ is defined as

$$f_{p_\theta}(\boldsymbol{\theta}) = \gamma \frac{N}{2} |p_\theta|^2 \tag{3}$$

where $\boldsymbol{\theta} = [\theta_1, \ldots, \theta_N]^T$ and $\gamma = \pm 1$. Each element of $\boldsymbol{\theta}$ is the phase angle of each robot. From the definitions of the synchronized and balanced phase configuration we know that $f_{p_\theta}(\boldsymbol{\theta})$ reaches its unique minimum when all phase angles of robots are synchronized (for $\gamma = -1$) or balanced (for $\gamma = 1$). Thus, the phase control problem is successfully transformed into an unconstrained nonlinear optimization problem by defining the above objective function, i.e., the synchronized and balanced phase configuration can be achieved by minimizing $f_{p_\theta}$.

The second issue can be tackled by utilizing particular control laws in RCFA. RCFA is designed based on $BFGS$ method to minimize the predefined objective function. $BFGS$ is an iterative method for solving unconstrained nonlinear optimization problem. It begins with an initial state of the parameter $\boldsymbol{\theta}$ and generates a sequence of improved estimates along descent direction. Finally, they will stop iterating when the minimum of the objective function is found. Thus, the main work of us is that designing appropriate descent directions, control laws and the rule of updating step length according to the principles of the general optimization methods.

The descent direction of the $BFGS$ method at the $k$th iteration is defined as

$$\boldsymbol{d}_k = -\boldsymbol{B}_k^{-1} \boldsymbol{g}_k \tag{4}$$

where $\boldsymbol{B}_k$ is an approximation of the true Hessian $\boldsymbol{H}_k$ of the objective function and is a symmetric positive definite matrix and $\boldsymbol{g}_k$ is the gradient of $f_{p_\theta}(\boldsymbol{\theta}_k)$. $\boldsymbol{d}_k$ is a descent direction ($\boldsymbol{g}_k^T \boldsymbol{d}_k < 0$). $\boldsymbol{B}_k$ will be updated at every iteration by the following

$$\boldsymbol{B}_{k+1} = \boldsymbol{B}_k - \frac{\boldsymbol{B}_k \boldsymbol{s}_k \boldsymbol{s}_k^T \boldsymbol{B}_k}{\boldsymbol{s}_k^T \boldsymbol{B}_k \boldsymbol{s}_k} + \frac{\boldsymbol{y}_k \boldsymbol{y}_k^T}{\boldsymbol{y}_k^T \boldsymbol{s}_k} \tag{5}$$

where $\boldsymbol{s}_k = \boldsymbol{\theta}_{k+1} - \boldsymbol{\theta}_k$, $\boldsymbol{y}_k = \boldsymbol{g}_{k+1} - \boldsymbol{g}_k$.

In order to achieve rectilinear formation with synchronized and balanced phase configuration, the control law is designed as

$$\boldsymbol{u}_k = \alpha_k \boldsymbol{d}_k \tag{6}$$

where $\alpha_k$ is the step length (control gain). Thus, when the objective function $f_{p_\theta}$ reaches its unique minimum, the control inputs will be zero, i.e., all the robots will move in straight lines and meanwhile synchronized and balanced phase configuration can be achieved by setting $f_{p_\theta}$ with $\gamma = -1$ and $\gamma = 1$, respectively.

For the circular formation pattern, we change the control law into

$$u_k = \alpha_k \widetilde{d}_k = \alpha_k d_k + \mathbb{1}_{(-\infty,0]}(\|g_k\|)\omega_d \tag{7}$$

where $\mathbb{1}_\Omega(x)$ is the indicator function, i.e., $\mathbb{1}_\Omega(x) = 1$ if $x \in \Omega$, and $\mathbb{1}_\Omega(x) = 0$, otherwise. $\omega_d = [\omega_1, \ldots, \omega_N]^T$, the elements of $\omega_d$ denote desired angular frequencies of the robots and are constant values. It is not difficult to show that $\widetilde{d}_k$ is always a descent direction. When $f_{p_\theta}$ reaches its unique minimum, we have $u_k = \omega_0$, i.e., all of the robots will move in circles with synchronized (for $\gamma = -1$) and balanced (for $\gamma = 1$) phase configuration.

## 3.2   Achieving Rectilinear/Circular Formation with Stabilization of the Average Linear Momentum

The average linear momentum is needed to measure the average position of the robots, which is defined as

$$\dot{Z} = \frac{1}{N} \sum_{m=1}^{N} v_m e^{i\theta_m}. \tag{8}$$

Apparently, the average position of all the robots stabilize on a fixed point if and only if $\dot{Z} = 0$.

In order to make a group of $N$ robots stabilize on a fixed average position, the objective function $f_{\dot{Z}}(\theta) : \mathcal{R}^N \to \mathcal{R}$ is given by

$$f_{\dot{Z}}(\theta) = \frac{N}{2}|\dot{Z}|^2. \tag{9}$$

Thus, $f_{\dot{Z}}(\theta)$ reaches its unique minimum when the stabilization of the average linear momentum is achieved.

Based on the same principles described in Sect. 3.1, the control laws (6) and (7) can also be utilized to achieve the rectilinear and circular formation with stabilization of the average linear momentum.

**Remark 1:** If all the robots have a common unit speeds, all three types of phase configurations can be achieved by utilizing the order parameter, since $p_\theta$ can be interpreted as the derivative of the average position $Z = \frac{1}{N} \sum_{m=1}^{N} x_m + iy_m$. However, we assume that all the robots have nonidentical constant speeds so that $p_\theta$ and $\dot{Z}$ are treated separately in this paper. Note that synchronized phase configuration can also be reached by maximizing $f_{\dot{Z}}$.

## 3.3   RCFA: Rectilinear/Circular Formation Algorithm

Based on the discussion above, all the robots can move in straight lines or circular orbits with three types of phase configurations by minimizing the corresponding

---

**Algorithm 1.** RCFA

---

1  Set $\boldsymbol{v}$ and $\boldsymbol{\omega}_0$ to be constant vectors and $\boldsymbol{B}_0 = \boldsymbol{I}_n$.
2  Define the objective function by (3) or (9) ;
3  **for** $k = 1, 2, \ldots$ **do**
4  |    Compute the descent direction $\boldsymbol{d}_k$ by (4);
5  |    Seek appropriate step length $\alpha_k$ according to **Algorithm 2**;
6  |    Compute control input according to (6) or (7) ;
7  |    Compute $\boldsymbol{B}_{k+1}$ by means of (5);
8  **end**

---

objective functions with particular control laws. The implementation details of the RCFA are presented in the Algorithm 1.

In order to make the control input acquires variable control gains. The control gain is searched by using the following Algorithm 2. Note that two types of descent directions $\boldsymbol{d}_k$ and $\widetilde{\boldsymbol{d}}_k$ are denoted by $\mathbf{d}_k$ in this algorithm.

---

**Algorithm 2.** Step length search algorithm

---

1  Choose $a_1 \in (0, 1)$, $\alpha \in (0, 1)$; Set $j = 0$;
2  **repeat** until $f(\boldsymbol{\theta}_k + \alpha^j \boldsymbol{d}_k) \leq f(\boldsymbol{\theta}_k) + a_1 \alpha^j \boldsymbol{g}^T(\boldsymbol{\theta}_k) \boldsymbol{d}_k$
3  $j = j + 1$;
4  **end repeat**
5  Terminate with $\alpha_k = \alpha^j$.

---

To obtain global convergence of RCFA, the angle $\widetilde{\theta}_k$ between $\mathbf{d}_k$ and $-\boldsymbol{g}_k$ is needed, which is defined by

$$\cos \widetilde{\theta}_k = \frac{-\boldsymbol{g}_k^T \mathbf{d}_k}{\|\boldsymbol{g}_k\| \|\mathbf{d}_k\|}. \tag{10}$$

We first show the convergence of Algorithm 2 in the RCFA by the following lemma.

**Lemma 1.** *Suppose that $\{\boldsymbol{\theta}_k\}$ is a sequence generated by (6) or (7) with Algorithm 1, and $f(\boldsymbol{\theta}_k)$ is bounded below in $\mathcal{R}^N$. Assume also that the gradient $\boldsymbol{g}_k$ of $f(\boldsymbol{\theta}_k)$ exists and is uniformly continuous in the level set*

$$\mathcal{L}(\boldsymbol{\theta}_0) = \{\boldsymbol{\theta}_k \in \mathcal{R}^N | f(\boldsymbol{\theta}_k) \leq f(\boldsymbol{\theta}_0)\} \tag{11}$$

*for an arbitrary staring point $\boldsymbol{\theta}_0 \in \mathcal{R}^N$. If the descent direction $\boldsymbol{d}_k$ satisfies that*

$$0 \leq \widetilde{\theta}_k \leq \frac{\pi}{2} - \eta, \quad \eta \in (0, \frac{\pi}{2}), \tag{12}$$

*we have $\|\boldsymbol{g}(\boldsymbol{\theta}^*)\| = 0$ for any limit point $\boldsymbol{\theta}^*$ of the sequence $\{\boldsymbol{\theta}_k\}$.*

*Proof.* We prove this lemma by contradiction and suppose that $\boldsymbol{\theta}^*$ is a limit point of the sequence $\{\boldsymbol{\theta}_k\}$ and $\|\boldsymbol{g}(\boldsymbol{\theta}^*)\| \neq 0$. Since $\mathbf{d}_k$ is a descent direction, $\{f(\boldsymbol{\theta}_k)\}$ is monotonically decreasing. Also because $f(\boldsymbol{\theta}_k)$ is bounded below, the limit of $f(\boldsymbol{\theta}_k)$ exists and we obtain that $f(\boldsymbol{\theta}_k) - f(\boldsymbol{\theta}_{k+1}) \to 0, \quad f(\boldsymbol{\theta}_k) \to f(\boldsymbol{\theta}^*)$. From Algorithm 2, we have that $-a_1 \boldsymbol{g}_k^T \widetilde{\boldsymbol{s}}_k \to 0, \boldsymbol{g}_k^T \widetilde{\boldsymbol{s}}_k \to 0$ where $\widetilde{\boldsymbol{s}}_k = \alpha^j \mathbf{d}_k$. In addition, we can also show that $\cos \widetilde{\theta}_k = \frac{-\boldsymbol{g}_k^T \widetilde{\boldsymbol{s}}_k}{\|\boldsymbol{g}_k\| \|\widetilde{\boldsymbol{s}}_k\|}$, and since $\cos \widetilde{\theta}_k > 0$ according to (12), we have $\|\widetilde{\boldsymbol{s}}_k\| \to 0$.

The parameter $j$ used in Algorithm 2 is the least nonnegative integer which makes the inequality true. Thus for $\alpha^{j-1} = \alpha^j / \alpha$, we have

$$f(\boldsymbol{\theta}_k + \alpha^{j-1} \mathbf{d}_k) - f(\boldsymbol{\theta}_k) > a_1 \alpha^{j-1} \boldsymbol{g}_k^T \mathbf{d}_k. \tag{13}$$

Note that $\alpha^{j-1} \mathbf{d}_k = \widetilde{\boldsymbol{s}}_k / \alpha$, we can rewrite (13) as

$$f(\boldsymbol{\theta}_k + \frac{\widetilde{\boldsymbol{s}}_k}{\alpha}) - f(\boldsymbol{\theta}_k) > a_1 \boldsymbol{g}_k^T \frac{\widetilde{\boldsymbol{s}}_k}{\alpha}. \tag{14}$$

Let $\boldsymbol{p}_k = \frac{\widetilde{\boldsymbol{s}}_k}{\|\widetilde{\boldsymbol{s}}_k\|}$, then $\frac{\widetilde{\boldsymbol{s}}_k}{\alpha} = \frac{\|\widetilde{\boldsymbol{s}}_k\|}{\alpha} \boldsymbol{p}_k$. According to $\|\widetilde{\boldsymbol{s}}_k\| \to 0$, we have that $\alpha'_k = \frac{\|\widetilde{\boldsymbol{s}}_k\|}{\alpha} \to 0$ and (14) can be rewritten as

$$\frac{f(\boldsymbol{\theta}_k + \alpha'_k \boldsymbol{p}_k) - f(\boldsymbol{\theta}_k)}{\alpha'_k} > a_1 \boldsymbol{g}_k^T \boldsymbol{p}_k. \tag{15}$$

On the one hand, since $\|\boldsymbol{p}_k\| = 1$, $\{\|\boldsymbol{p}_k\|\}$ is bounded. Therefore, there exists a convergent subsequence $\{\|\boldsymbol{p}_k\|\}$ and $\{\|\boldsymbol{p}_k\|\} \to \|\boldsymbol{p}^*\| = 1$. By taking the limit on both sides of (15), we obtain that $\boldsymbol{g}^T(\boldsymbol{\theta}^*) \boldsymbol{p}^* \geq a_1 \boldsymbol{g}^T(\boldsymbol{\theta}^*) \boldsymbol{p}^*$. Thus,

$$\boldsymbol{g}^T(\boldsymbol{\theta}^*) \boldsymbol{p}^* \geq 0. \tag{16}$$

On the other hand, $\boldsymbol{p}_k = \frac{\widetilde{\boldsymbol{s}}_k}{\|\widetilde{\boldsymbol{s}}_k\|} = \frac{\mathbf{d}_k}{\|\mathbf{d}_k\|}$, we have

$$-\boldsymbol{g}_k^T \boldsymbol{p}_k = -\boldsymbol{g}_k^T (\frac{\mathbf{d}_k}{\|\mathbf{d}_k\|}) = \|\boldsymbol{g}_k\| \cos \widetilde{\theta}_k \geq \|\boldsymbol{g}_k\| \sin \eta. \tag{17}$$

By taking the limit of (17), we have $-\boldsymbol{g}(\boldsymbol{\theta}^*)^T \boldsymbol{p}^* \geq \|\boldsymbol{g}(\boldsymbol{\theta}^*)\| \sin \eta > 0$, i.e.,

$$\boldsymbol{g}(\boldsymbol{\theta}^*)^T \boldsymbol{p}^* < 0, \tag{18}$$

which gives a contradiction to (16). Therefore, $\|\boldsymbol{g}(\boldsymbol{\theta}^*)\| = 0$. ∎

In the end, the global convergence of RCFA is presented in the following theorem.

**Theorem 1.** *Let $\boldsymbol{B}_0$ be any symmetric positive definite initial matrix, and let $\boldsymbol{\theta}_0$ be a staring point for which Assumption 6.1 proposed in [13] is satisfied. Then the sequence $\{\boldsymbol{\theta}_k\}$ generated by the RCFA converges to the minimizer $\boldsymbol{\theta}^*$ of $f_{p_\theta}$.*

*Proof.* The theorem follows from the results of Theorem 6.5 in [13] and its proof is omitted here. ∎

# 4  Simulations and Experiments

In this section, simulations and experiments are performed on a group of WMRs to verify the feasibility and effectiveness of RCFA.

## 4.1  Simulation Results

First, a group of five robots modeled by (1) are configured with speeds $v_1 = 0.7, v_2 = 0.5, v_3 = 0.5, v_4 = 0.3$ and $v_5 = 0.3$. The initial phase angles are shown in Figs. 1(a) and 2(a), in which $\theta_1 = 0, \theta_2 = -\pi/4, \theta_3 = \pi/4, \theta_4 = -\pi/2$ and $\theta_5 = \pi/2$. If we set the objective function as $f_{p_\theta}$ (with $\gamma = -1$), rectilinear formation with synchronized phase configuration is achieved by utilizing RCFA with the control law (6). If we set the objective function to $f_{\dot{Z}}$, as expected, all the robots can keep moving in rectilinear formation with a fixed average position. The motion of robots in the two simulations are shown in Figs. 1(b) and 2(b) (the average position of robots is marked by $\oplus$), respectively.

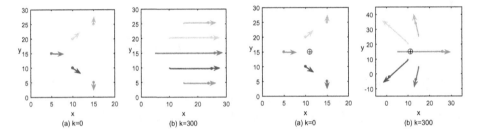

**Fig. 1.** Rectilinear formation with synchronized phase configuration

**Fig. 2.** Rectilinear formation with stabilization of the average linear momentum

Then, a group of three robots are configured with speeds $v_1 = 1.5, v_2 = 1$ and $v_3 = 0.8$ to achieve the circular formation pattern by using RCFA with the control law (7). The initial phase angles are shown in Figs. 3(a) and 4(a), in which $\theta_1 = 3\pi/4, \theta_2 = \pi/2$ and $\theta_3 = \pi/4$. By setting $f_{p_\theta}$ as the objective function with $\gamma = -1$ and $\gamma = 1$, circular formation with synchronized and balanced phase configurations are achieved, respectively. The motion of the robots are shown in Figs. 3(b) and 4(b).

## 4.2  Experimental Results

Two typical formation experiments are carried out on a group of two PIONEER 3DX mobile robots installed with the Robot Operating System (ROS). The PIONEER 3DX is equipped with two processors, a laptop (2.2-GHz CPU and 8-GB memory) for running the algorithm and communicating with other robots via WIFI, and a 32-bit microprocessor (44.2368 MHz) for motion control.

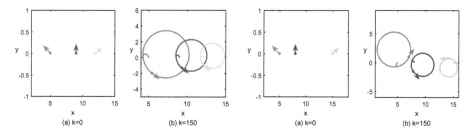

**Fig. 3.** Circular formation with synchronized phase configuration

**Fig. 4.** Circular formation with balanced phase configuration

The position of each robot is estimated by using its on-board motor encoder. The updating frequency of the algorithm is set as 10 Hz. In the first experiment, we initialize the phase angles of the two robots to 0 and $\pi/2$, respectively. Similarly, we set the initial phase angles of the two robots to 0 and $\pi/4$ in the second experiment, respectively. Rectilinear and circular formation with synchronized phase configuration are shown in Figs. 5(a) and 6(a), respectively. The complete moving trajectories of the two robots are recorded in an odometer log and plotted in Figs. 5(b) and 6(b).

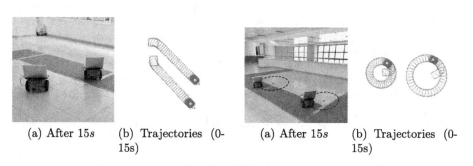

(a) After 15$s$     (b) Trajectories (0-15s)

(a) After 15$s$     (b) Trajectories (0-15s)

**Fig. 5.** The snapshots and trajectories of the rectilinear formation with synchronized phase for a group of two robots.

**Fig. 6.** The snapshots and trajectories of the circular formation with synchronized phase for a group of two robots.

## 5    Conclusion and Future Work

This paper studies rectilinear and circular formation control problems for multiple unicycle type vehicles. A novel variable-gain RCFA is proposed based on *BFGS* method and global convergence properties can be guaranteed under some mild assumptions. Furthermore, the control inputs in present paper contain only the orientations of the robots and the speeds of the robots can be nonidentical.

Simulation and experiment results verified the feasibility and effectiveness of the proposed algorithm.

In our future work, we aim to investigate another formation pattern, in which all the robots are required to move in concentric circular formation (a common circle or different circles around a common center) with phase synchronization/balancing. This type of formation control problem is more complicated since it requires both circular orbit control and phase control.

# References

1. Loria, A., Dasdemir, J., Jarquin, N.A.: Leader-follower formation and tracking control of mobile robots along straight paths. IEEE Trans. Control Syst. Technol. **24**, 727–732 (2016)
2. Bin, X., et al.: Distributed multi-robot motion planning for cooperative multi-area coverage. In: 13th IEEE International Conference on Control and Automation, Ohrid, pp. 361–366. IEEE Press (2017)
3. Briñón-Arranz, L., Schenato, L., Seuret, A.: Distributed source seeking via a circular formation of agents under communication constraints. IEEE Trans. Control Syst. Technol. **3**, 104–115 (2016)
4. Zheng, R., Lin, Z., Yan, G.: Ring-coupled unicycles: boundedness convergence control. Automatica **45**, 2699–2706 (2009)
5. Lin, Z., Francis, B., Maggiore, M.: Necessary and sufficient graphical conditions for formation control of unicycles. IEEE Trans. Autom. Control **50**, 121–127 (2005)
6. Lalish, E., Morgansen, K.A., Tsukamaki, T.: Oscillatory control for constant-speed unicycle-type vehicles. In: 46th IEEE Conference on Decision and Control, Louisiana, pp. 5246–5251. IEEE Press (2007)
7. Alonso-Mora, J., Breitenmoser, A., Rufli, M., Siegwart, R., Beardsley, P.: Multi-robot system for artistic pattern formation. In: Proceedings of IEEE International Conference on Robotics and Automation, Shanghai, pp. 4512–4517. IEEE Press (2011)
8. Sepulchre, R., Paley, D.A., Leonard, N.E.: Stabilization of planar collective motion: all-to-all communication. IEEE Trans. Autom. Control **52**, 811–824 (2007)
9. Sepulchre, R., Paley, D., Leonard, N.E.: Stabilization of planar collective motion with limited communication. IEEE Trans. Autom. Control **53**, 706–719 (2008)
10. Napora, S., Paley, D.: Observer-based feedback control for stabilization of collective motion. IEEE Trans. Control Syst. Technol. **21**, 1846–1857 (2013)
11. Jain, A., Ghose, D.: Collective circular motion in synchronized and balanced formations with second-order rotational dynamics. Commun. Nonlinear Sci. Numer. Simul. **54**, 156–173 (2018)
12. Seyboth, G.S., Wu, J., Qin, J., Yu, C., Allgower, F.: Collective circular motion of unicycle type vehicles with nonidentical constant velocities. IEEE Trans. Control Netw. Syst. **1**, 167–176 (2014)
13. Wright, S., Nocedal, J.: Numerical Optimization. Springer, New York (2006). https://doi.org/10.1007/978-0-387-40065-5
14. Antonelli, G., Chiaverini, S., Fusco, G.: A calibration method for odometry of mobile robots based on the least-squares technique: theory and experimental validation. IEEE Trans. Robot. **21**, 994–1004 (2005)
15. Kuramoto, Y.: Chemical Oscillations, Waves, and Turbulence. Springer, New York (2003)

# State Estimation for Swarm UAVs Under Data Dropout Condition

Hongzhe Yu, Weifan Zhang, Xinjun Sheng$^{(\boxtimes)}$, and Wei Dong

Shanghai Jiao Tong University, 800 Dongchuan Road, Shanghai, China
{ben0107,zwf_jaccount,xjsheng,dr.dongwei}@sjtu.edu.cn

**Abstract.** In this work, a method based on position predicting, velocity filtering and self adaptive parameter tunning is addressed for state estimation and control for swarm of mini unmanned aerial vehicles (UAVs), in order to deal with random noise and data dropout appeared during flights. Under conditions of random data dropout rates and communication latencies, the presented algorithm gives position prediction based on filtered velocity estimation and it fuses the prediction with sensor data. At the same time it corrects the prediction by the error between prediction and measurement of the previous step. The algorithm is designed for tracking mini UAVs with identical marker configuration, and the principles refered is in potential of serving to state estimation in various circumstances. Based on this localization algorithm, a cascade nonlinear control model is developed for swarm UAV control. This work contributes mainly to the object localization and control in a multi-agent system in which all the agents are considered to be in an identical form, hoping that this work will be the testbed for more complicated swarm robot control experiments. Comparison results of state estimation are presented by implementing experiments with or without data dropout.

**Keywords:** State estimation · Data dropout compensation
UAV control

## 1 Introduction

### 1.1 Swarm UAV

In the recent years, unmanned aerial vehicles (UAVs) swarm has been attracting lots of attention. The deployment of an UAVs swarm is able to accomplish much more complex and difficult tasks than a single vehicle. Considerable promising applications are expanded in both military and civil areas. For example, UAV swarms can operate within military missions under a mission-based framework [1]. UAV swarms can also be deployed for patrolling and monitoring an area of people [2] or an area of wild-life [3]. As for post-disaster management,

---

H. Yu and W. Zhang—Equal contributors.

© Springer Nature Switzerland AG 2018
Z. Chen et al. (Eds.): ICIRA 2018, LNAI 10984, pp. 81–91, 2018.
https://doi.org/10.1007/978-3-319-97586-3_7

UAVs can play a crucial role in disaster response by mapping terrain, estimating damage, etc. [4] Several applications in agriculture domain are also presented recently. Loayza *et al.* performs a sowing seeding task using a centralized UAVs swarm [5], and a swarm of collaborating UAVs are designed for field coverage and weed mapping by Albani *et al.* [6]. In general, UAVs swarm includes various technologies and theories, such as robot sensing, data fusing, cooperative optimization, information network and so on. Especially with the development of drone markets, UAVs swarm are highly regarded by different research areas.

Over the last few years, many multi-robot systems have been presented, which can be classified into two typical types according to the global information accessibility [7]. One is the robotic swarms whose agents only have access to local information and limited communication ability. The other type is called general multi-robot system and its agents are able to obtain global information and all-to-all communication. Another sorting principle is based on the communication structure. Under this principle, multi-robot systems are divided into three subclasses, i.e., centralized systems, decentralized systems, and semi-centralized system (semi-centralized systems are also known as leader-follower structure [8]). Some quick conclusions can be drawn from two classifications above. 1. All of the robotic swarms are decentralized. It requires every robot to have at least one active position sensor so that they can build up the system. 2. A general multi-robot system can either be centralized or decentralized. Since decentralized algorithms can be transformed into a variety of centralized algorithms by employing a systematic methodology in the all-to-all communication condition [9], both active and passive position sensors are feasible for general multi-robot systems.

In this paper, we mainly discuss the centralized general multi-robot system. One major problem of this type of systems is data dropout and many compensation solutions have been put forward.

## 1.2  Related Work

For indoor state estimation, different methods and solutions have been proposed. A decentralized localization system using ultra-wideband radio triangulation [11] is introduced but with error in position (about 10 cm) that is not tolerable by swarm formation flights. James A. Preiss *et al.* introduce an Iterative closest point (ICP) frame-to-frame tracking method to track uavs [12], which provides a method to register a newly obtained point cloud to a previous one. This method considers the compensation for temporary communication latencies by implementing an Extended Kalman filter (EKF) to fuse data from motion capture system and IMU measurement, which requires the completeness of information in each frame obtained (with no point dropout) during flying. Though similar in key components, the work mentioned in this paper differs from previous works in the way that it utilizes the geometry characteristic of the markers and fuses the estimated velocity and acceleration to serve for prediction of the state estimation in the next frame by which it can compensate for possible noise and data dropouts. Random marker dropout (easily caused by overshadowing among agents in tightened formations) and other noisy conditions (random occurrence of reflections which are visible to the motion-capture cameras, etc.) reduces the robustness of the state estimations if not taken into considerations.

## 2    State Estimation

Compared to outdoor circumstances, indoor motion capture systems such as vicon provides much higher precision in position estimation [10], which allows the realization of complicated and precise control algorithms. However, typical tracking systems provides only the precisely measured markers. In presence of a multiple objects, most tracking systems fail to provide robust single-point object tracking algorithms. Furthermore, when facing with multiple objects, main exiting tracking software systems require different configurations for every single agent. These two aspects give rise to a mounting complexity of experiment implementation when the number of objects amounts. In this section an algorithm to track all the uavs is developed, basing on an uniform marker configuration used for every uav.

### 2.1    Method Overview

Based on identical marker configuration whose geometry characteristic is initially known, we present in this section an algorithm to track vehicles. The algorithm fuses the velocity informations to give predictions, in addition to the geometry characteristic and the relative positions of the markers. After each estimation, the algorithm uses the absolute value of the error of the previous estimation to correct the current prediction. The algorithm is offboard and is driven at 50 Hz on a PC by the ROS node who receives vicon marker ROS message.

### 2.2    Steps in the Method

The algorithm can be divided into several phases:

1. Prediction:
   In the prediction phase it is assumed that the velocity stays constant, and the prediction is modified in accordance the error of the prediction on last time step. The prediction phase is presented in the Eq. (1):

$$\widetilde{x_k} = x_{k-1} + v_{k-1} \times \Delta t + \rho \times \varepsilon_{k-1} \tag{1}$$

2. Registration of markers to corresponding vehicles[1]:
   In this step, it is assumed that changes in attitude between two time steps are negligible and there is no command on yaw angle $\psi$ and its rate $\omega$. Experiments show good performance of this assumption at moderate aggressive flight. When no more than 2 points around the predicted center position of an uav are detected, a most similar frame comparison (MSFC) algorithm is implemented. In order to find out in what direction the detected point is,

---

**Algorithm 1.** Marker registration algorithm

---

**Input:** Point cloud: vicon markers
**Output:** Vehicle positions $x_k$

1: algorithm to track object from a given point cloud
2: **if** the first time detect marker **then**
3:     Initialization: assign the positions $x_0$ and yaw angle $\psi$ for every vehicle, save the relative vectors $r$
4: **else:** at step k
5:     **for** vehicle $i$ **do**
6:         find the set of points close to $\widetilde{x_{k,i}}$: $Pts = \{pt_0, pt_1, ..., pt_n\}$
7:         **if** $|Pts| > 2$ **then**
8:             check the geometry relation of the vectors formed by points in $Pts$
9:             **if** right geometry relation **then**
10:                calculate $x_{k,i}$ using $Pts$
11:                find vehicle $i$, go to the $i+1$ vehicle
12:            **else**
13:                Cannot find vehicle $i$, go to $i+1$ vehicle
14:        **if** $|Pts| <= 2$ **then**
15:            **for** every $pt_j$ $in$ $Pts$ **do**
16:                $r_{i,j} = pt_j - \widetilde{x_{k,i}}$
17:                find the most similar $r$ (note as $r_{i,j}^*$) to $r_{i,j}$
18:            set the $x_k$ as $\overline{pt_j - r_{i,j}^*}$
19:        **else**
20:            Cannot find vehicle, go to the $i+1$ vehicle
21: **return** positions $x_k$

---

referring to the predicted center, the MSFC algorithm calculates the vector from the predicted center to the detected point, compares it with the four original vectors registered at initialization step and selects the most similar one as the direction the current detected point is in, relative to the predicted center.

3. Renew the error of the prediction in the step k: in this phase of the algorithm, the error of the predicted position is recalculated and is used to renew the weight it takes in the next estimation step.

$$\varepsilon_k = x_k - \widetilde{x_k} \tag{2}$$

$$\delta = \|\varepsilon\| \tag{3}$$

$$\rho = \frac{\delta}{\delta + \lambda} \tag{4}$$

where $\lambda$ is a hyper-parameter which depends on the average absolute of the error. We design the expression of the weight $\rho$ as in (4) because it is aimed to adapt the weight of the prediction in accordance to the norm of its error.

### 2.3 UAV Marker Configuration and Its Initialization

In [12] a method of initialization is mentioned with a guess of the yaw angle for each vehicle. This method requires iterations for the guess of the initial yaw angles begin from which they converge to their estimated values. This method provides possibilities of recognition failure and high computational cost. Within the scope of this paper a square configuration is used as the initialization configuration for uav markers, and the center of the squares represents the position of uavs. For considerations on precision and computational effectiveness, the initial yaw angles are manually assigned for every vehicle through an opencv interface each time the experiment is launched. This assignment successfully initializes the yaw angle of every vehicle in correspond with their true directions related to the coordinate system in the vicon software. With all these informations, the initial position of every vehicle and four vectors pointing from the center to the four markers of the vehicle is obtained. In later discussions, for the purpose of simplification, it is assumed that there is no command on yaw angle and its rate; further, it is also assumed that during the flight the changes of attitudes are negligible.

### 2.4 Velocity Filtering

As shown in Sect. 4, experiments show that when the rate of marker dropout is high, the estimation of velocities is accompanied with sharply oscillated noises. A filtering solution is proposed that, before it is used for prediction in the next estimation step, a first-order low pass filter with self-adapted parameters is designed and implemented to extract the low frequency component of the raw velocity $\hat{v}_k = x_k - x_{k-1}$ from the noise. The principle of first order low pass filter is presented in (5):

$$v_k = \alpha \times \hat{v}_k + (1 - \alpha) \times v_{k-1} \tag{5}$$

Self-adaptation design of the filter parameter $\alpha$ is based on the principles: (1) The filter is designed to track the input data when it varies rapidly (if the variation of the input data is larger than a threshold $\epsilon$). The agility of the filter augments with the varying rate of the input data. (2) The filter is designed to minimize the impact of sharp changes (if the direction of two consecutive variation changes, the counter and $\alpha$ are reset to zero). (3) The filter is designed to follow the consecutive augmentation or decrease of the input data (if two consecutive changes is towards the same direction, we tend to augment $\alpha$). Here the design of the self-adapted $\alpha$ is presented in Fig. 1.

## 3    Control

The controller used in this paper is based on [13], to which augmented a cascade design *(P-PID)* for position control with the inner loop is a proportion controller for position-velocity and the outer loop is a PID controller for velocity-acceleration. Both the inner and outer loop are added with feed forward terms

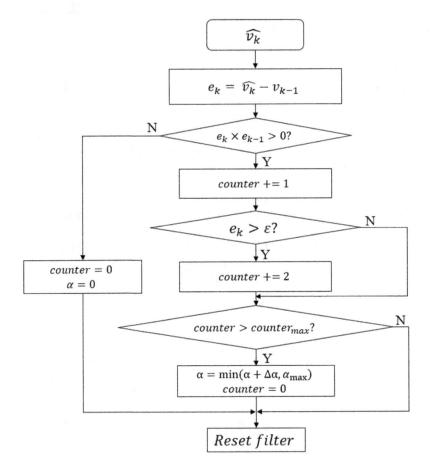

**Fig. 1.** Design of self-adapted low-pass filter

from the trajectory plan. The on-board control input is the attitude setpoints $(\phi, \theta, \dot{\psi})$, and the off-board commander setpoints are position $p_{des}$, velocity $v_{des}$, and acceleration $a_{des}$. Position error is defined as

$$e_p = p_{des} - p \tag{6}$$

In the inner loop the velocity is renewed with the position error, and is fused with the feed forward velocity from the planned trajectory:

$$\tilde{v} = K_{pp} \times e_p \tag{7}$$

$$v = \tilde{v} \times (1 - \rho) + v_{des} \times \rho \tag{8}$$

In the outer loop we define velocity error as

$$e_v = v - v_{des} \tag{9}$$

and the velocity error is utilized to renew the acceleration with a PID controller, fused with the feed forward acceleration from the trajectory plan:

$$\tilde{a} = K_p \times e_v + K_i \times \int e_v + K_d \times \Delta e_v \tag{10}$$

$$a = \tilde{a} \times (1 - \eta) + a_{des} \times \eta \tag{11}$$

where $K_{pp}, K_p, K_i, K_d$ are positive diagonal matrices.

## 4   Experiment

All experiments in this paper are conducted with the crazyflie micro uav [14]. A Vicon motion capture system is used for testing the state-estimation method and for estimating vehicles' position and velocity. Onboard gyros and IMU sensors are used for estimating the vehicles' attitude. Our software is written in C++ in the ROS kinetic environment. It is running on a PC with Ubuntu 16.04, TM i7-6700HQ, 2.60 Hz, and 16 GB RAM. We use ROS messages to transmit the point cloud information between motion-capture system and PC, and also the state estimation between PC and UAVs.

An experiment is conducted with the command of two vehicles to execute taking off (two phases: 0.2 m/s and 0.3 m/s) - hovering - circling (two uavs getting to the same radius of 0.8 m and beginning to circling at angular rate of 0.785 rad/s (equivalent to 8 s/r)). The experiment is conducted with the same command under different conditions: one with low marker dropout rate and the other with high marker dropout rate.

A. Performance without marker dropout:

In optimal conditions where marker dropout rate stays nearly to zero, the algorithm presented performs well in the test of circling as shown in Fig. 2. The results show that the position of the two vehicles follows well the curves of sinusoid and the velocities without much oscillations.

B. Velocity filtering in presence of data dropout and the position estimation result.

It is shown in Fig. 2 that the estimation of velocity is obtained with oscillations and noise. Experiments were conducted to try to find out the relation between noise presented in velocity and data dropout rate. In some extremely noisy conditions where the estimated velocity of vehicles oscillates violently added with marker dropout rate staying at a high level, vehicles are easily out of control even when they are commanded to track simple trajectories. Figures 3 and 4 show the impact of marker dropout rate on the oscillation of estimated velocity. When the two vehicles were commanded to follow the same trajectory with the same angular speed, vehicle0 who flies under severe marker dropout condition has considerable oscillations in the velocity curve while the other vehicle gives back tolerable estimation of velocity. In order to augment the precision of the position prediction, we implement the Low-Pass filter to the velocity calculated by difference of two consecutive positions, as mentioned in Sect. 2.4.

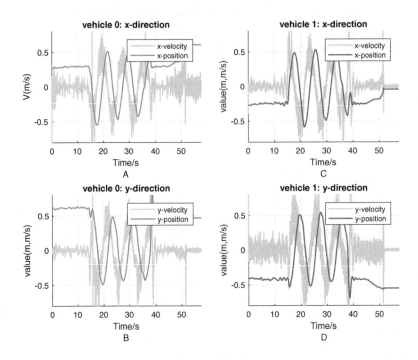

**Fig. 2.** Position of two vehicles commanded circling

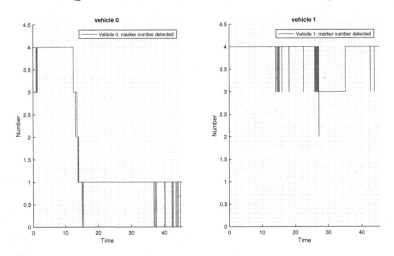

**Fig. 3.** Number of belonging markers detected for each of the two vehicles

From Figs. 3 and 4 we conclude that while different rates of marker recognition affects the calculated value of velocity, we can reduce the violent and sharp oscillation of velocities by filtering it with the filter proposed. State estimation with velocity filter added in the prediction step. Figure 5 shows the result of position estimation after filtering the velocity.

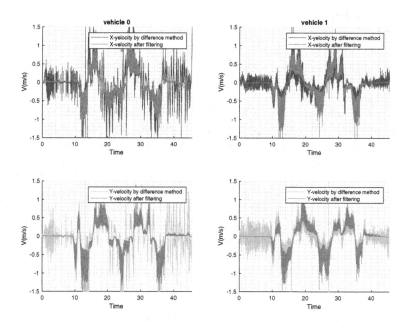

**Fig. 4.** Velocity oscillations under different data-dropout conditions

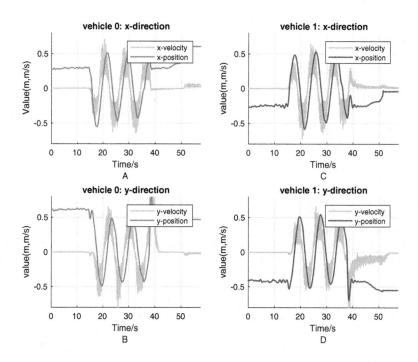

**Fig. 5.** Position and velocity of two vehicles after passing the low pass filter

## 5    Conclusion

State estimation is the base and one of the key components in robot control systems. We have described a method for indoor state estimation of swarm uavs under influences of random noise and data dropout. In this paper, the state estimation issue is simplified by utilizing the high precision of measurements provided by the motion capture systems. Although indoor motion capture systems provides high resolution in single-point position measurements, our algorithm takes into consideration complex situations met by swarm robot control in terms of state estimation and object tracking. To achieve robust performances, we fully utilize the geometry characteristic of the markers on one vehicle, together with precise prediction of positions based on smooth velocities filtered by a self-adaptive low pass filter. Experiments show good tracking and filtering results. The method mentioned in this paper is hoped to serve as the testbed for more complicated swarm robotic control systems in the future. In future work, it is planned to include the control on yaw angle and its rate into the prediction of the position, together with other attitude changes, in order to augment the capability and robustness of the algorithm in more aggressive flights.

**Acknowledgement.** This work was partially supported by the National Natural Science Foundation of China (Grant No. 51605282) and National Science and Technology Major Project.

## References

1. Giles, K., Giammarco, K.: Mission-based Architecture for Swarm Composability (MASC). Procedia Comput. Sci. **114**, 57–64 (2017)
2. Yatskin, D., Kalinov, I.: Principles of solving the space monitoring problem by multirotors swarm. In: International Conference on Engineering and Telecommunication (EnT), pp. 47–50 (2017)
3. Schneider, D.: Open season on drones? IEEE Spectr. **51**(1), 32–33 (2014)
4. Tanzi, T., Apvrille, L., Dugelay, J.L., et al.: UAVs for humanitarian missions: autonomy and reliability. In: Global Humanitarian Technology Conference, pp. 271–278. IEEE (2014)
5. Loayza, K., Lucas, P., Pelaez, E.: A centralized control of movements using a collision avoidance algorithm for a swarm of autonomous agents. In: IEEE Second Ecuador Technical Chapters Meeting, pp. 1–6. IEEE (2017)
6. Albani, D., Nardi, D., Trianni, V.: Field coverage and weed mapping by UAV swarms. IEEE/RSJ International Conference on Intelligent Robots and Systems, pp. 4319–4325. IEEE (2017)
7. Arpino, G., Morris, K., Nagavalli, S., Sycara, K.: Using information invariants to compare swarm algorithms and general multi-robot algorithms: a technical report. In: IEEE International Conference on Robotics and Automation (2018)
8. Wan, S., Lu, J., Fan, P.: Semi-centralized control for multi robot formation. In: International Conference on Robotics and Automation Engineering (ICRAE), pp. 31–36 (2017)

9. Xuan, P., Lesser, V.: Multi-agent policies: from centralized ones to decentralized ones. In: International Joint Conference on Autonomous Agents and Multiagent Systems, pp. 1098–1105. ACM (2002)
10. Lupashin, S., Hehn, M., Mueller, M.W., Schoellig, A.P., Sherback, M., D'Andrea, R.: A platform for aerial robotics research and demonstration: the flying machine arena. Mechatronics **24**(1), 41–54 (2014)
11. Ledergerber, A., Hamer, M., D'Andrea, R.: A robot self-localization system using one-way ultra-wideband communication, Mechatronics, In: IEEE/RSJ International Conference on Intelligent Robots and Systems, pp. 3131–3137 (2015)
12. Preiss, J.A., Honig, W., Sukhatme, G.S., Ayanian, N.: Crazyswarm: a large nano-quadcopter swarm. In: IEEE International Conference on Robotics and Automation, pp. 3299–3304 (2017)
13. Mellinger, D., Kumar, V.: Minimum snap trajectory generation and control for quadrotors. In: IEEE International Conference on Robotics and Automation, pp. 2520–2525 (2011). https://doi.org/10.1109/ICRA.2011.5980409
14. Bitcraze. https://www.bitcraze.io/. Accessed 4 Oct 2017

# Virtual-Datum Based Cooperative Kinematic Constraints Analysis for Dual-Robotic System

Qi Fan[1], Zeyu Gong[1], Bo Tao[1(✉)], and Jianlan Li[2]

[1] State Key Laboratory of Digital Manufacturing Equipment and Technology,
Huazhong University of Science and Technology, Wuhan 430074,
Hubei, The People's Republic of China
taobo@hust.edu.cn

[2] School of Energy and Power Engineering, Huazhong University of Science and Technology,
Wuhan 430074, Hubei, The People's Republic of China

**Abstract.** In the process of multi-robot collaboration, the movements of multiple robotic end effector usually have some certain cooperative kinematic constraints relationship. However, these kinematic constraints are often difficult to analyze, especially when the task is complicated. For this reason, a cooperative kinematic constraints analysis method based on a virtual datum is proposed in this paper, and the virtual datum is mainly constructed according to the geometrical characteristics of the object and the task requirements. As a result, the virtual datum is acting as a bridge to associate the position, orientation, and velocity of each robot's end effector at any moment. Further, the coordinated motion planning and control of dual robots will be easily implemented.

**Keywords:** Multi-robot collaboration · Kinematic constraints · Motion planning
Collaborative control

## 1 Introduction

At present, the multi-robotic (industrial) system has been widely used in various industrial applications [1–3], such as the handling of large-scale overweight objects, the spraying and welding of auto parts, and the drilling and assembly of aircraft skins. Compared to a single-robotic (industrial) system, the multi-robot system mainly has the following advantages: (1) a larger work space; (2) a greater load; (3) a higher flexibility. These advantages have greatly improved the operational capability and performance of the robot system, and have attracted many researchers to study the related technologies of multi-robot collaboration, especially the coordinated motion planning and control of the multi-robotic system [4, 5].

Multi-robot collaboration, that is, multiple robots cooperate to accomplish the same task, and can be divided into three categories: (1) parallel collaboration, multiple robots share workspaces, but the movements are independent of each other; (2) sequential collaboration, multiple robots work in sequence according to a certain order or time; (3) synchronous collaboration, multiple robots coordinate with each other and perform tasks that meet the certain constraints (position, orientation, and velocity, etc.) at the same

© Springer Nature Switzerland AG 2018
Z. Chen et al. (Eds.): ICIRA 2018, LNAI 10984, pp. 92–100, 2018.
https://doi.org/10.1007/978-3-319-97586-3_8

time. Since the first two types of the collaboration have no position or orientation constraints, this paper will focus on synchronous collaboration, and taking the dual-robotic system as a representative to analyze the cooperative kinematic constraints in collaboration process.

With regard to the analysis of cooperative kinematic constraints of the dual/multi-robotic system, some researchers have conducted relevant research [6–11]. In [6], two coordinated robots with the object they contact are considered as a closed kinematic chain. When the chain is in motion, the positions and orientations of the two robots must satisfy a set of holonomic constraints. In this way, the coordinated motion of dual robots can be realized simply by master-slave control. Similarly, the kinematic constraints equation for three robots manipulating a plate can be established, and the coordinated motion of the system is achieved by master-and-two-slaves control [7]. Further, a task-space regulation scheme for collaborative handling of two cooperative manipulators is proposed by analyzing the absolute and relative movement constraints [8]. In the above studies, there are no relative movement between the end effector of the robots and the object. But in some case, one or more robots in a multi-robotic system will have relative motion with the object. With the background of dual-robot cooperative welding, [9] analyzes the kinematic constraints of a system when there is a deterministically relative motion between the cooperative robots. In [10], an object-oriented general formulation for cooperative tasks is proposed, and the motion of the single arm in the system is computed via kinematic transformations between the relevant coordinate frames. In [11], kinematic constraint for curved-surface nondestructive testing is analyzed, which is special in that the constraint relationship is time-varying and closely related to the geometry of the object.

Overall, kinematic constraints analysis is an effective way to achieve coordinated motion planning and control of multi-robotic system. However, existing research are mostly for specific scenarios. So, this paper will aim to propose a general method for kinematic constraints analysis of dual-robotic system. The proposed method will first construct a virtual datum, which will be served as a bridge to achieve the analysis of coordinated kinematic constraints.

The rest of this paper is organized as follows. Section 2 gives the problem statement of the coordinated kinematic constraints. Next, the generation process of virtual datum is described in Sect. 3. And the mathematical analysis of cooperative kinematic constraints is given in Sect. 4. Finally, the conclusions are given in Sect. 5.

## 2  Problem Statement

In the process of synchronous collaboration of the dual-robotic system, the completion of the task mainly depends on the common movement (fully consistent movement, and there is no relative movement) or relative movement between the robot's end effector and the object (rigid body). Therefore, in order to complete the task better, there are two problems that need to be solved:

a. Analyze the movement relationship between the robot's end and the object.
b. Analyze the movement relationship between the ends of two robots.

It should be pointed out that the above-mentioned movement relationships mainly refer to relative position, relative orientation and relative velocity, etc.

To facilitate the subsequent description, we first give three definitions:

a. Effective action point (EAP), that is, the actual interaction point between the robot's end effector and the object, such as the contact center during handling or clamping.
b. Virtual datum point (VDP), which is a moving point used for kinematics constraint analysis at a certain time. In a sense, the virtual datum can represent the motion state of the object or the interaction state with the robot.
c. Virtual datum (VD), i.e. the collection of virtual reference points, with a time distribution or spatial distribution attribute properties.

Meanwhile, a typical dual-robot collaboration system as shown in Fig. 1, in which there are mainly contain two robots, one object, and some frames.

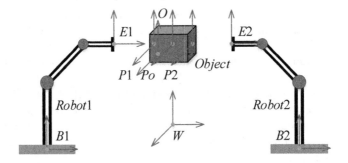

**Fig. 1.** Schematic diagram of a typical dual-robot collaboration system

The definition of these frames are as follows:

a. The world coordinate system {W}.
b. The base frame {B1} ({B2}) of the robot1 (robot2).
c. The end flange center's frame {E1} ({E2}) of the robot 1 (robot2).
d. The end effector's EAP frame {P1} ({P2}) of the robot 1 (robot2).
e. The VDP frame.
f. The workpiece frame {O}.

Here, we define a matrix $^{X}T_{Y}$, which is a $4 \times 4$ homogeneous transformation matrix, and represents the rotation and translation relation between the frame {X} and {Y}, and $^{X}T_{Y}(t)$ indicates that $^{X}T_{Y}$ is a matrix that changes over time. Then, the two problems mentioned above can be further described as:

a. Analysis the rotation and translation relation between the end flange center's frame {E1} ({E2}) and the VDP frame at any time. That is, solving the matrix $^{Po}T_{E1}(t)$ and $^{Po}T_{E2}(t)$.
b. Analysis the rotation and translation relation between the end flange center's frame {E1} and {E2} at any time. That is, solving the matrix, $^{E1}T_{E2}(t)$ or $^{E2}T_{E1}(t)$.

## 3   Generation of Virtual Datum

Some studies have shown that the kinematic constraints of synchronous collaboration for two robots are usually determined by the geometrical characteristics of the object itself and the task requirements. Therefore, we can mainly construct the virtual datum based on the characteristics of the object itself to facilitate the kinematic constraint analysis for the dual-robot collaboration.

In order to more clearly introduce the construction of the virtual datum, we classify the synchronous collaboration of the dual-robotic systems into three categories, which based on whether there is relative motion between the robot's end effector and the object. As shown in Fig. 2, the three types are: (a) two common movements; (b) a common movement (robot1) and a relative movement (robot2); (c) two relative movements.

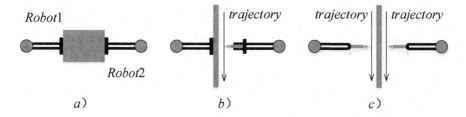

**Fig. 2.**   The classification of the synchronous collaboration for dual-robotic systems.

Then, the virtual datum that can simplify the kinematic constraint analysis are given for the above three types, respectively. As shown in Fig. 3, the distribution of VDPs is given in an intuitive way. Among them, the point P1 represents the robot1's EAP, the point P2 represents the robot2's EAP, and the point Po represents the VDP.

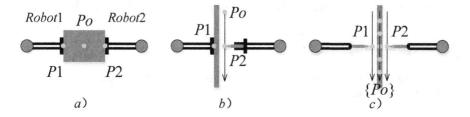

**Fig. 3.**   The distribution of VDPs

In type (a), since there is no relative motion between P1 (P2) and the object, which means the kinematic constraint between them is fixed. Therefore, the VDP used for the kinematic constraint analysis can be arbitrarily chosen as a point that can represent the movement of the object in space. In particular, the point is selected as the mass center of the object. This type is common in cooperative handling.

In type (b), since P1 has no relative movement with the object, the kinematic constraint between them is also fixed. Therefore, the simplification of kinematic constraint analysis between P2 and the object should be mainly considered in the construction of virtual datum. Generally, the relative movement path between P2 and the object in the task space is known. So, the starting point of the relative movement path between P2 and the object is a good choice of the VDP. This type is common in the process of dual-robot cooperative welding.

In type (c), since P1 and P2 both have relative movement with the object, in order to be able to associate the motions of P1 and P2, the VDP can't usually be selected as a specific point on the object, but a series of set of points on the object. The set of points is just the required virtual datum. As mentioned earlier in this section, we should establish a certain rule, which mainly consider the geometrical characteristics of the object itself and the task requirements, to construct the virtual datum. This type is common in the drilling and assembly of aircraft skin.

In order to clearly illustrate the construction rules of the virtual datum, a representative example is given. As shown in Fig. 4, a special object, whose surface is a freeform surface and has thickened asymmetric geometrical features, is the most complex type of common objects. Here, the virtual datum construction method for such an object will be given, and as follows:

(a) Solving the intersection between the object and m parallel and uniformly-distributed planes, that is, get the set of outlines of the object on the m cross-planes, and mark the set of outlines as $\{Lj\}$, $j \in N_+, 1 \le j \le m$, and Lj represents the intersection between the object and the j-th cross-plane.

(b) Select the outline Lj and straighten it so that the tangent of its lowest point is horizontal, and mark the adjusted outline as $\overline{L}_j$.

(c) Solving the intersection between $\overline{L}_j$ and n uniformly-distributed horizontal lines, and divide them into two point sets, that is $\{^jP_i^L\}$ and $\{^jP_i^R\}$, $i \in N_+, 1 \le i \le n$, and the $\{^jP_i^L\}$ ($\{^jP_i^R\}$) represents the point set of the intersection of the i-th horizontal line and the left (right) side of the outline $\overline{L}_j$. Then, obtain the new point set $\{^jP_i^M\}$ represents the middle point of the left and right side intersection, and there is,

$$^jP_i^M = \frac{1}{2}\left(^jP_i^L + {}^jP_i^R\right) \tag{1}$$

(d) Curve fitting for the point sets $^jP^M$ respectively and to obtain m lines, and each line is a virtual datum.

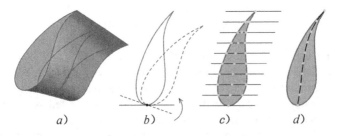

a)    b)    c)    d)

**Fig. 4.** The construction rules and process of virtual datum.

It should be noted that the construction rule of the virtual datum is task-oriented, i.e. it needs to consider the geometric characteristics of the object and the task requirements. Once the geometric and individual requirement information is given, the proposed method can be easily transplanted to solve different problem.

## 4    Analysis of Cooperative Kinematic Constraints

According to the previous analysis, it is not difficult to find that the constructed virtual datum can be used as a bridge to link the movement of the two robots' EAP. More importantly, the virtual datum can simplify the analysis of cooperative kinematic constraints. In the following, we will introduce constraints analysis based on virtual datum from the position, orientation, and velocity of two robots.

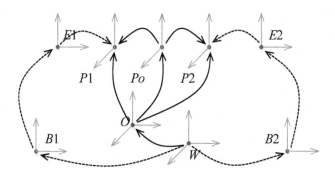

**Fig. 5.**  The frames topology diagram of a typical dual-robot collaboration system.

Based on the typical dual-robot collaboration system shown in Fig. 1, the topology diagram of system frames is given in Fig. 5. Generally, the transformation relationships between the frames in Fig. 5 mainly include the following types.

- $^W T_{B1}$ and $^W T_{B2}$: The transformation relation between the base frame of robot1 (robot2) and the world frame. It is usually a constant matrix.
- $^{B1} T_{E1}(t)$ and $^{B2} T_{E2}(t)$: The transformation relation between the end flange center's frame and the base frame of the robot. They are both the time-varying matrix, and determined by the robot's joint angular displacement.
- $^{E1} T_{P1}$ and $^{E2} T_{P2}$: The transformation relation between end effector's EAP frame and the end flange center's frame of the robot. They are usually also a constant matrix, and determined by the structural parameters of the end effector.
- $^W T_O(t)$: The transformation relation between the workpiece frame and the world frame. It is closely related to the task, sometimes (the object is fixed) it may be a constant matrix, that is, $^W T_O(t) \equiv {}^W T_O(0)$.
- $^O T_{P1}(t)$ and $^O T_{P2}(t)$: The transformation relation between the robot end effector's EAP frame and the workpiece frame. They are also closely related to the task. If there is no relative movement between the robot end effector's EAP and the object, two equations will be obtained, $^O T_{P1}(t) \equiv {}^O T_{P1}(0), {}^O T_{P2}(t) \equiv {}^O T_{P2}(0)$.

- $^{O}T_{Po}(t)$: The transformation relation between the VDP frame and the workpiece frame. It is not only closely related to the task, but also related to the object itself. If the VDP can be selected as a specific point on the object, the $^{O}T_{Po}(t)$ will become a constant matrix, that is, $^{O}T_{Po}(t) \equiv {}^{O}T_{Po}(0)$.
- $^{Po}T_{P1}(t)$ and $^{Po}T_{P2}(t)$: The transformation relation between the robot end effector's EAP frame and the VDP frame. They are two very important matrices that can reflect the interaction between the robot's end effector and the object. And if there is no relative movement between the robot end effector's EAP and the object, we can also think that $^{O}T_{P1}(t) \equiv {}^{O}T_{P1}(0), {}^{O}T_{P2}(t) \equiv {}^{O}T_{P2}(0)$.

### 4.1   Constraints Analysis for Position and Orientation

Based on the above transformation relationship, we can introduce a new transformation relation, that is,

- $^{W}T_{Po}(t)$: The transformation relation between the VDP frame and the world frame. And it can be obtained from Eq. (2).

$$^{W}T_{Po}(t) = {}^{W}T_{O}(t) \cdot {}^{O}T_{Po}(t) \tag{2}$$

Further analysis of the Eq. (2), we can find an interesting phenomenon: in practical applications, the matrices $^{W}T_{O}(t)$ and $^{O}T_{Po}(t)$ usually only one of is time-varying, which means that the matrices $^{W}T_{Po}(t)$ can better reflect the essence of the task.

Then, the joint motion of robot1 and robot2 will be associated with the motion of the VDP, and as follows,

$$^{B1}T_{E1}(t) = ({}^{W}T_{B1})^{-1} \cdot {}^{W}T_{Po}(t) \cdot {}^{Po}T_{P1}(t) \cdot ({}^{E1}T_{P1})^{-1} \tag{3}$$

$$^{B2}T_{E2}(t) = ({}^{W}T_{B2})^{-1} \cdot {}^{W}T_{Po}(t) \cdot {}^{Po}T_{P2}(t) \cdot ({}^{E2}T_{P2})^{-1} \tag{4}$$

Combining Eqs. (2) and (3), the motions of the end of robot1 and robot2 will be associated with virtual datum, which serves as a bridge, and their relative motion relationship $^{E1}T_{E2}(t)$ will be obtained.

$$^{E1}T_{E2}(t) = {}^{E1}T_{P1} \cdot ({}^{Po}T_{P1}(t))^{-1} \cdot {}^{Po}T_{P2}(t) \cdot ({}^{E2}T_{P2})^{-1} \tag{5}$$

Equation (5) represents the holonomic constraints for the position and orientation between the end flange center of robot1 and robot2 during their cooperative motion. And it also implies that the motion of two robots in synchronous collaboration usually has a definite corresponding relation.

Obviously, the $4 \times 4$ homogeneous transformation matrix $^{X}T_{Y}$ can be represented by a $3 \times 3$ rotation matrix $^{X}R_{Y}$ and a $3 \times 1$ translational vector $^{X}P_{Y}$. Then,

$$^{X}T_{Y} = \begin{bmatrix} ^{X}R_{Y} & ^{X}P_{Y} \\ 0 & 1 \end{bmatrix} \tag{6}$$

Where $^X R_Y$ and $^X P_Y$ represents the translation relation and rotation relation between $\{X\}$ and $\{Y\}$, respectively.

Combining Eqs. (5) and (6), the constraints determined by Eq. (5) can be divided into two parts, one for position constraint and the other for orientation constraint.

$$^{E1}P_{E2}(t) = {}^{E1}R_{P2}(t) \cdot \left({}^{E2}R_{P2}\right)^{-1} + {}^{E1}R_{Po}(t) \cdot {}^{Po}R_{P2}(t) + {}^{E1}R_{P1} \cdot \left({}^{Po}P_{P1}(t)\right)^{-1} + {}^{E1}P_{P1} \qquad (7)$$

$$^{E1}R_{E2}(t) = {}^{E1}R_{P1} \cdot \left({}^{Po}R_{P1}(t)\right)^{-1} \cdot {}^{Po}R_{P2}(t) \cdot \left({}^{E2}R_{P2}\right)^{-1} \qquad (8)$$

### 4.2   Constraints Analysis for Velocity

The translational velocity and angular velocity of a rigid body are analyzed in [12], and their relationship to translation vectors and rotation matrices is given,

$$v = \dot{P} \qquad (9)$$

$$\begin{cases} [\omega\times] \equiv \dot{R}R^{-1} = \begin{bmatrix} 0 & -\omega_Z & \omega_Y \\ \omega_Z & 0 & -\omega_X \\ -\omega_Y & \omega_X & 0 \end{bmatrix} \\ \omega = \begin{bmatrix} \omega_X & \omega_Y & \omega_Z \end{bmatrix}^T \end{cases} \qquad (10)$$

Where $[\omega\times]$ is a $3 \times 3$ antisymmetric matrix.

Combining Eqs. (7) and (9), the translational velocity relationship between the end of robot1 and robot2 at the task space will be obtained,

$$^{E1}v_{E2}(t) = {}^{E1}\dot{P}_{E2}(t) \qquad (11)$$

Combining Eqs. (8) and (10), the angular velocity relationship between the end of robot1 and robot2 at the task space will be obtained,

$$\left[{}^{E1}\omega_{E2}(t)\times\right] = {}^{E1}\dot{R}_{E2}(t) \cdot \left({}^{E1}R_{E2}(t)\right)^{-1} \qquad (12)$$

Equations (11) and (12) represents the holonomic constraints for the translational velocity and angular velocity in Cartesian space between the end of robot1 and robot2 during their cooperative motion.

By now, constraint relations of the position, orientation and Cartesian velocity for the dual-robotic system robots have been definitely proposed, and can apply to the three types of synchronous collaboration mentioned above. Further, these analysis results can contribute to the realization of collaborative motion planning and control for a dual or multi-robotic systems.

## 5   Conclusion

This paper focuses on the research of multi-industrial robot collaboration and presents a novel method for collaborative kinematic constraints analysis. The core of the proposed method is to construct a virtual datum for constraints analysis. Represented by a dual-robotic system, the common collaborations are divided into three categories and the construction of virtual datum is introduced in order. The geometric characteristics of the object itself and task requirements is the main factor in determining the method of construction. Using the virtual benchmark as a bridge, the holonomic constraints for position, orientation, and velocity of the dual robots can be obtained. The analysis results can closely correlate the motion of the two robots, and to improve the collaborative performance of the multi-robotic system.

**Acknowledgments.** This work is supported in part by the National Science Foundation of China under Grant 91748204 and 51575215, and the National Key Research and Development Program of China under Grant 2017YFB1301504.

## References

1. von Albrichsfeld, C.: A self-adjusting active compliance controller for multiple robots handling an object. Control Eng. Pract. **2**(10), 165–173 (2002)
2. Pellegrinelli, S.: Multi-robot spot-welding cells: an integrated approach to cell design and motion planning. CIRP Annals **1**(63), 17–20 (2014)
3. Honglun, H.: Dynamic modeling and sensitivity analysis of dual-robot pneumatic riveting system for fuselage panel assembly. Ind. Robot Int. J. Robot. Res. Appl. **2**(43), 221–230 (2016)
4. Arai, T.: Editorial: advances in multi-robot systems. IEEE Trans. Robot. Autom. **5**(18), 655–661 (2002)
5. Caccavale, F., Uchiyama, M.: Cooperative manipulators. In: Siciliano, B., Khatib, O. (eds.) Springer Handbook of Robotics, pp. 701–718. Springer, Heidelberg (2008). https://doi.org/10.1007/978-3-540-30301-5_30
6. Luh, J.Y.S.: Constrained relations between two coordinated industrial robots for motion control. Int. J. Robot. Res. **3**(6), 60–70 (1987)
7. Tzafestas, C.S.: Path planning and control of a cooperative three-robot system manipulating large objects. J. Intell. Robot. Syst. **2**(22), 99–116 (1998)
8. Caccavale, F.: Task-space regulation of cooperative manipulators. Automatica **6**(36), 879–887 (2000)
9. Yahui, G.: Cooperative path planning and constraints analysis for master-slave industrial robots. Int. J. Adv. Robot. Syst. **3**(9), 88–100 (2012)
10. Basile, F.: Task-oriented motion planning for multi-arm robotic systems. Robot. Comput. Integr. Manuf. **5**(28), 569–582 (2012)
11. Zongxing, L.: Kinematic constraint analysis in a twin-robot system for curved-surface nondestructive testing. Ind. Robot Int. J. Robot. Res. Appl. **2**(43), 172–180 (2016)
12. Craig, J.: Introduction to Robotics: Mechanics and Control, 3rd edn. Prentice Hall, Englewood Cliffs (2004)

# Distributed Hunting for Multi USVs Based on Cyclic Estimation and Pursuit

Binbin Hu[1,2], Bin Liu[1,2(✉)], Zhecheng Xu[1,2], Tao Geng[1], Ye Yuan[2], and Hai-Tao Zhang[1,2(✉)]

[1] Guangdong HUST Industrial Technology Research Institute, Dongguan 523000, Guangdong, China
`binliu92@hust.edu.cn`, `zht@mail.hust.edu.cn`
[2] The State Key Lab of Digital Manufacturing Equipment and Technology, Image Processing and Intelligent Control Key Laboratory of Education Ministry of China, School of Automation, Huazhong University of Science and Technology, Wuhan 430074, People's Republic of China

**Abstract.** In this brief, we study the hunting problem for a group of underactuated surface vehicles (USVs), in which the vehicles converge to the target as the center, as well as maintain the desired relative distance to the target when rotating around the target at the same speed. A approach based on the cyclic matrix is delivered. The overall control objectives are divided into two subobjectives, where the first is target circling that all vehicles rotate a circle around the target, and the second is that the vehicles are eventually evenly spaced on the circle. The former part is based on the cyclic estimation of target to get close to the target, the latter is designed by also cyclic pursuit strategy using the relative angle between the neighbors. An important feature of the controller is that not all vehicles know the target's position. For hunting with obstacle avoidance, artificial potential method and label's change strategy between neighbors are also applied to guarantee obstacle avoidance. Numerical simulations are given to verify the effectiveness of the proposed controller.

**Keywords:** Distributed hunting · Cyclic estimation and pursuit Multi underactuated surface vehicles

## 1 Introduction

Recent decades witnesses the rapid development of distributed control of multi-agent systems. It is partially due to the increasing need to perform more difficult and complex tasks, where it contributes to increasing efficiency, reducing the system cost and providing the redundancy against individual failure. In particular, hunting behavior where multi agents enclose the target in a certain area,

Supported by the Guangdong Innovative and Entrepreneurial Research Team Program under Grant 2014ZT05G304.

Z. Chen et al. (Eds.): ICIRA 2018, LNAI 10984, pp. 101–112, 2018.
https://doi.org/10.1007/978-3-319-97586-3_9

has attracted much attention recently. For example, In [15], a feedback control method was first taken into account to make multi robots round up the target robot into a certain area. In [11], the neural network method and the methods of dynamic alliance was studied for the pursuit and capture behavior. In [12,14], the author performed the case where the unicycles form a circular formation under all-to-all communication with unit constant velocity or nonidentical constant velocity.

Then, formation under the cyclic pursuit strategy was further studied. In [9], a cyclic controller for the desired pursuit pattern of moving target in 3D space was developed. In particular, there are many case where the target is given and known to all the vehicles. In [2], the limited visibility of the onboard sensors were took into account in the cyclic pursuit strategy. In [7], the author studied the case where a rigidity of graphs was utilized in the spaced formation of circle. A hybrid control law of the cyclic enclosing formation was introduced in [10]. It was shown in [16] that a distributed dynamic control law for circle formation of unicycles when the target is just known to one cycles was developed. In [17], the cycle formation is only based on the bearing-only measurement. Furthermore, it is some work of circle formation with more general networks. In [13], a balanced graph condition was considered in designing the dynamic controller. A controller is studied for circle formation with even a jointly connected network was proposed in [3,4]. In [5], the author developed the case of which cyclic pursuit formation with a hierarchical controller.

Compared with the existing result, the main contribution of this brief are listed in the following four aspects. First, the aforementioned results on hunting problem almost consider the case of unicycles where the target is known to all the vehicles. Our proposed controller is based on the information of estimation and neighbors via cyclic communication network $\mathbb{G}$. Second, a hierarchical structure of the controller is to solve the underactuated and nonlinear characters. Third, artificial method and label-change strategy are considered to avoid collision, space evenly and rotate around the target.

The rest of this paper is organized as follows: Sect. 2 introduces some preliminaries and gives definition of hunting problem. Section 3 presents the distributed hunting controller based on the cyclic estimation and pursuit strategy. Some computer simulation results are presented in Sect. 4. Section 5 concludes the article.

*Notation:* Throughout the paper, $\mathbb{R}^n$ denotes the $n$-dimensional Euclidean Space. $\|\cdot\|$ denotes the Euclidean norms. $(\cdot)_{ij}$ denotes the element of $(\cdot)$ in row i, column j. $\lambda_{min}(\cdot)$ denotes the smallest eigenvalues of a square matrix $(\cdot)$.

## 2    Problem Formulation

First, consider a group of N underactuated surface vehicles represented by the dynamics found in [6] with kinematics and kinetics:

$$\begin{cases} \dot{\boldsymbol{\eta}}_i = \mathbf{J}(\psi_i)\mathbf{v}_i \\ \mathbf{M}_i\dot{\mathbf{v}}_i + \mathbf{C}_i\mathbf{v}_i + \mathbf{D}_i\mathbf{v}_i = \boldsymbol{\tau}_i \end{cases} \tag{1}$$

with

$$\boldsymbol{\eta}_i = [x_i, y_i, \psi_i]^T, \mathbf{v_i} = [u_i, v_i, r_i]^T, \boldsymbol{\tau}_i = [\tau_{iu}, 0, \tau_{ir}]^T$$

$$\mathbf{M}_i = \begin{bmatrix} m_{11i} & 0 & 0 \\ 0 & m_{22i} & 0 \\ 0 & 0 & m_{33i} \end{bmatrix}, \mathbf{C}_i = \begin{bmatrix} 0 & 0 & c_{13i} \\ 0 & 0 & c_{23i} \\ c_{31i} & c_{32i} & 0 \end{bmatrix}$$

$$\mathbf{D}_i = \begin{bmatrix} d_{11i} & 0 & 0 \\ 0 & d_{22i} & 0 \\ 0 & 0 & d_{33i} \end{bmatrix}, \mathbf{J} = \begin{bmatrix} \cos(\psi_i) & -\sin(\psi_i) & 0 \\ \sin(\psi_i) & \cos(\psi_i) & 0 \\ 0 & 0 & 1 \end{bmatrix}.$$

$\boldsymbol{\eta}_i \in \mathbb{R}^3$ is the position vector in the earth-fixed reference frame; $\mathbf{v}_i \in \mathbb{R}^3$ is the velocity vector in the body-fixed reference frame; $\mathbf{M}_i \in \mathbb{R}^{3 \times 3}$ is the inertia matrix; $\mathbf{C}_i \in \mathbb{R}^{3 \times 3}, \mathbf{D}_i \in \mathbb{R}^{3 \times 3}$ denote the coriolis and centripetal matrix damping matrix, respectively; $\boldsymbol{\tau}_i \in \mathbb{R}^3$ is control vector with $\tau_{iu}$ the surge force and $\tau_{ir}$ the yaw moment.

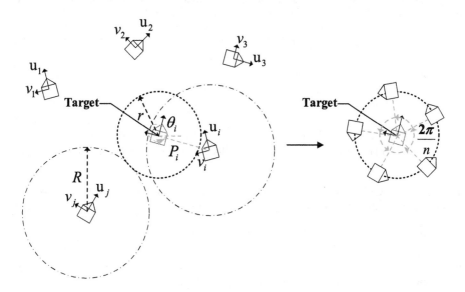

**Fig. 1.** Distributed hunting behavior (USVs which are initially located in plane forms a circle formation and rotate the target).

Figure 1 illustrates a group of $N$, underactuated surface vehicles perform the distributed hunting behavior for a preset target $P^*$. Given sense radius $R > 0$, if $||P_i - P_*|| \leq R$, vessel $i$ get the position $P_*$. For ease of expression, we label the vehicles as follows:

*Remark 1.* The label are sorted first in ascending order in a counterclockwise manner based on the angle $\theta_i, 0 \leq \theta_i \leq 2\pi$ and $\theta_i \leq \theta_{i+1}, i = 1, 2, \ldots, N - 1$ between their position and estimation of target.

$$\begin{aligned} \theta_i &= \mathrm{atan2}(x_i^P, y_i^P) \\ &= \mathrm{atan2}\big((x_i^Q - x_i^P), (y_i^Q - y_i^P)\big) \end{aligned} \tag{2}$$

where $P_i(x_i^P, y_i^P)^T$, $Q_i(x_i^Q, y_i^Q)^T$ is the position of vessel $i$ and the estimation $i$ of the target. $D_i \triangleq \|P_i - Q_i\|$ is the relative position from vehicle $i$ to estimator $i$.

Consider the vehicles' communication networks are described by $\theta_i$ as $\mathbb{G} = (\nu, \varepsilon)$, where $\nu = \{1, 2, \ldots, N\}$ and $\varepsilon = \{(1, 2), (2, 3), \ldots, (N-1, N), (N, 1)\}$. It means that vehicle $i$ only gets the information from neighbor $i + 1$ that are in front of itself.

Now, we are ready to provide two problems of distributed hunting problem as below:

**Problem 1:** Collective Hunting Problem: Given n vessels defined as (1), design a distributed control law:

$$u_i = f(P_i, Q_i, \theta_i, \theta_{i+1}), i \in \nu$$

such that $n$ vehicles perform the collective hunting behavior by spacing evenly on the same circle and rotating around the target, as follows.

$$\begin{cases} D_i = r, \\ (\theta_{i+1} - \theta_i) mod(2\pi) = \frac{2\pi}{n} \\ \dot{\theta}_i = \dot{\theta}_j, \end{cases} \tag{3}$$

where $i \neq j$ and $i, j \in \nu$. When $i = n, i + 1 = 1$. $r$ is defined as the hunting radius.

Note that, in real application, USV is a rigid-body system with its body and length, the control problem becomes a collective control problem with obstacle avoidance, as follows.

**Problem 2:** Collective Hunting with Obstacle Avoidance Problem: Given n vessels defined as (1), design a distributed control law:

$$u_i = f(P_i, P_j, Q_i, \theta_i, \theta_{i+1}), \ i \in \nu, j \in N_i$$

where $N_i = \{j, j \in \nu | (\|P_i - P_j\| \leq \frac{1}{2}R)\}$. Such that $n$ vehicles achieve not only the above objectives but also the obstacle avoidance as below.

$$P_i(t) \neq P_j(t) \neq P_*(t) \tag{4}$$

where $i \neq j, i, j \in \nu$ at any time $t$.

In this brief, we focus on the problem of collective hunting with obstacle avoidance. The following section will give the control law to achieve the above all objectives.

## 3   Cyclic Estimation and Pursuit Hunting Controller Design

From the practical viewpoint, it is important to achieve the desired global hunting behavior through only local information. Figure 2 illustrates the structure of distributed hunting controller, which consists two parts, namely, hunting behavior control and vehicle kinetic control. The immediate control signal $u_i^d, \psi_i^d$ are only based on the estimator $Q_i$ and relative angle $\theta_{i+1}, \theta_i$. Then discuss and prove the stability of the closed system. Finally, the surge force $\tau_{iu}$ and the sway force $\tau_{ir}$ of the hunter vehicle $i$ are derived via the PID controller. Some assumptions are given before designing the control law.

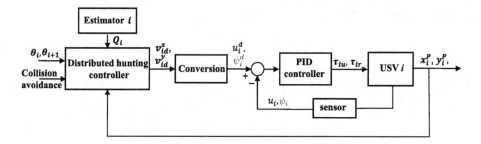

**Fig. 2.** Structure of distributed hunting controller

*Assumption 1.* $R \geq r$ and vessel $i$ get the global position of $P_*$ if target is in the sense radius $R$.

*Remark 2.* $R \geq r$ means that all vehicles eventually sense the position of position of target, which is a necessary condition for solving the collective hunting problem. USV equips with the differential GPS and millimeter wave radar, which can calculate the global position of $P_*$ if $||D_i|| \leq R$.

*Assumption 2.* The vehicle $i$ needs to know the initial label of its own.

*Remark 3.* Label-change strategy contribute to the obstacle avoidance between the vessels during the hunting. The following section give the reason.

### 3.1   Cyclic Estimation of the Target for Balanced Hunting

In this section, a decentralized estimator of the target's position $Q_i$ is designed for vehicle $i$ to get close to the target. The vehicles exchange their estimation via the cyclic communication graph $\mathbb{G}$. Before giving the form of estimator, a lemma of circular matrix is introduced.

**Lemma 1.** *Every circular matrix* $\mathbf{C} \in \mathbb{R}^{n \times n}$ *can be represented as in [9]:*

$$\mathbf{C} = circ(c_1, c_2, \ldots, c_n)$$
$$= c_1 I_n + c_2 \Pi_n + c_3 \Pi_n^2 + \cdots + c_n \Pi_n^{n-1} \tag{5}$$

*where* $\Pi_n = circ(0, 1, 0 \ldots, 0) \in \mathbb{R}^{n \times n}$. *Further, the circulant's representer is defined as:*

$$p_c(\lambda) = c_1 + c_2 \lambda + c_3 \lambda^2 + \cdots + c_n \lambda^{n-1} \tag{6}$$

*Since* $\mathbf{C} = p_c(\Pi_n)$. *Then, the eigenvalues of* $\mathbf{C}$ *are* $\lambda_i = p_c(\omega^{n-1})$, *where* $\omega = e^{j2\pi/n}$ *with* $j = \sqrt{-1}$ *and* $i = 1, 2, \ldots, n$. ∎

From the lemma above, we will have the dynamic equation of estimator $Q$ as follows:

$$\dot{Q}_i = k_1((Q_{i+1} - Q_i) + sgn^+(\Delta_i)(P^* - Q_i)) \tag{7}$$

where $Q_i$ is vehicle $i$'s estimation of target when $i = 1, 2, \ldots, n-1$. When $i = n$, estimator $Q_{i+1}$ is equal to $Q_1$. $k_1 > 0$ is a positive constant, which determines the convergence speed of estimators. $\Delta_i = R - \|D_i\|_2$ represents difference between the sense radius $R$ and relative distance $\|D_i\|_2$ to the target. $sgn^+(\cdot)$ is a sign function defined as follows:

$$sgn^+(x) = \begin{cases} 1, x \geq 0 \\ 0, x < 0 \end{cases}$$

Further, the initial value of the $Q_i$ is based on the sense radius $R$ of the equipment, just like radar, which is calculated as follows:

$$Q_i(0) = sgn^+(\Delta_i)P^*(0) \tag{8}$$

for all $i \in n$. $P^*(0)$ denotes the initial position of the target.

In order to analysis the overall estimation of the multi underactuated surface vehicles, we will rewrite the (7) in the following vector form:

$$\dot{Q}(t) = AQ(t) + B(t)(\mathbf{1} \otimes P^* - Q(t)) \tag{9}$$

with $A = circ(-k_1, k_1, 0, \ldots, 0) \in \mathbb{R}^{n \times n}$, $B \in \mathbb{R}^{n \times n}$ is a matrix where the diagonal terms $b_{ii} = 0, or, k_1$, the other term $b_{ij} = 0, i \neq j$. Where $Q = [Q_1, Q_2, \ldots, Q_n]^T \in \mathbb{R}^n$ and $cir$ denotes the circular matrix. $\mathbf{1}$ represents a column vector $[1, 1, \ldots, 1]^T$. Then we will give the theorem for the estimation.

**Theorem 1.** *Suppose that estimation $Q$ of the target based on the cyclic communication network $\mathbb{G}$ uses the form of (9) with the initial conditions (8), one has $\lim_{t \to \infty}(Q(t) - \mathbf{1} \otimes P^*) = 0, \forall i \in n$.* ∎

*Proof:* First, we introduce the estimation error $\mathbf{e}^Q = Q - \mathbf{1} \otimes P^*$, which is also a column represented $\mathbf{e}^Q = [e_1^Q, e_2^Q, \ldots, e_n^Q]$. Take the derivative of $\mathbf{e}^Q$ and substitute the (9), we will have the following dynamics:

$$\begin{aligned}
\dot{\mathbf{e}}^Q(t) &= \dot{Q} \\
&= AQ(t) + B(\mathbf{1} \otimes P^* - Q(t)) \\
&= (A - B)(\mathbf{e}^Q + \mathbf{1} \otimes P^*) + B(\mathbf{1} \otimes P^*) \\
&= (A - B)\mathbf{e}^Q + A(\mathbf{1} \otimes P^*) - B(\mathbf{1} \otimes P^*) + B(\mathbf{1} \otimes P^*) \\
&= (A - B)\mathbf{e}^Q
\end{aligned} \tag{10}$$

Since $A$ is circular matrix, its representer is $p_A(\lambda) = -k_1(1 + \lambda)$ and the eigenvalues are given as $\lambda_i = p_A(\omega^{i-1}), i = 1, 2, \ldots, n$, from Lemma 1. Then, the eigenvalues can be rewritten in complex form as:

$$\lambda_i = k_1[\cos(\tfrac{2\pi(i-1)}{n}) - 1] + jk_1 \sin(\tfrac{2\pi(i-1)}{n})$$

where $i = 1, 2, \ldots, n$. Since $k_1 > 0$. The matrix A always has a zero eigenvalue, $\lambda_1$, while the remaining $n - 1$ eigenvalue $\lambda_i, i = 2, 3, \ldots, n$, lies in the left-half complex plane. Based on the Gersǧorin disk theorem [8], all the eigenvalues of $A = [a_{ij}]$ are located in the disks as follows:

$$D_i = \{z \in \mathbb{C} : |z - a_{ii}| \leq \sum_{j \in A, j \neq i} a_{ij}\} \tag{11}$$

Then the eigenvalues of matrix $A$ are all located in the circle centered at $(-k_1, 0)$ with the radius of $k_1$. Since matrix $B$ only changes the center of the disk for $A$. The eigenvalues of matrix $(A - B)$ are also located in the disk as follows.

$$D'_i = \{z \in \mathbb{C} : |z - a_{ii} - b_{ii}| \leq \sum_{j \in A, j \neq i} a_{ij}\} \tag{12}$$

where $b_{ii}$ is equal to 0 or $k_1$, which is determined by whether the vehicle $i$ can sense the target's position or not. So the eigenvalues of $(A - B)$ are located in the circle $D'$, where the radius is the same as $A$ and center of the disk may moves left a distance of $k_1$.

Therefore, $Q(t)$ converges to the stable and invariant. Based on (9), it is easy to get that $Q_1(t_1) = Q_2(t_1) = \cdots = Q_n(t_1)$ when $t = t_1$. From *Assumption 1*, if the estimation $Q$ converge stable in the circle but not the center of the circle, which means that there is s static error between the estimation $Q_i$ and the target $P^*$. There must be at least one vehicle can sense the target's position, which means that the feedback of the target's position for estimation. Then estimation exchange their information via cyclic communication network to finally eliminate the error between estimation and target. Finally, *Theorem 1* is proved. ∎

## 3.2 Rotation and Cyclic Pursuit

In this section, we design the hierarchical controller for the hunting problem with obstacle avoidance via the information of estimation of target and neighborhood. The first step is to derive the immediate signal to get close to the target and rotate around the target.

$$\mathbf{u} = \mathbf{u}^r + \mathbf{u}^\theta \tag{13}$$

where $\mathbf{u} \in \mathbb{R}^n$ is a column vector controller defined as $[\mathbf{u}_1, \mathbf{u}_2, \mathbf{u}_3, \ldots, \mathbf{u}_n]^T$. It is consists of two parts: $\mathbf{u}^r \in \mathbb{R}^n$ is a feedback control law determining $D$. The second part $\mathbf{u}^\theta \in \mathbb{R}^n$ denotes the cyclic pursuit law determining $\theta$, which is to space evenly and rotate around the target.

Then, $Q_i$ and $D_i$ are seen as the center of the circle and radius respectively, the immediate signal directly determine $D_i$ and $\theta_i$, given the kinematics between the vehicle $i$ and estimator $i$ as below.

$$\begin{cases} \dot{D}_i = \mathbf{u}_i^r \\ \dot{\theta}_i = \mathbf{u}_i^\theta \end{cases} \tag{14}$$

Then we give the detailed control law for $\mathbf{u}^r$ and $\mathbf{u}^\theta$ to control $D$ and $\theta$.

$$\mathbf{u}^r = k_2(\mathbf{1} \otimes r - D); \tag{15}$$

where $k_2$ is the positive constant. $D = [D_1, D_2, D_3, \ldots, D_n]^T$ is a column denoting the relative distance from vehicles to the estimation.

**Theorem 2.** *Suppose that the control part $\mathbf{u}^r$ based on the estimation of target uses the form of (15), all vehicles asymptotically converge to the circular obits of the target.* ■

*Proof:* Define a *Lyapunov* function $V = \sum\limits_{i=1}^{n} \frac{1}{2}(r - D_i)^2$, take derivative of $V$ and substitute (14) and (15):

$$
\begin{aligned}
\dot{V} &= -\sum_{i=1}^{n}(r - D_i)\dot{D}_i \\
&= -(\mathbf{1} \otimes r - D)^T \dot{D} \\
&= -k_2(\mathbf{1} \otimes r - D)^T(\mathbf{1} \otimes r - D) \\
&\leq 0
\end{aligned}
\tag{16}
$$

where it notes that $V$ is equal to zero only when $D_1, D_2, D_3, \ldots, D_n$ is equal to $r$. Since $V_i = \frac{1}{2}(r - D_i)^2 \geq 0$, so from the *Lyapunov* theorem, It is easily to see that $D$ asymptotically converges to $\mathbf{1} \otimes r$ with control law $\mathbf{u}^r$. ■

The second step is to make vehicles rotate around the target and space evenly by the control law $\mathbf{u}^\theta$, which is in the form:

$$
\mathbf{u}^\theta = k_3(C\boldsymbol{\theta} + P)
\tag{17}
$$

where $C = circ(-k_3, k_3) \in \mathbb{R}^{n \times n}$, $\boldsymbol{\theta} = [\theta_1, \theta_2, \ldots, \theta_n]^T \in \mathbb{R}^n$, $P = [0, 0, \ldots, 2\pi]^T \in \mathbb{R}^n$.

**Theorem 3.** *Suppose that the control part $\mathbf{u}^\theta$ based on cyclic pursuit strategy uses the form of (17), all vehicles asymptotically space evenly and rotate around the target.*

*Proof:* The theorem is achieved based on a similar idea used in [9]. The previous parts implies $\dot{\boldsymbol{\theta}} = \mathbf{u}^\theta$. Then substitute (17) and take the derivative of $\dot{\boldsymbol{\theta}}$, then we have $\ddot{\boldsymbol{\theta}} = C\dot{\boldsymbol{\theta}}$. Since $C$ is circular matrix, its representer is $p_C(\lambda) = -k_3(1 + \lambda)$ and the eigenvalues are given as $\lambda_i = p_C(\omega^{i-1}), i = 1, 2, \ldots, n$, from Lemma 1. Then, the eigenvalues can be rewritten in complex form as matrix $A$. Therefore, the matrix C always has a zero eigenvalue, $\lambda_1$, while the remaining $n - 1$ eigenvalue $\lambda_i, i = 2, 3, \ldots, n$, lies in the left-half complex plane, which means that $\dot{\boldsymbol{\theta}}$ converges to the null space $\{\boldsymbol{\sigma} | \sigma I_n, \sigma \in \mathbb{R}\}$, $I_n = [1, 1, \ldots, 1]^T \in \mathbb{R}^n$, which corresponds to the $\lambda_1 = 0$; i.e.. $\ddot{\boldsymbol{\theta}}(t_1) = [\dot{\theta}_1(t_1), \dot{\theta}_2(t_1), \ldots, \dot{\theta}_n(t_1)]^T$ satisfies $\dot{\theta}_1(t_1) = \dot{\theta}_2(t_1) = \cdots = \dot{\theta}_n(t_1) = \sigma$ in the steady state. Furthermore, we have:

$$
\sum_{i=1}^{n} \dot{\theta}_i(t) = 2k_3\pi \quad for \ all \ t \geq 0
\tag{18}
$$

Therefore, we have $\dot{\theta}_i(t_1) = \frac{2k_3\pi}{n} = k_3(\theta_{i+1}(t_1) - \theta_i(t_1))$. From (17), we will derive that:

$$
\begin{cases}
\theta_{i+1}(t_1) - \theta_i(t_1) = \frac{2\pi}{n}; & i = 1, 2, \ldots, n - 1 \\
\theta_1(t_1) - \theta_i(t_1) + 2\pi = \frac{2\pi}{n}; & i = n
\end{cases}
\tag{19}
$$

Therefore, we proof the *Theorem* 3. Finally, the circular formation of vehicles and space evenly and rotation around the target are formed.

## 3.3   Collision Avoidance and Change the Label Between Vehicles

In the previous part, we achieve the objectives of the hunting problem. Note that during the forming of the circular formation, the label of vehicles is set by the initial different estimations of the target, it maybe break rule of Remark 1. For example, the label of vehicles maybe first set as $1, 2, 4, 3$, but the actual label around the target should be $1, 2, 3, 4$. It may cause serious problems such as the communication network $\mathbb{G}$ is not cyclic network again. Here, we propose the artificial potential method and the strategy of changing label between neighbors. The artificial potential methods are stated as follows:

$$\zeta_i = \begin{cases} (\frac{1}{2}R - D_i)^2, & D_i \le \frac{1}{2}R \\ 0, & D_i > \frac{1}{2}R \end{cases} \tag{20}$$

where $\zeta_i$ is a potential function. $R$ is the sense radius of the vehicle. $\zeta_i$ is set as repulsive function added to the immediate signal.

Combining the kinetics and kinematics of USV (1), the desired surge velocity $u_i^d$ and sway velocity $v_i^d$ by the immediate signal. Transform it into the earth framework, we can finally get the desired velocity of vehicle $i$:

$$\begin{cases} v_{id}^x = \mathbf{u_i^r} \cos \theta_i - \mathbf{u_i^\theta} \sin \theta_i + \sum_{j \in N_i}^{n} \zeta_j \cos \varrho_j \\ v_{id}^y = \mathbf{u_i^r} \sin \theta_i - \mathbf{u_i^\theta} \cos \theta_i + \sum_{j \in N_i}^{n} \zeta_j \sin \varrho_j \end{cases} \tag{21}$$

where $v_{id}^x, v_{id}^y$ is the desired velocity in the earth framework. $\varrho_j$ denotes the vehicles whose angle from vehicle $i$ to vehicle $j$, which is in the following form:

$$\varrho_i = \text{atan2}((y_i^P - y_j^P), (x_i^P - x_j^P))$$

For the USV vehicle, the surge velocity $u_i^d$ and $\psi_i^d$ can be derived from (refControl.20),

$$u_i^d = \sqrt{v_{id}^x \times v_{id}^x + v_{id}^y \times v_{id}^y}, \ \psi_i^d = \text{atan2}(v_{id}^y, v_{id}^x) \tag{22}$$

## 3.4   Vehicle Kinetic Control

The following work is to design a control law $\tau_{iu}$ and $\tau_{ir}$ to make the surge velocity $u_i$ and the sway velocity $v_i$ of the hunter $i$ converges to the desired immediate control signal $u_i^d$ and $v_i^d$. Analyse the kinetic equation of the under-actuated hunter vehicles, the surge velocity $u_i$ and the heading angle $\psi_i$ are mainly influenced by control input $\tau_{iu}$ and $\tau_{ir}$ respectively. So for simplicity of achievement, a PID controller is proposed to make the goal. Define the surge velocity error $e_i^u$ and the phase angle error $e_i^\psi$ as:

$$\begin{cases} e_i^u = u_i - u_i^d \\ e_i^\psi = \psi_i - \psi_i^d \end{cases} \tag{23}$$

where $e_i^u$ and $e_i^\psi$ are the surge velocity error and heading angle error between vehicle states and the desired immediate control signal. Based on the PID controller scheme, the control law of hunter vehicle $i$ is obtained as follows:

$$\begin{cases} \tau_{iu} = k_i^{pu} e_i^u + k_i^{iu} \int_0^t e_i^u \, dt + k_i^{du} \dot{e}_i^u \\ \tau_{ir} = k_i^{pp} e_i^\psi + k_i^{ip} \int_0^t e_i^\psi \, dt + k_i^{dp} \dot{e}_i^\psi \end{cases} \tag{24}$$

where $(k_i^{pu}, k_i^{iu}, k_i^{du})$ and $(k_i^{pp}, k_i^{ip}, k_i^{dp})$ are positive constant, which represent proportionality coefficient, integral coefficient and differential coefficient of the surge velocity and phase angle of hunter vehicle $i$.

## 4   Simulation

In this section, we carry out some numerical simulations to demonstrate the performance of the controller (13) for hunting problem. Consider an underactuated surface vehicle with model parameters just as in [1]: $m_{11} = 1.956, m_{22} = 2.405, m_{33} = 0.043, d_{11} = 2.436, d_{22} = 12.992, d_{33} = 0.0564$. For the purpose of comparisons, we suppose the controller is in two cases, which contains label-change strategy and which is not.

For simulation use, we define one target and five hunters, making the sense radius is $R = 8$ m and hunting radius is $r = 5$ m. Then we give the initial conditions for the hunting problem as follows: the target position is set $P_* = (-5, -3)^T$, the hunter vehicles' positions and states are set without loss of generality, only one vessel initially knows the position of target:

$$\mathbf{P} = \begin{bmatrix} 2 & 8 \\ 8 & -2 \\ -2 & 10 \\ -2 & -7.5 \\ 4 & -10 \end{bmatrix}, \psi = \begin{bmatrix} \arctan 4 \\ \arctan \frac{1}{7} \\ \arctan \frac{1}{5} \\ \arctan \frac{1}{5} \\ \arctan \frac{1}{5} \end{bmatrix}, \mathbf{v} = \begin{bmatrix} 0 & 0 & 0 \\ 0 & 0 & 0 \\ 0 & 0 & 0 \\ 0 & 0 & 0 \\ 0 & 0 & 0 \end{bmatrix}$$

The control parameter of hunter vehicle $i$ are taken as $k_1 = 0.5, k_2 = 1, k_3 = 0.15, k_i^{pu} = 12, k_i^{iu} = 0.7, k_i^{du} = 0, k_i^{pp} = 0.1, k_i^{ip} = 0, k_i^{du} = 5$. Simulation results are shown in the following figures, and the simulation time is set 50 s.

Figures 3 and 4 show the distributed hunting behavior with label changing between neighbors. Figure 3 shows that vehicles finally surround and rotate

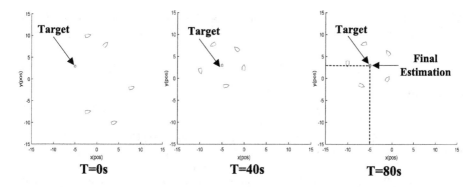

**Fig. 3.** Distributed hunting behavior where label can change between neighbors.

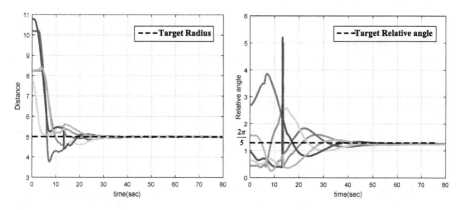

**Fig. 4.** Distance and relative angle between the angle where label can change between neighbors

around the target, and the vehicles space evenly around the target. Figure 4 indicates that distance from target and relative angle have oscillations at the beginning, because vehicles change its label between neighbors. But it can asymptotically converges to the desired value. One can conclude the effectiveness of the controller for distributed hunting by multi USVs.

## 5    Conclusions

In this paper, we have studied collective hunting problem with obstacle avoidance for a group of USVs. The problem includes two subobjectives of hunting target, for which each vehicle maintains the desired distance from target, spaces evenly and rotates around the target. The controller are based on the cyclic estimation of target and feedback control law to get close to the target. It is designed by also cyclic pursuit strategy using the relative angle between the neighbors. Artificial potential method and strategy of changing label between neighbors are also applied to guarantee obstacle avoidance and achieve the goal. It is worth to mention that the control law guarantee almost collision during the forming the formation.

# References

1. Ashrafiuon, H., Muske, K.R., McNinch, L.C., Soltan, R.A.: Sliding-mode tracking control of surface vessels. IEEE Trans. Ind. Electron. **55**(11), 4004–4012 (2008)
2. Ceccarelli, N., Di Marco, M., Garulli, A., Giannitrapani, A.: Collective circular motion of multi-vehicle systems. Automatica **44**(12), 3025–3035 (2008)
3. Chen, Z., Zhang, H.T.: No-beacon collective circular motion of jointly connected multi-agents. Automatica **47**(9), 1929–1937 (2011)
4. Chen, Z., Zhang, H.T.: A remark on collective circular motion of heterogeneous multi-agents. Automatica **49**(5), 1236–1241 (2013)
5. El-Hawwary, M.I., Maggiore, M.: Distributed circular formation stabilization for dynamic unicycles. IEEE Trans. Autom. Control **58**(1), 149–162 (2013)
6. Fossen, T.I.: Marine control system-guidance, navigation and control of ships, rigs and underwater vehicles. Marine Cybemetics (2002)
7. Frew, E.W., Lawrence, D.A., Morris, S.: Coordinated standoff tracking of moving targets using lyapunov guidance vector fields. J. Guidance Control Dyn. **31**(2), 290–306 (2008)
8. Horn, R.A., Horn, R.A., Johnson, C.R.: Matrix Analysis. Cambridge University Press, Cambridge (1990)
9. Kim, T.H., Sugie, T.: Cooperative control for target-capturing task based on a cyclic pursuit strategy. Automatica **43**(8), 1426–1431 (2007)
10. Lan, Y., Yan, G., Lin, Z.: Distributed control of cooperative target enclosing based on reachability and invariance analysis. Syst. Control Lett. **59**(7), 381–389 (2010)
11. Ni, J., Yang, S.X.: Bioinspired neural network for real-time cooperative hunting by multirobots in unknown environments. IEEE Trans. Neural Netw. **22**(12), 2062–2077 (2011)
12. Sepulchre, R., Paley, D.A., Leonard, N.E.: Stabilization of planar collective motion: all-to-all communication. IEEE Trans. Autom. Control **52**(5), 811–824 (2007)
13. Sepulchre, R., Paley, D.A., Leonard, N.E.: Stabilization of planar collective motion with limited communication. IEEE Trans. Autom. Control **53**(3), 706–719 (2008)
14. Seyboth, G.S., Wu, J., Qin, J., Yu, C., Allgöwer, F.: Collective circular motion of unicycle type vehicles with nonidentical constant velocities. IEEE Trans. Control Netw. Syst. **1**(2), 167–176 (2014)
15. Yamaguchi, H.: A cooperative hunting behavior by multiple nonholonomic mobile robots. In: 1998 IEEE International Conference on Systems, Man, and Cybernetics, vol. 4, pp. 3347–3352. IEEE (1998)
16. Yu, X., Liu, L.: Distributed circular formation control of ring-networked nonholonomic vehicles. Automatica **68**, 92–99 (2016)
17. Zheng, R., Liu, Y., Sun, D.: Enclosing a target by nonholonomic mobile robots with bearing-only measurements. Automatica **53**, 400–407 (2015)

# The Wireless Communications for Unmanned Surface Vehicle: An Overview

Junfeng Ge[1,2(✉)], Tao Li[1,2], and Tao Geng[1]

[1] Guangdong HUST Industrial Technology Research Institute,
Guangdong Province Key Lab of Digital Manufacturing Equipment,
Dongguan 523808, Guangdong, China
gejf@hust.edu.cn
[2] Key Laboratory of Image Processing and Intelligent Control of Education Ministry of China,
School of Automation, Huazhong University of Science & Technology,
Wuhan 430074, Hubei, China

**Abstract.** With growing interest in commercial, scientific and environmental issues on oceans, lakes and rivers, there has been a corresponding growth in demand for the development of unmanned surface vehicles (USVs). The wireless communication module is one of the key components of USV because of the data and control information communication with ground control stations and other vehicles to perform cooperative control. This paper presents a survey on the wireless communication techniques that can be used in USVs and a detailed discussion about the advantages and disadvantages of these techniques for data and control information transmission. The paper first provides an overview of both typical and recent wireless communication techniques, along with some principles and parameters. Next, wireless communication techniques used in USVs are outlined and classified according to single USV and multi-USVs. Finally, some advices on choosing wireless communication techniques and some general challenges towards high-speed and swarming USVs are highlighted.

**Keywords:** Unmanned surface vehicle (USV) · Wireless communication
Multi-USV networks

## 1 Introduction

Unmanned Surface Vehicle (USV) is an intelligent and unmanned water surface platform, which can navigate autonomously in the marine, lake and river environment and complete various tasks such as environmental perception, target detection. The applications of USV include harbor surveillance, water quality sampling, hydrologic survey, maritime search and rescue, anti-submarine warfare [1–4].

Depending on practical applications, USVs may have a variety of appearances and functionalities. However, the basic elements such as propulsion and power unit, perception and navigation unit, guidance and control unit, communication unit, must be included in every USV. The fundamental architecture of a typical USV is shown in Fig. 1.

© Springer Nature Switzerland AG 2018
Z. Chen et al. (Eds.): ICIRA 2018, LNAI 10984, pp. 113–119, 2018.
https://doi.org/10.1007/978-3-319-97586-3_10

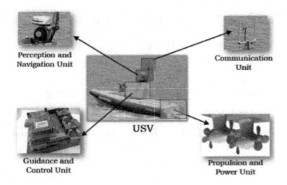

**Fig. 1.** The fundamental architecture of a typical USV

The propulsion and power unit provides the speed and heading of USV. Most existing USVs use rudder and propeller (or water jet) propulsion systems. The perception and navigation unit concentrates on identifying the USV's current and future states according to its surrounding environment information. The guidance and control unit focuses on determining the proper control forces and moments to be generated with instruction provided by the navigation unit and the control objectives. The communication unit has wireless communication with ground control stations and exchange data including receiving commands from the ground station and reporting the operating status and videos. Reliability of communication systems is therefore of great importance [5–8].

In order to enhance robustness and reliability of USV, improve mission performance, increase their spatiotemporal capacity, and enlarge the coverage of surveillance and measurement, the USVs are clustered into swarm application under cooperative control. Consequently, there are four types of wireless communication for USVs as shown in Fig. 2.

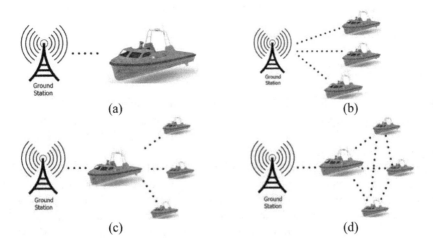

**Fig. 2.** Four types of wireless communication for USVs

The four types of communication configurations are point-point, point-multipoint, point-point and point-multipoint and mesh network. Different configuration requires different communication performance. Decentralized cooperative control shares information among the USVs, which brings about numerous challenges including limited communication bandwidth, transmission noise, and communication delays, dropouts and failures [1, 9–13].

This paper presents a survey on the wireless communication techniques that can be used in USVs and a detailed discussion about the performance according to single USV and multi-USVs. Finally, some advices on choosing wireless communication techniques and some general challenges towards high-speed and swarming USVs are presented.

## 2   The Wireless Communication Techniques

According to the different coverage ability, the wireless communication techniques can be divided into short-range and long-range wireless communication methods. The short-range wireless communication methods contains Bluetooth, Zigbee, Ultra WideBand, Wi-Fi, and Radio Frequency (RF). The long-range wireless communication methods includes High-gain Wi-Fi (Microwave bridge), GPRS, 3G, WiMAX, and LTE [14–18].

Bluetooth is a wireless technology standard for exchanging data over short distances from fixed and mobile devices, and building personal area networks (PANs). Invented by telecom vendor Ericsson in 1994. The IEEE standardized Bluetooth as IEEE 802.15.1, but no longer maintains the standard. The Bluetooth SIG oversees development of the specification, manages the qualification program, and protects the trademarks. The transmission frequency band of Bluetooth is the 2.4 GHz ISM band that is universally available to the public worldwide, providing a transmission rate of 1 Mbps and a transmission distance of 10 m.

Zigbee is an IEEE 802.15.4-based specification for a suite of high-level communication protocols used to create personal area networks with small, low-power digital radios. Zigbee is a low-power, low data rate, and close proximity wireless ad hoc network. The technology defined by the Zigbee specification is intended to be simpler and less expensive than other wireless personal area networks (WPANs), such as Bluetooth or more general wireless networking such as Wi-Fi. Applications include home energy monitors, traffic management systems, and other consumer and industrial equipment that requires short-range low-rate wireless data transfer.

Ultra-wideband (also known as UWB, ultra-wide band and ultraband) is a radio technology that can use a very low energy level for short-range, high-bandwidth communications over a large portion of the radio spectrum. UWB has traditional applications in non-cooperative radar imaging. Most recent applications target sensor data collection, precision locating and tracking applications. UWB can achieve data transfer rates of hundreds of Mbit/s to several Gbit/s in a range of about 10 m.

Wi-Fi or WiFi is a technology for wireless local area networking with devices based on the IEEE 802.11 standards. Wi-Fi most commonly uses the 2.4 GHz UHF and 5.8 GHz SHF ISM radio bands. Wi-Fi also allows communications directly from one device to another without an access point intermediary. This is called ad hoc Wi-Fi transmission which can be used for multi-USV communication. The Wi-Fi signal range depends on the frequency band, radio power output, antenna gain and antenna type as well as the modulation technique. An access point compliant with either 802.11b or 802.11g, using the stock antenna might have a range of 100 m. The same radio with an external semi parabolic antenna (15 dB gain) might have a range over 20 miles, which is also called high-gain Wi-Fi. The newest 802.11ac can achieve the Gigabit throughput.

RF communications incorporate a transmitter and a receiver. They are of various types and ranges. Several carrier frequencies are commonly used in commercially available RF modules, including those in the industrial, scientific and medical (ISM) radio bands such as 433.92 MHz, 915 MHz, and 2400 MHz. These frequencies are used because of national and international regulations governing the use of radio for communication.

GPRS (General Packet Radio Service) is a packet oriented mobile data service on the 2G and 3G cellular communication system's global system for mobile communications (GSM). In 2G systems, GPRS provides data rates of 56–114 Kbit/second. 2G cellular technology combined with GPRS is sometimes described as 2.5G, that is, a technology between the second (2G) and third (3G) generations of mobile telephony. It provides moderate-speed data transfer, by using unused time division multiple access (TDMA) channels.

3G, short for third generation, is the third generation of wireless mobile telecommunications technology. It is the upgrade for 2G and 2.5G GPRS networks, for faster internet speed. 3G telecommunication networks support services that provide an information transfer rate of at least 0.2 Mbit/s. Later 3G releases, often denoted 3.5G and 3.75G, also provide mobile broadband access of several Mbit/s. It is expected that 3G will provide transmission rates at a minimum data rate of 2 Mbit/s for stationary or walking users, and 348 Kbit/s in a moving vehicle.

WiMAX (Worldwide Interoperability for Microwave Access) is a family of wireless communication standards based on the IEEE 802.16 set of standards. WiMAX was initially designed to provide 30 to 40 megabit-per-second data rates, with the 2011 update providing up to 1 Gbit/s for fixed stations. WiMAX is a long range system, covering many kilometers that uses licensed or unlicensed spectrum to deliver connection to a network, in most cases the Internet.

Long-Term Evolution (LTE) is a standard for high-speed wireless communication for mobile devices and data terminals, based on the GSM/EDGE and UMTS/HSPA technologies. It increases the capacity and speed using a different radio interface together with core network improvements. LTE is commonly marketed as 4G LTE and can provides downlink peak rates of 300 Mbit/s, uplink peak rates of 75 Mbit/s and QoS provisions permitting a transfer latency of less than 5 ms in the radio access network. LTE has the ability to manage fast-moving mobiles and supports multi-cast and broadcast streams.

The current typical communication techniques can be summarized in the Table 1. The methods can be used in USV are listed too. All the available methods are proper for Multi-USV communication using point-multipoint mode, especially, the Zigbee, Wi-Fi and RF can be used to set up the ad hoc network for swarm control.

**Table 1.** The typical wireless communication techniques

| ID | Name | Range type | Range | Speed | Network structure | Can be used by USV |
|----|------|-----------|-------|-------|-------------------|--------------------|
| 1 | Bluetooth | Short | 10 m | 1 Mbps | Direct | No |
| 2 | Zigbee | Short | 100 m | 250 kbps | Direct/Ad hoc | Yes |
| 3 | Ultra WideBand | Short | 10 m | 100 Mbps | Direct | No |
| 4 | Wi-Fi | Short | 100 m | 600 Mbps | Direct/Ad hoc | Yes |
| 5 | RF | Short | 1 km | 2k–2 Mbps | Direct | Yes |
| 6 | High-gain Wi-Fi | Long | 5 km | 600 Mbps | Direct/Ad hoc | Yes |
| 7 | RF | Long | 5 km | 20 Mbps | Direct/Ad hoc | Yes |
| 8 | GPRS | Long | 10 km | 100 kbps | Direct | Yes |
| 9 | 3G | Long | 10 km | 2 Mbps | Direct | Yes |
| 10 | WiMAX | Long | 10 km | 1 Gbps | Direct | Yes |
| 11 | 4G | Long | 10 km | 300 Mbps | Direct | Yes |

## 3  The Wireless Communication in USV

If the USV adopts a real-time data transmission scheme, its operating range will be limited by the distance of wireless data communication. Generally, a point-to-point communication method can transmit up to several km, and a 4G network can cover about 10 km of the coastline, which is the maximum operating range of the scheme. If non-real-time data transmission and artificial intelligence obstacle avoidance schemes are used, the scope of operation is theoretically limited by the power supply of USV.

The performance of the USVs in domestic market is listed in Table 2. Most domestic USVs choose 2.4G Wi-Fi or RF for short-range communication and 4G for long range communication.

For short-range communication, we can use the 2.4G Wi-Fi modules. Each USV is connected to the Wi-Fi access point in the same local area network. The Wi-Fi access point can be installed on the shore or on the leader USV. This ensures that each USV can use the TCP/IP protocol to transfer data.

For long-distance communications, we can use 4G modules. USVs send data to the ground control center or cloud server for data storing through 4G modules. The cloud server can forward the data to the ground control center. After that, the USV receives the information from the ground control center forwarded by the cloud server to autonomously navigate, and sends the real-time heading, speed, and location data to the ground control center during the sailing process.

**Table 2.** The performance of the USVs in domestic market

| Name | | TianXing NO.1 | JingHai NO.3 | M75 | Huawei NO.5 |
|---|---|---|---|---|---|
| Picture | | | | | |
| Company | | Harbin Engineering University | Shanghai University | Zhuhai Yunzhou | Hua Ce |
| Parameters | Type | Monomer | Monomer | Monomer | Three-body |
| | Length | 13m | 6.28m | 5m | 1.6m |
| | Max speed | ≥ 50 knots | ≥ 10 knots | ≥ 30 knots | ≥ 10 knots |
| | Propulsion | Double surface slurry | Unknown | Jet Pump | Propeller |
| | Distance | 1000km | 200 nautical miles | 150 nautical mile | Unknown |
| Short-range Comm. | | 2.4G | Unknown | 2.4G | Radio / Network Bridge |
| Long-range Comm. | | Unknown | 4G/ Satellite | Unknown | Unknown |

## 4    Conclusions

This paper presents a survey on the wireless communication techniques that can be used in USVs and the discussion about the advantages and disadvantages of these techniques for data and control information transmission. The Zigbee, Wi-Fi and Radio Frequency (RF) can be used for short-range wireless communication. The high-gain Wi-Fi (Microwave bridge), RF, GPRS, 3G, WiMAX, and LTE can be used in long-range wireless communication for single or multiple USV networks. The Zigbee, Wi-Fi and RF can also be used to form the ad hoc USV networks.

With the rapid development of 5G wireless communication techniques [19], USVs can transmit video and control or status information through faster broadband channel and more convenient narrow band techniques such as LoRa and NB-IoT with acceptable delay. At that moment, the wireless communication challenges in USV swarm may not be in the way.

**Acknowledgements.** This work was supported by the Guangdong Innovative and Entrepreneurial Research Team Program under Grant 2014ZT05G304.

# References

1. Liu, Z., Zhang, Y., Yu, X., et al.: Unmanned surface vehicles: an overview of developments and challenges. Ann. Rev. Control **41**, 71–93 (2016)
2. Pinto, E., Santana, P., Marques, F., Mendonça, R., Lourenço, A., Barata, J.: On the design of a robotic system composed of an unmanned surface vehicle and a piggybacked VTOL. In: Camarinha-Matos, L.M., Barrento, N.S., Mendonça, R. (eds.) DoCEIS 2014. IAICT, vol. 423, pp. 193–200. Springer, Heidelberg (2014). https://doi.org/10.1007/978-3-642-54734-8_22
3. Qi, J., Peng, Y., Wang, H., et al.: Design and implement of a trimaran unmanned surface vehicle system. In: International Conference on Information Acquisition, pp. 361–365. IEEE (2017)
4. Shi, X., Wang, S., Xu, Z., et al.: Hardware system for unmanned surface vehicle using IPC. Adv. Mater. Res. **971**, 507–510 (2014)
5. Christensen, A., Oliveira, S., Postolache, O., et al.: Design of communication and control for swarms of aquatic surface drones. In: International Conference on Agents and Artificial Intelligence, Lisbon, Portugal (2015)
6. Heo, J., Kim, J., Kwon, Y.: Analysis of design directions for unmanned surface vehicles (USVs). J. Comput. Commun. **5**(7), 92–100 (2017)
7. Ferreira, B., Coelho, A., Lopes, M., et al.: Flexible unmanned surface vehicles enabling future internet experimentally-driven research. In: Oceans, Aberdeen, UK. IEEE (2017)
8. Powers, C., Hanlon, R., Schmale, D.: Remote collection of microorganisms at two depths in a freshwater lake using an unmanned surface vehicle (USV). PeerJ **6**(5), e4290 (2018)
9. Campbell, S., Naeem, W., Irwin, G.W.: A review on improving the autonomy of unmanned surface vehicles through intelligent collision avoidance manoeuvres. Ann. Rev. Control **36**(2), 267–283 (2012)
10. Sánchez-García, J., García-Campos, J.M., Arzamendia, M., et al.: A survey on unmanned aerial and aquatic vehicle multi-hop networks: wireless communications, evaluation tools and applications. Comput. Commun. **119**, 43–65 (2018)
11. Yanmaz, E., Yahyanejad, S., Rinner, B., et al.: Drone networks: communications, coordination, and sensing. Ad Hoc Netw. **68**, 1–15 (2017)
12. Manley, J.E.: Unmanned surface vehicles, 15 years of development. In: Oceans, Quebec, Canada. IEEE (2008)
13. Dunbabin, M., Grinham, A., Udy, J.: An autonomous surface vehicle for water quality monitoring. In: Australian Conference on Robotics and Automation, Sydney Australia (2009)
14. Curcio, J., Leonard, J., Patrikalakis, A.: SCOUT a low cost autonomous surface platform for research in cooperative autonomy. In: Oceans, Washington, DC, USA, pp. 725–729. IEEE (2016)
15. Sousa, J., Gonçalves, G.A.: Unmanned vehicles for environmental data collection. Clean Technol. Environ. Policy **13**(2), 369–380 (2011)
16. Zeng, Y., Zhang, R., Teng, J.L.: Wireless communications with unmanned aerial vehicles: opportunities and challenges. IEEE Commun. Mag. **54**(5), 36–42 (2016)
17. Zhan, P., Yu, K., Swindlehurst, A.L.: Wireless relay communications with unmanned aerial vehicles: performance and optimization. IEEE Trans. Aerosp. Electron. Syst. **47**(3), 2068–2085 (2011)
18. Wiki Homepage. https://en.wikipedia.org. Accessed 10 Apr 2018
19. Zhang, S., Zhang, H., Di, B., et al.: Cellular UAV-to-X Communications: Design and Optimization for Multi-UAV Networks in 5G (2018). arXiv:1801.05000v1

# PolyMap: A 2D Polygon-Based Map Format for Multi-robot Autonomous Indoor Localization and Mapping

Johann Dichtl[(✉)], Luc Fabresse, Guillaume Lozenguez, and Noury Bouraqadi

IMT Lille-Douai, Douai, France
johann.dichtl@gmail.com

**Abstract.** Autonomous exploration is an important tasks in many robotic fields such as disaster response scenarios. In time critical situations, the use of multiple robots can reduce the time to create a complete map of the environment. However among the most popular map formats in use today, none are ideal for the multi-robot autonomous indoor localization. In terms of memory usage, visualization, and usability in navigation and exploration tasks, all formats have some strengths and weaknesses.

In this paper we introduce PolyMap, a map format that is based on *simple* polygons. Since the polygons are based on line segments, this is a special case of vector-based map formats. This format provides advantages in terms of memory footprint over occupancy grids, while not falling behind in visualization. Its sparse nature is also an advantage for navigation tasks, in particular when the map needs to be shared over a wireless network connection. Additionally the explicit modeling of frontiers helps with autonomous exploration.

**Keywords:** Vector maps · Indoor mapping · Exploration
Multi-robot systems

## 1 Introduction

In many indoor applications, robots have to autonomously map an environment without human assistance. For example, fire fighters can use robots to build a map before entering a building that caught fire. Rapid exploration is critical in such scenarios to minimize damage and maximize survival chances of victims. This is why it is interesting to rely on multiple robots to explore the environment and collaboratively build a map.

Currently popular 2D map formats have various shortcomings for this tasks. Occupancy grids, the most commonly used format, have a large memory footprint, which is particularly limiting when relying on wireless network connectivity to share maps between robots. Feature-based maps tend to not model the shape of obstacles, making navigation and visualization for human use difficult or impossible. Furthermore they don't define frontiers, making autonomous

© Springer Nature Switzerland AG 2018
Z. Chen et al. (Eds.): ICIRA 2018, LNAI 10984, pp. 120–131, 2018.
https://doi.org/10.1007/978-3-319-97586-3_11

exploration more difficult. Vector-based map formats also don't have frontiers, and often have gaps between vectors, making it impossible to clearly distinguish explored space from unknown space.

In this paper, we investigate 2D map formats that are the most appropriate for such applications. Starting from our reference scenario we draw a list of requirements (Sect. 2), that exhibit the shortcomings of the state of existing map formats (Sect. 3). We then introduce PolyMap (Sect. 4), a 2D map format that represent the environment as a collection of polygons. We also evaluate PolyMap by showing how it addresses our requirements. The paper ends with a conclusion (Sect. 5) that summarizes our contributions and sketches some future work.

## 2 Requirements

To speed up mapping by using a robotic fleet, robots have to spread and explore different parts of the environment. This means that they have to somehow decide which one goes where. This decision relies on *frontiers* [14] between explored space and the unknown one. Each reachable frontier is assigned to one or more robots. Robots then navigate towards their respective target frontiers, which lead to exploring new areas of the environment. Each robot then shares its local map with others to build a bigger map gathering all explored areas. The list of frontiers is updated and again assigned to robots. This process is repeated until we get a map with no reachable frontier left.

From the above description, we can infer that a map format for autonomous exploration and mapping should meet the following requirements:

**Explicit Exploration Frontiers.** Explicit frontiers enable collaboration. They materialize the exploration tasks assigned to each robot. This is why the map format should allow representing frontiers.

**Support for Path Planning.** To evaluate frontiers' reachability, as well as the cost to reach them, robots should be able to perform path planning. The map format should then unambiguously distinguish free navigable areas from obstacles.

**Lightweight.** To cover large areas, robotic fleets can include many robots, possibly dozens. All these robots need to share their local maps over a wireless network, to achieve collaborative exploration. The map format should be lightweight memory wise to save network bandwidth and support large fleets of robots.

**Visualization for Human Use.** In many scenarios, a human supervises the robots' mission or uses the built map to perform some task. For example, responders might use the map built by robots to go rescue trapped humans. The map format should then be usable by humans to identify different areas of the explored environment.

## 3   State of the Art

The purpose of a map varies from use case to use case. Therefore it is of no surprise, that different map formats emerged to satisfy different use cases. The two major classifications that are of interest to us are *metric* and *topological* map formats. Metric maps model geometric information of the environment such as obstacles and free space, in some form of coordinates that allow to measure distances. Topological maps model connectivity between different locations. As shown in Fig. 1, metric maps can be further split into three sub-groups: occupancy grids, feature-based, and vector-based.

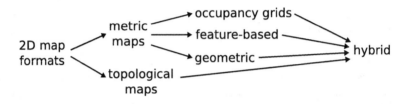

**Fig. 1.** Map formats families

*Occupancy Grid.* The dominating 2D map format is the *occupancy grid*, also called *grid map* [12]. An occupancy grid is a matrix, where each cell contains the estimated probability that the space it represents is traversable. In this context *traversable* only means that the space has been observed (i.e. it is not unknown space) and is not occupied by an obstacle. This probability is typically quantized into three possible states: traversable/free, occupied, and unknown/unexplored.

Figure 2 shows an example of a three-state grid map. This quantization reduces the required resources to create the map and distribute it in a network.

This format is easy to visualize for human use (as seen in Fig. 2), and can also be used for navigation purposes. This makes the map format very versatile. It is however relatively expensive in terms of CPU and memory requirements, which limits the maximum size of the created maps. The map resolution (i.e. how big a single cell is) influences both resources as well. The overall size of the matrix storing the grid is limited by the hardware, and if it needs to be shared between multiple robots/computers, by the available network bandwidth.

Topological information is embedded only implicitly, by treating the grid as a big graph. Here we have edges between two neighboring cells if and only if both cells are marked as traversable space. This results in a relatively large graph that is not optimized for navigation tasks. Therefore, occupancy grids are only used for navigation tasks in relatively small environments, e.g. in an office building.

*Feature-Based Maps.* The second map format family is the *feature-* or *landmark-based* map format. This map format stores distinct features or landmarks of the environment (e.g. corners or artificial beacons) with their relative position. For

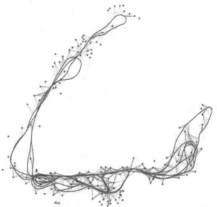

**Fig. 2.** A 3-state occupancy grid created from the *Intel Research Lab* data set. The cell states are: free (white), unexplored (gray), and occupied (black). Image and data set source: http://www2.informatik. uni-freiburg.de/~stachnis/datasets.html

**Fig. 3.** An example of a landmark-based map from the *Victoria Park* data set. The red crosses show the detected landmarks, while the robot's estimated trajectory is displayed in blue. The black lines indicate at which position (i.e. which keyframe) the landmark was observed. Image source: https://sourceforge.net/p/ slam-plus-plus/wiki/Compiling%20and %20running/?version=14 (Color figure online)

example Castellanos et al. [2] uses line segments as features. Unlike the *occupancy grid*, this is a sparse representation of the environment (see Fig. 3).

The advantages of this format stem from the sparse nature of the landmarks, typically translating into significantly smaller memory footprints and computation power. This in return allows larger areas to be covered by this map format before being limited by the computational hardware.

The main disadvantage of this format is, that the transition from traversable space to unexplored space is not modeled. In particular, frontiers are not part of the map format, so frontier-based exploration is not possible. Furthermore, features don't necessarily model obstacles, even though some examples exist that use vectors as features, e.g. [2,8,10]. If visualization is meaningful with this map format depends on whether the chosen feature models obstacle shapes.

*Vector Maps.* The third and currently least used map format family of the three is a parametric representation of the environment. Parametric in this context refers to parametric curves (splines and bezier curves), and line segments.

In the context of SLAM line segments are often also called vectors. They are the only ones used for SLAM [5], since they are significantly easier to handle for computations such as collision control.

Line segment based maps model the borders between traversable space and obstacles via line segments. This representation has been used since early robotic Localization and Mapping research [3]. The resulting maps is very usable by humans, since it allows them to identify obstacles in the environment.

Compared with occupancy grids, line segment based maps are sparse in nature, since only the boundaries are explicitly modeled. This sparse nature makes vector map lightweight and hence very appropriate for sharing over wireless.

Existing vector-based map formats [1,4,6,8–11] suffer from two major limitations. First, they lack a representation of frontiers between explored and unexplored spaces. Besides, the resulting map is not always navigable because of possible gaps between vectors representing boundaries of the same obstacle. A notable exception is the map format of TvSLAM [4], where segments delimiting a given obstacle are always connected. However, likewise all other vector-based format, TvSLAM maps don't distinguish between unexplored areas from explored traversable ones.

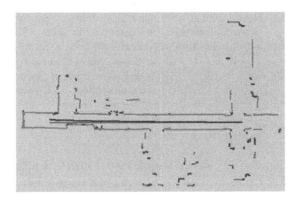

**Fig. 4.** A small vector map created with edge-extraction from point clouds. Long vectors are colored red, shorter vectors are blue. Source: [6]. (Color figure online)

*Hybrid Maps.* Hybrid map formats, as indicated in Fig. 1, are a mix of two or more maps formats. They are not explicitly included in Table 1, because their properties can be deduced from the type of maps that they are composed of. Two examples for hybrid map formats are: Ravankar et al. [10] and Lv et al. [8], both combining feature-based and vector-based maps.

*Summary.* Table 1 summarizes our evaluation of map format families with regard to requirements of Sect. 2.

**Explicit Exploration Frontiers.** Occupancy grids make it easy to make frontiers explicit. Unexplored space is well differentiated from free space and obstacles. Thus, frontiers are *free* cells that have neighboring *unexplored* cells.

**Table 1.** Evaluation of existing map formats

|  | Occupancy grids | Feature-based | Vector-based |
|---|---|---|---|
| Exploration frontiers | Yes | No | No |
| Path planning | Yes | Not always[a] | Not always[b] |
| Lightweight | No | Yes | Yes |
| Visualization | Yes | Not always[c] | Yes |

[a]The map needs to embed topological information to support path planning since features alone typically are not sufficient to model traversable space.

[b]Path planning can be challenging if there are gaps between vectors.

[c]In general, feature-based maps don't model obstacle shapes, and are not suitable for visualization. However exceptions exist, e.g. [8,10]

The classic feature-based map format does not contain any frontiers. Vector-based map formats could model frontiers via vectors that are marked as such, but we are currently not aware of any implementations that make use of this.

**Support for Path Planning.** Typically, the first step for path planning is to create a topological graph of the environment. Building this graph from an occupancy grid is technically not needed since the format already implicitly contains such a graph: each free cell represents a node in the graph, and two neighboring cells are connected if both are free. This method however is relatively expensive in terms of computational and memory requirements for path finding, since the graph is relatively large. Hence, often a dedicated topological map is created from the metric map instead. Feature based maps do not allow to create topological maps. Vector maps do not allow to build topological maps. One reason are the gaps between the vectors (as shown for example in Fig. 4). Path planning is also challenging with vector maps, since there is no clear distinction between free space and unexplored space.

**Lightweight.** Occupancy grids have the highest memory requirements among the discussed formats. The memory depends only on the size of the area covered by the map and the chosen resolution. It does not depend on the number of obstacles, or the shape of the obstacles. Feature-based maps have a low memory consumption which correlates with the number of features in the environment/map. As a result, the memory consumption can be altered by choosing a different feature detector, or tweaking the parameters of the selected feature detector. Vector-based map formats are sparse representation of the environment, and as such have low memory requirements. Furthermore, the number of line segments in both formats depends on the amount of obstacles (and their shape), but not on the size of the environment.

**Visualization for Human Use.** All presented map formats except feature-based maps are considered suitable for visualization. For feature-based maps, it depends on the type of feature: those that do model the position and shape of obstacles are suitable for visualization, the rest is not. Occupancy grids can easily be exported as bitmaps (typically in the *portable network graphic*

(PNG) format), vector maps are either rasterized as bitmaps or exported in a vector format such as *scaleable vector graphics* (SVG). Either provide an easy-to-understand top-down-view of the covered area.

# 4  PolyMap

Looking at the previous section, we can see that vector-based map format is promising. Vector-based maps are both lightweight and appropriate for human use. In this section, we introduce PolyMap, a vector-based map format that addresses the two limitations of vector maps. We show that PolyMap is appropriate for path planning and for exploration.

## 4.1  Polygon-Based Map Format

PolyMap represents the environment using simple polygons. *Simple* polygons are polygons that consist of non-intersecting line segments that are joined pair-wise to form a closed path [13].

Formally, a polygon-based map $M$ is defined as:

$$type: \quad T = \{border, frontier\} \tag{1}$$

$$vector: \quad V = \{(x_i, t)|1 \leq i \leq 2; x_i \in \mathbb{R}^2; x_1 \neq x_2\} \tag{2}$$

$$polygon: \quad P = \{v_i|i \geq 3; i \in V\} \tag{3}$$

$$map: \quad M = \{p_i|1 \leq i \leq n; p \in P\} \tag{4}$$

with $P$ forming a *simple* polygon, and $n$ being the number of polygons in the map.

We define two types of line segments:

**Obstacles:** they represent the outline of obstacles.

**Frontiers:** they model the transition from free/traversable space to unknown/ unexplored space.

Every line segment has exactly one type, polygons can contain line segments of different types. The direction of line segments is chosen, so that clockwise oriented polygons contain traversable space inside. Figure 5 shows an example of map containing two different types of line segments: red line segments represent obstacles and green line segments represent frontiers. A map is composed of a single (clockwise oriented) large polygon that encompasses the entire explored area. Inside this polygon are typically multiple other polygons with a counter-clockwise orientation, modeling obstacles.

Using directed line segment like this, has the advantage that it is easy to determine whether a given point is inside free or unexplored space; a property that feature-based and vector-based map formats are missing. For this, we only need to find the closest line segment to the point and look at which side of the polygon the point is located. Figure 6 illustrates this with two points: one is inside the explored area and the other is outside. This not only helps for visualization, but also allows localization with particle filters to discard particles that are inside obstacles or otherwise in unexplored space.

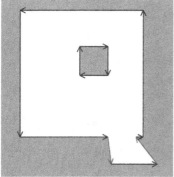

**Fig. 5.** An example of a map consisting of only simple polygons. Obstacles are red and frontiers are green. (Color figure online)

**Fig. 6.** Two points, one inside free space, the other in unexplored space.

## 4.2    Evaluation of PolyMap

PolyMap does address all the requirements presented in Sect. 3. Being based on vectors grouped into polygons, a PolyMap has by definition the advantages of vector-based map format. It is lightweight and can be easily visualized by humans. The remaining of this section first discusses how PolyMap meets the two other requirements and then how to build polygon-based maps.

*Explicit Exploration Frontiers.* The types of line segments of a polygon are either obstacles or frontiers. Outside a polygon is unexplored space, while inside represents the explored space. Passages between the two spaces, if any, are represented thanks to frontier segments.

*Support for Path Planning.* PolyMaps are suitable for navigation because polygons separate traversable areas and unexplored ones. The sparse nature of this map format makes it easy to create a topological graph using visibility graphs, or random sampling points in traversable space [7]. Such topological graphs are then suitable for navigation and path planning.

*How to Build a PolyMap?* To use PolyMaps in practice, we need to build such maps. One idea is to build it from an existing occupancy grid map. Doing so can be achieved in several ways, for example by extending the approach of Baizid *et al.* [1] to form closed polygons. Another possibility would be to utilize a wall-following strategy to outline the borders of traversable space, creating closed polygons in the process. Yet another method is to create a vector for every transition from free space into non-free space, and build polygons from the collected vectors. We used this idea in our implementation (programmed in

**Data:** occupancy grid map $G$
**Result:** PolyMap $M$
create dictionary $D$
/* create vectors                                                    */
**foreach** *cell* $c \in G$ **where** *c is free space* **do**

> **foreach** *neighbor* $n$ *of* $c$ **where** *n is not free space* **do**
>> /* vector orientation is CCW with respect to $c$              */
>> **if** *n is obstacle* **then**
>>> | create vector $v$ at border of $c$ and $n$ of type obstacle
>>
>> **else if** *n is unexplored* **then**
>>> | create vector $v$ at border of $c$ and $n$ of type frontier
>>
>> **end**
>> add vector $v$ to dictionary D with $v$ start point as key
>> in case of two vectors sharing the same start point, both are stored
>>> alongside
>
> **end**

**end**
/* create polygons                                                   */
**while** $D$ *is not empty* **do**

> create empty polygon $P$
> take random vector $v$ from $D$
> remove $v$ from $D$
> w := v
> **while** $w$ *end point* $\neq v$ *start point* **do**
>> P add $w$
>> $w := D$ at key $w$ end point
>> remove $w$ from D
>
> **end**
> $M$ add $P$

**end**
/* aggregate vectors                                                 */
**foreach** *polygon* $P \in M$ **do**

> **foreach** *vector* $v \in P$ **do**
>> w := vector in $P$ **where** $w$ start point = $v$ end point
>> **if** $w$ *orientation* $= v$ *orientation* **and** $w$ *type* $= v$ *type* **then**
>>> set $v$ end point to $w$ end point
>>> remove $w$ from $P$
>>
>> **end**
>
> **end**

**end**

**Algorithm 1:** How to build a PolyMap from a grid map

Pharo[1]) to build PolyMaps from grid maps. The pseudo-code for this is shown in Algorithm 1.

Figure 7 shows the result on applying our algorithm to convert a grid map of the Intel Lab data set[2] to a PolyMap. Obstacles are shown with red vectors and

---

[1] https://pharo.org/.

[2] http://ais.informatik.uni-freiburg.de/slamevaluation/datasets.php.

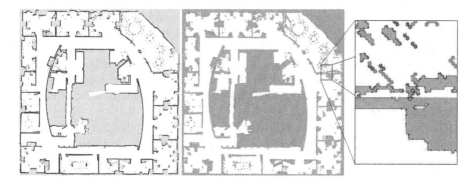

**Fig. 7.** Occupancy grid (left), full converted PolyMap (middle) and zoom in the PolyMap (right) (Color figure online)

frontiers with green vectors. The resulting PolyMap in this example consists of 829 polygons, with a total of 18054 vectors. The right-most figure is a zoomed-in section of the PolyMap, displaying individual vectors. Interestingly, the size of the full PolyMap is approximately the size of the compressed grid map. In this particular example, the grid map (in PNG format) takes about 100 kB, and the uncompressed PolyMap requires about 160 kB. Compressed, the PolyMap shrinks to approximately 44 kB (ZIP[3]) and 20 kB (7z[4]). Nevertheless, this algorithm only creates either horizontal or vertical vectors. Further optimization that allows diagonal vectors could result in even smaller memory footprints of the maps. Such optimizations will be considered in future work.

*Multi-robot Exploration Using PolyMaps.* Being a lightweight map format with a sparse representation of the environment, PolyMaps have low network bandwidth requirements. In multi-robot exploration scenarios, this allows us to use more robots, cover larger areas, or exchange maps more frequently than when compared with occupancy grids. Also, due to the sparse nature of the map format combined with explicit frontiers, frontiers can be selected easily and assigned to individual robots as exploration goals. Priority can be based on size of the frontier (i.e. length of the vectors) and proximity to the robot. The process of merging multiple PolyMaps from different robot is an interesting topic for future work.

## 5 Conclusion

Polygon-based map formats provide advantages by employing a sparse representation of the environment while still modeling obstacles with an accuracy that is comparable to occupancy grids. The reduction in memory consumption allows

---

[3] ISO/IEC 21320-1:2015, https://www.iso.org/standard/60101.html.
[4] https://www.7-zip.org/.

to cover larger areas compared to occupancy grids, and helps to do deal with the bandwidth bottleneck when broadcasting the map over a network. This is particularly helpful if maps need to be distributed within a robot fleet.

Explicitly modeling frontiers and enforcing closed polygons makes the maps much more useful for navigation and exploration tasks. Using directed line segments also makes it easy to determine whether a given point in located in free or unexplored space.

For future work, we plan to implement a full SLAM application that utilizes the PolyMap format and to test it in autonomous exploration scenarios. We further want to use the PolyMap format in multi-robot systems to rapidly explore an environment with a fleet of robots. This includes merging PolyMaps that have been created bu different robots in the fleet.

**Acknowledgment.** This work is part of the CPER DATA project that is supported by Région Hauts de France, and the French state.

# References

1. Baizid, K., Lozenguez, G., Fabresse, L., Bouraqadi, N.: Vector maps: a lightweight and accurate map format for multi-robot systems. In: Kubota, N., Kiguchi, K., Liu, H., Obo, T. (eds.) ICIRA 2016. LNCS (LNAI), vol. 9834, pp. 418–429. Springer, Cham (2016). https://doi.org/10.1007/978-3-319-43506-0_37
2. Castellanos, J.A., Montiel, J., Neira, J., Tardós, J.D.: The SPmap: a probabilistic framework for simultaneous localization and map building. IEEE Trans. Robot. Autom. **15**(5), 948–952 (1999)
3. Chatila, R., Laumond, J.P.: Position referencing and consistent world modeling for mobile robots. In: Proceedings of the 1985 IEEE International Conference on Robotics and Automation, vol. 2, pp. 138–145. IEEE (1985)
4. Chen, Y., Qu, C., Wang, Q., Jin, Z., Shen, M., Shen, J.: TVSLAM: an efficient topological-vector based SLAM algorithm for home cleaning robots. In: Huang, Y.A., Wu, H., Liu, H., Yin, Z. (eds.) ICIRA 2017. LNCS (LNAI), vol. 10464, pp. 166–178. Springer, Cham (2017). https://doi.org/10.1007/978-3-319-65298-6_16
5. The Robot Map Data Representation (MDR) Working Group: IEEE standard for robot map data representation for navigation, sponsor: IEEE robotics and automation society, June 2016. http://standards.ieee.org/findstds/standard/1873-2015.html
6. Jelinek, A.: Vector maps in mobile robotics. Acta Polytech. CTU Proc. **2**(2), 22–28 (2015)
7. LaValle, S.M.: Planning Algorithms. Cambridge University Press, New York (2006)
8. Lv, J., Kobayashi, Y., Ravankar, A.A., Emaru, T.: Straight line segments extraction and EKF-SLAM in indoor environment. J. Autom. Control Eng. **2**(3), 270–276 (2014)
9. Pfister, S.T., Roumeliotis, S.I., Burdick, J.W.: Weighted line fitting algorithms for mobile robot map building and efficient data representation. In: Proceedings of the IEEE International Conference on Robotics and Automation, 2003, ICRA 2003, vol. 1, pp. 1304–1311. IEEE (2003)
10. Ravankar, A., Ravankar, A.A., Hoshino, Y., Emaru, T., Kobayashi, Y.: On a hopping-points SVD and hough transform-based line detection algorithm for robot localization and mapping. Int. J. Adv. Robot. Syst. **13**(3), 98 (2016)

11. Sohn, H.J., Kim, B.K.: VecSLAM: an efficient vector-based SLAM algorithm for indoor environments. J. Intell. Rob. Syst. **56**(3), 301–318 (2009). https://doi.org/10.1007/s10846-009-9313-2

12. Stachniss, C., Grisetti, G., Hähnel, D., Burgard, W.: Improved rao-blackwellized mapping by adaptive sampling and active loop-closure. In: Proceedings of the Workshop on Self-Organization of AdaptiVE behavior (SOAVE) (2004)

13. Toussaint, G.: Efficient triangulation of simple polygons. Visual Comput. **7**(5–6), 280–295 (1991)

14. Yamauchi, B.: A frontier-based approach for autonomous exploration. In: Proceedings of CIRA 1997 (1997)

# Improved Input-to-State Stability Analysis of Discrete-Time Time-Varying Systems

Tianrui Zhao[1]([⊠]) (iD), Lili Jia[2] (iD), and Weiwei Luo[1] (iD)

[1] Center for Control Theory and Guidance Technology,
Harbin Institute of Technology, Harbin 150006, China
hittianruizhao@163.com, luo425wei@163.com
[2] Yunnan University Dianchi College, Kunming 650228, China
lilijiadianchi@163.com

**Abstract.** This note is concerned with input-to-state stability analysis of discrete-time switched nonlinear time-varying (DSNTV) systems. Some sufficient conditions are derived for testing input-to-state stability (ISS) of discrete-time nonlinear time-varying (DNTV) systems. ISS of DSNTV systems is further studied. The main feature of these obtained results is that the time-difference of Lyapunov functions of subsystems is allowed to be indefinite. Globally uniformly asymptotically stability (GUAS) and globally uniformly exponentially stability (GUES) concepts are utilized for analysis stability of general DNTV and DSNTV systems. An numerical example demonstrate the effectiveness of the proposed approaches.

**Keywords:** Discrete-time systems · Input-to-state stability
Nonlinear time-varying systems · Indefinite time-difference
Switched systems

## 1 Introduction

Switched systems are a class of important systems composed of many subsystems, which have received extensive attention from researchers in the past decades. The application of the switched systems are very extensive, such as artificial intelligence systems, power systems, and economic systems. Because switching is widespread in practical systems, switched systems have received great attention in the control community and a large number of results have been reported (see [3,6,10,17,18,20,21] and the reference therein). Particularly, stability analysis of switched system is the most important and fundamental problem in these results. Various stability concepts such as asymptotic stability [23], input-to-state stability (ISS) [13], and finite time stability have been thoroughly investigated for both continuous and discrete systems by exploiting Lyapunov's

---

The authors would like to thank Professor Bin Zhou for fruitful suggestions.

Z. Chen et al. (Eds.): ICIRA 2018, LNAI 10984, pp. 132–143, 2018.
https://doi.org/10.1007/978-3-319-97586-3_12

second method. For this stability analysis method, unlike the well-established time-invariant system stability theory, stability analysis of time-varying systems is full of challenges. This is also true for discrete nonlinear time-varying (DNTV) systems and discrete switched nonlinear time-varying (DSNTV) systems.

ISS means that the bound of the system response can be determined by the bounds of disturbance input and system initial state. ISS has proved to be very effective in describing the external disturbance input of the control system, and has attracted more and more attention of researchers since the pioneering result [15]. As mentioned earlier, ISS analysis can be applied on both continuous-time and discrete-time systems. For continuous-time systems, ISS analysis was studied in [15,16,19]. However, when $V(t,x) \geq \rho(|u|)$, the time derivative of Lyapunov functions along the trajectories of the considered system in these documents must be negative. Recently, an improved ISS analysis method based on Lyapunov function is studied in [3,12,14,22,24], which allows the time derivative of Lyapunov functions to be indefinite. For discrete-time systems, ISS analysis was studied in [6,8,10,11]. Whereas, the time-shift of the ISS-Lyapunov function needs to be negative under some additional condition on $u$. [7] improved these results, and proposed two methods. The advantage of this method is that the time-difference of the Lyapunov functions are allowed to be indefinite. Inspired by these existing work, in this paper, we continue to extend the method in [3,7,22,24] to the ISS analysis of DNTV and DSNTV systems.

Lyapunov's second method has been recognized as a very powerful tool in stability theory, especially the time-invariant systems. However, stability analysis for time-varying systems is more challenging than time-invariant systems. Therefore, stability analysis methods for time-varying systems still need to be further studied. For both continuous systems and discrete systems, the classical Lyapunov stability analysis method requires the derivative or difference of the Lyapunov function to be negative (see [2,4,5] and the reference therein). Recently, we have established a stability analysis method for several kinds of time-varying systems in [22–25] that allows the time derivative (time difference) of Lyapunov functions to be indefinite. Meanwhile, a similar stability analysis method was also used in the ISS analysis for continuous nonlinear time-delay systems by Ning et al. [13] and switched systems by Chen and Yang [3], respectively.

The main purpose of this paper is to establish stability theorems to discrete-time time-varying systems by using improved Lyapunov functions developed in [3,24]. The main contributions of this paper are highlighted as follows: (1). With the help of the scalar stable function [23,25] and an improved comparison lemma, some sufficient conditions are derived for testing ISS of DNTV systems. (2). Extend the results from Item 1 to DSNTV systems. (3). Studied the globally uniformly asymptotically stability (GUAS) and globally uniformly exponentially stability (GUES) of the previous two kinds of systems. The time-difference of Lyapunov functions in all these methods is allowed to be indefinite under some additional condition, which is more easily satisfied than the conditions in [6,8].

**Notation:** In this paper, $\mathbf{R}$, $\mathbf{R}^+$, and $\mathbf{R}^n$ denote the set of real numbers, non-negative real numbers, and $n$-dimensional Euclidean space respectively. For $p$, $q \in J$ with $p \leq q$, $\mathbf{I}[p,q]$ is the set $\{p, p+1, \ldots, q\}$ and $\mathbf{I}[p,\infty)$ is the set $\{p, p+1, \ldots\}$. Let $J = \mathbf{I}[0,\infty)$ and $\Pi_{j=k_0}^k H(k) = I_n$, $\forall k < k_0$, $k_0 \in J$. Denote $l_\infty^m = \{f(k) : J \to \mathbf{R}^m, |\sup_{k \in J}\{|f(k)|\} < \infty\}$. For $a$, $b \in J$, and a function $f$, we denote $|f|_{\mathbf{I}[a,b]} = \sup\{|f(s)|, s \in \mathbf{I}[a,b]\}$. The function $\psi : \mathbf{R}^+ \to \mathbf{R}^+$ is a $\mathcal{K}$ function, if $\psi$ is continuous and strictly increasing and $\psi(0) = 0$. If $\psi \in \mathcal{K}$ and moreover $\lim_{s \to \infty} \psi(s) = \infty$, then it is denoted by $\psi \in \mathcal{K}_\infty$. The function $\psi(s,k) : \mathbf{R}^+ \times \mathbf{R}^+ \to \mathbf{R}^+$ is a $\mathcal{KL}$ function, if $\psi(\cdot,k) \in \mathcal{K}$ for a fixed $k$, and $\psi(s,\cdot)$ is decreasing with $\lim_{k \to \infty} \psi(s,k) = 0$ for a fixed $s$. For a positive constant $x$, the function $\lfloor x \rfloor$ denotes the largest integer not greater than $x$.

## 2    Problem Formulation and Preliminaries

### 2.1    Systems Description

We first consider the following discrete nonlinear time-varying system (DNTV)

$$x(k+1) = f(k, x(k), u(k)), x(k_0) = x_0, k \geq k_0, \tag{1}$$

where initial time $k_0 \in J$, the state of the system $x(k) : J \to \mathbf{R}^n$, the input $u(k) : J \to \mathbf{R}^m$ is assumed to be locally essentially bounded. We also assume that $f : \mathbf{R}^+ \times \mathbf{R}^n \times \mathbf{R}^m \to \mathbf{R}^n$ is locally essentially bounded with $f(k,0,0) = 0$.

For futher use, we introduce the following definition, which characterize several stability notions for system (1).

**Definition 1.** *The system (1) is said to be:*

1. *Input-to-state stable (ISS) if there exist a $\mathcal{KL}$-function $\sigma$ and a $\mathcal{K}$-function $\gamma$ such that, for each $u \in l_\infty^m$, and $k \geq k_0$, $k_0 \in J$, (see [8])*

$$|x(k)| \leq \sigma(|x_0|, k - k_0) + \gamma(|u|_{\mathbf{I}[k_0, k-1]});$$

2. *Globally uniformly asymptotically stable (GUAS) (with $u \equiv 0$), if there exists a $\mathcal{KL}$-function $\sigma$ such that, for any $k \geq k_0$, $k_0 \in J$, (see [1,9])*

$$|x(k)| \leq \sigma(|x_0|, k - k_0);$$

3. *Globally uniformly exponentially stable (GUES) (with $u \equiv 0$), if there exist two positive constants $\alpha$ and $\beta < 1$ such that, for any $k \geq k_0$, $k_0 \in J$, (see [1,9])*

$$|x(k)| \leq \alpha \beta^{k - k_0} |x_0|.$$

We next consider the following discrete-time switched nonlinear time-varying (DSNTV) systems:

$$x(k+1) = f_{\vartheta(k)}(k, x(k), u(k)), x(k_0) = x_0, k \geq k_0, \tag{2}$$

where initial time $k_0 \in J$, the state of the system $x(k) \in \mathbf{R}^n$, the input $u(k) \in \mathbf{R}^m$ is assumed to be locally essentially bounded, and switching signal $\vartheta : J \to Q = \{1, 2, \ldots, q\}$. Let $k_1 < k_2 < \cdots < k_t$, $t \geq 1$, denote the switching instants of $\vartheta(\tau)$ for $k_0 < \tau < k$ and $\{x(k_0) : (i_0, k_0), (i_1, k_1), \ldots, (i_t, k_t), \ldots, | i_t \in Q, t \in J\}$ denote the switching sequence. When $k \in [k_j, k_{j+1})$, the $i_j$th subsystem is active. For each $i \in Q$, $f_i : \mathbf{R}^+ \times \mathbf{R}^n \times \mathbf{R}^m \to \mathbf{R}^n$ is locally essentially bounded with $f_i(k, 0, 0) = 0$. Moreover, system (1) and system (2) have the same stability definition. Let $\varkappa(a, b)$ denote the number of switches occurring in the interval $\mathbf{I}[a, b]$. Throughout this paper, we assume that $x(k)$ is a single-valued function.

## 2.2  Scalar Stable Functions and Comparison Lemma

To build our results, we need the following basic concepts recalled from [23, 25]. Consider the following scalar discrete linear time-varying (DLTV) system

$$y(k + 1) = \mu(k) y(k), k \in J, \tag{3}$$

where $\mu(k) : J \to \mathbf{R}^+$ and $y \in \mathbf{R}$.

**Definition 2.** *[23, 25] The function $\mu(k)$ is a uniformly stable (US) function if system (3) is US, a asymptotically stable (AS) function if system (3) is AS, and a uniformly asymptotically stable (UAS) function if system (3) is UAS.*

By noting that the state transition matrix $\phi_1(k, k_0)$ for system (3) is [23, 25]

$$\phi_1(k, k_0) = \prod_{i=k_0}^{k-1} \mu(i), \forall k \geq k_0, k_0 \in J. \tag{4}$$

We have the following result.

**Lemma 1.** *[23, 25] The function $\mu(k) : J \to \mathbf{R}^+$ is*

1. *US if and only if there exists a number $\chi \geq 1$ such that, for any $k \geq k_0, k_0 \in J$, $\phi_1(k, k_0) \leq \chi$ is satisfied;*
2. *AS if and only if $\lim_{k \to \infty} \phi_1(k, k_0) = 0$;*
3. *UAS if and only if there exist two numbers $\chi \geq 1$ and $\lambda \in (0, 1)$ such that*

$$\phi_1(k, k_0) \leq \chi \lambda^{k-k_0}, \forall k \geq k_0, k_0 \in J. \tag{5}$$

To obtain the ISS stability theorem, we need the following comparison lemma.

**Lemma 2.** *Let $y(k)$ be a function satisfying*

$$y(k + 1) \leq \mu(k) y(k), \quad \text{whenever } y(k + 1) \geq \psi(k), \tag{6}$$

*where $\mu(k)$ is a US function and $\psi(k) : J \to \mathbf{R}$ is a sequence (which can be dependent on $y$). Then, the following inequality is established*

$$y(k) \leq \max \left\{ \phi_1(k, k_0) y(k_0), \chi |\psi|_{\mathbf{I}[k_0, k-1]} \right\}. \tag{7}$$

*Proof.* We consider the following inequality in two cases,

$$y(s+1) \geq \psi(s). \tag{8}$$

**Case 1:** For all $s \in \mathbf{I}[k_0, k-1]$, (8) holds. By using (4), (6) and Gronwall inequality, we have

$$y(k) \leq \phi_1(k, k_0) y(k_0). \tag{9}$$

**Case 2:** For some $s \in \mathbf{I}[k_0, k-1]$, (8) does not hold true. Let $k^* \in \mathbf{I}[k_0, k-1]$ be the maximal number such that $y(k^*+1) < \psi(k^*)$. Then we have either $k^* < k-1$ or $k^* = k-1$. If $k^* < k-1$, then (8) holds for all $s \in \mathbf{I}[k^*+1, k-1]$, which implies $y(s+1) \leq \mu(s) y(s)$, and thus

$$y(k) \leq \phi_1(k, k^*+1) y(k^*+1) < \chi\psi(k^*) \leq \chi |\psi|_{\mathbf{I}[k_0, k-1]}. \tag{10}$$

If $k^* = k-1$, then by the definition of $k^*$ and (6), we have

$$y(k) = y(k^*+1) < \psi(k^*) \leq \chi |\psi|_{\mathbf{I}[k_0, k-1]}. \tag{11}$$

Now, combining (9), (10) and (11), we conclude that (7) is true.

## 3    Main Results

### 3.1    Improved ISS Analysis for DNTV Systems

We first establish an improved ISS analysis for DNTV system (1).

**Theorem 1.** *Assume that there exist two functions* $u_1, u_2 \in \mathcal{K}_\infty$, *a function* $\rho \in \mathcal{K}$, *a AS and US function* $\mu(k)$, *and a function* $V(k, x) : J \times \mathbf{R}^n \rightarrow \mathbf{R}$, *such that, for all* $k \in J$ *and* $x \in \mathbf{R}^n$, *the following conditions hold:*

(A).   $u_1(|x|) \leq V(k, x(k)) \leq u_2(|x|)$.
(B).   $V(k+1, x(k+1)) \leq \mu(k) V(k, x(k))$ *if* $V(k+1, x(k+1)) \geq \rho(|u(k)|)$.

*Then the DNTV system (1) is ISS.*

*Proof.* We choose $y(k) = V(k, x(k)) = |y(k)|$, $\psi(k) = \rho(|u(k)|)$. Then (B) is just in the form of (6) and it follows from Lemma 2 that, for any $k \geq k_0 \in J$,

$$y(k) \leq \max\left\{\Theta(k), \chi |\psi|_{\mathbf{I}[k_0, k-1]}\right\} \leq \Theta(k) + \chi |\psi|_{\mathbf{I}[k_0, k-1]},$$

where $\Theta(k) = \phi_1(k, k_0) y(k_0)$. Hence, by using $\alpha(a+b) \leq \alpha(2a) + \alpha(2b), \alpha \in \mathcal{K}, a \geq 0, b \geq 0$ and condition (A), we get, for any $k, k_0 \in J$, with $k \geq k_0$,

$$|x(k)| \leq u_1^{-1}(y(k)) \leq u_1^{-1}(2\Theta(k)) + u_1^{-1}\left(2\chi |\psi|_{\mathbf{I}[k_0, k-1]}\right). \tag{12}$$

We next construct a $\mathcal{KL}$-function $\sigma_1(|x_0|, k-k_0)$ satisfying $2\Theta(k) \leq \sigma_1(|x_0|, k-k_0)$. Since $\mu(k)$ is AS, namely, $\lim_{k\rightarrow\infty} \phi_1(k, k_0) = 0$, there exists a sequence

$\{k_i\}$, satisfying, for $k, i \in J$, $k_i \to \infty, i \to \infty$, such that $2\Theta(k) \leq \frac{2\chi y(k_0)}{i+1} \leq \frac{2\chi u_2(|x_0|)}{i+1}$, $k \geq k_i$. Then $\mathcal{KL}$-function $\sigma_1$ is defined as, for all $j \geq 1$,

$$\sigma_1 = \begin{cases} \frac{2\chi u_2(|x_0|)}{j} \left(1 - \frac{k-k_j}{(j+1)(k_{j+1}-k_j)}\right) & , k \in \mathbf{I}\,[k_j, k_{j+1}), \\ 2\chi u_2(|x_0|) & , k \in \mathbf{I}\,[k_0, k_1). \end{cases}$$

From which and (12), we have

$$|x(k)| \leq u_1^{-1}\left(\sigma_1(|x_0|, k - k_0)\right) + u_1^{-1}\left(2\chi\,|\psi|_{\mathbf{I}[k_0,k-1]}\right),$$

which shows that the system (1) is ISS. The proof is finished.

From Example 1 in [23], it is easy to know that we can find a scalar stable function $\mu(k)$ that is US and AS but is not UAS. But, if scalar stable function $\mu(k)$ is UAS, then it is a AS and US function. Therefore, we can get the following corollary, which seems more concise.

**Corollary 1.** *[7] Assume that there exist two functions $u_1, u_2 \in \mathcal{K}_\infty$, a function $\rho \in \mathcal{K}$, a UAS function $\mu(k)$, and a function $V(k,x) : J \times \mathbf{R}^n \to \mathbf{R}$, such that, for all $k \in J$ and $x \in \mathbf{R}^n$, (A) and (B) are satisfied, then the DNTV system (1) is ISS.*

We mention that Theorem 1 and Corollary 1 can be viewed as the discretization of Theorem 2 in [3] and Theorem 3 in [24] respectively. In addition, Theorem 1 and Corollary 1 are an extension of Lemma 3.5 in [8]. Because, under condition $V(k+1, x(k+1)) \geq \rho(|u(k)|)$, the time-shift of Lyapunov functions can take both negative and positive values. DLTV (researched in [z]) is a special case of DNTV.

## 3.2   Improved ISS Analysis for DSNTV Systems

We then can provide the following ISS stability theorems for system (2).

**Theorem 2.** *Assume that there exist two functions $u_1, u_2 \in \mathcal{K}_\infty$, a function $\rho \in \mathcal{K}$, a US function $\mu_i(k)$, two positive constants $b \geq 1, B$, and a function $V_i(k,x) : J \times \mathbf{R}^n \to \mathbf{R}$, such that, for all $k \in J$ and $x \in \mathbf{R}^n$,*

(C)   $u_1(|x|) \leq V_{i_j}(k, x(k)) \leq u_2(|x|)$.

(D)   $V_{i_j}(k+1, x(k+1)) \leq \mu_{i_j}(k) V_{i_j}(k, x(k))$ *if* $V_{i_j}(k+1, x(k+1)) \geq \rho(|u(k)|)$.

(E)   $V_{i_{j+1}}(k, x(k)) \leq b V_{i_j}(k, x(k))$, $\forall i_j \in Q$.

(F)   $\lim_{k \to \infty} m(k) = 0$ *and* $n(k) < B$,

*where* $m(k) = b^{\varkappa(k_0, k)} \phi_2(k, k_0)$ *and* $n(k) = \Sigma_{k_0 < k_s < k}(b^{\varkappa(k_s, k)} \phi_2(k, k_s))$. *Then the system (2) is ISS.*

*Proof.* We also choose $y_{i_j}(k) = V_{i_j}(k, x(k)) = |y_{i_j}(k)|$, $\psi(k) = \rho(|u(k)|)$. The state transition matrix for $\mu_i(k)$, denoted by $\phi_2(k, k_0)$, is given by

$$\phi_2(k, k_0) = \prod_{s=k_0}^{k-1} \mu_{\vartheta(s)}(s), \forall k \geq k_0, k_0 \in J. \tag{13}$$

Meanwhile, Lemma 1 and Lemma 2 remains true for $\phi_2$. By using Lemma 2 agian, we have, for any $k \in \mathbf{I}[k_j + 1, k_{j+1}]$, $j \in J$,

$$y_{i_j}(k) \leq \phi_2(k, k_j) y_{i_j}(k_j) + \chi|\psi|_{\mathbf{I}[k_j, k-1]}. \tag{14}$$

Now, for proving ISS, we need to prove that, for all $k \geq k_0$,

$$y_{\vartheta(k)}(k) \leq m(k)y_{i_0}(k_0) + (1 + n(k))\chi|\psi|_{\mathbf{I}[k_0, k-1]}. \tag{15}$$

We prove this by mathematical induction. For $k \in \mathbf{I}[k_0 + 1, k_1]$, (15) follows from (14), namely,

$$y_{i_0}(k) \leq b\phi_2(k, k_0) y_{i_0}(k_0) + \chi|\psi|_{\mathbf{I}[k_0, k-1]}.$$

We now assume that (15) is true for $k \in \mathbf{I}[k_j + 1, k_{j+1}]$, namely,

$$y_{i_j}(k) \leq m(k)y_{i_0}(k_0) + (1 + n(k))\chi|\psi|_{\mathbf{I}[k_0, k-1]}, \tag{16}$$

and want to show that (15) is still satisfied for $k \in \mathbf{I}[k_{j+1} + 1, k_{j+2}]$. By using (E), (14), and (16), we have

$$\begin{aligned}
y_{i_{j+1}}(k) &\leq \phi_2(k, k_{j+1}) y_{i_{j+1}}(k_{j+1}) + \chi|\psi|_{\mathbf{I}[k_{j+1}, k-1]} \\
&\leq b\phi_2(k, k_{j+1}) y_{i_j}(k_{j+1}) + \chi|\psi|_{\mathbf{I}[k_{j+1}, k-1]} \\
&\leq b\phi_2(k, k_{j+1}) \left( m(k_{j+1})y_{i_0}(k_0) + (1 + n(k_{j+1}))\chi|\psi|_{\mathbf{I}[k_0, k_{j+1}-1]} \right) \\
&\quad + \chi|\psi|_{\mathbf{I}[k_{j+1}, k-1]} \\
&\leq b^{\varkappa(k_0, k)}\phi_2(k, k_0) y_{i_0}(k_0) + \chi|\psi|_{\mathbf{I}[k_0, k-1]} \\
&\quad \times \left( 1 + b\phi_2(k, k_{j+1}) + \sum_{k_0 < k_s < k_{j+1}} \left( b^{\varkappa(k_s, k)}\phi_2(k, k_s) \right) \right) \\
&= m(k)y_{i_0}(k_0) + (1 + n(k))\chi|\psi|_{\mathbf{I}[k_0, k-1]}.
\end{aligned}$$

Therefore, (15) is satisfied for all $k \geq k_0, k_0 \in J$. Then, by using $\alpha(a + b) \leq \alpha(2a) + \alpha(2b), \alpha \in \mathcal{K}, a \geq 0, b \geq 0$, (C), and (15), we get, for all $k \geq k_0$

$$\begin{aligned}
|x(k)| &\leq u_1^{-1}(2u_2(|x_0|)m(k)) + u_1^{-1}\left( 2(1 + n(k))\chi|\psi|_{\mathbf{I}[k_0, k-1]} \right) \\
&\leq u_1^{-1}(2u_2(|x_0|)m(k)) + u_1^{-1}\left( 2(1 + B)\chi|\psi|_{\mathbf{I}[k_0, k-1]} \right).
\end{aligned}$$

Since $\lim_{k \to \infty} m(k) = 0$, using a similar method in Theorem 1, we can construct a $\mathcal{KL}$-function $\sigma_2$ such that, for all $k \in J$, $2u_2(|x_0|)m(k) \leq \sigma_2(|x_0|, k - k_0)$. Hence, the system (2) is ISS. The proof is finished.

**Corollary 2.** *Assume that there exist an integer $m \geq 1$ and a constant $\bar{\delta}_1$ satisfying $\delta_j \triangleq k_{j+1} - k_j \geq \bar{\delta}_1, \forall j \geq m$, with $j \in J$, two functions $u_1, u_2 \in \mathcal{K}_\infty$, a function $\rho \in \mathcal{K}$, a UAS function $\mu_i(k)$, a positive constant $b \geq 1$ and a function $V_i(k, x) : J \times \mathbf{R}^n \to \mathbf{R}$, such that, for all $k \in J$ and $x \in \mathbf{R}^n$, (C), (D), (E) and*

$$\sqrt[\bar{\delta}_1]{b}\lambda < 1, \tag{17}$$

*are satisfied. Then the system (2) is ISS.*

*Proof.* We also choose $y_{i_j}(k) = V_{i_j}(k, x(k)) = \left|y_{i_j}(k)\right|$, $\psi(k) = \rho\left(|u(k)|\right)$. Moreover, the state transition matrix for $\mu_i(k)$ is defined in (13). According to the analysis of the Theorem 2, we have, for all $k \geq k_0$

$$|x(k)| \leq u_1^{-1}\left(2u_2\left(|x_0|\right)m(k)\right) + u_1^{-1}\left(2\left(1 + n(k)\right)\chi\left|\psi\right|_{\mathbf{I}[k_0, k-1]}\right), \tag{18}$$

where $m(k) = b^{\varkappa(k_0, k)}\phi_2(k, k_0)$ and $n(k) = \Sigma_{k_0 < k_s < k}(b^{\varkappa(k_s, k)}\phi_2(k, k_s))$. We consider two cases:

**Case 1:** A finite number of switching for system (2).

**Case 2:** An infinite number of switching for system (2).

In Case 1, there exist an integer $q$ and a constant $n_1$ such that the $q$-th subsystem is active on $\mathbf{I}[k_q, \infty)$, and

$$m(k) = b^{\varkappa(k_0, k)}\phi_2(k, k_0) \leq b^{q+1}\chi\lambda^{k-k_0},$$

$$n(k) = \sum_{k_0 < k_s < k}\left(b^{\varkappa(k_s, k)}\phi_2(k, k_s)\right) = n_1 < \infty,$$

from which and (18) it follows that, for all $k \geq k_0$,

$$|x(k)| \leq u_1^{-1}\left(m_1 u_2\left(|x_0|\right)\lambda^{k-k_0}\right) + u_1^{-1}\left(2(1 + n_1)\chi\left|\psi\right|_{\mathbf{I}[k_0, k-1]}\right),$$

where $m_1 = 2\chi b^{q+1}$. Hence, system (2) is ISS in Case 1.

Now consider Case 2. Let $k \in \mathbf{I}[k_t + 1, k_{t+1}], t \in J$. Then we have either $t \geq m$ or $t < m$. If $t < m$, then we have system (2) is ISS by using the same approach as Case 1. If $t \geq m$, then we have,

$$\varkappa(k_0, k) = m + \sum_{j=m}^{t}\frac{k_{j+1} - k_j}{\delta_j} \leq \frac{1}{\bar{\delta}_1}\sum_{j=0}^{t-1}(k_{j+1} - k_j) + 2m \leq \frac{1}{\bar{\delta}_1}(k - k_0) + 2m,$$

from which and condition (17) it follows that, for all $k \geq k_0$

$$m(k) = b^{\varkappa(k_0, k)}\phi_2(k, k_0) \leq \chi b^{2m}\left(\sqrt[\bar{\delta}_1]{b}\lambda\right)^{k-k_0},$$

$$n(k) = \sum_{k_0 < k_s < k}(b^{\varkappa(k_s, k)}\phi_2(k, k_s)) \leq \sum_{k_0 < k_s < k}\left(b^{\varkappa(k_s, k)}\chi\lambda^{k-k_s}\right)$$

$$= n_2 + \sum_{k_m \leq k_s < k}(b^{\varkappa(k_s, k)}\chi\lambda^{k-k_s}) \leq n_2 + \frac{b\chi}{1 - b\lambda^{\bar{\delta}_1}},$$

where $n_2 = \Sigma_{k_0 < k_s < k_m} (b^{\varkappa(k_s,k)} \chi \lambda^{k-k_s})$. Hence, all the conditions in Theorem 2 are satisfied, the system (2) is ISS. The proof is finished.

**Theorem 3.** *Assume that system (2) has an infinite number of switching. Assume that there exist two functions $u_1, u_2 \in \mathcal{K}_\infty$, a function $\rho \in \mathcal{K}$, a US function $\mu_i(k)$, a positive constant $b \geq 1$, a positive integer $p$, a constant $\bar{\lambda} \in (0,1)$ and a function $V_i(k, x) : J \times \mathbf{R}^n \to \mathbf{R}$, such that, for all $k \in J$ and $x \in \mathbf{R}^n$, (C), (D), (E) and*

$$b^p \phi_2 \left( k_{p(j+1)}, k_{pj} \right) \leq \bar{\lambda} < 1 \tag{19}$$

*are satisfied. Then the system (2) is ISS.*

*Proof.* According to the analysis of the Theorem 2, we also have, for all $k \geq k_0$, (18) holds, where $\psi(k), m(k)$ and $n(k)$ are given in Theorem 2 and Corollary 2. By using (D), (E) and (19), we have $\lim_{k \to \infty} m(k) = \lim_{k \to \infty} b^{\varkappa(k_0,k)} \phi_2(k, k_0) \leq \lim_{k,z \to \infty} b^{p(z+1)} \phi_2(k, k_{pz}) \phi_2(k_{pz}, k_0) \leq \lim_{z \to \infty} \chi b^p \bar{\lambda}^z = 0$, where $z = \lfloor \varkappa(k_0, k)/p \rfloor$. Next, we claim that there exists a constant $L$ such that,

$$n(k) \leq \chi \left( 1 + \frac{b^p \chi}{1 - \bar{\lambda}} \right) \sum_{s=0}^{p-1} b^s = L < \infty. \tag{20}$$

For all $k \geq k_0$, there exist two positive integers $j$ and $l$, such that $k \in \mathbf{I}[k_j + 1, k_{j+1}] \subseteq \mathbf{I}[k_{lp} + 1, k_{(l+1)p}]$. Similarly to the proof of Theorem 5 in [3], we next consider two cases.

**Case 1:** $l = 0$. In this case, we have

$$n(k) = \sum_{k_0 < k_s < k} \left( b^{\varkappa(k_s,k)} \phi_2 \left( k, k_s \right) \right) \leq \chi \sum_{s=1}^{p-1} b^s. \tag{21}$$

**Case 2:** $l > 0$. In this case, we have

$$n(k) = \sum_{k_0 < k_s < k} \left( b^{\varkappa(k_s,k)} \phi_2 \left( k, k_s \right) \right)$$

$$\leq \chi \sum_{s=1}^{p-1} b^s + b^{\varkappa(k_{lp},k)} \chi^2 \sum_{s=0}^{p-1} b^s \sum_{i=0}^{l-1} \bar{\lambda}^i$$

$$\leq \chi \left( 1 + \frac{b^p \chi}{1 - \bar{\lambda}} \right) \sum_{s=0}^{p-1} b^s. \tag{22}$$

Combining (21) and (22), we can get (20) and $n(k)$ is bounded. Hence, the system (2) is ISS.

### 3.3    Asymptotic Stability Analysis

We finally introduce some crateria for system (1) and (2) with $u \equiv 0$.

**Corollary 3.** *[25] Then the system (1) with $u \equiv 0$ is:*

1. *GUAS if there exist two functions $u_1, u_2 \in \mathcal{K}_\infty$, a UAS function $\mu(k)$ and a function $V(k, x) : J \times \mathbf{R}^n \to \mathbf{R}$ such that, for all $x \in \mathbf{R}^n$, (A), and*

$$V(k + 1, x(k + 1)) \leq \mu(k) V(k, x(k)), \forall k \in J, \tag{23}$$

   *are satisfied.*

2. *GUES if there exist three positive constants $u_1, u_2, \pi$, a UAS function $\mu(k)$ and a function $V(k, x) : J \times \mathbf{R}^n \to \mathbf{R}$ such that, for all $k \in J$ and $x \in \mathbf{R}^n$, $u_1 |x|^\pi \leq V(k, x(k)) \leq u_2 |x|^\pi$, and (23) are satisfied.*

**Corollary 4.** *The assumption of $\delta$, $b$, $\mu_i(k)$ and $V_i(k, x) : J \times \mathbf{R}^n \to \mathbf{R}$ in Corollary 2 is still valid. Then the system (2) with $u \equiv 0$ is:*

1. *GUAS if there exist two functions $u_1, u_2 \in \mathcal{K}_\infty$ such that, for all $k \in J$ and $x \in \mathbf{R}^n$, (C), (E), (17), and*

$$V_i(k + 1, x(k + 1)) \leq \mu_i(k) V_i(k, x(k)), \tag{24}$$

   *are satisfied.*

2. *GUES if there exist three positive constants $u_1, u_2, \pi$, such that, for all $k \in J$ and $x \in \mathbf{R}^n$, $u_1 |x|^\pi \leq V_i(k, x(k)) \leq u_2 |x|^\pi$, (E), (17), and (24) are satisfied.*

## 4   An Numerical Example

Consider system (2) with $x(k) \in \mathbf{R}^2$, $u(k) \in \mathbf{R}$, $\sigma(k) \in Q = \{1, 2\}$, and

$$f_1(k, x(k), u(k)) = (2.1 - 2M(k)) x(k) + u(k),$$
$$f_2(k, x(k), u(k)) = (1.3 + M(k)) x(k) + u(k).$$

where $M(k) = \cos(k\pi/10)$. Let $V_1(k, x(k)) = V_2(k, x(k)) = |x(k)|$, then we have $V_1(k + 1, x(k + 1)) \leq (2.2 - 2M(k))V_1(k, x(k))$, $V_2(k + 1, x(k + 1)) \leq (1.4 + M(k))V_2(k, x(k))$, for $V_i(k, x(k)) \geq 100|u(k)|^2$, $i = 1, 2$. Let $\mu_1(k) = (2.2 - 2M(k))$, $\mu_2(k) = (1.4 + M(k))$, $\rho(s) = 100s^2$, $u_1(|x(k)|) = u_2(|x(k)|) = |x(k)|$, and $b = 1$, we can verify that (C), (D) and (E) are satisfied. Since the equations $\mu_1(k)$ and $\mu_2(k)$ are periodic functions with period 20, we have, for $k \in \mathbf{I}[0, 19]$,

$$\begin{cases} \mu_2(k) \leq \mu_1(k), \ k \in \mathbf{I}[5, 15], \\ \mu_1(k) < \mu_2(k), \ otherwise. \end{cases}$$

Assume that $\vartheta(k) = 2$ over $\mathbf{I}[20m + 5, 20m + 16]$, $\forall m \in J$ and $\vartheta(k) = 1$ in other case, then $\mu_i(k)$ is UAS, which implies $\sqrt[\delta]{b}\lambda < 1$. Hence the system is ISS.

## 5    Conclusion

This paper has studied the input-to-state (ISS) stability analysis of discrete-time switched nonlinear time-varying (DSNTV) systems by using Lyapunov's second method. The existing ISS approaches were improved and some sufficient conditions have been proposed to analyze the ISS of discrete-time nonlinear time-varying (DNTV) systems with the help of the concept of scalar stable function and an improved comparison lemma. The improved input-to-state stability analysis approache was applied on a class of DSNTV systems and some criteria are obtained. The advantage of the conditions obtained is that the time-difference of improved Lyapunov functions are allowed to be indefinite. Finally, globally uniformly asymptotically stability (GUAS) and globally uniformly exponentially stability (GUES) concepts were considered for analysis stability of general DNTV and DSNTV systems.

## References

1. Agarwal, R.P.: Difference Equations and Inequalities: Theory, Methods, and Applications. CRC Press, Boca Raton (2000)
2. Bai, X., Li, H.: Input-to-state stability of discrete-time switched systems. In: 2011 Second International Conference on Digital Manufacturing and Automation (ICDMA), pp. 652–653. IEEE, August 2011
3. Chen, G., Yang, Y.: Relaxed conditions for the input-to-state stability of switched nonlinear time-varying systems. IEEE Trans. Autom. Control **62**(9), 4706–4712 (2017)
4. Feng, W., Zhang, J.F.: Input-to-state stability of switched nonlinear systems. IFAC Proc. Vol. **38**(1), 324–329 (2005)
5. Guiyuan, L., Ping, Z.: Input-to-state stability of discrete-time switched nonlinear systems. In Control And Decision Conference (CCDC), 2017 29th Chinese, pp. 3296–3300. IEEE, May 2017
6. Huang, M., Ma, L., Zhao, G., Wang, X., Wang, Z.: Input-to-state stability of discrete-time switched systems and switching supervisory control. In: 2017 IEEE Conference on Control Technology and Applications (CCTA), pp. 910–915. IEEE, August 2017
7. Li, H., Liu, A., Zhang, L.: Input-to-state stability of time-varying nonlinear discrete-time systems via indefinite difference Lyapunov functions. ISA Trans. **77**, 71–76 (2018)
8. Jiang, Z.P., Wang, Y.: Input-to-state stability for discrete-time nonlinear systems. Automatica **37**(6), 857–869 (2001)
9. Jiang, Z.P., Wang, Y.: A converse Lyapunov theorem for discrete-time systems with disturbances. Syst. Control Lett. **45**(1), 49–58 (2002)
10. Lian, J., Li, C., Liu, D.: Input-to-state stability for discrete-time non-linear switched singular systems. IET Control Theory Appl. **11**(16), 2893–2899 (2017)
11. Liu, Y., Kao, Y., Karimi, H.R., Gao, Z.: Input-to-state stability for discrete-time nonlinear switched singular systems. Inf. Sci. **358**, 18–28 (2016)
12. Ning, C., He, Y., Wu, M., Liu, Q., She, J.: Input-to-state stability of nonlinear systems based on an indefinite Lyapunov function. Syst. Control Lett. **61**(12), 1254–1259 (2012)

13. Ning, C., He, Y., Wu, M., She, J.: Improved Razumikhin-type theorem for input-to-state stability of nonlinear time-delay systems. IEEE Trans. Autom. Control **59**(7), 1983–1988 (2014)
14. Peng, S.: Lyapunov-Krasovskii-type criteria on ISS and iISS for impulsive time-varying delayed systems. IET Control Theory Appl. **12**, 1649–1657 (2018)
15. Sontag, E.D.: Smooth stabilization implies coprime factorization. IEEE Trans. Autom. Control **34**(4), 435–443 (1989)
16. Sontag, E.D., Wang, Y.: On characterizations of the input-to-state stability property. Syst. Control Lett. **24**(5), 351–359 (1995)
17. Wang, Y.E., Sun, X.M., Mazenc, F.: Stability of switched nonlinear systems with delay and disturbance. Automatica **69**, 78–86 (2016)
18. Wu, X., Tang, Y., Cao, J., Mao, X.: Stability Analysis for Continuous-Time Switched Systems with Stochastic Switching Signals. IEEE Transactions on Automatic Control (2017)
19. Yu, Y., Zeng, Z., Li, Z., Wang, X., Shen, L.: Event-triggered encirclement control of multi-agent systems with bearing rigidity. Sci. China(Inf. Sci.) **60**(11), 110–203 (2017)
20. Zhang, W.A., Yu, L.: Stability analysis for discrete-time switched time-delay systems. Automatica **45**(10), 2265–2271 (2009)
21. Zhang, H., Qin, C., Luo, Y.: Neural-network-based constrained optimal control scheme for discrete-time switched nonlinear system using dual heuristic programming. IEEE Trans. Autom. Sci. Eng. **11**(3), 839–849 (2014)
22. Zhou, B., Luo, W.: Improved Razumikhin and Krasovskii stability criteria for time-varying stochastic time-delay systems. Automatica **89**, 382–391 (2016)
23. Zhou, B., Zhao, T.: On asymptotic stability of discrete-time linear time-varying systems. IEEE Trans. Autom. Control **62**, 4274–4281 (2017)
24. Zhou, B.: Stability analysis of non-linear time-varying systems by Lyapunov functions with indefinite derivatives. IET Control Theory Appl. **11**(9), 1434–1442 (2017)
25. Zhou, B.: Improved Razumikhin and Krasovskii approaches for discrete-time time-varying time-delay systems. Automatica **91**, 256–269 (2018)

# Optimal Control Problem
# for Discrete-Time Markov Jump Systems
# with Indefinite Weight Costs

Hongdan Li[1], Chunyan Han[2], and Huanshui Zhang[1($\boxtimes$)]

[1] School of Control Science and Engineering, Shandong University,
Jinan 250061, Shandong, China
lhd200908@163.com, hszhang@sdu.edu.cn
[2] School of Electrical Engineering, University of Jinan,
Jinan 250022, Shandong, China
cyhan823@hotmail.com

**Abstract.** In this article, the optimal control problem with indefinite state and control weighting matrices in the cost function for discrete-time systems involving Markov jump and multiplicative noise is discussed. Necessary and sufficient conditions of the solvability of indefinite optimal control problem in finite-horizon are obtained by solving the forward-backward stochastic difference equations with Markov jump (FBSDEs-MJ) derived from the maximum principle, whose method is different from most previous works [12], etc.

**Keywords:** Optimal control · FBSDEs-MJ · Indefinite
Markov jump system

## 1 Introduction

There are many factors to give rise to abrupt changes such as abrupt environmental disturbances, component failures or repairs and these changes often occur in many control systems, for instance, economic systems and aircraft control systems. This phenomenon can be modeled as Markov jump linear systems (MJLS). Owing to its widely application in practice, in recent years, the subject of MJLS is by now huge and is growing rapidly, see [1–7], and reference therein. Seeing that the importance of the linear quadratic (LQ) control problem in the study of control system, there are also many results about these problems with Markov jump. [8] considered the optimal control problems for discrete-time linear systems subject to Markov jump with two cases that the one without noise and the other with an additive noise in model. In [9], they illustrated the equivalence between the stability of the optimal control and positiveness of the coupled algebraic Riccati equation via the concept of weak detectability.

This work is supported by the National Natural Science Foundation of China under Grants 61573221, 61633014, 61473134.

© Springer Nature Switzerland AG 2018
Z. Chen et al. (Eds.): ICIRA 2018, LNAI 10984, pp. 144–152, 2018.
https://doi.org/10.1007/978-3-319-97586-3_13

It is noteworthy that all the above results are obtained under the common assumption that the weighting matrices of state and control in the quadratic performance index are required to be positive semi-definite even positive definite. However, when the weighting matrices have the requirement of symmetry only, the stochastic LQ problem may be still well posed. This case is called indefinite stochastic problem which often appear in economic fields such as portfolio selection problem. As regard to the problem, [10] and [11] investigated an indefinite stochastic LQ control problem for continuous-time linear systems subject to Markov jump in finite and infinite time horizon, respectively. [12] derived the necessary and sufficient condition for the well posedness of the indefinite LQ problem and the optimal control law were given in terms of a set of coupled generalized Riccati difference equations interconnected with a set of coupled linear recursive equations.

Inspired by the above literatures, in this paper, we study the optimal control problem for discrete-time systems involving Markov jump and multiplicative noise in which the state and control weighting matrices in the cost function are indefinite. The main contribution of this paper is that an optimal controller is explicitly shown by a generalized difference Riccati equation with Markov jump (GDRE-MJ) which is derived from the solution to the FBSDEs-MJ, which is a new method compared with the previous works studied the linear quadratic optimal problem involving Markov jump. The rest of this article is made up of the following sections. Section 2 mainly provides some results about optimal control with finite horizon. And Sect. 3 makes a summary.

The related notations in this article are expressed as follows:

$\mathbb{R}^n$ : the $n$-dimensional Euclidean space;
$\mathbb{R}^{m \times n}$ : the norm bounded linear space of all $m \times n$ matrices;
$Y'$ : the transposition of $Y$;
$Y \geq 0 (Y > 0)$: the symmetric matrix $Y \in \mathbb{R}^{n \times n}$ is positive semi-definite (positive definite);
$Y^\dagger$ : the Moore-Penrose pseudo-inverse of $Y$;
$\mathbf{Ker}(Y)$ : the kernel of a matrix $Y$;
$(\Omega, \mathcal{G}, \mathcal{G}_k, \mathcal{P})$: a complete probability space with the $\sigma$-field generated by $\{x(0), \theta(0), \cdots, x(k), \theta(k)\}$;
$E[\cdot|\mathcal{G}_k]$: the conditional expectation with respect to $\mathcal{G}_k$ and $\mathcal{G}_{-1}$ is understood as $\{\emptyset, \Omega\}$.

## 2  Preliminaries

Considering the following discrete-time Markov jump linear system with multiplicative noise:

$$x(k+1) = (A_{\theta(k)} + B_{\theta(k)}\omega(k))x(k) + (C_{\theta(k)} + D_{\theta(k)}\omega(k))u(k), \qquad (1)$$

where $x(k) \in \mathbb{R}^n$ denotes the state, $u(k) \in \mathbb{R}^m$ denotes control process and $\omega(k)$ is scalar valued random white noise with zero mean and variance $\sigma^2$. $\theta(k)$ is a discrete-time Markov chain with finite state space $\{1, 2, \cdots, L\}$ and transition probability $\rho_{i,j} = \mathrm{P}(\theta(k+1) = j|\theta(k) = i)(i, j = 1, 2, \cdots, L)$. We set $\pi_i(k) = \mathrm{P}(\theta(k) = i)(i = 1, 2, \cdots, L)$, while $A_i, B_i, C_i, D_i(i = 1, \cdots, L)$ are matrices of appropriate dimensions. The initial value $x_0$ is known. We assume that $\theta(k)$ is independent of $x_0$.

The cost function with finite horizon is as the following description.

$$J_N = \mathrm{E}\bigg\{\sum_{k=0}^{N}\big[x(k)'Q_{\theta(k)}x(k) + u(k)'R_{\theta(k)}u(k)\big]$$
$$+ x(N+1)'P_{\theta(N+1)}x(N+1)\bigg\}, \tag{2}$$

where $N > 0$ is an integer, $x(N+1)$ is the terminal state, $P_{\theta(N+1)}$ reflects the penalty on the terminal state, the matrix functions $R_{\theta(k)}$ and $Q_{\theta(k)}$ are symmetric matrices.

**Problem 1.** Find a $\mathcal{G}_k$-measurable controller $u(k)$ to minimize (2) subject to (1).

On the ground of the indefiniteness of weighting matrices, the above problem may be ill-posed. Hence, we should introduce next definitions and lemmas.

**Definition 1:** Problem 1 is called well posed if $\inf\limits_{u_0,\cdots,u_N} J_N > -\infty$ for any random variables $x_0$.

**Definition 2:** Problem 1 is called solvable if there exists an admissible control $(u_0^*, \cdots, u_N^*)$ such that (2) is minimized for any $x_0$.

**Remark 1:** From Theorem 4.3 in [15], the equivalence between the well-posedness and the solvability of Problem* can be obtained.

Due to the dependence of $\theta(k)$ on its past values, an extended version of the stochastic maximum principle which is suitable for the MJLS (1) is established in the sequel.

**Lemma 1 (Maximum Principle involving Markov Jump).** According to the linear system (1) and the performance index (2). If the linear quadratic problem min $J_N$ is solvable, then the optimal $\mathcal{G}_k$-measurable control $u(k)$ satisfies the following equilibrium condition

$$0 = \mathrm{E}[(C_{\theta(k)} + D_{\theta(k)}\omega(k))'\lambda_k + R_{\theta(k)}u(k)|\mathcal{G}_k], k = 0, \cdots, N, \tag{3}$$

where the costate $\lambda_k$ satisfies the following equation

$$\lambda_N = \mathrm{E}[P_{\theta(N+1)}x(N+1)|\mathcal{G}_N], \tag{4}$$
$$\lambda_{k-1} = \mathrm{E}[(A_{\theta(k)} + B_{\theta(k)}\omega(k))'\lambda_k + Q_{\theta(k)}x(k)|\mathcal{G}_{k-1}], k = 0, \cdots, N, \tag{5}$$

together the costate Eqs. (4)–(5) with state Eq. (1), the FBSDEs-MJ is established, which play a vital role in this paper.

*Proof.* Similar to the derivation for Maximum Principle (MP) as in [13,17], the MP (3)–(5) follows directly, the aforementioned conclusion can be derived using an analogous step, so its proof is omitted.

Now we will show the following theorem which is expressed the result of Problem 1.

**Theorem 1.** *Problem 1 is solvable if and only if the following generalized difference Riccati equations with Markov jump*

$$\begin{cases} P_i(k) = A_i'(\sum_{j=1}^{L} \rho_{i,j} P_j(k+1)) A_i + \sigma^2 B_i'(\sum_{j=1}^{L} \rho_{i,j} P_j(k+1)) B_i + Q_i \\ -M_i(k)' \Upsilon_i(k)^\dagger M_i(k), \\ \Upsilon_i(k) \Upsilon_i(k)^\dagger M_i(k) - M_i(k) = 0, \\ \Upsilon_i(k) \geq 0, \end{cases} \tag{6}$$

*in which*

$$\Upsilon_i(k) = C_i'(\sum_{j=1}^{L} \rho_{i,j} P_j(k+1)) C_i + \sigma^2 D_i'(\sum_{j=1}^{L} \rho_{i,j} P_j(k+1)) D_i + R_i, \tag{7}$$

$$M_i(k) = C_i'(\sum_{j=1}^{L} \rho_{i,j} P_j(k+1)) A_i + \sigma^2 D_i'(\sum_{j=1}^{L} \rho_{i,j} P_j(k+1)) B_i, \tag{8}$$

*has a solution. If this condition is satisfied, the analytical solution to the optimal control can be given as*

$$u^*(k) = -\Upsilon_i(k)^\dagger M_i(k) x(k), i = 1, \cdots, L, \tag{9}$$

*for $k = N, \cdots, 0$. The corresponding optimal performance index is given by*

$$J_N^* = E[x(0)' P_{\theta(0)}(0) x(0)]. \tag{10}$$

*The relationship of the costate $\lambda_{k-1}$ and the state $x(k)$ is given as*

$$\lambda_{k-1} = (\sum_{j=1}^{L} \rho_{i,j} P_j(k)) x(k), i = 1, \cdots, L. \tag{11}$$

*Proof* (Necessity). Assume that Problem 1 is solvable, we will investigate that there exist symmetric matrices $P_i(0), \cdots, P_i(N)$, $i = 1, \cdots, L$ satisfying the GDRE-MJ (6) by induction. To this end, we first set the following formula as

$$\underline{J}(k) = \inf_{u_k, \cdots, u_N} E\left[ \sum_{i=k}^{N} (x(i)' Q_{\theta(i)} x(i) + u(i)' R_{\theta(i)} u(i)) \right.$$

$$\left. + x(N+1)' P_{\theta(N+1)} x(N+1) | \mathcal{G}_{k-1} \right]. \tag{12}$$

It is obvious to know that for any $k_1 < k_2$, when $\underline{J}(k_1)$ is finite then $\underline{J}(k_2)$ is also finite by the stochastic optimality principle. Since Problem* is supposed to be solvable, we can see that $\underline{J}(k)$ is finite for any $0 \leq k \leq N$.

Firstly, we let $k = N$, from system (1), we know that

$$
\underline{J}(N) = \inf_{u_N} \mathrm{E}\Big\{ x(N)'[Q_i + A_i'(\sum_{j=1}^{L} \rho_{i,j} P_j(N+1))A_i + \sigma^2 B_i'
$$

$$
\cdot(\sum_{j=1}^{L} \rho_{i,j} P_j(N+1))B_i]x(N) + 2x(N)'[A_i'(\sum_{j=1}^{L} \rho_{i,j} P_j(N+1))C_i
$$

$$
+\sigma^2 B_i'(\sum_{j=1}^{L} \rho_{i,j} P_j(N+1))D_i]u(N) + u(N)'[R_i + C_i'
$$

$$
\cdot(\sum_{j=1}^{L} \rho_{i,j} P_j(N+1))C_i + \sigma^2 D_i'(\sum_{j=1}^{L} \rho_{i,j} P_j(N+1))D_i]u(N)|\mathcal{G}_{N-1}\Big\}
$$

By Lemma 4.3 in [15] and the finiteness of $\underline{J}(N)$, it yields that there indeed exist symmetric matrix $P_i(N)$ satisfying

$$
\underline{J}(N) = \mathrm{E}[x(N)'P_i(N)x(N)],
$$

and furthermore,

$$
P_i(N) = A_i'(\sum_{j=1}^{L} \rho_{i,j} P_j(N+1))A_i + \sigma^2 B_i'(\sum_{j=1}^{L} \rho_{i,j} P_j(N+1))B_i + Q_i
$$

$$
- M_i(N)'\Upsilon_i(N)^\dagger M_i(N), \tag{13}
$$

$$
\Upsilon_i(N)\Upsilon_i(N)^\dagger M_i(N) - M_i(N) = 0, \tag{14}
$$

$$
\Upsilon_i(N) \geq 0, \tag{15}
$$

in which

$$
\Upsilon_i(N) = C_i'(\sum_{j=1}^{L} \rho_{i,j} P_j(N+1))C_i + \sigma^2 D_i'(\sum_{j=1}^{L} \rho_{i,j} P_j(N+1))D_i + R_i, \tag{16}
$$

$$
M_i(N) = C_i'(\sum_{j=1}^{L} \rho_{i,j} P_j(N+1))A_i + \sigma^2 D_i'(\sum_{j=1}^{L} \rho_{i,j} P_j(N+1))B_i. \tag{17}
$$

The optimal controller $u(N)$ will be calculated from (1), (3) and (4).

$$
0 = E[(C_{\theta(N)} + D_{\theta(N)}\omega(N))'\lambda(N) + R_{\theta(N)}u(N)|\mathcal{G}_N]
$$

$$
= \Big[C_i'(\sum_{j=1}^{L} \rho_{i,j} P_j(N+1))A_i + \sigma^2 D_i'(\sum_{j=1}^{L} \rho_{i,j} P_j(N+1))B_i\Big]x(N)
$$

$$
+\Big[C_i'(\sum_{j=1}^{L} \rho_{i,j} P_j(N+1))C_i + \sigma^2 D_i'(\sum_{j=1}^{L} \rho_{i,j} P_j(N+1))D_i + R_i\Big]u(N). \tag{18}
$$

So, from (16) and (17), we have that

$$u(N) = -\Upsilon_i(N)^\dagger M_i(N)x(N), \tag{19}$$

which is as (9) in the case of $k = N$.

As to $\lambda_{N-1}$, from (1), (4), (5) and (19), it yields that

$$\lambda_{N-1} = \mathrm{E}[(A_{\theta(N)} + B_{\theta(N)}\omega(N))'\mathrm{E}[P_{\theta(N+1)}x(N+1)|\mathcal{G}_N] + Q_{\theta(N)}x(N)|\mathcal{G}_{N-1}]$$

$$= \mathrm{E}\Big[A_i'(\sum_{j=1}^{L}\rho_{i,j}P_j(N+1))A_i + B_i'(\sum_{j=1}^{L}\rho_{i,j}P_j(N+1))B_i + Q_i$$

$$- M_i(N)'\Upsilon_i(N)^\dagger M_i(N)|\mathcal{G}_{N-1}\Big]x(N)$$

$$= (\sum_{i=1}^{L}\rho_{s,i}P_i(N))x(N), s = 1, \cdots, L, \tag{20}$$

which is satisfied (11) with $k = N$.

Now we assume that GDRE-MJ (6) has a solution $P_i(m)$, $k+1 \leq m \leq N$ and satisfying $\underline{J}(m) = \mathrm{E}[x(m)'P_i(m)x(m)]$ and $u(m)$, $\lambda(m-1)$ are as (9), (11), respectively, thus for $k$, we have

$$\underline{J}(k) = \inf_{u_k} \mathrm{E}\Big\{x(k)'Q_{\theta(k)}x(k) + u(k)'R_{\theta(k)}u(k) + \mathrm{E}[x(k+1)'P_i(k+1)$$

$$\cdot x(k+1)]|\mathcal{G}_{k-1}\Big\}$$

$$= \inf_{u_k} \mathrm{E}\Big\{x(k)'[Q_i + A_i'(\sum_{j=1}^{L}\rho_{i,j}P_j(k+1))A_i + \sigma^2 B_i'(\sum_{j=1}^{L}\rho_{i,j}P_j(k+1))B_i]$$

$$\cdot x(k) + 2x(k)'[A_i'(\sum_{j=1}^{L}\rho_{i,j}P_j(k+1))C_i + \sigma^2 B_i'(\sum_{j=1}^{L}\rho_{i,j}P_j(k+1))D_i]u(k)$$

$$+ u(k)'[R_i + C_i'(\sum_{j=1}^{L}\rho_{i,j}P_j(k+1))C_i + \sigma^2 D_i'(\sum_{j=1}^{L}\rho_{i,j}P_j(k+1))D_i]$$

$$\cdot u(k)|\mathcal{G}_{k-1}\Big\}.$$

Similarly, from Lemma 4.3 in [15] and the finiteness of $\underline{J}(k)$, we can obtain that there exist $P_i(k)$ satisfying GDRE-MJ (6). Furthermore, $\underline{J}(k) = \mathrm{E}[x(k)'P_i(k)x(k)]$. From now on by mathematical induction we obtain that GDRE-MJ (6) exists a solution.

In the case that GDRE-MJ (6) exists a solution and the inductive hypothesis, the optimal controller $u(k)$ can be obtained from (1) and (3).

$$0 = E[(C_{\theta(k)} + D_{\theta(k)}\omega(k))'(\sum_{j=1}^{L} \rho_{i,j}P_j(k+1))x(k+1) + R_{\theta(k)}u(k)|\mathcal{G}_k]$$

$$= \left[ C_i'(\sum_{j=1}^{L} \rho_{i,j}P_j(k+1))A_i + \sigma^2 D_i'(\sum_{j=1}^{L} \rho_{i,j}P_j(k+1))B_i \right] x(k)$$

$$+ \left[ C_i'(\sum_{j=1}^{L} \rho_{i,j}P_j(k+1))C_i + \sigma^2 D_i'(\sum_{j=1}^{L} \rho_{i,j}P_j(k+1))D_i + R_i \right] u(k), (21)$$

i.e.,

$$u(k) = -\Upsilon_i(k)^\dagger M_i(k)x(k). \tag{22}$$

From (1), (5) and (22), $\lambda_{k-1}$ can be derived as that

$$\lambda_{k-1} = E[(A_{\theta(k)} + B_{\theta(k)}\omega(k))'(\sum_{j=1}^{L} \rho_{i,j}P_j(k+1))x(k+1)] + Q_{\theta(k)}x(k)|\mathcal{G}_{k-1}]$$

$$= E\left[ A_i'(\sum_{j=1}^{L} \rho_{i,j}P_j(k+1))A_i + B_i'(\sum_{j=1}^{L} \rho_{i,j}P_j(k+1))B_i + Q_i \right.$$

$$\left. - M_i(k)'\Upsilon_i(k)^\dagger M_i(k)|\mathcal{G}_{k-1} \right]x(k)$$

$$= (\sum_{i=1}^{L} \rho_{s,i}P_i(k))x(k), s = 1, \cdots, L. \tag{23}$$

The proof about necessity is end.

(Sufficiency): When the GDRE-MJ (6) has a solution, we will show that Problem 1 is solvable.

Denote $V_N(k, x(k)) \triangleq E[x(k)'P_{\theta(k)}(k)x(k)]$. From (1) we deduce that

$$V_N(k, x(k)) - V_N(k+1, x(k+1))$$

$$= E[x(k)'P_{\theta(k)}(k)x(k) - x(k+1)'P_{\theta(k+1)}(k+1)x(k+1)]$$

$$= E\Big\{ x(k)'[Q_i - M_i(k)'\Upsilon_i(k)^\dagger M_i(k)]x(k) - x(k)'M_i'(k)u(k)$$

$$- u(k)'M_i(k)x(k) - u(k)'\Upsilon_i(k)u(k) + u(k)'R_iu(k) \Big\}$$

$$= E\Big\{ x(k)'Q_ix(k) + u(k)'R_iu(k) - [u(k) + \Upsilon_i(k)^\dagger M_i(k)x(k)]'\Upsilon_i(k)[u(k)$$

$$+ \Upsilon_i(k)^\dagger M_i(k)x(k)] \Big\}$$

Adding from $k = 0$ to $k = N$ on both sides of the above equation, we have that

$$V_N(0, x(0)) - V_N(N + 1, x(N + 1))$$

$$= \mathrm{E} \sum_{k=0}^{N} \left\{ x(k)'Q_i x(k) + u(k)'R_i u(k) \right.$$

$$\left. - [u(k) + \Upsilon_i(k)^{\dagger} M_i(k)x(k)]'\Upsilon_i(k)[u(k) + \Upsilon_i(k)^{\dagger} M_i(k)x(k)] \right\}. \tag{24}$$

The above mentioned equation implies that

$$J_N = \mathrm{E}[x_0' P_{\theta(0)} x_0] + \sum_{k=0}^{N} [u(k) + \Upsilon_i(k)^{\dagger} M_i(k)x(k)]'\Upsilon_i(k)[u(k) + \Upsilon_i(k)^{\dagger} M_i(k)x(k)].$$

Considering $\Upsilon_i(k) \geq 0$, we have $J_N \geq \mathrm{E}[x_0' P_{\theta(0)} x_0]$. Therefore, the optimal controller can be given by $u(k) = -\Upsilon_i(k)^{\dagger} M_i(k)x(k)$ and the optimal cost is given by $J_N = \mathrm{E}[x_0' P_{\theta(0)} x_0]$.

This completes the proof.

## 3   Conclusions

This article mainly study the linear quadratic optimal control problem for discrete-time systems involving Markov jump and multiplicative noise. The state and control weighting matrices in the cost function are allowed to be indefinite. By solving the FBSDEs-MJ derived from the extended maximum principle, we conclude that the indefinite optimal control problem in finite-horizon is solvable if and only if the corresponding GDRE-MJ has a solution, which is an easy verifiable conclusion compared with operator type results.

## References

1. Costa, O.L.V., Fragoso, M.D., Marques, R.P.: Discrete Time Markov Jump Linear Systems. Springer, New York (2005). https://doi.org/10.1007/b138575
2. Tugnait, J.K.: Adaptive estimation and identification for discrete systems with Markov jump parameters. IEEE Trans. Autom. Control **27**, 1054–1064 (1982)
3. Shi, P., Boukas, E.K., Agarwal, R.K.: Control of Markovian jump discrete-time systems with norm bounded uncertainty and unknown delay. IEEE Trans. Autom. Control **44**(11), 2139–2144 (1999)
4. Wu, L., Shi, P., Gao, H.: State estimation and sliding mode control of Markovian jump singular systems. IEEE Trans. Autom. Control **55**(5), 1213–1219 (2010)
5. Costa, O.L.V., Assumpia Filho, E.O., Boukas, E.K., Marques, R.P.: Constrained quadratic state feedback control of discrete-time Markovian jump linear systems. Automatica **35**, 617–626 (1999)
6. Costa, O.L.V.: Linear minimum mean square error estimation for discrete-time Markovian jump linear systems. IEEE Trans. Autom. Control **39**(8), 1685–1689 (1994)

7. Costa, O.L.V., do Val, J.B.R.: Full information H control for discrete-time infinite Markovjump parameter systems. J. Math. Anal. Appl. **202**, 578–603 (1996)
8. Costa, O.L.V., Fragoso, M.D.: Discrete-time LQ-optimal control problems for infnite Markov jump parameter systems. IEEE Trans. Autom. Control **40**(12), 2076–2088 (1995)
9. do Val, J.B.R., Costa, E.F.: Stabilizability and positiveness of solutions of the jump linear quadratic problem and the coupled algebraic Riccati equation. Trans. Autom. Control **50**(5), 691–695 (2005)
10. Li, X., Zhou, X.: Indefinite stochastic LQ controls with Markovian jumps in a finite time horizon. Commun. Inf. Syst. **2**(3), 265–282 (2002)
11. Li, X., Zhou, X.: M. A. Rami.: Indefinite stochastic linear quadratic control with Markovian jumps in infinite time horizon. J. Glob. Optim. **27**, 149–175 (2003)
12. Costa, O.L.V., Wanderlei, L.P.: Indefinite quadratic with linear costs optimal control of Markov jump with multiplicative noise systems. Automatica **43**, 587–597 (2007)
13. Zhang, H., Li, L., Xu, J., Fu, M.: Linear quadratic regulation and stabilization of discrete- time systems with delay and multiplicative noise. IEEE Trans. Autom. Control **60**(10), 2599–2613 (2015)
14. Q. Qi, H. Zhang.: A Complete Solution to Optimal Control and Stabilization for Mean-field Systems: Part I, Discrete-time Case, pp. 1–20 (2016). arXiv preprint arXiv: 1608.06363
15. Aitrami, M., Chen, X., Zhou, X.: Discrete-time indefinite LQ control with state and control dependent noises. J. Global Optim. **23**(34), 245–265 (2002)
16. Zhang, W., Chen, B.S.: On stabilizability and exact observability of stochastic systems with their applications. Automatica **40**, 87–94 (2004)
17. Zhang, H., Wang, H., Li, L.: Adapted and casual maximum principle and analytical solution to optimal control for stochastic multiplicative- noise systems with multiple input-delays. In: Proceedings of the 51st IEEE Conference Decision Control, Maui, pp. 2122–2127, HI, USA (2012)

# Human-Machine Interaction

# Stretchable Tactile and Bio-potential Sensors for Human-Machine Interaction: A Review

Wentao Dong[1], YongAn Huang[2,3](✉), Zhouping Yin[2,3], Yuyu Zhou[4], and Jiankui Chen[2,3]

[1] School of Electrical and Automation Engineering, East China Jiaotong University, Nanchang 330013, China
[2] State Key Laboratory of Digital Manufacturing Equipment and Technology, Huazhong University of Science and Technology, Wuhan 430074, China
yahuang@hust.edu.cn
[3] Flexible Electronics Research Center, Huazhong University of Science and Technology, Wuhan 430074, China
[4] Department of Laboratory and Facility Management, Jinan University, Guangzhou 510632, China

**Abstract.** Human machine interaction (HMI) technologies have been widely applied to the fields of the complicated task assignment, biological health monitoring, prosthesis techniques, and clinical medicine. In this paper, different kinds of HMI modes are reviewed, such as tactile sensors, biological sensors, and multisensory data. Stretchable electronics integrated with multi-function sensors on the polydimethylsiloxane (PDMS) substrate are laminated onto the skin surface for collecting temperature, strain, pressure, biological signals simultaneously. More conformable and natural human-machine interaction methods would be realized, which will provide effective ways for human-robot interaction similar to human-to-human interaction, and finally drive the development of the coexisting-cooperative-cognitive robot (Tri-Co Robot) technology.

**Keywords:** Human machine interaction · Flexible/stretchable electronics
Tactile sensor · Biopotential sensor · Multi-sensory data

## 1 Introduction

Human machine interaction (HMI) is the field to study the information interaction between robots and humans, where the motion of the robots is controlled by the signals from human [1]. Conventional rigid sensor based on silicon wafer would lead to discomfort to body when it is laminated onto the skin surface. Flexible and stretchable electronics are the development direction for HMI [2, 3]. It is very important for the flexible and stretchable electronics to be designed with stretchability to satisfy the deformation of the soft skin and to control the robot more conveniently [4, 5].

The schematic graph for the human to communicate or interact with robots or machines is shown in Fig. 1. The HMI modes can be divided into several categories: (i) soft tactile sensors, (ii) biological sensors, and (iii) multi-sensory data. To understand

© Springer Nature Switzerland AG 2018
Z. Chen et al. (Eds.): ICIRA 2018, LNAI 10984, pp. 155–163, 2018.
https://doi.org/10.1007/978-3-319-97586-3_14

the development of the soft tactile sensing technologies, it is necessary to compare the four kinds of sensors: piezoelectric sensors, piezoresistive sensors, capacitive sensors, and acoustic sensors. Wang et al. designed and developed many piezoelectric materials and devices which transduce the human strain variation to the output voltage for HMI applications [2]. Bao et al. designed and fabricated an ultra-sensitive capacitive sensor array with pyramid-structured polydimethylsiloxan (PDMS) for pressure visualization [4, 6, 7]. Sound and audio systems were also widely applied to speech analysis and recognition, which are the hottest research areas [8]. Bio-potential signals from human bodies were also used to HMI application, such as, electromyography (EMG) [9, 10], electroencephalogram (EEG) [5, 11], and electrooculography (EOG) [12]. Features of the biological signals are extracted to control the motion of the external actuators. A prosthetic hand was designed to record multi-sensory data simultaneously, and data fusion methods were proposed to control the prosthesis [13].

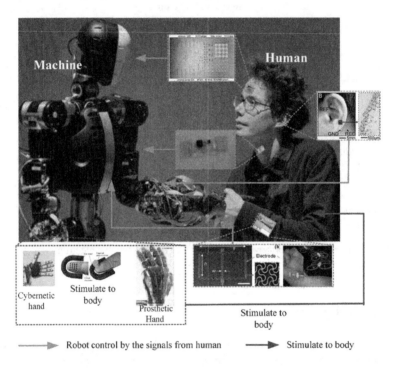

**Fig. 1.** Schematic graph for humans interacting with the robots

There are many review articles about the flexible and stretchable electronics which have been applied to biological healthy monitoring, electronic skin, and HMIs. Soft tactile sensors for HMI enable a generation of touch-sensitive health-care robots to attend the old and the disabled to collect human physiological signals for health monitoring [14]. However, all these reviews mainly focused on the design and development of the stretchable E-skin in different applications. The HMI modes should be also classified via the flexible and stretchable electronics. The rest of the review article is arranged as follows. In Sect. 2, soft tactile sensors are reviewed for HMI. Bio-potential signals for HMI are reviewed in Sect. 3. Multi-sensory data for HMI is introduced in Sect. 4. The whole review article is concluded in Sect. 5.

## 2    Tactile Sensors for HMI

It is important for the robot integrated with tactile sensors to interact with surroundings or humans for completing complicated and dynamic tasks. During the past decade, large numbers of soft tactile sensors were designed to HMI applications. As a branch of mechanical sensors, soft tactile sensors are categorized by the transduction methods, such as, piezoelectric sensors, piezoresistive sensors, capacitive sensors, and acoustic sensors.

Piezoelectricity refers to the production of electrical charging in certain materials under mechanical force due to the occurrence of electrical dipole moments [15]. This approach was used to convert mechanical stresses and vibrations into electrical signals via piezoelectric materials with high sensitivity, rapid response, and a high piezoelectric coefficient. A rapid-response flexible pressure sensor matrix which was based on the direct conversion of mechanical stress was developed to pressure visualization [16]. With external pressure applied to the flexible sensors, the sensor matrix could give the corresponding pressure distribution. Figure 2 depicted that transparent ZnO Nanowire sensor was designed for self-powered gesture recognition [17]. With the motion of different fingers, various control commands were generated to control the motion of the external actuators based on soft piezoelectricial sensors [18].

**Fig. 2.** HMI based on piezoelectricial sensors [17, 18]

Piezoresistive sensors enable transduction of force variations into changes in resistance which is easily detected by an electrical measuring system [19]. These sensors were widely used to detect strain information of human bodies. Stretchable silicon strain sensor integrated with serpentine structure was designed to control the motion of the external actuator as shown in Fig. 3 [20]. Piezoresistive sensors with liquid metals were encapsulated by elastomer channels for strain monitoring, which had excellent electrical and mechanical properties. Wu et al. designed a epidermal strain sensor with liquid metal for finger motion detection, and two features with the finger bending were recognized which demonstrated that the liquid metal strain sensor had the potential application in

HMI control [21]. Javey et al. developed a tactile sensing glove for capturing a variety of comprehensive hand motions, such as holding, gripping, grasping, squeezing and so on [19]. It clearly gives a promising application in the force visualization and HMI application.

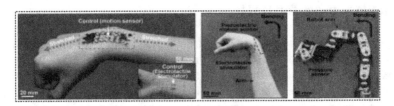

**Fig. 3.** HMI based on piezoresistive sensors [20]

The capacitance (C) of a parallel plate capacitor, denoted as the ability to store a charge, is described by $C = \varepsilon A/d$, where $\varepsilon$ is the dielectric constant, and A and d are the area and the distance between the two electrodes, respectively. Figure 4 depicted that flexible, pressure-sensitive active matrix on a plastic substrate with high-pressure sensitivity and rapid response time based on microstructured rubber was designed and fabricated, which could be used to precisely map the static pressure distribution and health monitoring. Bao et al. demonstrated a transparent and stretchable capacitive sensor array based on carbon nanotube electrodes on elastic substrates that was sensitive to both pressure and strain [6]. The sensitivity of the capacitive sensor array depended on the pyramid-structured PDMS which increased the air voids between PDMS and organic semiconductor layer. When a little fly stood on the sensor, it would generate a plus signal and give pressure visualization with high sensitivity.

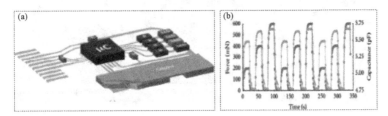

**Fig. 4.** HMI based on soft capacitive sensors.

The epidermal mechano-acoustic sensors were compatible to the soft curvilinear skin for capturing acoustic vibrations for local words and the speech signals, and the human voices were recognized to control the motion of the robot [22]. The mechano-acoustic vibrations were recorded while speaking "left," "right," "up," and "down." The spectrogram highlighted the unique time-frequency characteristics of each of the four words. The intimate contact between the sensors and the skin rendered their operation almost unaffected by ambient acoustic noise. These features could allow the epidermal acoustic sensor to be used for communication in loud environments by first responders, ground controllers, or security agents.

## 3   Bio-potential Sensors for HMI

The contraction of muscles could generate EMG signals, which were acquired by the surface EMG electrodes noninvasively [23]. The stretchable multichannel sEMG patches were designed for robot manipulation via eight gestures of the hand [24]. Combined with the large-scale size, the multichannel noninvasive HMI via stretchable μm thick sEMG patches successfully manipulated the robot hand with eight different gestures, whose precision was as high as conventional gel electrodes array. Figure 5(a) depicted the stretchable EMG electrode was designed to control the motion of mobile robot via different human gestures [9].

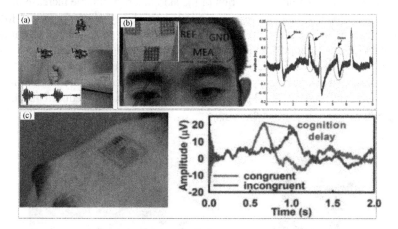

**Fig. 5.** HMI based on soft bio-potential sensors. (a) EMG sensor [9] (b) EOG sensor [27]; (c) EEG sensor [5].

The EOG sensor mounted on the forehead that was prepared by exfoliating the stratum corneum with tape yielded reproducible, high-quality results, as demonstrated in alpha rhythms recorded from awake subjects with their eyes closed [5, 25, 26]. The expected feature at 0–10 Hz appeared clearly in the Fourier-transformed data. This activity disappeared when the eyes were open. The EOG system for HMI applications consisted three modules: sensors for EOG recording, the signal collection instrument, and the computer control system (Fig. 5(b)) [27]. The signals were recorded by a commercial amplifier system to control the robots or computer games via different eye motion.

Brain–computer interface (BCI) technology was a radically communication option for those with neuromuscular impairments that prevent them from using conventional augmentative communication methods [5]. Current BCI recorded the EEG signal at the scalp or single-unit activity from cortex to control cursor movement, select letters or icons, or operate a neuroprosthesis [28]. The experimental setup for HMI based on EEG signals included a participant wearing the electrodes, an amplifier, an analog-to-digital converter, classifiers, and software as shown in Fig. 5(c). It was demonstrated that people played computer games with the EEG sensor for HMI application [29].

## 4    Multi-sensory Data for HMI

Multi-sensory data infusion methods were used to build a natural interface allowing humans to interact with robots or machines similar to human-to-human communication through speech or gestures, which required more information from both the human and machines.

A prosthetic hand integrated with artificial skin was highly compliant, and mechanically couples to the curvilinear surface of the prosthesis as shown in Fig. 6 [13]. The stacked layers highlighted the location of the embedded electronics, sensors and actuator. The thermal actuators were in fractal inspired formats to facilitate uniform heating during stretching and contraction of the skin layer. A notable increase in EMG activity for the loaded compared cases suggested an ability to detect loaded lifting. Stooping motions induced signature responses of the strain gauge, distinct from those associated with squatting [13]. In this circumstance, the temperature sensor provided a reliable indicator of muscle exertion, associated with increases in metabolic reactions and blood flow that lead to corresponding increases in skin temperature. Strain/pressure sensors positioned beneath the humidity sensors could exhibit reduced mechanical responses to external deformations.

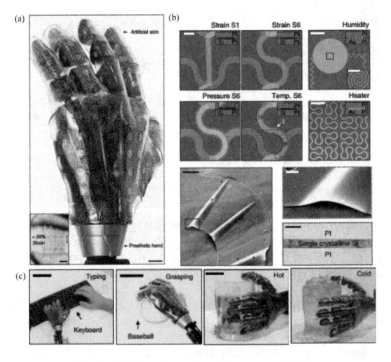

**Fig. 6.** HMI based on soft multi-sensory data [13].

Flexible multifunctional devices with EMG, temperature, and strain signal recording simultaneously were designed to a more practical HMI application for humans to interact with machines [30]. The temperature sensor consisted of a serpentine conductive trace (Cr/Au) with a width of 20 μm, thickness of 200 nm, and length of 26 mm in total. The strain sensor and the EMG sensor used the same metal pattered in different kinds of geometry. Together with the EMG sensor, the strain gauge, and temperature sensor, this platform offered important functionality intervention for lower back exertion.

## 5    Conclusion and Discussions

Robots integrated with soft sensors are also illustrated for disease treatment. Different modes of HMI are introduced, which includes tactile sensors, biological sensors, sand multi-sensory data. Soft tactile sensors for HMI will aid future development in robotics, biomedical devices, sports, automobile, textiles, and many other fields. Stretchable electronics integrated with multi-function sensors on the PDMS substrate are laminated onto the skin surface for collecting temperature, strain, pressure, and EMG signals simultaneously. Multi-source data fusion methods are adopted for the robot to accomplish the task and interact with human.

Although HMI applications have developed fast in recent years, there exist some difficult problems. (1) The production of devices with low-power consumption or self-powering ability remains a topic worthy of in-depth study, because the energy crisis is currently one of the largest challenges in our society. (2) The cost of tactile sensors is another problem for hindering the development and application in the field of service robot and health care. Low-cost materials and fabrication process for tactile and biological sensors are effective to promote HMI application.

Future HMI will also intelligently respond to variations in the external environment based on novel information transmission technology. With the rapid development of artificial intelligence technologies, intelligent HMI with smart sensing, computing and decision technologies would provide an effective way for people to interact with machines similar to human-to-human interaction.

**Acknowledgement.** The authors also acknowledge supports from the National Natural Science Foundation of China (51635007), Program for HUST Academic Frontier Youth Team, and Special Project of Technology Innovation of Hubei Province. (2017AAA002). The authors would like to thank Flexible Electronics Manufacturing Laboratory in Comprehensive Experiment Center for Advanced Manufacturing and Equipment Technology.

## References

1. Uchida, S., Mori, A., Kurazume, R., Rin-Ichiro, T., Hasegawa, T., Sakoe, H.: Early Recognition and Prediction of Gestures for Proactive Human-Machine Interface. Technical report of IEICE PRMU, vol. 104, pp. 7–12 (2004)
2. Wang, X., Dong, L., Zhang, H., Yu, R., Pan, C., Wang, Z.L.: Recent progress in electronic skin. Adv. Sci. **2**, 1–21 (2015)
3. Hammock, M.L., Chortos, A., Tee, B.C., Tok, J.B., Bao, Z.: 25th anniversary article: the evolution of electronic skin (e-skin): a brief history, design considerations, and recent progress. Adv. Mater. **25**, 5997–6038 (2013)

4. Lipomi, D.J., et al.: Skin-like pressure and strain sensors based on transparent elastic films of carbon nanotubes. Nat. Nanotechnol. **6**, 788–792 (2011)
5. Kim, D.H., et al.: Epidermal electronics. Science **333**, 838–843 (2011)
6. Mannsfeld, S.C., et al.: Highly sensitive flexible pressure sensors with microstructured rubber dielectric layers. Nat. Mater. **9**, 859–864 (2010)
7. Tee, B.C., Wang, C., Allen, R., Bao, Z.: An electrically and mechanically self-healing composite with pressure- and flexion-sensitive properties for electronic skin applications. Nat. Nanotechnol. **7**, 825–832 (2012)
8. Shriver, S., Toth, A., Zhu, X., Rudnicky, A., Rosenfeld, R.: A unified design for human-machine voice interaction, pp. 247–248 (2001)
9. Wentao, D., Chen, Z., Wei, H., Lin, X., Yong'an, H.: Stretchable human-machine interface based on skin-conformal sEMG electrodes with self-similar geometry. J. Semicond. **39**, 014001 (2018)
10. Dong, W., Zhu, C., Wang, Y., Xiao, L., Ye, D., Huang, Y.: Stretchable sEMG electrodes conformally laminated on skin for continuous electrophysiological monitoring. In: Huang, Y., Wu, H., Liu, H., Yin, Z. (eds.) ICIRA 2017. LNCS (LNAI), vol. 10464, pp. 77–86. Springer, Cham (2017). https://doi.org/10.1007/978-3-319-65298-6_8
11. Jeong, J.W., et al.: Capacitive epidermal electronics for electrically safe, long-term electrophysiological measurements. Adv. Healthc. Mater. **3**, 642–648 (2014)
12. Mishra, S., et al.: Soft, conformal bioelectronics for a wireless human-wheelchair interface. Biosens. Bioelectron. **91**, 796–803 (2017)
13. Kim, J., et al.: Stretchable silicon nanoribbon electronics for skin prosthesis. Nat. Commun. **5**, 5747 (2014)
14. Wang, F.: Soft tactile sensors for human-machine interaction. In: Tao, X. (ed.) Handbook of Smart Textiles. Springer, Singapore (2015). https://doi.org/10.1007/978-981-4451-45-1_26
15. Yu, P., Liu, W., Gu, C., Cheng, X., Fu, X.: Flexible piezoelectric tactile sensor array for dynamic three-axis force measurement. Sensors **16**, 819 (2016)
16. Gao, L., et al.: Epidermal photonic devices for quantitative imaging of temperature and thermal transport characteristics of the skin. Nat. Commun. **5**, 4938 (2014)
17. Pradel, K.C., Wu, W., Ding, Y., Wang, Z.L.: Solution-derived ZnO homojunction nanowire films on wearable substrates for energy conversion and self-powered gesture recognition. Nano Lett. **14**, 6897–6905 (2014)
18. Dong, W., Xiao, L., Hu, W., Zhu, C., Huang, Y., Yin, Z.: Wearable human–machine interface based on PVDF piezoelectric sensor. Trans. Instit. Meas. Control **39**, 398–403 (2017)
19. Takei, K., et al.: Nanowire active-matrix circuitry for low-voltage macroscale artificial skin. Nat. Mater. **9**, 821–826 (2010)
20. Ying, M., et al.: Silicon nanomembranes for fingertip electronics. Nanotechnology **23**, 344004 (2012)
21. Jeong, S.H., Zhang, S., Hjort, K., Hilborn, J., Wu, Z.: PDMS-based elastomer tuned soft, stretchable, and sticky for epidermal electronics. Adv. Mater. (2016)
22. Liu, Y., et al.: Epidermal mechano-acoustic sensing electronics for cardiovascular diagnostics and human-machine interfaces. Science Advances **2**, e1601185–e1601185 (2016)
23. Jeong, J.W., et al.: Materials and optimized designs for human-machine interfaces via epidermal electronics. Adv. Mater. **25**, 6839–6846 (2013)
24. Zhou, Y., Wang, Y., Liu. R., Lin, X., Zhang, Q., Huang, Y.: Multichannel noninvasive human–machine interface via stretchable μ m thick sEMG patches for robot manipulation. J.Micromech. Microeng. **28**, 014005 (2018)

25. Jin, L., Xian, H., Jiang, Y., Niu, Q., Xu, M., Yang, D.: Research on evaluation model for secondary task driving safety based on driver eye movements. Adv. Mech. Eng. **6**, 624561 (2015)
26. Sigari, M.-H., Pourshahabi, M.-R., Soryani, M., Fathy, M.: A review on driver face monitoring systems for fatigue and distraction detection. Int. J. Adv. Sci. Technol. **64**, 73–100 (2014)
27. Guo, X., et al.: A human-machine interface based on single channel EOG and patchable sensor. Biomed. Signal Process. Control **30**, 98–105 (2016)
28. Wolpaw, J.R., et al.: Brain-computer interface technology: a review of the first international meeting. IEEE Trans. Rehabil. Eng. **8**, 164–173 (2000)
29. Norton, J.J., et al.: Soft, curved electrode systems capable of integration on the auricle as a persistent brain-computer interface. Proc. Natl. Acad. Sci. U.S.A. **112**, 3920 (2015)
30. Xu, B., et al.: An epidermal stimulation and sensing platform for sensorimotor prosthetic control, management of lower back exertion, and electrical muscle activation. Adv. Mater. **28**, 4462–4471 (2015)

# Hand Detection and Location Based on Improved SSD for Space Human-Robot Interaction

Qing Gao[1,2], Jinguo Liu[1(✉)], Zhaojie Ju[3], Lu Zhang[4], Yangmin Li[5], and Yuwang Liu[1]

[1] The State Key Laboratory of Robotics, Shenyang Institute of Automation,
Chinese Academy of Sciences, Shenyang 110016, China
liujinguo@sia.cn
[2] University of the Chinese Academy of Science, Beijing 100049, China
[3] School of Computing, University of Portsmouth, Portsmouth, PO1 3HE, UK
[4] Key Laboratory of Space Utilization, Technology and Engineering Center for Space Utilization,
Chinese Academy of Sciences, Beijing 100094, China
[5] Department of Industrial and Systems Engineering, The Hong Kong Polytechnic University,
Hong Kong 999077, China

**Abstract.** In the astronaut-space robot interaction based on hand gestures, the detection and location of hands are the premise and basis of vision-based hand gesture recognition and hand tracking. In this paper, the SSD (Single Shot Multibox Detector) which is a kind of deep learning model is utilized to detect and locate astronaut's hands for space human-robot interaction (SHRI) based on hand gestures. First of all, in order to meet the needs of hand detection and location, an improved SSD model is designed to detect hands when they are shown as small targets in images. Then, a platform for SHRI is built and a set of hand gestures for SHRI are designed. Finally, the proposed SSD model is validated experimentally on a homemade hand gesture database for proving the superiority of this improved SSD model to small target hands detection.

**Keywords:** Human-robot interaction · Hands detection · SSD · Deep learning

## 1 Introduction

Nowadays, in the space missions, space robots are often controlled interactively by staff or astronauts because of the limited intelligence of space robots and safety. In general, the advantage of astronaut is that it has a strong sense of perception, decision-making and planning capabilities, while the advantage of the space robot is that it can achieve smooth, high-precision and wide-range operations. SHRI technology effectively combines the advantages of astronauts and space robots and plays an important role in space missions [1]. Among them, the hand gesture-based SHRI with its natural and intuitive, informative, non-contact advantages is very suitable for applications in SHRI tasks. At present, there have been some achievements in this field at home and abroad [2–4]. At the same time, the design of SHRI system and the design of interactive hand gestures are crucial to gesture-based SHRI. A set of reasonable and natural SHRI hand gestures can help astronauts to control space robots efficiently and conveniently.

© Springer Nature Switzerland AG 2018
Z. Chen et al. (Eds.): ICIRA 2018, LNAI 10984, pp. 164–175, 2018.
https://doi.org/10.1007/978-3-319-97586-3_15

Astronauts' hands detection and location are premise and basis of hand gestures recognition and hands tracking. They play a key role in the entire hand gesture interaction process. At present, significant efforts have been made in the field of hands detection and location [5–7]. Among them, deep learning has made breakthrough progress in visual-based hand gestures interaction [8–11]. For the target detection and location problems, the related deep learning models have R-CNN, Fast R-CNN, Faster R-CNN, YOLO and SSD. Among them, the SSD model has excellent performance in the real-time detection and location of targets. It uses end-to-end training method to achieve a balance between speed and accuracy (More accurate than Faster R-CNN and faster than YOLO) [12]. However, the SSD model also has the problem of inaccurate detection of small targets. For example, if the astronaut is far away from the camera during operation, the hand may be considered as a small target. In this situation, the SSD model can't detect the hand accurately. Therefore, the SSD model needs to be improved to increase the detection and location accuracy of hand when it considered as a small target.

In this paper, aiming at hand detection and location problems, the SSD model is improved. A feature-based SSD model is designed to detect the hand precisely when it is a small target. In addition, aiming at SHRI tasks, the second-generation astronaut assistant robot (AAR-2) which was developed by Shenyang Institute of Automation (SIA), Chinese Academy of Sciences (CAS) is chose to act as the space robot to set up the SHRI system platform. What's more, a set of hands-coordination interactive hand gestures which are called "left hand for instruction and right hand for operation" is designed. In the experiment part, a set of interactive astronaut hand gestures database is manufactured by ourselves. Under this database, the SSD model and the improved SSD model in this paper are trained and verified respectively, and their test results are compared.

The rest of this paper is structured as follows: In Sect. 2 the improved SSD model will be introduced. In Sect. 3 the SHRI system and the hands-coordination interactive hand gestures will be introduced. And in Sect. 4 the SSD model and the improved SSD model will be tested on a homemade astronaut hand gestures interactive database, and the experiment results will be compared to analyze the effectiveness and superiority of the improved SSD model. At last, the conclusion and the prospect of this article will be shown in Sect. 5.

## 2   Hand Detection and Location Method Based on Improved SSD Model

Based on the SHRI, the requirements for astronaut's hand detection and location are shown as follows:

(1) Ensure the real-time capability of hand detection and location;
(2) Can detect and locate different hand gestures;
(3) When the astronaut's hand is shown as a small target in the image, it can also be detected and located precisely.

## 2.1   Introduction to SSD Model

Based on the above requirements, using SSD model to detect and locate the astronaut's hand can learn the characteristics of different hand gestures and can guarantee the real-time capability. The network structure is shown in Fig. 1.

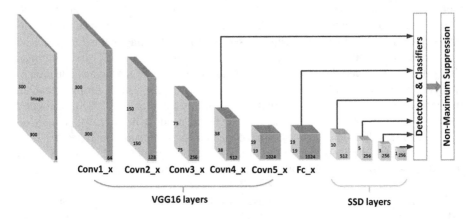

**Fig. 1.**  SSD architecture.

As shown in Fig. 1, the SSD is a full convolution neural network detector with different layers to detect objects of different sizes. However, the detection of small targets by the SSD model often does not perform well. The reason is that in shallower layers, the feature maps are large with more contextual information, but semantic information is not enough; in deeper layers, the semantic information is much enough, but after too many pooling layers, the feature maps are very small. When the detected hand is a small target, it needs a feature map which is large enough to provide sufficient contextual information and semantic information to distinguish hand from the background.

## 2.2   Introduction to Improved SSD Model

In order to solve the above problem, the information of the shallower layer can be combined with the information of the deeper layer to design a layer that has both enough contextual information and enough semantic information. Inspired by DSSD [13] and Feature-Fused SSD [14], the improved SSD model is designed as Fig. 2.

According to the principle of SSD model, shallower layers networks are used to detect small targets, and deeper layers networks are used to detect large targets. As shown in Fig. 2, aiming at the insufficient semantic information of shallower layers networks, the shallower layers should be fused with the deeper layers to increase the semantic information of shallower layers networks. It can increase the detection accuracy of the network to small targets. From [14], it can be seen that Conv4_3 layer has the best detection effect on small targets. Therefore, in this paper, Conv4_3 layer and Conv6_2 layer are utilized to fusion according to their features to improve the network detection accuracy of small target hands.

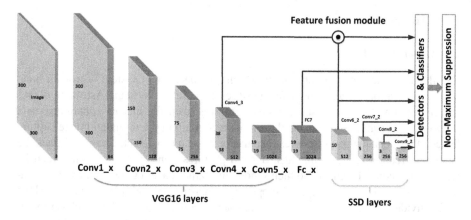

**Fig. 2.** Improved SSD architecture.

## 2.3 Feature Fusion Module

The feature fusion module is indicated by Fig. 3. The specific method is that conduct two deconvolution operations on the feature map of Conv6_2 layer to get Deconv6_2 layer which has the same size with Conv4_3 layer. Then two $3 \times 3$ convolutional layers are used after Conv4_3 layer and Deconv6_2 layer for learning the better features to fuse. After this, normalization layers are following with different scales respectively, i.e. 10, 20. Finally, from [13] it can be seen that element-wise product method can get the best accuracy result compared with concatenation and element-sum methods, so it can fuse the feature maps of the two layers according to element-wise product method to produce a new feature map Fusion_conv43_conv62 layer to detect and locate small targets.

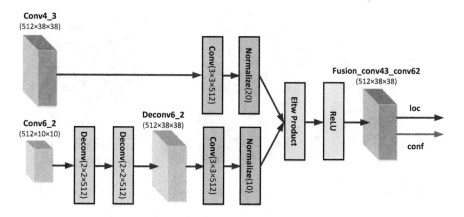

**Fig. 3.** Feature fusion module.

## 3   SHRI System

### 3.1   Experiment Platform

The SHRI system includes three parts which are space robot system, astronaut hand gestures operation system and hand gestures recognition system. Astronaut issues control commands through specific hand gestures, then the images containing hand gestures are transmitted to the hand gestures recognition system via an image capture device. After this, the hand gestures recognition system can detect, locate and track the hands and recognize the semantics of hand gestures. Finally, hand gestures can be decoded into control signals and then send them to the space robot to control its operation. According to the multi-spatial scope between astronaut and space robot, the SHRI can be divided into astronaut-robot shared space (shoulder-to-shoulder) and astronaut-robot non-shared space (Line-of-sight, Over-the-horizon, Interplanetary) [1]. In the astronaut-robot shared space mode, the image capture device and the gesture recognition system can be mounted on the space robot. While in the astronaut-robot non-shared space mode, the image capture device and gesture recognition system need to be installed on the console.

Select the AAR-2 as the space robot [15, 16]. This robot works at the space station and is utilized to assist astronauts to complete some space missions. It has six degrees of freedom, and can fly freely in the space station cabin. So its movement is very suitable for hand gestures interactive control. The schematic diagram that the astronaut uses hand gestures to control the AAR-2 is shown as Fig. 4.

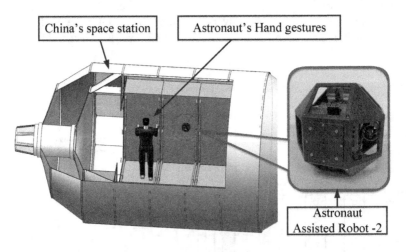

**Fig. 4.**  The schematic diagram that the astronaut uses hand gestures to control the AAR-2.

Using the FT-200 miniature simulation air-floating platform to simulate the space micro-gravity environment. The AAR robot can be mounted on it and can move on a marble platform with three degrees of freedom motion (translate along X axis and Y axis and rotate around z axis). The Kinect is selected as the image capture device to capture astronaut's hand gestures information. In terms of software, at the astronaut's hand gestures images recognition part, choose ROS (Robot Operating System) as the operating system; at the Deep Learning part, choose Caffe (Convolutional Architecture for Fast Feature Embedding) as the framework for deep learning models; at robot system part, the STM32 processor is selected as the control unit for controlling robot drivers and related motion control algorithm. The whole SHRI system experiment platform is illustrated in Fig. 5.

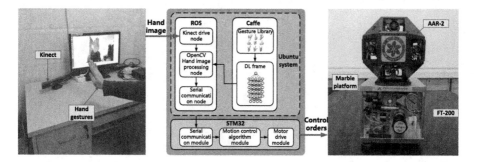

**Fig. 5.** SHRI system experiment platform.

## 3.2 Hand Gestures for SHRI

There are two types of hand gestures in SHRI: one is command-type hand gestures. It means that astronauts communicate certain deterministic semantic information to space robots through changing the morphology or spatial orientation of their hands. The other is control-type hand gestures. It means that astronauts transfer quantitative parameters to space robots by moving their hands.

Based on the above, a set of hands-coordination interactive hand gestures which are called "left hand for instruction and right hand for operation" is designed based on the ASL hand gestures [17]. The specific hand gestures semantic correspondence is shown as Table 1: astronaut can use the left hand to set up 8 semantics commands those are "Begin to control", "Stop control", "Finish control", "Path tracking", "Linear motion", "Rotational Motion", "Object approximation" and "Data transmission". When the left hand is recognized as the "Begin to control" gesture command, the robot performs a response to start control. After starting the control, when the left hand is recognized as the "Path tracking", the space robot will follow the track of right hand; when the left hand gesture is recognized as the "Linear motion", the space robot will make translate with right hand; and when the left hand gesture is recognized as the "Rotational Motion", the space robot will rotate with the right hand. when the left hand gesture is recognized as the "Object approximation", the space robot will move to the target; And when the left hand gesture is recognized as the "Data transmission", the robot will send position and attitude data and sensor test data to the host computer; Each time a hand gesture is completed, the left hand can be transformed into a "Finish control" gesture, then it will switch to a next gesture command. When the left hand is recognized as the "Stop control", interactive control between astronaut and space robot will end, and the gesture commands are no longer valid unless the next "Begin to control" hand gesture is detected.

**Table 1.** Chart of astronaut-space robot hands-coordination hand gestures.

| Hand gesture semantics | ASL letters | Left hand gestures | Right hand gestures |
|---|---|---|---|
| Begin to control | $B$ | | - |
| Stop control | $S$ | | - |
| Finish control | $F$ | | - |
| Path tracking | $P$ | | |
| Linear motion | $L$ | | |
| Rotational motion | $R$ | | |
| Object approximation | $O$ | | - |
| Data transmission | $D$ | | - |

As can be seen from Table 1, recognizing the left hand instructions requires the recognition of hand gesture semantics [18]. While recognizing the right hand operation requires tracking the hand in six degrees of freedom. And hand detection and location are crucial in both hand gestures semantic recognition and hand tracking. Therefore, detection and location of astronaut's hands are researched deeply in the following content.

## 4    Experiments

### 4.1    Hand Gestures Database for SHRI

The Hand gestures database for SHRI needs to meet the following requirements:

(1) Have the 8 hand gestures in the chart of astronaut-space robot hands-coordination hand gestures;
(2) Have images contain hands of different sizes.

Since the known hand gestures databases do not meet all the above requirements, a set of Space Robot Simple Sign Language (SRSSL) database was made by ourselves. This hand gestures database includes six different human hand gestures RGB images. Each of them contains the eight types of hand gestures in the chart of astronaut-space robot hands-coordination hand gestures, and includes five different sized hand gestures. Each person has 1000 hand gestures images, so the database contains a total of 6,000 hand gestures images. Select five people's hand gestures images (5000 images) as the training data, and another person's hand gestures images (1000 images) as the test data. A part of images ("Begin to control" hand gestures images) of SRSSL database are shown in Fig. 6. Where *PN* denotes the person number, *XS*, *S*, *M*, *L*, *XL* denote five different sizes hand targets. For example, *XS* denotes the minimum hand target and *XL* denotes the maximum hand target.

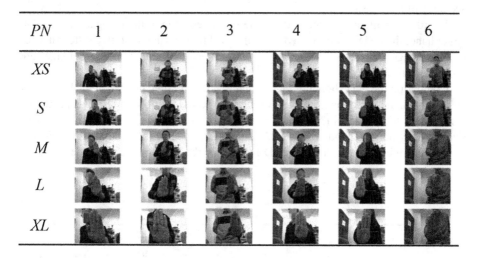

**Fig. 6.**  SRSSL database.

Use the above SRSSL database to evaluate the improved SSD model. The hand detection and location experiment will be conducted and the experiment results will be compared with the results of the SSD model.

### 4.2    Hand Detection and Location Experiment for Small Target Hands

The hardware equipment of these experiments is shown as follows: Intel Core I5-6400 CPU, NVIDIA GeForce GTX 1060 6 GB GDDR5, 16 GB ROM. And the experiments were conducted in Caffe environment under Ubuntu 14.04 64 bit OS system.

This article is primarily to solve the problem that the SSD model can't detect small targets precisely. In order to validate the improved SSD model proposed in this paper, the SSD model and the improved SSD model are trained and tested on the above SRSSL database respectively. The types of recognition include the 8 kinds of SHRI hand gestures.

Table 2 shows the average accuracies of the SSD method and the improved SSD method and the accuracies of each type of hand gesture.

**Table 2.** The comparison of detection results on SRSSL database.

| Methods | B | S | F | P | L | R | O | D | mAP |
|---|---|---|---|---|---|---|---|---|---|
| SSD | 79.1 | 83.5 | 75.6 | 70.4 | 86.7 | 85.8 | 86.8 | 86.9 | 81.8 |
| Improved SSD | 86.9 | 89.5 | 83.5 | 78.1 | 89.1 | 89.4 | 90.3 | 90.8 | 87.2 |

From Table 2, it can be seen that the mAP of improved SSD proposed in this paper is 87.2%. Compared with the SSD method, the accuracy improves 5.4%. What's more, the detection accuracy of each hand gesture of improved SSD is higher than SSD. Therefore, the improved SSD in this paper can contribute to improving the detection accuracy of hand gestures.

In order to validate the accuracy of the improved SSD to the recognition of small target hand, the SSD and improved SSD are used to experiment on five different sizes of hand images respectively. The detection accuracies of hands in different sizes are shown as follows.

Comparing the data shown in Figs. 7 and 8, it can be seen that when the hand sizes are *XS* and *S*, the detection accuracies of the SSD model are 46% and 83% respectively. While the detection accuracies of the improved SSD model are 63% and 92% respectively. And when the hand sizes are *M*, *L* and *XL*, the detection accuracy of improved SSD model are similar to that of SSD model. So the improved SSD model makes a

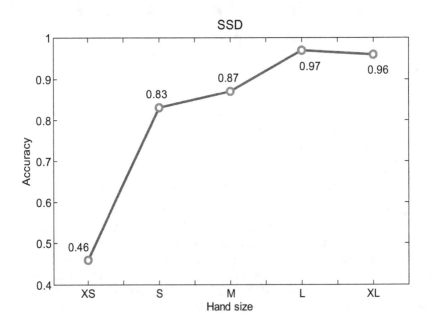

**Fig. 7.** The hand detection accuracies of SSD.

significant improvement on the precision of hand detection of small targets compare with the SSD model. And the detection accuracy of large target hand is also high.

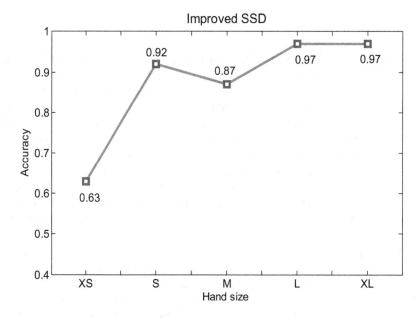

**Fig. 8.** The hand detection accuracies of improved SSD.

For a same RGB image with a small target hand, the detection and location effect pictures of SSD and improved SSD are shown in Fig. 9.

**Fig. 9.** The left picture is the detection and location effect picture of SSD, and the right picture is the detection and location effect picture of improved SSD. (Color figure online)

As can be observed in Fig. 9, the red boxes in the pictures are the detected gestures of networks. The blue words are the credibility of hands. Although both models can detect the hand successfully when it is displayed as a small target, but the improved SSD model predicts a higher credibility (66.4082%) than SSD model (50.1292%). Therefore.

In a word, the experiments fully demonstrate the effectiveness of the improved SSD model for small target hand detection and location.

## 5   Conclusion

In this paper, we mainly dealt with hand gestures interaction tasks in SHRI and researched on hands detection and location. Where the innovations of this paper includes: (a) Aiming at the problem that SSD model does not work well for small target detection, an improved SSD model was proposed. It fused the features of Conv4_3 layer and Conv6_2 layer to improve the detection of small target hands; (b) In the SHRI system, a set of hands-coordination interactive hand gestures which is called "left-handed instruction and right-handed operation" was designed; (c) Homemade a set of SRSSL database for experiments of hands detection and location.

The future work mainly includes: (a) Distinguishing and locating the right hand and the right hand and ensuring the real-time performance of the hand detection and location; (b) For the "right-handed operation", the six-DOF tracking of hand will be researched.

**Acknowledgments.** The authors would like to acknowledge the support from the Research Fund of China Manned Space Engineering (050102), the Key Research Program of the Chinese Academy of Sciences (Y4A3210301), the Natural Science Foundation of China under Grant No. 51775541, 51575412, 51575338 and 51575407, the EU Seventh Framework Programme (FP7)-ICT under Grant No. 611391, and the Research Project of State Key Lab of Digital Manufacturing Equipment & Technology of China under Grant No. DMETKF2017003.

## References

1. Fong, T., Nourbakhsh, I.: Interaction challenges in human-robot space exploration. Interactions **12**(2), 42–45 (2005)
2. http://www.pingwest.com/leap-motion-meets-nasa/. Accessed 31 Jan 2018
3. Wolf, M.T., Assad, C., Vernacchia, M.T., et al.: Gesture-based robot control with variable autonomy from the JPL BioSleeve. In: 2013 IEEE International Conference on Robotics and Automation (ICRA), pp. 1160–1165. IEEE, Karlsruhe, Germany (2013)
4. Liu, J.G., Luo, Y.F., Ju, Z.J.: An interactive astronaut-robot system with gesture control. Comput. Intell. Neurosci. **2016**, 11 (2016)
5. Grzejszczak, T., Kawulok, M., Galuszka, A.: Hand landmarks detection and localization in color images. Multimed. Tools Appl. **75**(23), 16363–16387 (2016)
6. Grzejszczak, T., Łegowski, A., Niezabitowski, M.: Application of hand detection algorithm in robot control. In: 17th International Carpathian Control Conference (ICCC). IEEE, Tatranska Lomnica, Slovakia (2016)
7. Raheja, J.L., Chaudhary, A., Maheshwari, S.: Hand gesture pointing location detection. Int. J. Light Electron Opt. **125**(3), 993–996 (2014)
8. Tompson, J., Stein, M., Lecun, Y., Perlin, K.: Real-time continuous pose recovery of human hands using convolutional networks. ACM Trans. Graph. **33**(5), 169 (2014)
9. Ge, L., Liang, H., Yuan, J., Thalmann, D.: Robust 3D hand pose estimation in single depth images: from single-view CNN to multi-view CNNs. In: 2016 IEEE Conference on Computer Vision and Pattern Recognition (CVPR), pp. 3593–3601. IEEE, Seattle, WA (2016)

10. Yamashita, T., Watasue, T.: Hand posture recognition based on bottom-up structured deep convolutional neural network with curriculum learning. In: 2014 IEEE International Conference on Image Processing (ICIP), pp. 853–857. IEEE, Paris, France (2014)
11. Molchanov, P., Gupta, S., Kim, K.: Hand Gesture Recognition with 3D Convolutional Neural Networks. In: CVPR 2015. IEEE, Boston, America (2015)
12. Liu, W., Anguelov, D., Erhan, D., Szegedy, C.: SSD: single shot multibox detector. In: 14th European Conference on Computer Vision (ECCV), pp. 21–37. IEEE, Amsterdam, The Netherlands (2016)
13. Fu, C.Y., Liu, W., Ranga, A., Tyagi, A., Berg, A.C.: DSSD: Deconvolutional Single Shot Detector. Computer Vision and Pattern Recognition. arXiv:1701.06659 (2017)
14. Cao, G.M., Xie, X.M., Yang, W.Z., et al.: Feature-Fused SSD: Fast Detection for Small Objects. Computer Vision and Pattern Recognition. arXiv:1709.05054 (2017)
15. Liu, J.G., Gao, Q., Liu, Z.W., Li, Y.M.: Attitude control for astronaut assisted robot in the space station. Int. J. Control Autom. Syst. **14**(4), 1082–1095 (2016)
16. Gao, Q., Liu, J.G., Tian, T.T., Li, Y.M.: Free-flying dynamics and control of an astronaut assistant robot based on fuzzy sliding mode algorithm. Acta Astronaut. **138**, 462–474 (2017)
17. Gattupalli, S., Ghaderi, A., Athitsos, V.: Evaluation of deep learning based pose estimation for sign language recognition. In: 9th ACM International Conference on PErvasive Technologies Related to Assistive Environments. IEEE, Greece (2016)
18. Gao, Q., Liu, J., Ju, Z., Li, Y., Zhang, T., Zhang, L.: Static hand gesture recognition with parallel CNNs for space human-robot interaction. In: Huang, Y., Wu, H., Liu, H., Yin, Z. (eds.) ICIRA 2017, Part I. LNCS (LNAI), vol. 10462, pp. 462–473. Springer, Cham (2017). https://doi.org/10.1007/978-3-319-65289-4_44

# An Intuitive Robot Learning
# from Human Demonstration

Uchenna Emeoha Ogenyi[1], Gongyue Zhang[1], Chenguang Yang[2],
Zhaojie Ju[1(✉)], and Honghai Liu[1]

[1] School of Computing, University of Portsmouth, Portsmouth, UK
{uchenna.ogenyi,gongyue.zhang,zhaojie.ju,honghai.liu}@port.ac.uk
[2] College of Engineering, Swansea University, Swansea, UK
chenguang.yang@swansea.ac.uk

**Abstract.** This paper presents a new way to teach a robot certain motions remotely from human demonstrator. The human and robot interface is built using a Kinect sensor which is connected directly to a remote computer that runs on processing software. The Cartesian coordinates are extracted, converted into joint angles and sent to the workstation for the control of the Sawyer robot. Kinesthetic teaching was used to correct the reproduced demonstrations while only valid resolved joint angles are recorded to ensure consistence in the sent data. The recorded dataset is encoded using GMM while GMR was employed to extract and reproduce generalised trajectory with respect to the associated time-step. To evaluate the proposed approach, an experiment for a robot to follow a human arm motion was performed. This proposed approach could help non-expert users to teach a robot how to perform assembling task in more human like ways.

**Keywords:** Programming by demonstration
Learning from demonstration · Learning by imitation

## 1 Introduction

Since the introduction of collaborative robots, humans can currently work in the same workspace with robots and the application of robot in human endeavour has continued to increase. However, the applicability of robot will be much diversified if it is much easier to teach a robot how to perform a task. The most viable technique that is currently used to teach a robot how to perform a task is Programming by Demonstration (PbD) which is also known as Learning from Demonstration (LfD) or Learning by Imitation. This technique enables a robot to learn from human teacher or other robots. The goal is to let users program a robot by simply giving it instructions on what to do in the form of demonstrations or by using other interactive means [5,10]. This method, of course reduces the cost of employing professionals to program the robot as it consequently allows the robot to learn from the users regardless of the users knowledge of robotics or robot programming.

© Springer Nature Switzerland AG 2018
Z. Chen et al. (Eds.): ICIRA 2018, LNAI 10984, pp. 176–185, 2018.
https://doi.org/10.1007/978-3-319-97586-3_16

In robotics, many researchers have employed various PbD techniques in order to transfer human skills to a robot. This includes the use of optical marker to track the motion of the human demonstrator [4]. This process involves fixing a tracking sensor on the human body to measure the skeletal movement of the part of the body of interest. However, the equipment causes inconvenience to the human user and sometimes results to data drift. Similar but more convenient technology was employed in [12] to control a robot via teleoperation. In the paper, the authors employed the use of a vision sensor (Kinect) to capture human joint information and then directly mapped it to the robot manipulator in order to control the robot in real time. By using this approach, human is free from the inconveniences that arise in the former. Another interesting approach is kinesthetic teaching which involves guiding the robot by the operator by moving the robot arm through the task to be implemented [6]. Kinesthetic teaching holds some advantages over other aforementioned approaches. Firstly, Kinesthetic approach does not resolve to singularity and external hardware devices that may cause inconveniences to human demonstrator are not required. Secondly, data recording more complete and accurate as a human physically guides the robot through the desired path or task. Some other approaches have been used and some have involved combining two approaches in other to utilize the advantages that both could offer. In [7], both kinesthetic teaching and the use of motion sensor were employed in the proposed human to robot gesture skill transfer.

In this paper we proposed a method that employed similar approach with [7] but with a different strategy. Unlike in [7] were both Kinesthetic teaching and motion sensor that could cause data drift and inconvenience to users were employed, in this paper, both kinesthetic teaching and vision based sensor which does not require users to wear any additional interface and permits for better orientation of the human arm were employed. During the demonstration of task, the Kinect sensor allows the user to record the human arm movement which is mapped directly to the robot controller to allow the robot reproduce the demonstrated gesture. On the other hand, Kinesthetic teaching is then applied to help the robot arm move correctly to the desired posture. During the process, valid trajectories from the demonstrations were extracted and thereafter Gaussian Mixture Model (GMM) and Gaussian Mixture Regression (GMR) were employed to encode and retrieve the generalized version of the trajectories respectively.

## 2   System Description

The framework comprises of two computer systems where one (remote) is connected directly to a Kinect sensor and the second (workstation) is connected to the Sawyer robot. Teaching the robot starts when the demonstrator stands in front of the Kinect sensor and maintains a certain distance and line of sight with the Kinect sensor. The human demonstrator moves his right arm to demonstrate sequences of steps to follow for the workings of the task to be taught. Then the Kinect sensor tracks the motion of the human arm and allows for human joint positions to be extracted. Employing vector approach [14], joint angles of the robot arm was calculated and passed to the workstation connected to Sawyer robot via User Datagram Protocol (UDP). By this means, human joint position is mapped to the robot arm and the robot arm is made to move by mimicking the human arm.

## 2.1 Hardware Setup

The platform for this research is centered on Sawyer robot. Sawyer is a single-arm robot with a total of eight degrees of freedom (8DOF) consisting of joint0 to joint6 (j0-j6) on the arm and one more joint on the head (see Fig. 1). Apart from j6 which has 540-degree of rotation that enables smooth positioning of the end-effector in different angles and also increase smooth maneuvering around obstacles; the rest of the joints have 350-degree of rotation. Sawyer is embedded with several sensors and actuators that are essential for research. Sawyer is provided with a motor encoder for every DOF, except for the head actuator. This makes it possible for joint angles to be extracted from each joint. Also embedded are force/torque sensor on each of its joints which accounts for collision detection and a Cognex camera attached to the wrist close to the end-effector which could be used to ensure a better view of the objected it interacts with. The Robotiq [1], recorded that the high precision of Sawyer is because the harmonic drive motors comes with zero backlash gear boxes.

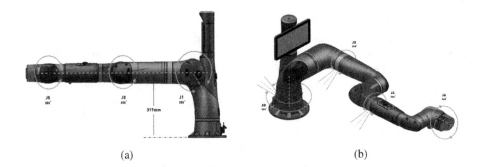

(a)                                                    (b)

**Fig. 1.** The figures shows Sawyer robot joints and link lenghts: (a) Show the Pitch joints while (b) Shows all the joints and their corresponding link lengths

The Kinect sensor can provide both RGB and depth information of a human teacher. The teacher performs demonstrations which is teleoperated to the robot via a Kinect sensor. During this process, human arm joint positions are mapped to the robot arm and the joint angle trajectories are recorded while robot arm is being moved by mimicking human motion.

## 2.2 Software Components

**Processing Software:** Processing software is an open source software that was built for teaching fundamentals of programming language. Processing comes with very simple syntax and a lot of graphical visual feedback to make programming easier for non-programmers. Processing is compatible with Windows and Linux and comes with a lot of libraries and functions that enables integration with various devices such as Kinect sensor.

**Robot Operating System and ROSPY:** Sawyer robot runs on Robot Operating System (ROS) platform. ROS has a lot of software libraries and tools that allows for development of robotic projects. Various research-oriented robots including Sawyer use this platform for the development of various kinds of robotic projects. With rospy, programmer can rapidly interface python with the ROS tools and parameters. Rospy pays a great interest on the developers' time over runtime which in turn creates enabling environment for speedy prototyping and testing of algorithm within ROS [2].

**Transmission Protocol:** The traditional Transmission Control Protocol (TCP) ensures no packet lost during data transmission but does this at the merit of slower transmission speed [11]. User Datagram Protocol (UDP) became a better candidate since fast speed of data transmission is of paramount importance in this system. In this paper, the UDP was used to transmit data from the remote computer to the workstation which is connected directly to the sawyer robot.

## 3  Probabilistic Model Selection

Several statistical models including Gaussian Mixture Model (GMM) have been used to successfully model human motion and retrieve generalized version of the motion in the past [3,13]. However, GMM is robust to noisy data, it is computationally inexpensive and can capture correlations between continuous features. All these characteristics make GMM a powerful tool for robotic data analysis. Therefore, the encoding, generalization and reproduction of the demonstrations are performed using GMM/GMR.

### 3.1  Gaussian Mixture Model

Considering a dataset of N data points with a dimensionality of D collected when a human demonstrator performed a task. Each sensory data gathered during the demonstrations contain joint angles and corresponding time-step. Therefore, the joint angle trajectories collected are $\chi = \{\chi\}_{j=1}^{N}$. Each datapoint consist of temporal and spatial variables which are donated by as t and s as in $\chi = \{\chi_t, \chi_s\}_{j=1}^{N}$ and has $\chi_t$ and $\chi_s$, for the temporal and spatial vectors respectively.

The dataset was used to encode the GMM of K components as defined by the probability density function:

$$p(\chi_j) = \sum_{k=1}^{K} p(k)p(\chi_j|k) \tag{1}$$

where $p(k)$ is the prior and $p(\chi_j|k)$ is a functional probability density function. For a mixture of K Gaussian distributions of dimensionality D, the parameters in Eq. (1) are defined as:

$$p(k) = \pi_k$$
$$p(\chi_j|k) = N(\chi_j; \mu_k, \Sigma_k) =$$

$$\frac{1}{\sqrt{(2\pi)^D|\Sigma_k|}} \ell^{\frac{((\chi_j - \mu_k)^T \Sigma_k^{-1}(\chi_j - \mu_k))}{2}} \tag{2}$$

where $N(\chi_j; \mu_k, \Sigma_k)$ represents the probability of a datapoint $\chi$ with respect to the normal distribution $N(\mu, \Sigma)$ and the Gaussian distributions defined by the prior probability $\pi_k$, mean vectors $\mu_k$ and covariance matrices $\Sigma_k$.

## 3.2   Gaussian Mixture Regression

After encoding the trajectories, GMR was employed to retrieve a smooth generalized version of the trajectories. Using GMR retrieval process is more advantageous over other stochastic methods as it can provide a more reliable way of reconstructing the Gaussian model [8]. The observed data $\chi = \{\chi_t, \chi_s\}$ is first modelled by the joint probability distribution P $\{\chi_t, \chi_s\}$. A generalized trajectory is then computed by estimating E $[p(\chi_s|\chi_t)]$ and cov $[p(\chi_s|\chi_t)]$ which is used to extract the constraints of the performed task. Just like in a regression problem, where a set of input variable X $\in \mathbb{R}^p$ and response variable Y $\in \mathbb{R}^q$ are given, where p and q are the dimensionality of the model input and output respectively. The aim of the regression, is to estimate the conditional expectation of Y given X from a set of observations $\{X, Y\}$. In the case of this paper, the regression aims at estimating the conditional probability of $\chi_s$ given $\chi_t$ where $\chi_s$ is a vector of positions at a given time $\chi_t$. Therefore, by computing the conditional expectation of $\chi_s$ at each time step, the generalized trajectories are obtained.

$$\mu_k = \{\mu_{k,k}, \mu_{s,k}\}, \Sigma_k = \begin{pmatrix} \Sigma_{tt,k} & \Sigma_{ts,k} \\ \Sigma_{st,k} & \Sigma_{ss,k} \end{pmatrix}$$

For each component of $k$, the expected distribution of $\chi_s, k$ given temporal value $\chi_t$ is

$$p(\chi_{s,k}|\chi_{t,k}) = N(\chi_{s,k}; \hat{\chi}_{s,k}, \hat{\Sigma}_{ss,k})$$
$$\hat{\chi}_{s,k} = \mu_{s,k} - \Sigma_{st,k}(\Sigma_{tt,k})^{-1}(\chi_t - \mu_{t,k})$$
$$\hat{\Sigma}_{ss,k} = \Sigma_{ss,k} - \Sigma_{st,k}(\Sigma_{tt,k})^{-1}\Sigma_{ts,k},$$

The mixture of $K$ component distributions $N(\hat{\chi}_{s,k}, \hat{\Sigma}_{ss,k})$ is done according to the prior $\beta_k$ , where $\beta_k = p(k|\chi_t)$ is determined by the probability of the component $k$.

$$p(\chi_s|\chi_t) = \sum_{k=1}^{K} \beta_k N(\chi_s; \hat{\chi}_{s,k}, \hat{\Sigma}_{ss,k}) \qquad (3)$$

$$\beta_k = \frac{p(k)p(\chi_t|k)}{\sum_{i=1}^{K} p(i)p(\chi_t|i)} = \frac{\pi_k N(\chi_t; \chi_{t,k}, \Sigma_{tt,k})}{\sum_{i=1}^{K} \pi_k N(\chi_t; \chi_{t,i}, \Sigma_{tt,i})}$$

Using Eq. (3) an estimation of the conditional expectation $\chi_s$ given $\chi_t$ is computed for a mixture of K components as show in Eq. (4). A generalized form of the trajectory and associated covariance matrix as used to reproduce the movement by evaluating the $(\hat{\chi}_s, \hat{\Sigma}_{ss})$ at different time steps $\chi_t$.

$$\hat{\chi}_s = \sum_{k=1}^{K} \beta_k \chi_{s,k}, \quad \hat{\Sigma}_{ss} = \sum_{k=1}^{K} \beta_k^2 \hat{\Sigma}_{ss,k}. \qquad (4)$$

### 3.3   Learning Parameter

Although, several approaches for estimating the parameters of GMM exist, however, the most popular, and the one used in this paper is the expectation-maximization (EM) algorithm [9], which iteratively optimizes the model using maximum likelihood estimates. Given that $p_{k,j}$ is defined as the posterior probability $p(k|\chi_j)$ computed using Bayes theorem $p(k|\chi_j) = \frac{p(k)p(\chi_j)}{\sum_{i=1}^{K} p(i)p(\chi_j|i)}$. Using a rough estimation by k-means segmentation, the parameter $\theta = \pi_k, \mu_k, \sum_k, E_k$ are iteratively computed until convergence.

$E - step$

$$p_{k,j}^{(r+1)} = \frac{\pi_k^{(r)} N(\varsigma_j; \mu_k^{(r)}, \Sigma_k^{(r)})}{\sum_{i=1}^{K} \pi_k^{(r)} N(\varsigma_j; \mu_i^{(r)}, \Sigma_i^{(r)})}$$

$$E_k^{(r+1)} = \sum_{j=1}^{N} p_{k,j}^{(r+1)}$$

$M - step$

$$\pi_k^{(r+1)} = \frac{E_k^{(r+1)}}{N}$$

$$\mu_k^{(r+1)} = \frac{\sum_{j=1}^{N} p_{k,j}^{(r+1)} \chi_j}{E_k^{(r+1)}}$$

$$\sum_k^{(r+1)} = \frac{\sum_{j=1}^{N} p_{k,j}^{(r+1)} (\chi_j - \mu_k^{(r+1)})(\chi_j - \mu_k^{(r+1)})^T}{E_k^{(r+1)}}$$

## 4    Experimental Results

For each given gesture, 3 demonstrations involving the use of the vision sensor and 3 demonstrations using Kinesthetic teaching are provided. Thereafter, the original dataset of 7DOFs was reduced to 2-demension but the important variance is retained. With this, the GMM parameter is estimated using the EM algorithm. From Fig. 2b, it could be observed that 3–5 GMM components could fit the encoded gestures. Using GMR a generalized trajectory among the demonstrated trajectories could be retrieved as presented in Fig. 2c.

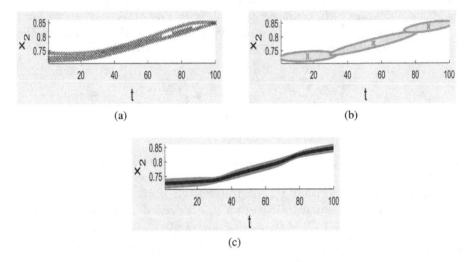

(a)                                                              (b)

(c)

**Fig. 2.** (a) A set of three sample demonstrations performed by a human. (b) The computed GMM for the demonstrated task in (a). Fig. 2c is the computed GMR model based on the computed GMM.

Reproduction of the 3 demonstrated gestures by the Sawyer robot was evaluated as shown in Fig. 3. The resulting trajectories of the robot correspond to the motion performed by the demonstrator. The trajectories of the robot represent the joint angles generated during task reproduction. Due to technical limitations, the reproduced demonstration by the robot is not exactly the posture of the demonstrator (see Fig. 3).

Another proof of consent conducted was asking some group of people to collaborate with the robot in a box stacking task. The experiment involved 10 persons with each allowed to stack 4 blocks together in not more than five trials. Five of them practiced how to perform the block stacking task while the other half were only instructed on how to perform the task and were never given a change to practice it before performing the task. It is expected that those who practiced before performing the task should finish much earlier than those who did not, but it turns out that the time difference is not as much as

**Fig. 3.** (a) The robot mimics the resulting trajectory of the human arm motion: (a) A gesture towards the right (b) A gesture towards the robot (c) A gesture pointing upward.

**Fig. 4.** The task was to stack the boxes in the order of 1–4. Figures (a), (b) and (c) are snapshots of a user performing the stacking task.

**Table 1.** Block stacking result

| Test category | Trained person | Untrained person |
|---|---|---|
| Stack completion time | 13.86 ± 2.729 | 26.3 ± 3.976 |

anticipated. The outcome of the experiment shows that the average completion time for trained persons is 13.86 min while that for untrained persons is 26.3 min as presented in Table 1. This result is an indication that with little practice, users can perfect on the use of the proposed system. Figure 4 shows the snapshot of the stacking task performed by one of the users.

## 5    Conclusion

This paper shows a new way to teach a robot how to perform a task via Programming by Demonstration. The approach proposed how to teach a robot to follow human arm motion based on GMM/GMR. The process commenced by acquiring the robot joint angles when it mimics human demonstrated gesture captured via a Kinect sensor. The acquired dataset is complemented using kinesthetic teaching aimed to correctly move the robot arm towards achieving a more accurate demonstration. Encoding of the recorded valid joint angles was performed using GMM while a smooth generalised trajectory of the demonstrated motion was achieved using GMR. The experiment shows that this dual means of data acquisition could be employed in training a robot via PbD.

**Acknowledgements.** The authors would like to acknowledge support from DREAM project of EU FP7-ICT (grant no. 611391), Research Project of State Key Lab of Digital Manufacturing Equipment & Technology, China (Grant no. DMETKF2017003), National Natural Science Foundation of China (grant no. 51575412) and the Tertiary Education Trust Fund (TETFund), Nigeria.

## References

1. COLLABORATIVE ROBOT EBOOK and COMPARATIVE CHART : 19 collaborative robots compared and analysed in this ebook. http://blog.robotiq.com/collaborative-robot-ebook. Accessed: 31 12 2016
2. Package Summary, rospy. http://wiki.ros.org/rospy. Accessed 23 03 2017
3. Bentivegna, D.C., Atkeson, C.G., Cheng, G.: Learning from observation and practice using primitives. In: AAAI 2004 Fall Symposium on Real-life Reinforcement Learning. Citeseer (2004)
4. Bestick, A.M., Burden, S.A., Willits, G., Naikal, N., Sastry, S.S., Bajcsy, R.: Personalized kinematics for human-robot collaborative manipulation. In: 2015 IEEE/RSJ International Conference on Intelligent Robots and Systems (IROS), pp. 1037–1044. IEEE (2015)
5. Billard, A., Calinon, S., Dillmann, R., Schaal, S.: Robot programming by demonstration. In: Siciliano, B., Khatib, O. (eds.) Springer handbook of robotics, pp. 1371–1394. Springer, Heidelberg (2008). https://doi.org/10.1007/978-3-540-30301-5_60
6. Calinon, S., Billard, A.: Active teaching in robot programming by demonstration. In: The 16th IEEE International Symposium on Robot and Human interactive Communication 2007, RO-MAN 2007, pp. 702–707. IEEE (2007)
7. Calinon, S., Billard, A.: Incremental learning of gestures by imitation in a humanoid robot. In: Proceedings of the ACM/IEEE International Conference on Human-robot Interaction, pp. 255–262. ACM (2007)
8. Calinon, S., Guenter, F., Billard, A.: On learning, representing, and generalizing a task in a humanoid robot. IEEE Trans. Syst. Man Cybern. Part B (Cybern.) **37**(2), 286–298 (2007)
9. Dempster, A.P., Laird, N.M., Rubin, D.B.: Maximum likelihood from incomplete data via the EM algorithm. J. Royal Stat. Soc. Series B (Methodol.) **39**, 1–38 (1977)

10. Ge, J.G.: Programming by demonstration by optical tracking system for dual arm robot. In: 2013 44th International Symposium on Robotics (ISR), pp. 1–7. IEEE (2013)

11. Masirap, M., Amaran, M.H., Yussoff, Y.M., Ab Rahman, R., Hashim, H.: Evaluation of reliable UDP-based transport protocols for internet of things (IOT). In: 2016 IEEE Symposium on Computer Applications & Industrial Electronics (ISCAIE), pp. 200–205. IEEE (2016)

12. Reddivari, H., Yang, C., Ju, Z., Liang, P., Li, Z., Xu, B.: Teleoperation control of Baxter robot using body motion tracking. In: 2014 International Conference on Multisensor Fusion and Information Integration for Intelligent Systems (MFI), pp. 1–6. IEEE (2014)

13. Saunders, J., Nehaniv, C.L., Dautenhahn, K.: Teaching robots by moulding behavior and scaffolding the environment. In: Proceedings of the 1st ACM SIGCHI/SIGART Conference on Human-Robot Interaction, pp. 118–125. ACM (2006)

14. Shiffman, D.: A Roadmap for U.S. Robotics vector is a collection of values that describe relative position in space. https://processing.org/tutorials/pvector/. Accessed 20 03 2018

# Deep Reinforcement Learning Based Collision Avoidance Algorithm for Differential Drive Robot

Xinglong Lu, Yiwen Cao, Zhonghua Zhao$^{(\boxtimes)}$, and Yilin Yan

Department of Instrument Science and Engineering, Shanghai Jiao Tong University,
Shanghai 200240, People's Republic of China
zhaozh@sjtu.edu.cn

**Abstract.** In this paper, collision avoidance problem is investigated for differential drive robot running in pedestrian environment, which requires for natural and safe interaction between robot and human. Based on deep reinforcement learning, a human-aware collision avoidance algorithm is proposed to find a smooth and collision-free path. A well designed reward function ensures the robot navigates without collision and obeys right-pass norm simultaneously. The slow convergence problem during training is addressed by pre-training the neural network using supervised learning. The simulation results show that the proposed algorithm can find a feasible and norm-obeyed path which achieves a natural human-robot interaction compared with traditional method.

**Keywords:** Deep reinforcement learning · Collision avoidance
Differential drive robot

## 1 Introduction

Collision avoidance is a crucial part of autonomous navigation under pedestrian environment for service robot, such as companion robot [1], guide robot [2] and intelligent wheelchair [3]. These robots usually use two-wheel differential drive as a chassis because of its simple structure and omnidirectional movement characteristics. The traditional collision avoidance algorithm that only seeks for finding a collision-free and time efficient path. However, collision avoidance between robot and pedestrian should also consider how to make robot's behavior acceptable, which means that the generated trajectory should be as smooth as possible with no sudden start/stop or sharp turn. Unfortunately, planning a collision-free, time efficient and smooth path under pedestrian environment is still challenging, because of the uncertain behavior of pedestrian. This task requires for precise perception of the human behavior and estimation of the pedestrian's intent (goal).

The common method to solve this problem is trying to model the control strategy or the motion of pedestrian, which can be considered as model-based

© Springer Nature Switzerland AG 2018
Z. Chen et al. (Eds.): ICIRA 2018, LNAI 10984, pp. 186–198, 2018.
https://doi.org/10.1007/978-3-319-97586-3_17

method. A Multi-Policy Decision Making (MPDM) framework [4] is proposed to make robot navigate amongst pedestrians by dynamically switching between *Go-Solo*, *Follow-other*, and *Stop* three policies. Social force model (SFM) [5] is designed for social force analysis when robot interacts with pedestrians and SFM has been implemented by many robot navigation applications [2,6]. But model-based method is difficult to generalize to unseen scenarios, because tuning a set of parameters that fit all environment is almost impossible. In addition, this method usually causes unnatural behavior, such as sudden start/stop and sharp turn.

Learning-based method is proposed to avoid time-consuming parameters tuning by off-line learning and on-line execution. An end-to-end framework is designed to map 2D-Lidar output to control command for multi-agent collision avoidance problem in [7]. Reinforcement learning is used to learn the optimal parameters of SFM [2]. To achieve autonomous navigating, the work in [8] teaches robot to imitate the behavior of pedestrians with inverse reinforcement learning. With the development of deep neural network (DNN), reinforcement learning uses DNN to approximate the value function or action-value function, which has been successfully implemented on playing Atari computer games [9] and controlling quadrotor [10]. The reinforcement learning method utilizes reward function to teach agent the rules that it should obey, rather than designing complicated control strategy. If adequate training data provided or exploring enough state space, deep reinforcement learning could obtain a optimal policy that maps current state to the best action.

In this work, we focus on collision avoidance when differential drive robot navigates under pedestrian environment and learn a policy that map the current state to best action using deep reinforcement leaning. The remainder of this paper is organized as follows. Section 2 introduces the detail of the problem formulation. The approach to train and test the deep reinforcement learning model is described in Sect. 3. Section 4 shows the simulation results and analysis. Finally, the conclusion of our work is delivered in Sect. 5.

## 2   Problem Formulation

In this section, the dynamics of differential drive robot and pedestrian is modeled at first, and then collision avoidance problem is represented in the format of Markov decision process (MDP). At last, deep reinforcement learning is formulated to solve the MDP problem.

### 2.1   Agent Modeling

Differential drive robot and pedestrian are considered as two types agent in this work. To get state space and action space in reinforcement learning, the dynamics of robot and pedestrian should be modeled first. As show in Fig. 1, the two-wheel geometry and gray circle represent differential drive robot and pedestrian respectively. The star gives the goal position for robot.

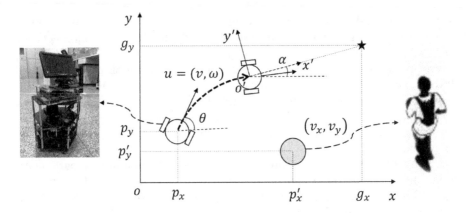

**Fig. 1.** Differential drive robot and pedestrian modeling

**Differential Drive Robot:** The state of differential drive robot contains position $(p_x, p_y)$, heading angle $\theta$, linear velocity $v$, angular velocity $\omega$, radius of robot's projection circle $r$. The goal position $(g_x, g_y)$ is also necessary in navigation task. To be compatible with ROS navigation package [13], velocity tuple $(v, \omega)$ is used as control command $u$. Suppose the differential drive robot runs from point $A(p_{x_t}, p_{y_t})$ to point $B(p_{x_{t+1}}, p_{y_{t+1}})$ with the command $u = (v_t, \omega_t), \omega_t \neq 0$ in $\Delta t$. The motion function is

$$p_{x_{t+1}} = p_{x_t} - \frac{v_t}{\omega_t} sin(\theta_t) + \frac{v_t}{\omega_t} sin(\theta_t + \omega_t \Delta t) \tag{1}$$

$$p_{y_{t+1}} = p_{y_t} + \frac{v_t}{\omega_t} cos(\theta_t) - \frac{v_t}{\omega_t} cos(\theta_t + \omega_t \Delta t) \tag{2}$$

$$\theta_t = \theta_t + \omega_t \Delta t \tag{3}$$

**Pedestrian:** Actually the state of pedestrian should be the same as robot, but the goal of pedestrian is not observable. Because it is hard to measure the heading angle of pedestrian, angular velocity can not be computed directly. For similarity, pedestrian's state includes position $(p'_x, p'_y)$ and velocity $(v_x, v_y)$ and its motion function is

$$p'_{x_{t+1}} = p'_{x_t} + v_x \Delta t \tag{4}$$

$$p'_{y_{t+1}} = p'_{y_t} + v_y \Delta t. \tag{5}$$

### 2.2   Collision Avoidance and MDP Representation

A collision avoidance problem under pedestrian environment can described as that a robot navigates from start point to goal position without collision and interacting with human safely and naturally. Generally a collision avoidance consist two part: perception module and decision module. Perception module is responsible for getting the state of robot and surrounding objects, such as static obstacle and moving obstacle. Decision module provides control command following a designed strategy based on the perception. In this work, only decision module is considered and moving obstacle is pedestrian. We are working on the assumption that only the nearest pedestrian along the path to goal is considered in multi-pedestrian environment. So the collision avoidance in pedestrian environment can be simplified to one robot interacting with one pedestrian situation.

To apply reinforcement learning to solve collision avoidance problem, a MDP representation should be abstracted. The MDP is a tuple $(S, A, R)$, where $S$ is state space, $A$ is action space, $R$ is reward function.

**State Space $S$:** The collision avoidance algorithm should only rely on the robot local coordination, which means the state need to be coordination invariant. So a robot-fixed coordinate $x'o'y'$ is used, which sets origin as robot's center point and x axis pointing to the robot's goal, as show in Fig. 1. Then the state is transformed to the robot-fixed coordination using Eqs. 6 and 7.

$$\begin{bmatrix} \tilde{p}'_x \\ \tilde{p}'_y \\ 1 \end{bmatrix} = \begin{bmatrix} cos(\alpha) & sin(\alpha) & -p_x \\ -sin(\alpha) & cos(\alpha) & -p_y \\ 0 & 0 & 1 \end{bmatrix} \begin{bmatrix} p'_x \\ p'_y \\ 1 \end{bmatrix} \tag{6}$$

$$\begin{bmatrix} \tilde{v}_x \\ \tilde{v}_y \end{bmatrix} = \begin{bmatrix} cos(\alpha) & sin(\alpha) \\ -sin(\alpha) & cos(\alpha) \end{bmatrix} \begin{bmatrix} v_x \\ v_y \end{bmatrix} \tag{7}$$

The state can be formulated as $s = [v, \omega, \theta, r, d_g, \tilde{p}'_x, \tilde{p}'_y, \tilde{v}_x, \tilde{v}_y, d_p] \in \mathbf{R}^{10}$, where $d_g$ is the distance between robot and its goal, $d_p$ is the distance between pedestrian and robot.

**Action Space $A$:** As velocity command is used to steer the service robot, action can be described as linear velocity $v$ and angular velocity $\omega$. The action space is discretized into finite tuples $(v, \omega)$.

**Reward Function $R$:** The reward function gives the reward feedback when robot executes an action from current state. It can be used to train the agent to obey some rules. Inspired by dynamic window approach (DWA) to collision avoidance algorithm [14], the reward function is set to reward reaching goal, high linear velocity and punish nearing pedestrian and collision.

* Reaching goal: reward 1;
* High linear velocity: if $|v| > \tau_v, 0 < \tau_v < v_{max}$, reward 0.5, where $\tau_v$ is the threshold of linear velocity, $v_{max}$ is the max linear velocity;
* Nearing Pedestrian: if $0 < d_p < \tau_d$, reward $-\tau_d + d_p$, where $\tau_d$ is the threshold of distance between robot and pedestrian, which can be used to set a safety margin for robot when navigates near pedestrian;
* Collision: if $d_p < 0$, reward $-0.5$.

To make the interaction between robot and human more acceptable, a right-pass norm is added into reward function. As shown in Fig. 2, the sad face and happy face means penalty and reward respectively. Specifically, the right-pass norm reward right hand passing, left hand overtaking and crossing behind. If the pedestrian does not appear in the norm-obeyed area, reward $-0.5$.

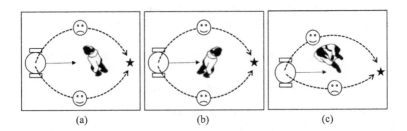

**Fig. 2.** Right-pass norm. (a) Passing; (b) Overtaking; (c) Crossing.

### 2.3   Deep Reinforcement Learning

The goal of deep reinforcement learning is to learn an optimal policy $\pi$ through exploring the environment with reward feedback. This means the agent chooses actions in a way that maximizes cumulative future reward. Deep Q-network (DQN) [9] uses a convolutional neural network (CNN) to approximate the optimal action-value function

$$Q^\star(s, a) = \max_\pi \mathbb{E}[r_t + \gamma r_{t+1} + \gamma^2 r_{t+2} + ... | s_t = s, a_t = a, \pi], \tag{8}$$

which is the sum of reward $r_t$ at time step $t$ and discounted by $\gamma$. And the reward is obtained by following the given policy $\pi$ and executing an action $a$ selected based on state $s$.

In this work, the action space is continuous and there is strong correlation between action and state. So it is hard to directly approximate action-value function. Instead, we approximate value function that denotes the expected time to reach goal based on the current state. A deep neural network (deep V-network) is used to approximate the value function, that is:

$$V^*(s) = \max_\pi \mathbb{E}[r_t + \gamma r_{t+1} + \gamma^2 r_{t+2} + ... | s_t = s, \pi], \tag{9}$$

where only state is considered. The optimal policy $\pi^*$ can be obtained by choosing the action that maximizes immediate reward $R(s, a)$ and next state's value,

$$\pi^*(s) = \max_a (R(s, a) + V^*(s'(s, a))), \tag{10}$$

where $s'(s, a)$ is the next state that can be computed by motion function of differential drive robot and pedestrian in Sect. 2.1.

The optimal value function $V^*(s; w)$ is obtained through off-policy deep reinforcement learning. $w$ is trainable parameters. We adopt two keypoints in DQN:

experience replay and soft parameter update. A dataset $D$ is setup as experience container that stores a tuple $(s_t, a_t, r_t, s'_t)$ at each time step $t$ and the state value. The value for each experience state is calculated using a target network $V(s'; w^-)$ and $w^-$ softly updated every $C$ episodes. The loss function of deep V-network is

$$L(w) = \mathbb{E}_{(s,a,r,s') \sim U(D)}[(r + \gamma \max_{a'} V(s'; w^-) - V(s; w))^2], \qquad (11)$$

which is the expectation of square differential value on samples of experience $(s, a, r, s') \sim U(D)$, uniformly selected at random from the dataset $D$ during learning.

## 3   DRL-Based Collision Avoidance

This section begins by describing the details about how to generate training dataset for pre-training the deep neural network utilizing ORCA algorithm [11]. Next, we elaborate the network architecture and training details about the deep reinforcement learning.

### 3.1   Dataset Generation

Different from the general reinforcement learning, off-policy reinforcement learning framework evaluates and updates different policies. So the training process can be accelerated using experience that produced by other collision avoidance algorithm. This work collects a sizable training data $D$ from ORCA, which is a multi-agent collision avoidance simulator. The setup for ORCA can be seen as Fig. 3, in which a two-agent system is simulated where circle with an arrow represents robot and rectangle with an arrow represents pedestrian. Three common scenarios are setup. Crossing scenario: change the angle $\alpha$ from 0 to $2\pi$. Overtaking scenario: change the vertical distance $d$. Passing scenario: change the vertical distance $d$.

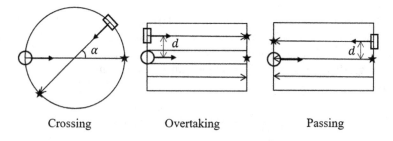

Crossing              Overtaking              Passing

**Fig. 3.** Our setup for collecting dataset from ORCA simulator.

Trajectory is generated at first, and then processed into state-value pairs $D = \{(s_k, y_k)\}_{k=0}^N$, where $y = \gamma^{t_g \cdot v_{max}}$, $t_g$ is the time to reach goal and $v_{max}$ is the maximum linear velocity for the robot. This work collects about 450 trajectories containing 10809 state-value pairs. Before feeding to the network, feature scaling is necessary to make each feature contribute equally to the state value. Otherwise, the pre-trained network make poor prediction and can not be used to explore environment efficiently.

## 3.2  Deep V-Network

The architecture of our Deep V-network is a multilayer perception with two hidden fully connected layers using ReLU activation function.

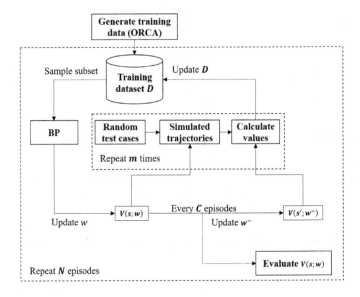

**Fig. 4.** The learning procedure of Deep V-Network.

**Pre-training:** To reduce agent randomly exploring time and accelerate the convergence of Deep V-network, the dataset $D$ is used to pre-train the network, which can be regarded as initialization procedure of deep reinforcement learning. Stochastic gradient descent (SGD) algorithm is used to minimize loss function $L_i(w_i)$ as

$$L_i(w_i) = \frac{1}{N} \sum_{k=1}^{N} (y_k - V(s_k, w_i))^2, \tag{12}$$

where $w_i$ is the parameters of Deep V-network at $i$th training iteration.

Here we make some notes about the pre-training process. First, the pre-training process essentially gives the robot a ability to explore environment more effectively, rather than explore randomly from scratch. And this process can also reduce the complexity of designing reward function, because there is no need to use continues reward to guide the robot reaching its goal. Second, the value function is no need to be optimal approximation after pre-training. We get the optimal policy through deep reinforcement learning rather than supervised regression. Third, this process is not just a copy of ORCA algorithm, which actually learns a nonlinear function to evaluate the state value that represents the expected time to reach goal.

**Deep Reinforcement Learning:** The deep V-network after pre-training can guide the robot to accomplish navigation. In this section, reinforcement learning is used to refine the network by $\epsilon$-greedy exploring environment (state space). The whole learning procedure can be seen as in Fig. 4.

In order to utilize the experience replay skill, the state generated by exploration during the training process is collected and stored into dataset $D$. To explore the environment, we first generate some random test cases consists of initial position, velocity, heading angle and goal position for robot and pedestrian. Then, $\epsilon-$greedy algorithm is iteratively used to choose current optimal action calculated using (10) or random action to guarantee explore enough space until reaching destination. To apply the soft parameter updating skill, the deep V-network $V(s; w)$ is duplicated as $V(s; w^-)$ after pre-training. $V(s; w^-)$ is used to process the exploring trajectories into state-value pairs and update the training data $D$ by replacing old data. The parameters of $V(s; w^-)$ is updated every $C$ episodes. The above procedures is repeated $m$ times and update the parameters of $V(s; w)$ by back-propagation with a batch of training data, meanwhile the episode counter increases one. The whole training process has $N$ episodes. To monitor the convergence of $V(s; w)$ network, some test cases are pre-defined and tested every $C$ episodes.

# 4 Simulation and Results

In this section, simulation setup is detailedly introduced at first. Then the simulation results and analysis is shown. Finally, we discuss the advantages and disadvantages of the deep reinforcement learning based collision avoidance algorithm.

## 4.1 Simulation Setup

A simulator is designed using OpenAI gym [12] and implemented with Python on Nvidia Jetson TX2 developer kit with dual-core NVIDIA Denver2 + quad-core ARM Cortex-A57 CPU, 256-core Pascal GPU and 8 GB of memory, which is shown in Fig. 5. The hidden layers of the deep value network consist of 100, 100 nodes respectively. Learning rate is set as 0.002 in pre training process and deep reinforcement leaning process. The training process lasts about 2 h when whole training episode $N$ set as 500. Refer to Table 1 for more parameter setting details.

(a)                    (b)

**Fig. 5.** Nvidia Jetson TX2. (a) TX2 with a monitor; (b) TX2 developer kit.

**Table 1.** Deep reinforcement learning parameters setting

| States dimension | 15 |
|---|---|
| Number of actions | 35 |
| Learning rate | 0.002 |
| Pre-training epoch | 20 |
| DRL epoch $N$ | 500 |
| Discount factor $\gamma$ | 0.9 |
| Initial $\epsilon$ | 0.5 |
| Exploring times $m$ | 10 |
| Update period $C$ | 20 |

### 4.2 Simulation Results

The following presents the simulation results under different test scenarios as shown in Fig. 6, where circle and rectangle with a line represent robot and pedestrian respectively. The stars give the positions of their goals. Simple geometric shape is used, because it is hard to draw complicated models in OpenAI gym environment.

**Swap**: In this scenario, robot and pedestrian move in opposite directions and swap their position at last, as shown in Fig. 6(a). The figure shows that the robot passes right under the right-pass norm while keeping a safe margin with pedestrian.

**Crossing**: In this scenario, robot and pedestrian move to their goals respectively and their path will interact in the middle area, which is responsible for the most collision cases, as shown in Fig. 6(b). As can be seen that the robot slow down when it knows collision will happen if it still keep current velocity and turn a little bit right to keep a distance with the pedestrian. This test can prove that our algorithm can make collision avoidance effectively.

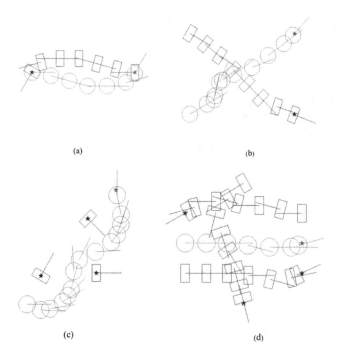

(a)                                    (b)

(c)                                    (d)

**Fig. 6.** Simulation results on four test scenarios. (a) Swap; (b) Crossing; (c) Static obstacle; (d) Multi-pedestrian.

**Static Obstacle**: In this scenario, pedestrians' current position is the same as their goals, which can be regarded as static obstacles, as shown in Fig. 6(c). As can be observed that the robot navigates around the obstacle smoothly. This test is used to demonstrate the generalization capability of our algorithm, because the deep V-network has never been trained on such scenarios.

**Multi-pedestrian**: To test robot navigates under multi-pedestrian environment, a three-pedestrians scenario is setup, in which one pedestrian passing the robot, one pedestrian running side by side to the sample direction with the robot and the last pedestrian navigating from up to down crosses with both robot and pedestrians, as shown in Fig. 6(d). The robot successfully finds a collision-free path and reaching its goal. The pedestrians are also guided by the same collision avoidance policy and reach their goals eventually.

### 4.3   Quantitative Analysis

To quantitatively evaluate our algorithm, here three performance metrics are defined:

(1) *safety margin*, the minimum distance between robot and pedestrians during traveling time. This metric denotes the collision avoidance performance, which is reflected in reward function.

(2) *right-pass obey proportion*, this can be measured by pre-defining some test cases and counting the proportion of the cases obeyed right-pass norm.
(3) *success rate*, successful reaching goal without collision. This metric can be used to estimate the collision avoidance effectiveness.

We set crossing, passing, overtaking and random test scenarios separately to estimate the above three metrics. And modified DWA like [16] is used to compare with our algorithm. The ideal safety margin should be $\tau_d$ meters, because collision avoidance reward function only cares about margin less $\tau_d$, here set as 0.3.

**Table 2.** Quantitative analysis

| Test scenarios configuration | | Safety margin (m) | | Right-pass rate | | Success rate | |
|---|---|---|---|---|---|---|---|
| Scenarios | Times | DWA | Ours | DWA | Ours | DWA | Ours |
| Crossing | 20 | 0.94 | 0.39 | NA | 15/20 | 13/20 | 18/20 |
| Passing | 20 | 1.00 | 0.25 | NA | 19/20 | 11/20 | 20/20 |
| Overtaking | 20 | 0.99 | 0.31 | NA | 14/20 | 14/20 | 19/20 |
| Random | 100 | 1.69 | 1.28 | NA | 81/100 | 87/100 | 93/100 |

As shown in Table 2, our algorithm achieved that right-pass norm obey proportion is 70% to 95% and success rate is 90% to 93% under all scenarios. While DWA has only 55% to 81% success rate and has no right-pass norm aware. Safety margin is about 0.3 m under crossing, passing and overtaking scenarios and is a little large under random test cases. The reason is randomly generated test cases are simple, requiring robot and pedestrian to travel strait towards their goals. DWA obviously has large safety margin. So compared to DWA, it can be concluded that the robot can successfully find a collision-free path that keeps a proper safety margin with pedestrians and also makes robot obey right-pass norm using our algorithm.

## 5    Conclusion and Discussion

We explored a deep reinforcement learning based collision avoidance algorithm for differential drive robot. By designing a set of reward and penalty items, the learned policy can guide a robot to navigate under pedestrian environment safely. Meanwhile, the robot is taught to obey right-pass norm that pedestrians always follow, which achieved a nature interaction between robot and human. The details of the architecture and training process of the deep V-network is presented. We demonstrate our algorithm is effective to navigate under pedestrian environment in simulation and perform better than DWA on safety margin, right-pass norm obey proportion and success rate.

Although the simulation results show the promising of learning-based collision avoidance method, some issues should be discussed to analyze the shortage of this work and guide the future work. It is hard to obtain the optimal policy. We repeat the learning procedure several times and the value function convergence to different target. The reason is that training data is not enough to represent the true state space. In the future, some exploring strategy during learning will be tried to get more broadly distributed samples like [10] does. The performance of the proposed algorithm on real robot need to be test. There are two ways to improve the deployment on real robot, considering state noise and using real robot to train. [15] uses real robot to do deep reinforcement learning and get a good result. In the future, we will try to compare the performance of the pure simulation model, model with noise estimation and real robot model.

# References

1. Ferrer, G., Garrell, A., Sanfeliu, A.: Robot companion: a social-force based approach with human awareness-navigation in crowded environments. In: 2013 IEEE/RSJ International Conference on Intelligent Robots and Systems (IROS), vol. 40, no. 6, pp. 1688–1694 (2013)
2. Dewantara, B.S.B., Miura, J.: Generation of a socially aware behavior of a guide robot using reinforcement learning. In: Electronics Symposium (2017)
3. Chao, W., Matveev, A.S., Savkin, A.V., Nguyen, T.N.: A collision avoidance strategy for safe autonomous navigation of an intelligent electric-powered wheelchair in dynamic uncertain environments with moving obstacles. In: 2013 European Control Conference, pp. 4382–4387 (2013)
4. Mehta, D., Ferrer, G., Olson, E.: Autonomous navigation in dynamic social environments using Multi-Policy Decision Making. In: IEEE/RSJ International Conference on Intelligent Robots and Systems (IROS), pp. 1190–1197 (2016)
5. Helbingp, D., Molnor, P.: Social force model for pedestrian dynamics. Phys. Rev. E **51**(5), 4282 (1995)
6. Ferrer, G., Garrell, A., Sanfeliu, A.: Social-aware robot navigation in urban environments. In: 2013 European Conference on Mobile Robots (ECMR), pp. 331–336 (2013)
7. Pinxin, L., Wenxi, L., Jia, P.: Deep-learned collision avoidance policy for distributed multiagent navigation. IEEE Robot. Autom. Lett. **2**(2), 656–663 (2017)
8. Kretzschmar, H., Spies, M., Sprunk, C., Burgard, W.: Socially compliant mobile robot navigation via inverse reinforcement learning. Int. J. Robot. Res. (2016)
9. Mnih, V., et al.: Human-level control through deep reinforcement learning. Nature **518**(7540), 529–533 (2015)
10. Hwangbo, J., Sa, I., Siegwart, R., Hutter, M.: Control of a quadrotor with reinforcement learning. IEEE Robot. Autom. Lett. **PP**(99), 1 (2017)
11. Van den Berg, J., Guy, S.J., Lin, M., Manocha, D.: Reciprocal n-body collision avoidance. In: Pradalier, C., Siegwart, R., Hirzinger, G. (eds.) Robotics Research. Springer Tracts in Advanced Robotics, vol. 70, pp. 3–19. Springer, Heidelberg (2011). https://doi.org/10.1007/978-3-642-19457-3_1
12. Brockman, G., et al.: OpenAI gym (2016)

13. ROS navigation package. http://wiki.ros.org/navigation
14. Fox, D., Burgard, W., Thrun, S.: The dynamic window approach to collision avoidance. IEEE Robot. Autom. Mag. **4**(1), 23–33 (1997)
15. Kahn, G., et al.: Uncertainty-aware reinforcement learning for collision avoidance. arXiv preprint arXiv:1702.01182 (2017)
16. Chen, Y.: Service robot navigation in large dynamic and complex indoor environments. Diss. University of Science and Technology of China (2017)

# A Novel Redundant Continuum Manipulator with Variable Geometry Trusses

Wenxiang Zhao and Wenzeng Zhang[✉]

Department of Mechanical Engineering, Tsinghua University, Beijing 100084, China
wenzeng@tsinghua.edu.cn

**Abstract.** This paper proposed a kind of variable geometry trussed (VGT) redundant continuum manipulators, which are a type of super-redundant robot. Compared to traditional robots, this robot has many advantages such as high rigidity, modularity, and flexibility. Due to these advantages, the VGT redundant continuum manipulators can be applied in remote control of space, underwater detection, and an industrial environment with obstacles. Besides, due to its deformable characteristics, it can also be used as a logging tractor to adaptively move in unstructured pipeline installations. This paper introduces a modular concept and considers the symmetry of the single-module robot during the construction process, which simplifies the kinematics calculation of a complex robot into a kinematics calculation of a single module, and makes this robot's kinematics' solutions be derived from the geometric relationships of the model explicitly. At the same time, due to the modular design, the robotic arm has a wide range of applications in the future.

**Keywords:** Manipulator · Ultra-redundant · Variable Geometry Truss Modular

## 1 Introduction

Since the first industrial robot was built in the United States in 1952 [1], the robot has developed very rapidly and has been widely used in many fields such as the automobile, the electronics, the nuclear and the service industry. Nowadays the robot configuration can generally be divided into series configurations and parallel configurations, and both of them have advantages and disadvantages.

The first to use in industrial production was a serial robot. In 1958 [2], Joseph F. Engel Berger, known as the "father of industrial robots," created the world's first robot company, Unimentation (Universal Automation), and participated in the design of the first Unimate robot. This is a 5-axis hydraulically driven robot for die casting and the control system is operated by just one computer [3]. It is mainly used for material transport between machines, which can rotate around the base and move up and down in the vertical direction. The Unimate robots are generally

© Springer Nature Switzerland AG 2018
Z. Chen et al. (Eds.): ICIRA 2018, LNAI 10984, pp. 199–211, 2018.
https://doi.org/10.1007/978-3-319-97586-3_18

considered to be the earliest industrial robots in the world. And now the series robots are also on the basis of these two robots on evolved.

In 1978, Hunt, a renowned institutional professor from Australia, proposed the Setwart platform mechanism which has six-degrees of freedom as a robotic mechanism, which is the origin of the concept of parallel robots [4]. A typical Stewart parallel mechanism consists of six independent and freely retractable struts, which are connected by a ball joint and a Hooke hinge. In this way, the upper platform and the lower platform have 6 independent motions and 6 degrees of freedom, which makes it can be moved in any direction and rotated about any axis in any direction or position.

With the development of science and technology, both of the two structures have been widely used. However, both configurations have their inevitable drawbacks.

For example, the axes of the series robots need to be controlled independently with corresponding encoders and sensors, which makes theirs control is difficult, and they are prone to have dynamic errors and cumulative errors [5]. In addition, the large working space would cause a large inertia. Although the parallel robot does not have the above problems, its motion analysis is complicated and there is a coupling effect between the drive axes. The most serious problem is that the working space of parallel robots is relatively small which makes them mostly used for a small working table and cannot handle large-sized parts [6].

At the end of the 1990s, the research of the hybrid mechanism gradually attracted people's attention. The hybrid mechanism is expected to integrate the advantages of parallel robots and series robots, which is a new development direction of robot. In 1994, Husain conducted a research on the positional dynamics of a three-branch hybrid platform mechanism. In 1999, the Shenyang Institute of Automation also successfully developed a five-axis machine tool (four-axis parallel, one-axis series). Research on hybrid machine has become a hot topic [7].

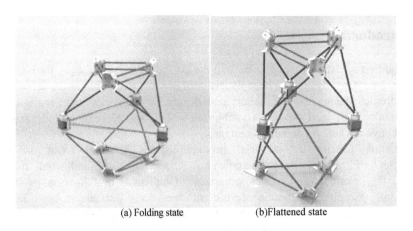

(a) Folding state                (b)Flattened state

**Fig. 1.** The different forms of VGT manipulators unit modular

Based on this background, this paper hopes to overcome the above-mentioned problems through the modular design based on the two structures. Utilizing the advantages of the truss mechanism and the axial force characteristics and the high stiffness to mass ratio, the robot can significantly increase the load capacity and positioning accuracy. VGT continuum manipulator came into being (Fig. 1).

## 2    The Principle of Design and Motion

### 2.1    The Principle of Design

The overall structure is similar to that of the Stewart platform, but the driving is to change the position and direction of the target plane by changing the side length of the triangle of the active layer to drive the movement of the mechanism. In contrast, the traditional Stewart platform does not have an efficient inverse motion solution currently. However, this design can use the length of each side of the active layer to accurately solve the position and direction of the work plane, so as to achieve accurate motion control [10].

The active layer changes the length of each side and drives the linkage movement. The passive layer realizes the series connection of each unit body, and the active layer and the passive layer are arranged alternately. When one actuator of the active layer is extended, the two triangles connected to it are changed to an obtuse triangle, and the height of the base is reduced, so that the unit body is bent toward the actuator, and the two passive planes are opposite to each other as shown in Fig. 2.

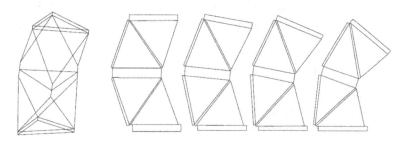

**Fig. 2.** The formation of a modular and .manipulators

The adjustment range of a unit body is relatively limited, but in the case of multiple unit bodies connected in series, not only has a large movement reachable space, but also can achieve a variety of functions such as motion path planning, as shown in Fig. 3.

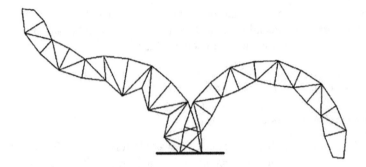

**Fig. 3.** The formation of the series manipulators

## 2.2 The Principle of Motion

Based on the above motion analysis, the original multi-degree-of-freedom spherical hinge can be simplified accordingly, taking into account that the scope of the sphere hinge is mechanically smaller as well as a certain gap, so that in the actual design, we dismantled the ball joint into a rotating hinge and a universal joint to achieve three rotational degrees of freedom. The degree of freedom is calculated that $S = 3$ which is consistent with the original design goals using the GrUbler-Kutzbach formula, where component N is taken as 14, and each rod in the active layer can be split into two components. The number of motion pairs and the degree of freedom are calculated as shown in Table 1 below [8, 9].

**Table 1.** The calculation of degree-of-freedom

| Kinetic pair | Number | Degrees of freedom |
|---|---|---|
| Revolute pair in passive layer | 6 | 1 |
| Sphere-trough pair in active layer | 6 | 3 |
| Linear pair in active layer | 3 | 2 |
| Revolute pair in active layer | 3 | 1 |

Due to the high requirements for accuracy and light weight, we finally choose the driving pattern of the motor plus screw, which achieves the purpose of reducing the weight (Fig. 4).

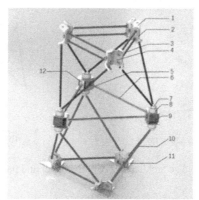

1-revolute pair; 2-mounting plate; 3-mounting rod; 4-connecting plate; 5-screw; 6-active rod; 7, 8-spherical hinge; 9-step motor; 10-connecting rod; 11-base panel; 12-active layer revolute pair

**Fig. 4.** The main assembly of VGT manipulators

# 3   Architecture of the VGT Manipulators

## 3.1   Main Assembly

According to actual demand in applications, a complete VGT redundant continuum manipulators can be composed of a different number of unit modules, so this paper only analyze one unit module. A unit module is divided into two passive layers and one active layer. The active layer consists of three stepping motors connected to three leadscrew to form a triangle. Each stepper motor uses 12 V DC to work independently. The screw rotates the motor into a telescopic motion, causing the two triangles connected to it to change. The unit body bends while the two passive layers rotate relative to each other. The three motors are independent of each other, so that the two passive planes have three relative generalized degrees of freedom.

## 3.2   The Architecture of Active Layer

The active layer consists of three drive units that form a triangle. Each drive unit includes a stepper motor, a motor bracket, a screw, a sleeve, and two universal joints. The motor bracket is mainly used to fix the motor and connect the universal joint and the sleeve connector together. The stepper motor connects to the lead screw and provides torque to rotate the lead screw. One end of the sleeve connects to the motor bracket through a connector that can rotate relative to the bracket, and the other end sleeves on the screw. The two revolute pair on each universal joint provide two degrees of freedom. The universal joint connects to the passive layer by two linkages, so that the two linkages and the passive layer form an equilateral triangle with a fixed side length (Fig. 5).

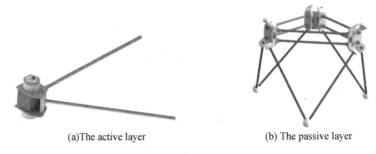

(a)The active layer                    (b) The passive layer

**Fig. 5.**   The architecture of active layer and passive layer

### 3.3   The Architecture of Passive Layer

The passive layer includes three fixed-length passive layer linkages and three connectors. The two adjacent linkages and the passive layer linkages on the two connecting members form an equilateral triangle, which can rotate about the passive layer connecting members under the action of the rotating pair on the connectors.

### 3.4   Motion Control System

In terms of motion control, we make it through communication between *Arduino* and *Matlab*. By giving the length of three rods, we can achieve the positive kinematics control. And by giving the end of the three coordinate values, we can also achieve inverse kinematics control. In addition, it also achieve the round path, straight path, polygonal path and other functions of the end coordinates. After calculation, Matlab pass predetermined three lengths of information to the Arduino to control. On the one hand, reducing the Arduino's computational burden, on the other hand, it makes it possible to implement multiple motion modes of the module.

The prime mover is stepper motors whose step angle is 1.8° with 8 mm lead screw, which make the control precision to 0.04 mm, achieving the desired accuracy. At the same time, among the modules of multiple units, we select the motor with the larger torque in the bottom module, while the motor with the small torque but light is in the top module to achieve a balanced state.

The final unit structure can achieve various movements and accurately reach any position in space, which achieve the desired goal, shown as Fig. 6.

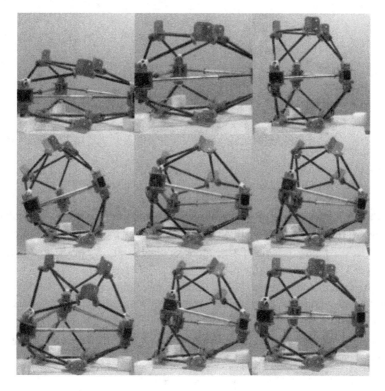

**Fig. 6.** The flexible deformation of a unit modular

# 4   Analysis of the VGT Manipulators

## 4.1   Kinematic Analysis

As shown in the Fig. 7 below, the module is regarded as a combination of two symmetrical octahedral truss structures [11]. The part in which the actuator directly control is a common symmetry plane M (active layer), including triangles $G_1, G_2, G_3$ composed of three length-adjustable sides $l_{12}, l_{23}, l_{31}$. The plane opposite to the plane M in each octahedral truss structure is defined as the bottom surface $B$ and the top surface $P$ (passive layer), and the centers of these two planar triangles are $O$ and $O_t$ respectively.

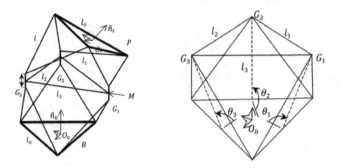

**Fig. 7.** The simplified octahedron model diagram

For the positive kinematics problem of the module unit, we need to calculate the coordinates of the center point $O_t$ of the end plane through the length of the three active bars. That is, the input variables for positive kinematics are $l_{12}, l_{23}, l_{31}$, and the output variable is $OO_t$.

We introduce the plane angles $\theta_1, \theta_2, \theta_3$ of the three planes and the bottom plane as the intermediate variables that link the input variables with the output variables, as shown in the Fig. 7 above.

To simplify the expression, noting that $N = \sqrt{l_s^2 - l^2/4}$, then the coordinates of the bottom three sides can be expressed as

$$O_1 = \left[ -\frac{l}{2\sqrt{3}}, 0, 0 \right] \tag{1}$$

$$O_2 = \left[ \frac{l}{4\sqrt{3}}, -\frac{l}{4}, 0 \right] \tag{2}$$

$$O_3 = \left[ \frac{l}{4\sqrt{3}}, \frac{l}{4}, 0 \right] \tag{3}$$

Then we can get the coordinates of the actuator plane's vertices $G_1, G_2, G_3$ :

$$G_1 = \left[ -\frac{l}{2\sqrt{3}} + Ncos\theta_1, 0, Nsin\theta_1 \right] \tag{4}$$

$$G_2 = \left[ \frac{l}{4\sqrt{3}} - \frac{N}{2}cos\theta_2, -\frac{l}{4} + \frac{\sqrt{3}Nc_2}{2}, Nsin\theta_2 \right] \tag{5}$$

$$G_3 = \left[ \frac{l}{4\sqrt{3}} - \frac{N}{2}cos\theta_3, \frac{l}{4} - \frac{\sqrt{3}Nc_3}{2}, Nsin\theta_3 \right] \qquad (6)$$

So that we can get the concrete expression

$$\left( 3k^2 + 2 - \left( \frac{l_{ij}}{N} \right)^2 \right) \sqrt{1 + tan\theta_i^2} \sqrt{1 + tan\theta_j^2}$$

$$= \pm 3k \left( \sqrt{1 + tan\theta_i^2} + \sqrt{1 + tan\theta_j^2} \right) - 1 + 2tan\theta_i tan\theta_j,$$

$$i,j = 1,2,3 \ldots i \neq j \qquad (7)$$

Among them

$$k = \frac{l}{2\sqrt{3}N}$$

According to the geometric relation of the following figure, under the condition of solving $\theta_1, \theta_2, \theta_3$, the coordinates of $O_t$ can be expressed as (Fig. 8):

$$O_t = \left( 2G_i \cdot \vec{n} + s \right)\vec{n} \qquad (8)$$

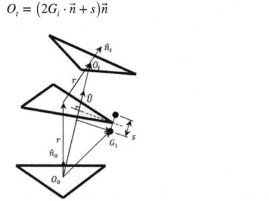

**Fig. 8.** The geometric relation of $O_0$ and $O_t$

## 4.2 Mechanical Analysis

After the design of the mechanism was completed, we used ANSYS to analyze the force under the action of its own weight and the load of 1 kg under the top. In order to simplify the model, we consider the nodes as rigid bodies and concentrate the mass of the rods on the nodes. The material settings correspond to the actual situation. According to the data, the bending strength and tensile strength of 3 K carbon fiber tubes are all 1500 MPa [12]. In order to check the carrying condition of the mechanism under the limit state, suppose there is no suspension system now, only the base support. The model and calculation results are realized in the form of a histogram, as shown in the Fig. 9.

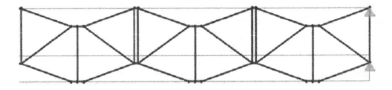

**Fig. 9.** The model of mechanical analysis.

We specify that in each small triangle, the left rod is the rod 1, the upper rod is the rod 2, and the lower rod is the rod 3. Due to the symmetrical structure, there are six different sets of members subjected to the force in the above figure. Now the six sets of members are subjected to the force in the form of a histogram as follows (Fig. 10):

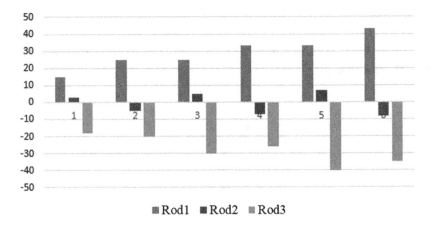

**Fig. 10.** Image of stress of each rod.

According to calculations, the force applied to the end member will not exceed 50 N. Since the diameter of the selected carbon fiber tube is 10 mm and the wall thickness is 1 mm, the maximum stress is calculated as

$$\delta = -\frac{M \times y}{I} = -\frac{F \times 1 \times \dfrac{d_1}{2}}{\dfrac{\pi(d_1^4 - d_2^4)}{64}} = 172 \, \text{Mpa} << [\delta] = 1500 \, \text{Mpa}$$

It can be seen that the strength of the structure meets the requirements at any time.

### 4.3   Experiments of Moduler Unit

In order to test the precision of the product when adjusting the position, we conducted a preliminary test on the manipulator. The test process are the following, a space rectangular coordinate system is established with the center of the ground triangle as the origin, and the coordinates of the center point of the top triangle are used as the measurement elements. We randomly select 9 sets of coordinates, and through theoretical calculations, we obtain the length of each side of the active layer needed to position the center point of the top triangle at this target. Using this calculation result as a parameter, combined with the plumb line and other tools to measure, get the actual coordinates of the center point, compared with the theoretical coordinates selected in advance, and analyze the error. Repeat 9 sets of experiments to get theoretical coordinates and actual coordinates and their errors are as Table 2.

**Table 2.**  The comparison of expected and actual coordinates

| Target coordinates/mm | | | Actual coordinates/mm | | | Relative error/% |
|---|---|---|---|---|---|---|
| X | Y | Z | X | Y | Z | |
| 0 | 0 | 253 | 0 | 0 | 250 | 1.186 |
| 0 | 0 | 340 | 0 | 3 | 334 | 1.973 |
| 0 | 0 | 400 | 0 | 6 | 394 | 2.121 |
| 30 | 0 | 330 | 26 | 4 | 319 | 3.733 |
| 0 | 60 | 350 | 3 | 55 | 342 | 2.788 |
| 20 | 60 | 350 | 16 | 52 | 344 | 3.028 |
| 100 | 0 | 300 | 89 | -6 | 289 | 5.273 |
| 0 | 120 | 290 | -6 | 110 | 278 | 5.332 |
| 100 | 100 | 300 | 92 | 108 | 292 | 4.178 |

The test results show that the relative error between the actual coordinates and the theoretical coordinates is within 6%, which is in an acceptable range.

The main causes of the error are:

(a)   The accuracy of the connector obtained by 3D printing is low, and there is a certain error in the assembly process;
(b)   The stiffness of the connector by 3D printing is small, and when the force is applied, a certain deformation and bending would occur;
(c)   Some measurement errors occurs;
(d)   Stepper motors have lost motion during operation.

## 5   The Application of VGT Manipulators

The structure maintains an indeterminate structure at any location, with high stiffness and light weight. After measurement, the total weight of one module is only 1.8 kg, the weight of two modules will not exceed 5 kg, and the movement range of the two modules also reaches 0.6 m. Compared with traditional series robots, the manipulator can achieve a greater range of motion [13]. From the above-mentioned mechanical analysis, it can be seen that in all motion ranges, a certain degree of strength can be ensured, which has a wide range of application values.

(1) The manipulators can be used as a support for a solar-powered windsurfing space deployment mechanism, which can be folded and saved space before it occurs. In space, fine-tuning the motor parameters can make the solar-powered windsurfing always stay at the best position.

(2) The manipulator can be developed as a pipeline robot [14]. The pipeline robot is composed of multiple modules. Rollers are added to the nodes of several active layers and the length of each side of the active layer can be freely adjusted within a certain range to ensure that the rollers are in close proximity to the pipe wall. Because the adjustment of the length of the active layer is relatively simple, this product can be automatically adapted to pipes of different diameters after modification. Refitted pipe robots can be used in pipeline inspection and maintenance operations. Traditional pipeline robots are difficult to adapt to complex pipelines with large changes in pipe diameter, but pipeline robots made with VGT structures can better adapt to complex pipelines.

(3) The manipulator has a strong scalability and large potential in development. The unit modules can not only design as octahedron VGTs, but also can be extended to decahedrons, dodecahedrons, etc. When the number of facets of the unit modules is larger, the movement that the manipulator can achieve becomes more complicated.

## 6    Conclusions

This paper proposed a variable geometry trussed redundant continuum manipulator with the advantages of serial robots and parallel robots, which effectively solved the inherent defects of various types of robots in modern industrial production. Using a modular parallel mechanism and connecting multiple modules in series, achieves complex space motion while ensuring high accuracy [15].

As mentioned by related scholars above, the possibility of ultra-redundant robots can only be limited by imagination.

**Acknowledgement.** This Research was supported by National Natural Science Foundation of China (No. 51575302).

## References

1. Bo, C.: The development and use of machine tools in historical perspective. J. Econ. Behav. Organ. **5**(1), 91–114 (1983)
2. Robotworx: History of industrial robots [EB/OL], 22 November 2013. http://www.robots.com/education/industrialhistory
3. Jone, M.: A brief history of awesome robots [EB/OL], 02 June 2013. www.motherjones.com/media/2013/05/robots-modern-unimate-watson-roomba-timeline

4. Reinholtz, V.A., Watson, L.T.: Enumeration and Analysis of Variable Geometry Truss Manipulators (1990)
5. Murthy, V., Waldron, K.J.: Position kinematics of the generalized lobster arm and its series parallel dual. Trans. ASME J. Mech. Des. **114**, 406–413 (1992)
6. Kraus, J.: Variable structure position control of an industrial robotic manipulator. J. Braz. Soc. Mech. Sci. **8**, 23–25 (2002)
7. Lee, H.Y., Roth, B.A.: Closed-from solution of the forward displacement analysis of a parallel mechanisms. In: Proceedings of the IEEE International Conference on Robotics and Automation (1993)
8. Miura, K.: Variable Geometry Truss Concept. The Institute of Space and Astronautical Science. Report NO. 614, September 1984
9. Suzuki, K., Maeda, Y., Ishibashi, Y., Fukushima, N.: Improvement of operability in remote robot control with force feedback. In: 2015 IEEE 4th Global Conference on Consumer Electronics (GCCE) (2015)
10. Fang, H., Fang, Y., Hu, M.: Forward position analysis of a novel three DOF parallel mechanism. In: Proceedings of the 11th World Congress in Mechanism and Machine Science, Tianjin, China, 1–4 April 2004, pp. 154–157. China Machinery Press, Beijing (2004)
11. Wei, G., Chen, Y., Dai, J.S.: Synthesis, mobility, and multifurcation of deployable polyhedral mechanisms with radially reciprocating motion. J. Mech. Des. Trans. ASME **136**(9), 091003 (2014)
12. Zhao, J.S., Chu, F., Feng, Z.J.: The mechanism theory and application of deployable structures based on SLE. Mech. Mach. Theor. **44**(2), 324–335 (2009)
13. Yang, Y., Jiang, B., Hu, S.: Fast trajectory planning for VGT manipulator via convex optimization. Int. J. Adv. Robot. Syst. **12**(9), 1 (2015)
14. Chettibi, T.: Synthesis of dynamic motions for robotic manipulators with geometric path constraints. Mechatronics **16**(9), 547–563 (2006)
15. Lin, F.Y.: Optimal motion planning for robotic manipulators. Adv. Mater. Res. **902**, 262–266 (2014)

# Rehabilitation Robotics

# Signal Compression Method Based Heart Rate Model Estimation and SMCSPO Control for Cardiac Rehabilitation with Treadmill

Hyun Hee Kim[1], Hwan Young Kim[2], Hong Ying Lee[2], and Min Cheol Lee[2(✉)]

[1] Division of Robotics Convergence, Pusan National University, Busan, Korea
sleepingjongmo@nate.com
[2] School of Mechanical Engineering, Pusan National University, Busan, Korea
reddish0621@naver.com, {lihongying,mclee}@pusan.ac.kr

**Abstract.** Cardiac rehabilitation with treadmill is widely used in heart disease medical centers in developed countries including Korea because it significantly reduces the prevalence of heart disease and is effective in treating mild and severe patients. For the automation of cardiac rehabilitation exercise, dynamic model of patient heart rate according to treadmill speed has been proposed several times in developed countries where heart rehabilitation studies are active. The heart rate model of the human body differs from person to person because it is greatly influenced by various factors such as age, weight, health condition, and heart disease severity. However, existing studies on human heart rate modeling do not reflect these individual differences. In this paper, we introduce a method of estimating the heart rate model by using only the exercise that can be easily performed before the treadmill cardiac rehabilitation by using the signal compression method. In the experiment, the heart rate model was estimated as a 2-order transfer function closest to the equivalent impulse response. Based on the estimated model, we simulated in MATLAB using sliding mode control with sliding perturbation observer (SMCSPO), which is a robust controller for system nonlinearity, parameter uncertainty and disturbance.

**Keywords:** Heart rate modeling · Cardiac rehabilitation
Signal compression method · SMCSPO

## 1 Introduction

Cardiac rehabilitation is very important for patients with heart disease. Rehabilitation exercise can restore the patient's ability to exercise and prevent further cardiovascular disease. Furthermore, it not only improves the quality of life but also reduces the mortality rate of all causes. The rehabilitation exercise is performed considering frequency, intensity, time(duration), and time, and mainly carries out aerobic exercise using treadmill and ergometer [1–3]. Knowing how heart rate changes according to exercise intensity will help many areas. This allows for more efficient rehabilitation exercise and can be beneficial to athletes, health trainer, and rehabilitation therapist. For this reason, various studies have been conducted to model heart rate according to

© Springer Nature Switzerland AG 2018
Z. Chen et al. (Eds.): ICIRA 2018, LNAI 10984, pp. 215–224, 2018.
https://doi.org/10.1007/978-3-319-97586-3_19

exercise intensity [4–8]. Hajek et al. modeled the HR response with a second order system using feedforward and feedback components [4]. Su et al. modeled the HR response by Hammerstein system, a system that consists of a static nonlinearity cascaded at the input of a linear system [5]. And, Nonlinear modeling was performed in [6] considering nonlinear factors due to human hormones and metabolism. A lot of studies have been done to design the controller to follow the preset heart rate trajectory when rehabilitating using these models [9–14]. This can be used to make rehabilitation exercise more efficient, and an autonomous system can be a great help if the therapist is responsible for many patients. However, their models were mainly obtained from normal males aged 20~30 years. In actual cardiac rehabilitation patients, heart rate characteristics may be different from those of normal people due to medication. In addition, the modeling was carried out for more than 30 min including warm up, exercise period, and rest period. This paper introduces a method of modeling the heart rate of patients with cardiac rehabilitation by using the signal compression method [15, 16]. Using this method, heart rate modeling can be obtained with only walking for about 10 min.

The heart rate model estimated by the signal compression method is used to implement the controller. Based on the estimated model, a robust control algorithm, sliding mode control with sliding perturbation observer (SMCSPO [17]) was simulated in MATLAB. SMCSPO controller is robust to disturbance, nonlinearity of the system, parameter uncertainty, and it is considered to be suitable for control of heart rate affected by many disturbances other than treadmill speed.

## 2    Heart Rate Model Estimation

### 2.1    Signal Compression Method

Signal compression is a kind of modeling theory for obtaining the transfer function of linear part of target system by converting the impulse signal to equivalent impulse signal. In the time domain, the impulse signal has an infinity value in a very short time near 0 time and the area of the signal is 1. when the signal is converted to a frequency domain (Laplace transform), the signal has a magnitude of 1 and a phase of 0. Thus, the pulse signal is input into the heart rate transfer function G(s), the output Y(s) about heart rate can be obtained directly. But in the heart rate system, the realization of an ideal impulse signal is impossible. Even it is a similar signal, an unlimited signal for a short time can also make the patient's heart under a very dangerous situation.

In order to obtain the dynamics of system, a full frequency response is required. The signal compression method is used to generate a similar impulse signal which has a flat power spectrum in a designed range of possible frequency to get a credible frequency response. The designed impulse is used to pass a time delay filter which has a certain phase margin in bode diagram. The lengthen signal can be used to give the real system caused by its finite magnitude and stretched time. After inputting this equivalent signal into the system, by using the product of the inverse function of filter, the transfer function of system can be calculated [15]. And the Fig. 1 shows the block diagram of signal compression method.

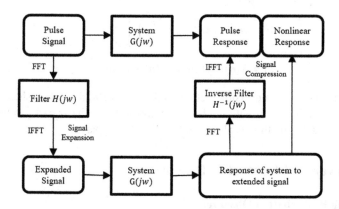

**Fig. 1.** Block diagram of signal compression method. The impulse signal is extended with a time delay filter and applied to the target system. Estimates the linear component of the system using an inverse filter on the response obtained from the system.

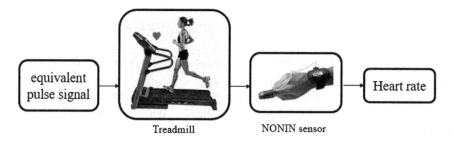

**Fig. 2.** Experiment system structure

Principle of the signal compression method is:

$$X(n) + jY(n) = P(n) \times H(jn)\left(0 \le n \le \frac{N}{2} - 1\right) \tag{1.1}$$

$$X(n) + jY(n) = P(n) \times H(jn)\left(n = \frac{N}{2}\right) \tag{1.2}$$

$$X(n) + Y(n) = X(N - n) + jY(N - n)\left(\frac{N}{2} + 1 \le n \le N - 1\right) \tag{1.3}$$

Here, P(n) is a function shows the power spectral density of test signal, H(jn) is a function of the filter. When signal compression method (filter) is designed, the equivalent pulse signal to process is important. The equivalent pulse signal $\delta_e(s)$ and the method to design a filter is displayed in the following way.

$$P(n) = M \, exp\left[-\left(\frac{n}{a}\right)^{12}\right] \tag{2.1}$$

$$H(jn) = \exp\left[-\left(\frac{12n^2}{b}\right)j\right] \tag{2.2}$$

$$\delta_e(s) = X(n) + jY(n) = P(n) \times H(jn) \tag{2.3}$$

If apply the signal compression method in linear system, it is established. When the method is used in a system with nonlinear component, the linear part can be separate from system and this method is also established in linear part.

## 3    Observer and Controller Design

### 3.1    Sliding Perturbation Observer

In this section, the structure of sliding perturbation observer is introduced briefly [17]. The governing equation of a general 2-order dynamics containing disturbance is defined as (3).

$$\ddot{x} = f(x) + \Delta f(x) + \sum_{i=1}^{n} \left[(b_i(x) + \Delta b_i(x))u_i\right] + d(t) \tag{3}$$

where $x$, input argument, is the state vector. The terms $f(x)$ is the linear dynamics terms and $\Delta f(x)$ is the nonlinear terms and uncertainties. The components $b_i(x)$ and $\Delta b_i(x)$ represent the elements of the control gain matrix and their uncertainties, while $d(t)$ is the external disturbance and $u_i$ is the control input. The terms $f(x)$ and $b_i(x)$ are known continuous functions of state.

Perturbation is defined as the combination of all the uncertainties of (3). The perturbation is given by:

$$\Psi(x, t) = \Delta f(x) + \sum_{i=1}^{n} \left[(\Delta b_i(x)u_i\right] + d(t) \tag{4}$$

It is assumed that the perturbations are upper bounded by a known continuous function of the state. An arbitrary positive number $\alpha_3$ and $\bar{u}$ is the new control variable to decouple the control variable using the following equation.

$$f(\hat{x}) + \sum_{i=1}^{n} \left[(\Delta b_i(\hat{x})u_i\right] = \alpha_3 \bar{u} \tag{5}$$

where "^" symbolizes the estimated quantity. The decoupled linear term is controlled by $\bar{u}$, and the nonlinear dynamics can be compensated by estimating perturbation. The state space representation is derived by:

$$\dot{x}_1 = x_2 \tag{6.1}$$

$$\dot{x}_1 = \alpha_3 \bar{u} + \Psi \tag{6.2}$$

$$y = x_1 \tag{6.3}$$

A new variable $x_3$ is defined as following equation to derive the perturbation from state sliding observer.

$$x_3 = \alpha_3 x_2 - \Psi / \alpha_3 \tag{7}$$

The perturbation is calculated using (7). The overall SPO structure to observe the system is derived as (8.1), (8.2) and (8.3).

$$\dot{\hat{x}}_1 = \hat{x}_2 - k_1 sat(\tilde{x}_1) - \alpha_1 \tilde{x}_1 \tag{8.1}$$

$$\dot{\hat{x}}_2 = \alpha_3 \bar{u} - k_2 sat(\tilde{x}_1) - \alpha_2 \tilde{x}_1 + \hat{\Psi} \tag{8.2}$$

$$\dot{\hat{x}}_3 = \alpha_3^2 \left( -\hat{x}_3 + \alpha_3 \hat{x}_2 + \bar{u} \right) \tag{8.3}$$

where perturbation $\hat{\psi}$ is can be calculated as (9).

$$\hat{\Psi} = \alpha_3 \left( -\hat{x}_3 + \alpha_3 \hat{x}_2 \right) \tag{9}$$

where $k_1$, $k_2$, $\alpha_1$, $\alpha_2$ and $\alpha_3$ are positive value and $\tilde{x}_1 = \hat{x}_1 - x_1$ is estimated error of state. The saturation function is used to reduce chattering in sliding surface.

$$sat(\tilde{x}_1) = \begin{cases} \tilde{x}_1 / |\tilde{x}_1|, & |\tilde{x}_1| \geq \epsilon_o \\ \tilde{x}_1 / \epsilon_o, & |\tilde{x}_1| \leq \epsilon_o \end{cases} \tag{10}$$

where $\epsilon_o$ is the boundary layer of SPO.

## 3.2 Sliding Mode Control with Sliding Perturbation Observer

This section explains the SMCSPO controller design. The estimated sliding function s is defined as (11).

$$\hat{s} = \dot{\hat{e}} + c\hat{e} \tag{11}$$

where $c$ is a slope of switching line and $e$ is the estimated position tracking error. The control $\bar{u}$ is selected to enforce $\hat{s}^T \hat{s} < 0$, outside a prescribed manifold. A desired $\hat{s}$ is like as

$$\dot{\hat{s}} = -Ksat(\hat{s}) \tag{12}$$

Using (8.1), (8.2), (8.3), (9) and (11), it is possible to compute $\hat{s}$ as

$$\dot{\hat{s}} = \alpha_3 \bar{u} - \left[ \frac{k_2}{\epsilon_o} + c\left( \frac{k_1}{\epsilon_o} \right) - \left( \frac{k_1}{\epsilon_o} \right)^2 \right] \tilde{x}_1 + \ddot{x}_{1d} + c\left( \hat{x}_2 - \dot{x}_{1d} \right) + \hat{\Psi} \tag{13}$$

From (12) and (13), a new control law is selected as

$$\bar{u} = \frac{1}{\alpha_3} \left\{ -Ksat(\hat{s}) + \left[ \frac{k_2}{\epsilon_o} + c\left(\frac{k_1}{\epsilon_o}\right) - \left(\frac{k_1}{\epsilon_o}\right)^2 \right] \tilde{x}_1 - \ddot{x}_{1d} - c\left(\hat{x}_2 - \dot{x}_{1d}\right) - \hat{\Psi} \right\} \quad (14)$$

## 4 Experiment

### 4.1 Experimental Equipment

In the experimental equipment section, the treadmill will used as test equipment whose model number is HERA-9000(I) manufactured by HEALTH ONE CO., LTD, Korea [18]. The communication between the computer and treadmill can be achieved through an RS232 serial port. And the part of heart rate data that we obtain from the test is measured by a NONIN sensor [19]. The total system structure is shown in Fig. 2.

The equivalent signal is a kind of test signal with a certain frequency. So, before we do the experiment, it is important that to know the cut-off frequency of inverter. The Fig. 3 shows the bode diagram of inverter and from this diagram it is can be known that the cut-off frequency is about 0.068 Hz. So when we select the equivalent pulse signal with a certain frequency characteristic, the frequency have to be smaller than the cut-off frequency of inverter. In this paper, the frequency of equivalent pulse signal always be taken as 0.065 Hz.

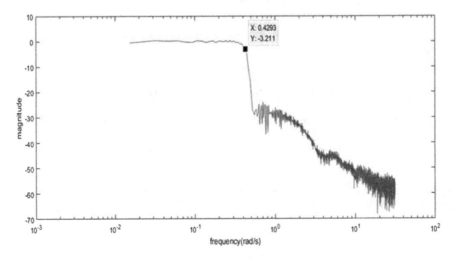

**Fig. 3.** Bode diagram of inverter

As the part of heart rate sensor communication, NONIN sensor use the Bluetooth communication to transfer information with computer. The sensor send 3 sets of packets per second.

## 4.2    Experimental Procedure and System Modeling

Through selecting a certain power density of equivalent pulse signal and designing a filter, here, the parameters in (2.1), (2.2) and (2.3) can be calculate as a = 30 and b = 180. Then we can get an equivalent signal with a maximum speed of 6 km/h as the Fig. 4(a) shows.

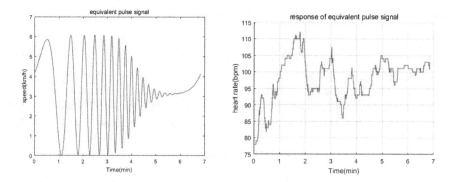

**Fig. 4.**  (a) Equivalent pulse signal (b) response of equivalent pulse signal

After we design this kind of equivalent test signal, it can be input into the treadmill use RS232 serial port. And then there will be a person accept the testing. This person who wearing a heart rate sensor, here is NONIN sensor, will do the treadmill exercise under the speed as equivalent pulse signal shows. After the tester running until the equivalent pulse signal stop, the data of heart rate sensor can be saved. Of course, the total process that transfer data, receive data and save data, is achieved by using program developed in the C# language ourselves.

The Fig. 4(b) shows the heart rate response when the tester do the treadmill exercise under the speed of equivalent pulse signal. It can be seen from this figure that, to a certain extent, the heart rate has a great relationship with the speed of treadmill. This is the response of time domain, here, we have to transform the time domain response into frequency domain and then use the inverse filter. The FFT and IFFT function from MATLAB will be used in transform the signal form time domain into frequency domain and frequency domain into time domain.

We suppose two unknown parameters $\omega_n$ and $\zeta$ in 2-order system,

$$G(s) = \frac{\omega_n^2}{s^2 + 2\zeta\omega_n + \omega_n^2} \tag{15}$$

And the MATLAB function corrcoef (A, B) will be used to obtain this two parameters when the 2-order system is most similar to the actual compression signal of heart rate. This function returns coefficients between two random variables A and B. If each variable has N scalar observations, then the Pearson correlation coefficient is defined as,

$$\rho(A, B) = \frac{1}{N-1} \sum_{i=1}^{N} \left( \frac{\overline{A_i - \mu_A}}{\sigma_A} \right) \left( \frac{B_i - \mu_B}{\sigma_B} \right) \tag{16}$$

where $\mu_A$ and $\sigma_A$ are the mean and standard deviation of A, and $\mu_B$ and $\sigma_B$ are the mean and standard deviation of B.

The correlation coefficient matrix of two random variables is the matrix of correlation coefficient for each pairwise variable combination,

$$R = \begin{pmatrix} \rho(A, A) & \rho(A, B) \\ \rho(B, A) & \rho(B, B) \end{pmatrix} \tag{17}$$

By using this function, parameters $\omega_n$ and $\zeta$ will be calculated when the correlation between compression signal of heart rate and the equivalent pulse signal response of the 2-order system is maximum. The parameters of 2-order system are shown in Table 1.

**Table 1.** Parameters of system

| Parameters | Values |
|------------|--------|
| $\omega_n$ | 5 |
| $\zeta$ | 1 |
| Correlation | 0.8613 |

## 5   Simulation

After using the signal compression method, the 2nd-order system can be obtained to describe the human heart rate linear part. At this section, this model will be used to design SMCSPO controller. To represent the human heart rate model in MATLAB, the heart rate model equation of Cheng was used [9]. This model expressed not only direct heart rate changes due to treadmill speed, but also slow heart rate changes due to human metabolism and hormone secretion. In addition, to confirm the robustness of the SMCSPO controller to disturbance, a sinusoidal signal of 1 rad/s period and a random signal of 6 s intervals were added to the heart rate model as a disturbance.

Usually, the good rehabilitation exercise required the tester do the treadmill exercise about 1 h and let the stable heart rate keep around at 130 bpm. So in the simulation, the target heart rate was set as follow. The first part is warm-up exercise, it is about 10 min long, the middle part is normal exercise, it is about 40 min, and the final part will take about 10 min to let heart rate calm down. Assuming a resting heart rate of 74 bpm, the target heart rate during exercise was set at 134 bpm, which is increased by 60 bpm.

Figure 5 shows the control results of the heart rate model by SMCSPO. In Fig. 5(a), we can see that the target heart rate is well followed despite the system nonlinearity and the effects of disturbances. However, the phenomenon that the heart rate does not fall properly in the cam-down section appears because the negative control input cannot be applied due to the characteristics of the heart rate model. Figure 5(b) shows the difference between target heart rate and output heart rate. There was no significant error in the main exercise period, which is the most important in cardiac rehabilitation exercise.

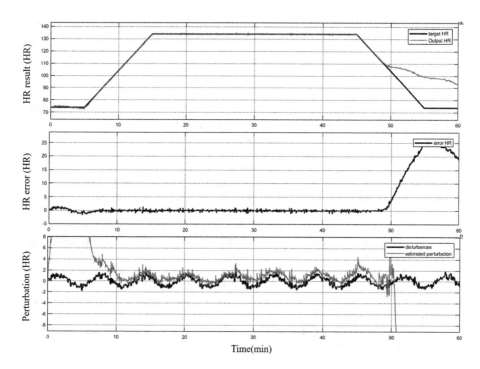

**Fig. 5.** (a) Heart rate control result (b) heart rate control error (c) estimated perturbation

The perturbation estimated from the sliding perturbation observer are shown in Fig. 5(c) along with external disturbances. The two signals are almost similar in appearance, but the reason for some offset and error is that the perturbation estimates both system nonlinearity and modeling error as well as disturbance. Also, it can be seen that the perturbation diverges in the cam-down section that is not properly controlled.

# 6    Conclusion

In this paper, the heart rate model of the human body is estimated by the signal compression method and the heart rate is controlled by the robust controller SMCSPO. The cross correlation coefficient of the model estimated by the signal compression method is 0.8613, which is not as good as that of the other systems having the characteristic of the 2-order system transfer function. In the next study, we need to estimate the model to the n-order system for more accurate parameter estimation. Also, a study on the experiment of SMCSPO control for actual human heart rate will be conducted.

**Acknowledgement.** This research was supported by the Ministry of Trade, Industry & Energy and the Korea Evaluation Institute of Industrial Technology (KEIT) with the program number of "10038660", the MOTIE (Ministry of Trade, Industry & Energy), Korea, under the Industry Convergence Liaison Robotics Creative Graduates Education Program supervised by the KIAT

(N0001126), and Basic Science Research Program through the National Research Foundation of Korea (NRF) funded by the Ministry of Education (1345253125).

# References

1. Kim, C.: Overview of cardiac rehabilitation. J. Korean Med. Assoc. **59**(12), 938–946 (2016)
2. Hong, K.P.: Cardiac rehabilitation. Korean Circ. J. **28**(3), 484–491 (1998)
3. Yang, Y.J.: Exercise testing and prescription. Hanyang Med. Rev. **29**(1), 20–27 (2009)
4. Hajek, M., Potucek, J., Brodan, V.: Mathematical model of heart rate regulation during exercise. Automatica **16**(2), 191–195 (1980)
5. Su, S.W., Wang, L., Celler, B.G., Savkin, A.V., Guo, Y.: Identification and control for heart rate regulation during treadmill exercise. IEEE Trans. Biomed. Eng. **BE-54**(7), 1238–1246 (2007)
6. Cheng, T., Savkin, A., Celler, B., Su, S.: Nonlinear modeling and control of human heart rate response during exercise with various work load intensities. IEEE Trans. Biomed. Eng. **55**(11), 2499–2508 (2008)
7. Lefever, J., Berckmans, D.: Time-variant modelling of heart rate responses to exercise intensity during road cycling. Eur. J. Sport Sci. **14**(S1), S406–S412 (2014)
8. Brodan, V., Hajek, M., Kuhn, E.: An analog model of pulse rate during physical load and recovery. Physiol. Bohemoslov. **20**, 189–198 (1971)
9. Cheng, T., Savkin, A.: A robust control design for heart rate tracking during exercise. In: IEEE EMBS 2008 (2008)
10. Su, S.W., Huang, S.: Optimizing heart rate regulation for safe exercise. Ann. Biomed. Eng. **38**(3), 758–768 (2010)
11. Argha, A., Su, S.W.: Designing adaptive integral sliding mode control for heart rate regulation during cycle-ergometer exercise using bio-feedback. In: EMBC (2015)
12. Mohammad, S., Guerra, T.M.: Heart rate control during cycling exercise using Takagi-Sugeno models. IFAC **44**, 12783–12788 (2011)
13. Mazenc, F., Malisoff, M.: Tracking control and robustness analysis for a nonlinear model of human heart rate during exercise. Automatica **47**(5), 968–974 (2011)
14. Scalzi, S., Tomei, P.: Nonlinear control techniques for the heart rate regulation in treadmill exercise. IEEE Trans. Biomed. Eng. **59**(3), 599–603 (2012)
15. Jin, S.Y., Lee, M.C.: Study on the identification of dynamic system with nonlinear terms using signal compression method and correlation coefficient. Korean Soc. Mech. Eng. **1**(1), 519–523 (1993)
16. Lee, M.C., Aoshima, N.: Identification and its evaluation of the system with a nonlinear element by signal compression method. Trans. SICE **25**(7), 729–736 (1989)
17. Moura, J.T., Elmali, H., Olgac, N.: Sliding mode control with sliding perturbation observer. J. Dyn. Syst. Measur. Control **119**(4), 657–665 (1997)
18. http://www.ehealth-one.com
19. http://www.nonin.com

# Measurement and Analysis of Upper Limb Reachable Workspace for Post-stroke Patients

Jing Bai and Aiguo Song[(✉)]

The State Key Laboratory of Bioelectronics, Jiangsu Key Lab of Remote
Measurement and Control, School of Instrument Science and Engineering,
Southeast University, Nanjing 210096, Jiangsu, People's Republic of China
a.g.song@seu.edu.cn

**Abstract.** The range of reachable workspace is related to the activity and motor function of the upper limbs. In this paper the upper limb reachable workspace of stroke patient was analyzed, and compared with the upper limb Fugl-Meyer score assessed by the therapist. In the experiment, the subject did the movement protocol by following the conductor. Different protocol was selected adaptively according to the arm activity. The avatar in the virtual environment was controlled synchronously to increase the fun of measurement. According to the movement trajectory of the upper extremity, reachable workspace sphere was fitted and relative surface area (RSA) was calculated to evaluate the performance of the upper limb. This study indicates that the RSA of upper limbs based on Kinect virtual environment has great potential in the assessment of upper limb performance of stroke patients and can be helpful for clinical evaluation.

**Keywords:** Adaptive · Reachable workspace · Post-stroke · Fugl-Meyer

## 1 Introduction

The "Report on the Chinese Stroke Prevention 2017" shows that the number of stroke patients who are over 40 years of age in China is 12.24 million. Also the trend of young patients with stroke is obvious. Stroke is one of the three highest fatal diseases in our country, with 75% of survivors having varying degrees of disability and loss of ability to work. However, the number of professional therapists for stroke patients is relatively small. Because there are more and more patients, the evaluation of upper limb movement function is arduous.

Traditional Upper limb motor function assessment methods usually include Rom [3], the Wolf motor function test [2], Fugl-Meyer [1], Brunnstrom stage [4] and so on. However, these methods are highly dependent on occupational therapists and need more time [5]. During rehabilitation, the rehabilitation recipe depends on the physician's assessment; because the assessment of rehabilitation is not quantitative, different doctors may make different assessment decisions [4].

The Kinect 3D Somatosensory Camera developed by Microsoft has been used to capture the movement of players in three-dimensional space [6]. Using Kinect Zhao propose a rule-based human motion tracking rehabilitation exercise that provides automated real-time assessment, feedback, and guidance for users performing

© Springer Nature Switzerland AG 2018
Z. Chen et al. (Eds.): ICIRA 2018, LNAI 10984, pp. 225–234, 2018.
https://doi.org/10.1007/978-3-319-97586-3_20

rehabilitation exercises at home without physical therapist supervision [7]. Su et al. developed a Kinect-enabled system for the patient to do the exercise and recorded as a base for evaluating the patient's rehabilitation exercise at home [8].

The exercise space of the upper extremities was measured, to evaluate the functional impairment of the upper extremities caused by some neurological diseases [9, 10]. Kurillo et al. focus on the technical aspect and accuracy of using Microsoft Kinect for assessment of reachable workspace as a potential outcome measure for the upper extremity [11].

Rapid advance in virtual reality technologies has been gaining a wide field in the motor rehabilitation process. Virtual reality technologies could assist the patients during unsupervised rehabilitation by providing an empathic feedback to improve their adherence to the treatment. Virtual reality [12] can create a highly interactive and immersive virtual environment, and Kinect can provide real-time motion status feedback for subjects. Virtual reality will improve the user's enthusiasm for the experiment; reduce the dependence on the doctor [13, 14].

In this paper, virtual environment technology is used to set up vivid virtual scenes for evaluating test experiments. Visual feedback and auditory feedback increase the interactivity and immersion of the assessment. Different protocol was selected adaptively according to the arm activity to eliminate patients' psychological stress, which comes from the mismatched movement protocols during the measurement. Subjects follow the instructions of the instructor in different directions to wave the limbs, while Kinect collects hand joint information and shoulder joint information, then the upper limb reachable workspace of the stroke patient was evaluated and analyzed.

**Fig. 1.** The experiment platform for upper limb

## 2  Method

### 2.1  Platform

As shown in Fig. 1, the experiment platform contains Microsoft Kinect camera (version 1) [15], tripod, laptop, 31.5-in. display monitor, loudspeaker and so on.

Kinect has three autofocus cameras: two infrared cameras optimized for depth detection and one standard visual-spectrum camera used for visual recognition. The Kinect SDK (Software Development Kit) for Windows provides detailed location and orientation information for up to two players standing in front of the Kinect sensor array. Previous devices have difficulty tracking human motion using a camera without body sensors; Kinect is a noninvasive and markerless method for motion tracking. In this paper Kinect 1.0 was used to collect the movement information of subjects; the tripod was used to place Kinect; the monitor provided a large field of view for the patient to watch easily. Unity 3d was used to design the reachable workspace experimental platform on the computer.

Overall, when a subject came to use the system to do experiment, Kinect built a visual character (avatar) to match him. The protocol level was selected adaptively. Then the instructor in the video began to show what actions the subject should follow. Moreover, the virtual environment will give subject a visual stimulation by the avatar standing next to the video that will copy the subject's actions when he follows the instructor. Finally, the joint data was recorded and analyzed to evaluate the upper limb reachable workspace.

## 2.2   Human Skeleton Matching

The skeleton of the subject collected by Kinect is bound to the skeleton of the virtual character model. Figure 2 shows the skeleton point collected by Kinect and the vivid virtual character model. So that the subject can control the character model to move in real time. This increases human-computer interaction and improves patient participation in the trial. Considering the age of patient, the mirror control was used for convenient. Just like the subject stood in front of the mirror and waved his arm. Soothing background music was played while measurement.

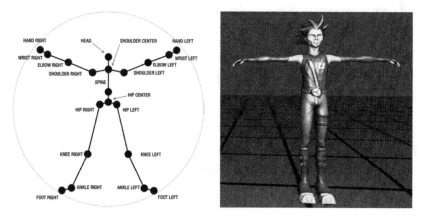

**Fig. 2.** Human skeleton matches virtual characters

## 2.3   Protocol

According to the range of the shoulder flexion extension, outreach adduction, internal rotation and external rotation. Moreover, depending on the patient's condition, the upper limb reachable workspace motion protocol was designed in three levels, as shown in Table 1, where the range and speed of the first level are the smallest for the severe dyskinesia patient, the range and speed of the third-level are maximum. The upper limb was moved in different ranges by different azimuth and elevation angles in the horizontal and vertical directions.

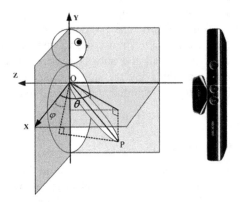

**Fig. 3.** Movement protocol demonstration

The starting position is the arm naturally put down and arm straight palm inward. The upper limb moves in the horizontal and vertical directions with the shoulder joint as the origin. One of the movement position is shown in Fig. 3. This paper chose to define a coordinate system similar to that used by the Microsoft Kinect. The Kinect is placed on the right side, and the subject is standing facing the Kinect sensor. The shoulder joint O is the origin, P is the recording point-the position of the hand, $\theta$ is the azimuth, $\varphi$ is the altitude. Different angles are set to measure the movement of upper limbs. When the altitude is 30, 60, 90 or 120, the Cheers sound is played to increase the patient's enthusiasm.

## 2.4   Adaptive Selection of Protocol Level

In order to reduce the psychological burden of the patient and avoid the negative emotion of the patient during the measurement, different levels of motion protocols were set adaptively.

Before the start of the experiment, the subject did the maximal arm outreach three times with the azimuth is zero. The reachable workspace coordinate system of right limb is shown in the Fig. 4(a). The stretching angle θ is shown in Fig. 4(b). It is calculated by the following formula.

$$\theta = \cos^{-1} \frac{\overrightarrow{CP} \cdot \overrightarrow{SH}}{|\overrightarrow{CP}||\overrightarrow{SH}|} \tag{1}$$

$$\overrightarrow{CP} = (Px - Cx, Py - Cy)$$

$$\overrightarrow{SH} = (Hx - Sx, Hy - Sy)$$

C stands for shoulder center, P stands for Spine, S stands for Shoulder right, and H stands for Right hand.

The system recorded the data automatically and calculated the average of the three abduction angles. The exercise protocol level was selected adaptively according to the average value. When the angle is in the range of 0 to 60°, the motion protocol is the first level, the speed of the demonstrator's arm is slow. When the angle is in the range of 60–120°, the motion protocol is the second level, and the arm speed is Medium speed. When the angle is within 120–180°, the action protocol level is level 3, and the demonstrator's arm speed is fast.

Table 1 shows the key motion of the experiment. The movement mainly in vertical and horizontal directions, the path is set by different azimuth angle and altitude angle. The experiment requires subject's hand to stretch as far as possible and to keep the elbow outstretched. When the subject's movement is good, the cheers sound is played, giving auditory feedback and encouraging the subject to continue the exercise.

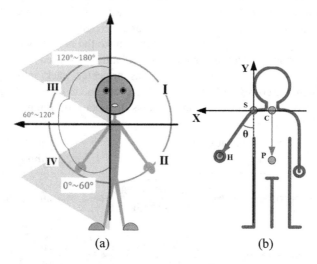

(a)                              (b)

**Fig. 4.** The reachable workspace coordinate system and four quadrants I, II, III, and IV of right limb

**Table 1.** Upper extremity reachable workspace protocol

| Level | Direction | Space angle | Path (°) 1 | 2 | 3 | 4 | 5 | Speed |
|---|---|---|---|---|---|---|---|---|
| 1 | Vertical | Azimuth | 0 | 45 | 90 | – | – | Slow |
|  |  | Altitude | 0–60 | 0–60 | 0–60 | – | – |  |
|  | Horizontal | Azimuth | 0–90 | 0–90 | – | – | – |  |
|  |  | Altitude | 30 | 90 | – | – | – |  |
| 2 | Vertical | Azimuth | 0 | 45 | 90 | – | – | Medium |
|  |  | Altitude | 0–120 | 0–120 | 0–120 | – | – |  |
|  | Horizontal | Azimuth | 0–120 | 0–120 | 0–120 | 0–120 | – |  |
|  |  | Altitude | 30 | 90 | 120 | 0–120 | – |  |
| 3 | Vertical | Azimuth | 0 | 45 | 90 | 135 | – | Fast |
|  |  | Altitude | 0–180 | 0–180 | 0–180 | 0–180 | – |  |
|  | Horizontal | Azimuth | 0–135 | 0–135 | 0–135 | 0–135 | 90 |  |
|  |  | Altitude | 30 | 90 | 120 | 0–180 | 0–90 |  |

## 3   Data Analysis

With the shoulder joint as the center of the sphere and the arm length as the radius, the upper limb stretches and draws arcs in different planes. The reachable workspace of upper limb is part of the sphere. The workspace is divided into four quadrants (I, II, III and IV) as shown in the Fig. 4, each quadrant corresponds to 1/4 sphere. The right upper limb is taken as an example.

The trajectory data that was collected by Kinect was filtered with a 6th-order low pass Butterworth filter with a cut-off frequency of 30 Hz. The data was rejected when the movement speed (tangential velocity) was lower than 50 mm/s. The nonlinear least squares method was used to fit the trajectory; then, the fitted trajectory data was projected into the spherical coordinate system, and the α-shape geometry was used to extract contour edges of the trajectory. In order to make the boundary of the reachable workspace more smooth, Hermite spline interpolating was used to interpolate the boundary points. Next, the trajectory data was projected back into the Cartesian coordinates. Then, the corresponding accessible surface patches were extracted. At last, the surface area was normalized. Figure 5 shows the data processing.

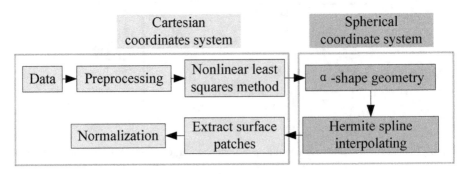

**Fig. 5.** Data processing

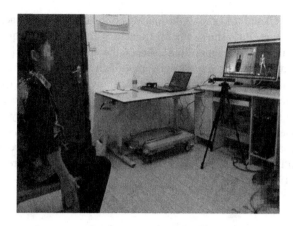

**Fig. 6.** Experiment scene

**Table 2.** Patient Clinical Characteristics

| ID | Gender (F/M) | Age | Affected Arm (L/R) | Month | Category | Upper limb Fugl-Meyer score | Level |
|----|-------------|-----|--------------------|-------|----------|-----------------------------|-------|
| 1 | F | 46 | Left | 1 | Infarction | 15 | I |
| 2 | M | 59 | Right | 2 | Hemorrhage | 19 | I |
| 3 | M | 62 | Right | 8 | Infarction | 37 | II |
| 4 | M | 57 | Left | 11 | Infarction | 43 | II |
| 5 | M | 45 | Right | 4 | Infarction | 57 | III |
| 6 | F | 53 | Right | 30 | Hemorrhage | 59 | III |
| 7 | M | 54 | Left | 10 | Hemorrhage | 65 | IV |
| 8 | M | 38 | Right | 13 | Hemorrhage | 64 | IV |

## 4   Results

Patient clinical characteristics are described in Table 2. It shows the affected arm, time after stroke, stroke category (cerebral hemorrhage or infarction) and Fugl-Meyer score.

Fugl-Meyer score (100) is equal to the total score of upper extremity function (66) plus the total score of lower extremity function (34). According to the clinical significance of the Fugl-Meyer assessment (FMA), human motor function is divided into four stages: severe dyskinesia (<50), significant dyskinesia (50–84), moderate dyskinesia (85–95), and mild dyskinesia (96–99). According to the proportion of upper extremity motor function scores, the clinical significance of dividing the upper extremity FMA score is: severe dyskinesia (<33), obvious dyskinesia (33–55), moderate dyskinesia (56–62) and mild dyskinesia (63–65). The experimental scene in Nanjing Tongren hospital is shown in Fig. 6.

Figure 7(a1), (a2) and (a3) shows the analysis of reachable workspace obtained in a healthy subject. Figure 7(a1) shows the 3D trajectory data after least squares fitting, the red dot is the ball center. Figure 7(a2) shows the trajectory data projected to the spherical coordinates, and the outer boundaries of the concave bounding polygon is obtained as shown by the blue line. Figure 7(a3) shows the envelope of the reachable workspace obtained by back projecting trajectory data to three dimensional space and fitting a spherical surface. Figure 7(b1), (b2) and (b3) shows the analysis of reachable workspace obtained in the patient 4. Figure 7(c1), (c2) and (c3) shows the analysis of reachable workspace obtained in the patient 2. The Relative surface area (RSA) is represented in different colors in each quadrant, respectively, area1 red, area2 green, area3 purple, and area4 pink.

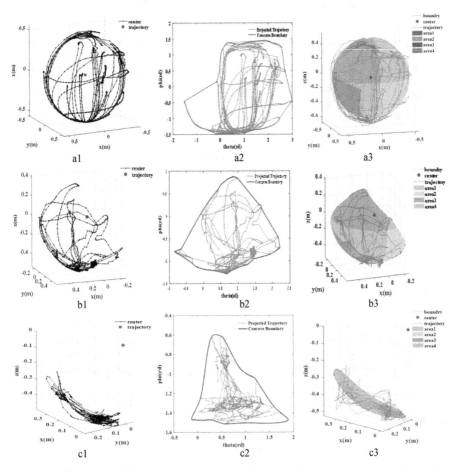

**Fig. 7.** Reachable workspace (a1) 3D trajectory after least squares method of healthy subject (a2) Boundary of healthy subject (a3) Graphical visualization of 3d reachable workspace of healthy subject. (b1) 3D trajectory after least squares method of patient 4 (b2) Boundary of patient 4 (b3) Graphical visualization of 3d reachable workspace of patient 4. (c1) 3D trajectory after least squares method of patient 2 (c2) Boundary of patient 2 (c3) Graphical visualization of 3d reachable workspace of patient 2. (Color figure online)

It can be seen from RSA in Fig. 7(a3), (b3) and (c3), healthy arm could naturally move to each quadrant; the upper limb of the patient 4 could reach the fourth quadrant, and its active space in the second and third quadrants was less, but it could hardly move to the first quadrant. As can be seen from the figure, as the injury of the upper extremities increases, the smoothness of the trajectory decreases, which indicates that the movement quality of the upper limb is decline and the movement becomes more and more awkward.

The average RSA plot by Fugl-Meyer level is shown in Fig. 8. The value of RSA is the average of each RSA for patients in the same level. As can be seen from the figure, with the decline in the Fugl-Meyer level, the RSA is gradually reduced. The upper limb of patients in level I are only active in the fourth quadrant. The patients in level IV are not much different from the RSA of healthy people.

There is a partial absence of the patient's upper limb working space. Overall, with the increase of upper limb injury, the RSA of upper limb reachable workspace was decreased, the smoothness of space trajectory was reduced, and the irregularity of path was increased. These figures suggest that compared with healthy subject, stroke patients have less RSA, both in the quadrant and in the total area. Patients in different Fugl-Meyer score also has different RSA.

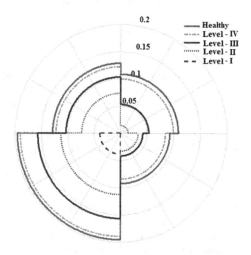

**Fig. 8.** The RSA plot by Fugl-Meyer level

## 5    Conclusion

The reachable workspace RSA in stroke patients based on Kinect was measured and analyzed. Patients with different Fugl-Meyer score were participated in this assessment.

The protocol with virtual environment can increase the interactivity and immersion of the assessment. The adaptive selection of protocol can improve the efficiency of the test, and prevent the patient from producing psychological pressure effectively.

Reachable workspace RSA has a certain relationship with the motor function disorder classification based on Fugl-Meyer score, and can be used as a potential quantitative assessment method to assess the motor function of the upper limb, and provide further basis for clinical treatment of post-stroke patients.

**Acknowledgment.** The authors would like to thank the anonymous reviewers for their valuable comments and helpful suggestions. In addition, we would like to thank all of the subjects who participated in the study.

This work has been supported by National Key R&D Plan (2016YFB1001301), The National Natural Science Foundation of China (91648206).

# References

1. Gladstone, D.J., Danells, C.J., Black, S.E.: The Fugl-Meyer assessment of motor recovery after stroke: a critical review of its measurement properties. Neurorehabilitation Neural Repair **16**(3), 232–240 (2002)
2. Wolf, S.L., Catlin, P.A., Ellis, M., et al.: Assessing Wolf motor function Test as outcome measure for research in patients after stroke. Stroke **32**(7), 1635–1639 (2001)
3. Macedo, L.G., Magee, D.J.: Differences in range of motion between dominant and nondominant sides of upper and lower extremities. J. Manipulative Physiol. Ther. **31**(8), 577–582 (2008)
4. Yu, L., Wang, J.P., Fang, Q., et al.: Brunnstrom stage automatic evaluation for stroke patients using extreme learning machine. In: IEEE Biomedical Circuits and Systems Conference, pp. 380–383. IEEE, Taiwan (2012)
5. Hondori, H.M., Khademi, M.: A review on technical and clinical impact of microsoft Kinect on physical therapy and rehabilitation. J. Med. Eng. **2014**(1), 846514 (2014)
6. Bai, J., Song, A., Xu, B., et al.: A novel human-robot cooperative method for upper extremity rehabilitation. Int. J. Soc. Robot. **9**(2), 265–275 (2017)
7. Zhao, W., Reinthal, M.A., Espy, D.D., et al.: Rule-based human motion tracking for rehabilitation exercises: realtime assessment, feedback, and guidance. IEEE Access **5**, 21382–21394 (2017)
8. Su, C.J., Chiang, C.Y., Huang, J.Y.: Kinect-enabled home-based rehabilitation system using Dynamic Time Warping and fuzzy logic. Appl. Soft Comput. **22**(5), 652–666 (2014)
9. Han, J.J., Bie, E.D., Nicorici, A., et al.: Reachable workspace and performance of upper limb (PUL) in duchenne muscular dystrophy. Muscle Nerve **53**(4), 545–554 (2016)
10. Han, J.J., Kurillo, G., Abresch, R.T., et al.: Upper extremity 3D reachable workspace analysis in dystrophinopathy using Kinect. Muscle Nerve **52**(3), 344–355 (2015)
11. Kurillo, G., Chen, A., Bajcsy, R., et al.: Evaluation of upper extremity reachable workspace using Kinect camera. Technol. Health Care **21**(6), 641–656 (2013)
12. Taylor, M.J., McCormick, D., Shawis, T., et al.: Activity-promoting gaming systems in exercise and rehabilitation. J. Rehabil. Res. Dev. **48**(10), 1171–1186 (2011)
13. Pei, W., Xu, G., Li, M., et al.: A motion rehabilitation self-training and evaluation system using Kinect. In: 2016 13th International Conference on Ubiquitous Robots and Ambient Intelligence (URAI), pp. 353–357. IEEE, Xi'an (2016)
14. Mastropietro, A., et al.: Quantitative EEG and virtual reality to support post-stroke rehabilitation at home. In: Chen, Y.-W., Tanaka, S., Howlett, R.J., Jain, L.C. (eds.) Innovation in Medicine and Healthcare 2016. SIST, vol. 60, pp. 147–157. Springer, Cham (2016). https://doi.org/10.1007/978-3-319-39687-3_15
15. Yu, T.: Kinect Application Development Combat: The Most Natural Way to Dialogue with the Machine. China Machine Press, Beijing (2014)

# A Multi-channel EMG-Driven FES Solution for Stroke Rehabilitation

Yu Zhou[1(✉)], Yinfeng Fang[2], Jia Zeng[1], Kairu Li[2], and Honghai Liu[1]

[1] State Key Laboratory of Mechanical System and Vibration,
Shanghai Jiao Tong University, Shanghai, China
{hnllyu,jia.zeng,honghai.liu}@sjtu.edu.cn
[2] Group of Intelligent System and Biomedical Robotics,
School of Creative Technologies, University of Portsmouth, Portsmouth, UK
{yinfeng.fang,kairu.li}@port.ac.uk
http://bbl.sjtu.edu.cn/

**Abstract.** Functional electrical stimulation (FES) has been applied to stroke rehabilitation for many years. However, users are usually involved in open-loop fixed cycle FES systems in clinical, which is easy to cause muscle fatigue and reduce rehabilitation efficacy. This paper proposes a multi-surface EMG-driven FES integration solution for enhancing upper-limb stroke rehabilitation. This wireless portable system consists of sEMG data acquisition module and FES module, the former is used to capture sEMG signals, the latter of multi-channel FES output can be driven by the sEMG. Preliminary experiments proved that the system has outperformed existing similar systems and that sEMG can be effectively employed to achieve different FES intensity, demonstrating the potential for active stroke rehabilitation.

**Keywords:** Functional electrical stimulation (FES)
Surface electromyography (sEMG) · Integration system
Stroke rehabilitation

## 1 Introduction

FES uses short electrical pulses in specific motor neurons to generate contractions in paralyzed muscles [1]. Since first used for foot drop rehabilitation by Liberson in 1960 [2], FES has become one of the important and effective treatments for stroke rehabilitation [3,4]. Most FES systems used in clinical stroke rehabilitation settings are passive open-loop control systems, which output the stimulation current at a predefined fixed model. Thus, it is easy to induce muscle fatigue and requires the users' constant attention to operate them. Moreover, it excludes the active participation of subjects and decreases the neuromuscular activity and energy consumption [5,6]. The rehabilitation efficiency can be further improved when FES is applied with the EMG feedback, in which patients voluntary movement intention can be detected and stimulation process

© Springer Nature Switzerland AG 2018
Z. Chen et al. (Eds.): ICIRA 2018, LNAI 10984, pp. 235–243, 2018.
https://doi.org/10.1007/978-3-319-97586-3_21

is adjusted by the muscle state in real time. Thus, the muscle fatigue is alleviated because of the lower average stimulation intensity compared to the cycling fixed FES. The EMG-driven closed-loop FES has been studied and shows an advantage over the cycling FES [7–9].

There still exist problems: (1) The use of separate FES and EMG devices to form the closed loop not only faces the communication and real-time problems between the subsystems but also makes the whole system cumbersome and time-consuming to attach and remove, which becomes one of the main reasons why FES/EMG closed-loop control strategies have not been widely used in clinical applications [1]. (2) Most existing commercial FES systems, such as the ParaStep system (Sigmedics. USA), the RehaStim2 (HASOMED GmbH, Germany) and the NESS H200 (Bioness, USA), are open-loop. Some other devices, such as the Compex Motion stimulator (Compex SA, Swiss), the WalkAide foot drop stimulator (NeuroMotion, USA), the NESS L300 (BIONESS, USA), are equipped with sensor interfaces for force-sensitive-resistors (FSR), accelerometers or push buttons to determine when the stimulation is required according to the users' physical state. However, the sensor input of them performs more like an on/off trigger which does not represent or make full use of users' movement intention or muscle status; Thus, an integration EMG-driven rehabilitation system is desired to solve the problems above.

This study proposes an attempt of a wireless multichannel EMG/FES integration solution for upper-limb stroke rehabilitation. The sEMG from subjects can be used to trigger the FES module, and as an index measuring to which degree a stroke patient recovered.

## 2    Multi-channel EMG/FES Integration Solution

This section presents the overall design of the EMG/FES integration system. As shown in Fig. 1, the overall architecture is composed of seven parts: electronic switch module (ESM), signal acquisition module (SAM), microcontroller unit (MCU), electric stimulator output module (ESOM), power supply module (PSM), bluetooth module (BM) and graphical user interface (GUI). The channels work in a time-sharing way, which means they are used to capture EMG signals in signal sampling model and output FES current in electrical stimulation model. The ESM, controlled by MCU I/O ports, is used to switch the connection between electrodes-to-ESOM and electrodes-to-SAM, and so that to protect SAM from the heavy current damage from ESOM. The EMG-driven FES is based on the signal sampling model and the electrical stimulation model. In the signal sampling model, the analog raw signals are amplified, low-pass filtered and digitized by SAM and then delivered to MCU for further processing, after that the digital signals can be wirelessly transmitted to the GUI. In the electric stimulation model, the MCU makes ESOM output the stimulation current in a certain parameter configuration (such as amplitude, frequency and pulse width) according to the EMG processing results.

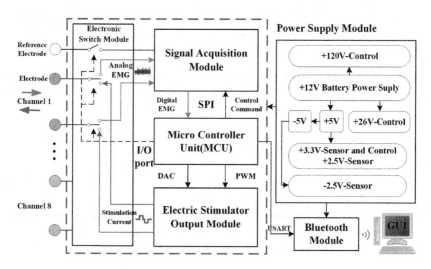

**Fig. 1.** The architecture of the sEMG/FES integration system.

## 2.1 sEMG Acquisition Module

The raw sEMG signals are always accompanied with various kinds of noises, such as physiological noise, ambient noise (50 or 60 Hz, power line radiation), electro-chemical noise from the skin-electrode interface and so on. So our previous study makes use of instrumentation amplifier to suppress common mode noise, band pass filter based on operational amplifiers to extract valid frequency band of EMG signal (20 Hz–500 Hz) and remove baseline noise as well as movement artifact, and comb filter to suppress the 50 Hz power line noise and multiples thereof. After all those filtering processes, the signals are digitized by the analog-to-digital converter (ADC) in MCU [10].

In this study, we chose the ADC integrated chip ADS1299 (Texas Instruments) as the main component of SAM. As shown in Fig. 2, the raw EMG signals are amplified, sampled and low-pass filtered in ADS1299 (the sampling rate is 1 kHz). Then the digitized sEMG signals are transmitted into MCU (STM32F103VCT6, STMicroelectronics) through the serial peripheral interface (SPI). The comb filter (center frequency, 50 Hz and multiples thereof) and high-pass filter (cut-off frequency, 20 Hz) towards the signals are realized in MCU to remove the baseline noise and movement artifact.

The low-pass filter is a on-chip digital third-order sinc filter whose Z-domain transfer function is (1). The Z-domain transfer function of the comb filter is (2). The high-pass filter is a six-order Butterworth filter defined by (3) in Z-domain. Compared to the design in [10], the hardware design in this study can not only extract clean sEMG but also save much space in PCB design.

$$H(Z)_{sinc} = \left| \frac{1 - Z^{-N}}{1 - Z^{-1}} \right|^3 \tag{1}$$

**Fig. 2.** The sEMG signal processing flow. RCF is the RC filter. DA, ADC and LF are the differential amplification, analog-to-digital conversion and low-pass filter within ADS1299. CF and HF are the comb filter and high-pass filter respectively in MCU.

where $N$ is the decimation ratio of the filter.

$$H(Z)_{comb} = \frac{i-1}{i} - \frac{1}{i}Z^{-T} - \frac{1}{i}Z^{-2T} - \cdots - \frac{1}{i}Z^{-iT} \tag{2}$$

where $T$ is a constant determined by sampling frequency (1 kHz) and the basic frequency of the power line noises (50 Hz); $i$ is the filter order, which determines the length of the previous signals being used to estimate noise.

$$H(Z)_{butterworth} = \frac{\sum_{k=0}^{M} b_k Z^{-k}}{1 - \sum_{l=1}^{N} a_l Z^{-l}} \tag{3}$$

where $a$ and $b$ are coefficients; $N$ and $M$ are determined by the filter order, meaning the length of the previous signals to be used.

## 2.2   Functional Electric Stimulation Module

The main parts of the functional electric stimulation module are the constant-current source circuit (Fig. 3a) and the bridge circuit (Fig. 3b). The magnitude of the stimulation current can be calculated by (4).

$$I_d = \frac{R_2 U_{REF}}{(R_2 + R_3)R_s} \tag{4}$$

where $I_d$ is the stimulation current applied to skin, $U_{REF}$ is the amplified DAC output from MCU, and $R_s$ is the sampling resistance to capture the stimulation magnitude of current in real time (shown in Fig. 3a). The current magnitude is the feedback to the circuit to maintain the constant current.

Under the control of the PWM (Pulse-Width Modulation) signals from the timers in MCU, the bridge circuit can realize the biphasic square wave stimulation current with desired frequency and duration. No matter symmetric biphasic pulses or asymmetric biphasic pulses, the area of the negative phase should be equal to that of the positive phase. The negative phase plays an important role in eliminating the charge accumulation in the skin and avoiding tissue damage. In order to avoid the two constant-current sources working at the same time, which will make the stimulation current into confusion, a short interval time (100 μs) is set between PWM1 and PWM2, which results in an interval between the positive and negative phase.

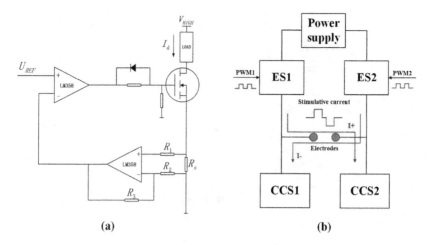

**Fig. 3.** The main functional part of the functional electric stimulation module: (a) Electrical schematic representation of constant-current source; (b) Schematic representation of the bridge circuit in ESOM. ES and CCS are electrical switch and constant-current source respectively. The positive phase current I+ is generated when the CCS 2 works and the negative phase current I− is generated when the CCS 1 works.

## 3  System Evaluation and Experiments

The function of sEMG acquisition, FES output and EMG-driven FES were evaluated respectively. It should be noted that all the experiment procedures in this study were approved by the SJTU School Ethics Committee and all subjects gave written informed consent and provided permission for publication of photographs for scientific and educational purposes.

### 3.1  EMG Acquisition Evaluation

In this part, two commercial sEMG sensors DataLOGMWX8 (Biometrics Ltd, UK) and Tringo Wireless (Delsys Inc, USA) were involved to be compared with the integration system with respect to the time-frequency domain characteristics. A healthy subject was asked to hold on hand close and hand open for 5 s in turn at a moderate level of effort according to the instructions on a computer screen, and the sEMG was recorded at the same time. The protocol was repeated on the same subject for testing different devices. And the electrodes/sensors were placed on the same belly position of flexor carpi ulnaris (FCU) after cleaning with alcohol [11]. As shown in Fig. 4a, the EMG/FES system could well detect EMG signals between 20 Hz and 350 Hz and showed better performance of the suppression for 50 Hz and multiples thereof than the two commercial devices.

In order to evaluate the signal quality of the integration system, signal-noise ratio (SNR) is compared among the integration system and the commercial devices mentioned before. A healthy subject without nerve and limb disease

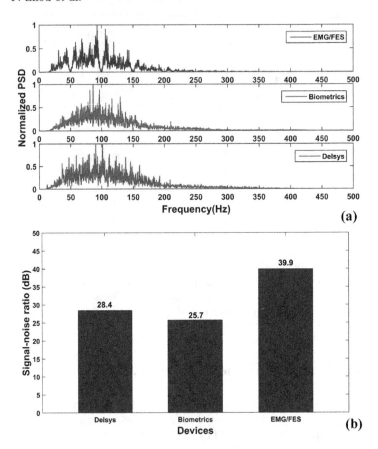

**Fig. 4.** Signals comparison between the proposed EMG/FES system and commercial sEMG acquisition systems in: (a) Normalized PSD of the acquired sEMG signals (b) Signal-noise ratio.

took part in the experiment. During one trial, the subject was asked to hold on the grip force sensor and keep the grip force at the level of 50% maximum volunteer contraction (MVC) with the vision feedback from computer for 10 s and then relax for 10 s, meanwhile the sEMG signals were detected with the electrodes/sensors placed on the belly position of FCU. The protocol was repeated on the same subject for testing different devices. The SNR was calculated by Eq. (5). And the result is shown in Fig. 4b. It indicates that the signal quality of the integration system is better than the two commercial devices.

$$SNR = 20log\frac{RMS_{signal}}{RMS_{noise}} \tag{5}$$

## 3.2    FES Output Evaluation

The parameters of the stimulation current consist of amplitude, pulse width and frequency. Figure 5 shows the waveform of the stimulation current generated by the proposed system with different parameters. It indicates that the system can output normal symmetrical and asymmetrical bipolar square wave for stimulation according to the design, though there is still small fluctuation appearing upon the positive phase when generating currents with certain amplitudes, which may cause uncomfortable acanthesthesia when applied to human skin.

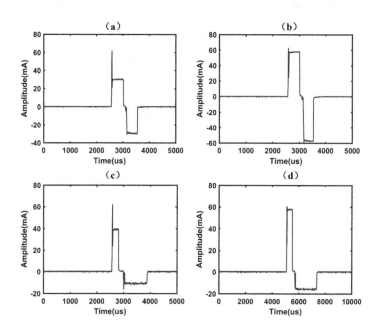

**Fig. 5.** Waveform of stimulation current with different parameters: (a) Amplitude-30 mA, pulse-400 µs, frequency-50 Hz; (b) Amplitude-60 mA, pulse-400 µs, frequency-50 Hz; (c) Amplitude-40 mA, pulse-200 µs (the positive phase), frequency-50 Hz; (d) Amplitude-60 mA, pulse-400 µs (the positive phase), frequency-50 Hz. (a) and (b) are symmetrical bipolar square wave. (c)–(d) are asymmetrical bipolar square wave.

## 3.3    EMG-Driven FES Experiments

Preliminary experiment was conducted to verify the performance of the sEMG-driven FES system developed in this study. A handgrip dynamometer (Biometrics Ltd, UK) was used to measure MVC force for the grasp of the subject. The experimental procedure was as follows. Firstly, the subject was asked to perform maximum MVC handgrip contractions for five times using his left hand with EMG signals of FCU stored. Two silica gel electrodes were placed on the belly of FCU and another two were placed on the wrist for reference. Each contraction was kept for 5 s with 20 s rest between two consecutive contractions. For every 5 s contraction EMG signals, the data of the first second and the last second were discarded and the rest 3 s signals were used for EMG root mean square (RMS) value calculation for one contraction. The mean value of the RMS for the five contractions was set as $RMS_{left}$; Secondly, the electrodes were moved and placed on the same position on the right hand, then the subject performed the handgrip with his right hand and controlled the grip force in realtime according to the visual feedback of the force value displayed on a computer screen in front of him and the EMG RMS was calculated every 300 ms to map the FES intensity. Equation 6 shows the mapping relationship between sEMG and FES.

$$\begin{cases} Q = \dfrac{RMS_{left} - RMS_{right}}{RMS_{left}} \\ I_R = \dfrac{I_{max}}{Q_{max} - Q_{min}} Q - \dfrac{I_{max}Q_{min}}{Q_{max} - Q_{min}} \end{cases} \tag{6}$$

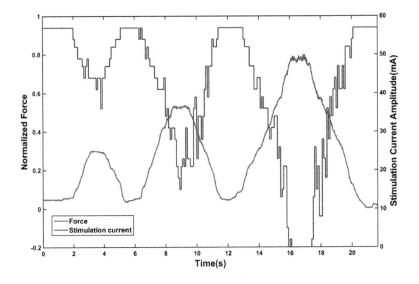

**Fig. 6.** The realtime FES output measurement based on the FCU EMG feedback when grip force varied. $Q_{max}$ and $Q_{min}$ were 100% and 20% respectively, $I_{max}$ was set as 57 mA. The three peaks of the force line represent $Q = 70\%$, $Q = 40\%$ and $Q = 20\%$ in order respectively. When $Q > 20\%$, the FES current amplitude changed from 0 mA to $I_{max}$ and be negative correlation with force. When $Q < 20\%$, the amplitude of the current is 0 mA.

Figure 6 shows the result of the experiment. It indicated that the FES intensity can be well controlled according to the EMG feedback in real time. The weaker the EMG is, the higher the FES intensity will be.

# 4    Conclusion

This paper presents a multi-channel EMG-driven FES integration system for stroke rehabilitation. The designed system has advantages of wireless communication, multiple channels, portable size and real-time capability. On one hand, the system can be used as sEMG acquisition device and FES device respectively; on the other hand, the FES can be optimized by the EMG, which demonstrates the potential of driving the FES treatment process using the patients' muscle state and movement intention for stroke rehabilitation.

**Acknowledgments.** This work is supported by the National Natural Science Foundation of China (No. 51575338, 51575407, 51475427) and the Fundamental Research Funds for the Central Universities (17JCYB03).

# References

1. Lynch, C.L., Popovic, M.R.: Functional electrical stimulation. IEEE Control Syst. **28**(2), 40–50 (2008)
2. Liberson, W.: Functional electrotherapy: stimulation of the peroneal nerve synchronized with the swing phase of the gait of hemiplegic patients. Arch. Phys. Med. Rehabil. **42**, 101 (1961)
3. Popović, D.B.: Advances in functional electrical stimulation (FES). J. Electromyogr. Kinesiol. **24**(6), 795–802 (2014)
4. Lyons, G.M., Sinkjær, T., Burridge, J.H., Wilcox, D.J.: A review of portable FES-based neural orthoses for the correction of drop foot. IEEE Trans. Neural Syst. Rehabil. Eng. **10**(4), 260–279 (2002)
5. Edgerton, V.R., Roy, R.R.: Robotic training and spinal cord plasticity. Brain Res. Bull. **78**(1), 4–12 (2009)
6. Lotze, M., Braun, C., Birbaumer, N., Anders, S., Cohen, L.G.: Motor learning elicited by voluntary drive. Brain **126**(4), 866–872 (2003)
7. Quandt, F., Hummel, F.C.: The influence of functional electrical stimulation on hand motor recovery in stroke patients: a review. Exp. Trans. Stroke Med. **6**(1), 9 (2014)
8. Hong, I.K., Choi, J.B., Lee, J.H.: Cortical changes after mental imagery training combined with electromyography-triggered electrical stimulation in patients with chronic stroke. Stroke **43**(9), 2506–2509 (2012)
9. Fujiwara, T.: Motor improvement and corticospinal modulation induced by hybrid assistive neuromuscular dynamic stimulation (hands) therapy in patients with chronic stroke. Neurorehabilitation Neural Repair **23**(2), 125–132 (2009)
10. Fang, Y., Zhu, X., Liu, H.: Development of a surface EMG acquisition system with novel electrodes configuration and signal representation. In: Lee, J., Lee, M.C., Liu, H., Ryu, J.-H. (eds.) ICIRA 2013. LNCS (LNAI), vol. 8102, pp. 405–414. Springer, Heidelberg (2013). https://doi.org/10.1007/978-3-642-40852-6_41
11. Forvi, E., et al.: Preliminary technological assessment of microneedles-based dry electrodes for biopotential monitoring in clinical examinations. Sens. Actuators A Phys. **180**, 177–186 (2012)

# A Common Prosthetic Rehabilitation Platform Based on Modular Design

Dehong Hao, Bo Lv, Sensen Liu, Xinjun Sheng$^{(\boxtimes)}$, and Xiangyang Zhu

State Key Laboratory of Mechanical System and Vibration,
Shanghai Jiao Tong University, 800 Dongchuan Road, Minhang District,
Shanghai, China
xjsheng@sjtu.edu.cn

**Abstract.** Virtual training environment for prosthetic rehabilitation is a promising research and commercial area. These years immersive virtual training environment is in rapid development, as Virtual reality (VR) and augmented reality (AR) technology as well as game-elements are embedded into the virtual training environment. While it is expected that these new technologies may raise the chance of recovery for amputees, it is also challenging for researchers to adopt these new technologies, as they require professional developing skills. That is why in this work, we present a common prosthetic rehabilitation platform based on modular design, to eliminate the gap between researchers and professional developers on VR/AR and video game. Two versions of the target achievement control (TAC) test are conducted, running on PC and HoloLens separately, to evaluate the feasibility and expansibility of this common platform and reveal the difference between 3D virtual training environment on HoloLens and the traditional one on PC.

**Keywords:** Prosthetic rehabilitation · Platform · Modular design
VR/AR · TAC test

## 1 Introduction

People who lose arms are faced with awkwardness of dysfunction of motor and sensory. Among all ways for amputees to return to normal life, prosthetic hand is the best choice, with a history that dates back to the seventeenth century. And in the middle of nineteenth century, Myoelectric prosthesis was introduced into market, and quickly became a hot issue in the research and commercial field. Nowadays, with the development of research on prosthesis design and myoelectric control, dexterous myoelectric prosthesis are available in the market [1,2], bringing great hope to amputees to regain control over daily life. These myoelectric prosthesis are usually capable of dexterous and natural control, thanks novel myoelectric control methods. These methods include pattern-recognition control [3] and simultaneous proportional control [4] and so on.

© Springer Nature Switzerland AG 2018
Z. Chen et al. (Eds.): ICIRA 2018, LNAI 10984, pp. 244–254, 2018.
https://doi.org/10.1007/978-3-319-97586-3_22

However, despite the advancement in the research field, myoelectric prosthesis is not widely used. It is even abandoned by amputees due to its high cost and relative poor functionality [5]. Additionally, boring and repetitive pre-prosthetic training also cause some amputees to give up in the very early stage. To address these problems, advanced training environment are gathering more and more attention from both the research and commercial field. In the early 21st century, with the development of computer science, virtual training environment was proposed in labs and then adopted to the commercial products, providing a much more efficient and comfortable rehabilitation experience with a lower cost. And to build a more immersive rehabilitation environment, virtual reality (VR) technology is utilized. These VR devices varies from prototypes in laboratories to mature commercial products [6,7]. Soon after that, Augmented reality (AR) technology was adopted to assist the amputees in interaction with the virtual surrounding, and has already shown promising result in prosthetic rehabilitation (such as treatment of phantom limb pain (PLP) [8]). Game-based rehabilitation is the blue ocean for commercial rehabilitation, because the entertainment feature it embedded promotes the engagement of amputees, and meanwhile the lower cost it required (compared to traditional rehabilitation equipment) prelieve the financial burden for amputees. Thus, game-based rehabilitation also becomes a promising area in research [9].

Though the novel environments introduced above have already surpass the previous environments, both by improving the tedious training as well as relieving patients' fatigue, problems also come along with these new technologies. For VR and AR, as discussed in the review [7], academic explanation for their effects is still insufficient. (Meanwhile more new VR/AR technologies and devices in development are making this question even more complex.) For game-based rehabilitation platform, which is popular in commercial application, it is still questionable whether it is based on a serious rehabilitation theory. For rehabilitation game designed in the lab, although some criteria have already been proposed [10], games developed by researchers are still too rudimentary for practical application. Contrastly, commercial developers usually lack basic knowledge on prosthetic rehabilitation. To address these problems, we present a common prosthetic rehabilitation platform based on modular design. This platform can eliminate the gap between researchers and developers by separating the rehabilitation method design and applications development process (while also allowing them to cooperate). We evaluate this platform by developing two versions of TAC tests [11], one 2-D version for computer and one 3-D version for HoloLens (an advanced AR device developed by Microsoft [12]). With the experiments of these 2D and 3D versions of TAC test on eight subjects, we investigate whether novel AR environment can bring improvement to the performance of rehabilitation training, and researched its advantages and disadvantages compared with traditional virtual training environment with computer and screen.

## 2   Method

### 2.1   Architecture

The overall architecture of this common platform is described in Fig. 1. The whole platform can be divided into two function units, the central server and the clients. The central server is on personal computer (PC), and contains basic modules for prosthetic rehabilitation. It is aimed for researchers to evaluate sensors, data processing algorithm and novel control methods. The clients are for multiple applications on different types of device. They are intended for professional developers, who may develop applications on clients. By separating the platform into two independent units, we facilitate the stand-alone development of researchers and developers.

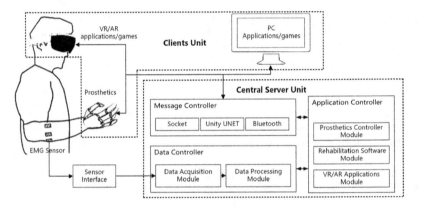

**Fig. 1.** Schematic diagram of the common prosthetic rehabilitation platform based on modular design.

**Central Server Unit.** The central server unit consists of three controllers, data controller, application control and message controller (as is shown in bottom-right in Fig. 1). The first controller is responsible for both the data acquisition and processing: In this controller, the data acquisition module acquires the electromyography (EMG) data from various types of EMG sensor, which each has unique manufacturer-provided interference (illustrated in the bottom-left in Fig. 1). Specially in this case, we use the Trigno sensors (Delsys Inc. Boston, MA, USA) for its high frequency of sample and wireless design. After data is acquired, the data processing module utilize various control methods to process data. The methods include direct control, pattern-recognition control and simultaneous proportional control. In this work, we implement pattern-recognition control with time domain (TD) features and linear discrimination analysis (LDA) classifier, which is a widely-used solution for dexterous prosthetic control [3]. The second controller, the application controller is implemented for specific clients and it is what controls the other two controllers: it controls data acquisition and processing in the first controller, then sends the command to clients via message controller (the third controller). The third controller, the message controller adapts to multiple clients on various device. It adopts three communication modules, i.e. socket, UNET, and Bluetooth. Socket and Bluetooth communication are the mainstream that covers most devices used for prosthetic rehabilitation, while UNET is relatively unfamiliar to researchers since its a unique built-in network in Unity3D, a popular game engine [13]. Each module can build a server, and thus this architecture supports multi-connection that simultaneously controls multiple devices. We design this communication solution so that VR/AR technology and game elements can be applied to prosthesis rehabilitation together. It should be noted that, the central server is independent of clients and thus can be used for testing tasks (e.g. testing of data processing algorithm). But essential user interface is recommended when central server is implemented to control specific clients.

**Clients Unit.** As introduced in the previous section, this platform supports multiple clients on different devices. These clients can either be specifically designed for this central server, or developed initially for other applications and later transplanted to this platform. This transplantation can be done by simply adding one of the three communication methods supported by the message controller.

In this work, we adopted SJT6 prosthetic hand [14], HoloLens and a desktop computer as three client devices. For SJT6 dexterous prosthetic hand, we implement a specific application controller to control the pattern-recognition process inside the data controller, and use the Bluetooth module in the message controller to send the control command to the prosthetic hand. In this way, the prosthesis hand can be controlled by the prosthetic controller module, which is further based on recognition result of data processing module. This prosthetic hand is able to perform multiple gestures like open/close hand, pronation/supination and movement of single finger. For HoloLens, which is the first self-contained AR headset that is able to present hologram fixed with the real world, it can provide more immersive experience with higher stability and portability, compared to other inferior AR products. An application and a game are designed, each can run on both the computer and the HoloLens. Later in tests, by comparing subjects' performance in the two training environment (PC and HoloLens), the effectiveness of these two types of rehabilitation training can be evaluated. And on this platform, a same application controller can be shared between the application and game, which saves much development work. For technical details, our application controller is based on TAC test while our data controller use pattern-recognition control method. And our game requires user to control a virtual 3D Mario to move in 3 DOFs, with a changing speed. The virtual Mario should be controlled to run from an original position to a specific destination within a certain time and reside for a brief time, which equals the expected experiment scheme of initial TAC test [11].

## 2.2  Test

Experiments are designed as shown in Fig. 2. We design two versions of rehabilitation application based on TAC test, one 2D TAC test on the PC and one 3D version on the HoloLens, to evaluate the feasibility and transportability of this platform, and research the trade-off between PC and HoloLens training environments. A user-friendly control panel is also designed, providing a convenient control of the experiment over subjects management, automatically data acquisition and storage.

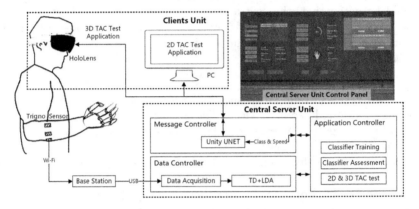

**Fig. 2.** The implementation of common platform for the experiment based on TAC test. The central server is on PC with a user-friendly control panel

**Subjects.** Eight able-bodied subjects participated in the experiment, all male with mean age of 24. All subjects took both 2D and 3D versions of TAC tests with more than one day between these two tests. They were informed of the whole experiment process. The experiment was approved by the Ethics Committee of Shanghai Jiao Tong University.

**Pattern Recognition.** Four Trigno Sensors are placed symmetrically around the muscle bulge on the forearm 2–3 cm below elbow, as is illustrated in the Fig. 2 [11]. EMG data are gathered to base station via Wi-Fi, then synchronized and transferred to the computer through USB cable, finally they are streamed to the data acquisition module with socket (built by utility software provided by Delsys). Seven motion classes (wrist flexion, wrist extension, wrist supination, wrist pronation, hand open, hand closed, and no movement) are included in this experiment. By performing these gestures, test subjects are required to control the 3-DOF movement of a virtual hand. In the training process, subjects can either see the virtual hand's movement on the PC screen (illustrated in Fig. 3) or hologram projected in the real world by the HoloLens, which is designed to assist them perform each movement comfortably and consistently without fatigue. For each movement, test subjects hold for 5 s, and after that there is a break of the same time between movements. With a sampling rate of 2000 Hz, in each movement the 6000 samples in the middle is segmented. Four TD features of the data (main absolute value, waveform length, slope sign changes and zero crossings) [3] are then extracted, with 150 ms-analysis window and 50 ms-window increment. An LDA classifier is then trained on the TD features. It should be noted that the root mean square (RMS) of EMG data of each motion is also used to determine moving speed of the virtual hand. The classifier is assessed by classification accuracy, namely the number of correct result by the total number. It is assessed after the training process. If the average classification accuracy of seven motion classes falls below a threshold (set as 85% in this experiment) [15], then the training process is repeated.

**TAC Test.** The TAC test (Target Achievement Control Test) is proposed by Rehabilitation Institute of Chicago in 2011 and designed to evaluate real-time myoelectric pattern-recognition control of multi-functional upper-limb prostheses [11]. In a TAC test, test subjects learn to control a virtual hand into a target posture using myoelectric pattern recognition. The virtual hand is shown on a PC screen and helps the subjects to learn to control the virtual hand naturally. The TAC test is more challenging compared to the previous tests, such as motion test, and also demands more complex experiment. We choose TAC test to evaluate the feasibility and stability of our platform, and investigate whether portable AR rehabilitation environment is superior to the 2D screen-based environment [16].

We design one 2-D version and one 3-D version of TAC test that each runs on PC and HoloLens, as is shown in Fig. 3(a), (b), (c), (d) respectively. In our TAC test, the virtual hand is controlled to move progressively in 3-DOFs. Its direction is determined by the result of classifier and moving speed is in proportional to RMS of EMG signal. In TAC test on this platform, the direction and speed is controlled by the application controller (as is illustrated in Fig. 2). Specifically, the application controller gives command to both HoloLens and the PC via UNET. In building this system, we first design the 2-D and 3-D version application controller for PC, and then transplanted it to run on HoloLens. In the transportation, only few code is needed to be changed. And this demonstrates the advantage of the architecture in this work, that only one control application is to coded, but can be adjusted to control both the 2D and 3D version of TAC test.

For the 2-D version of rehabilitation application on PC, there is one training process and one testing process. During the training process, test subjects learn to control the virtual hand, and once the classification accuracy exceeded a threshold, the TAC test interface with an additional semi-transparent virtual hand will appear. And then the testing process will start. Specifically, during the training period, text and movement of the virtual hand are displayed on the screen, assisting test subjects in learning to control in collaboration (as is shown in Fig. 3(a) and (b)). Subjects have enough time to familiarize themselves

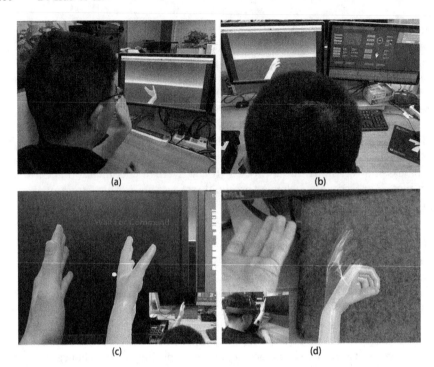

**Fig. 3.** Experiment record of the 2D and 3D version of TAC test. **(a)** training process of the 2D version of TAC test, **(b)** testing process of the 2D version of TAC test, **(c)** training process of the 3D version of TAC test in the first person perspective, **(d)** testing process of the 3D version of TAC test in the first person perspective. (Color figure online)

with the interface, until they think they can control the virtual hand to move progressively and naturally. They are also required to maintain for a dwelling time within a certain time-out period. And once the testing process begins, the semi-transparent hand will remain green if overlapped (within a certain tolerance) with the controlled virtual hand, to encourage test subjects to maintain at this state. And contrast to the 3-D version application, in the 2-D version the scream is fixed towards the test subjects.

For the 3-D version, the rehabilitation application on HoloLens employs the same experimental paradigm. But additionally, a hologram of virtual hand can be rotated and scaled, depending on the need of the test subjects, which provides a more immersive experience. The virtual hand is fixed relative to the surrounding, but subjects can observe it from any position and direction since it was a hologram. It should be noted that no marker is used to fix the virtual hand to the subjects' real hands, since real hand will often divert subjects attention from the virtual hand and sometimes confuse the subjects, according to the feedback from subjects in the pre-experiment. In the augmented reality environment, subjects are promoted to train and test classifier, and complete the TAC test under the instruction of the spatial sound from HoloLens, like from a real instructor aside and the motion of virtual hand with the same scale as their real hands.

For each version of TAC test, each subject has to perform 4 trails with 8 possible combinations of hand movement (open/lateral grip, extension/flexion, pronation/supination) in a random order. Each trial is conducted with a time-out of 45 s, a dwelling time of 2 s, and a tolerance of 5° for overlapping checking. The result of each trial is recorded as the complete time if completed successfully (including the 2 s-dwelling time), and if not successful, it is marked as failure. A questionnaire is conducted after each versions of TAC test is completed, to collect response on experience of experiment environment, especially differences between 2-D and 3-D environments.

## 3    Result and Discussion

Figure 4 shows the completion rate of each subject, the mean and standard deviation of all subjects' completion rate for 2D and 3D version. On average, the completion rate improved obviously, with an 76.953 ± 13.13% versus the 81.641 ± 13.15%, although two subjects (sub. 1 and sub. 6) showed a lower completion rate in the 3D test. Figure 5 shows the average completion time of each subject, the mean and standard deviation of all subjects' average completion time. The average of completion time decreased with 25.63 ± 3.00s vs. 24.68 ± 3.18s, showing the superior of the 3-D test. Three subjects (sub. 2, sub. 4 and sub. 5) spent more time in the 3D test.

The result of the questionnaire agree with the statistical analysis of the experiment data. All subjects agreed on that they got a more immersive experience in 3D environment, although six of them complaint that the HoloLens is a little heavy to wear through the whole experiment. And for the TAC test, five subjects thought 2D version is easier to learn. However, there is a diverted preference over 3D virtual hand four subjects thought it confused them with more information

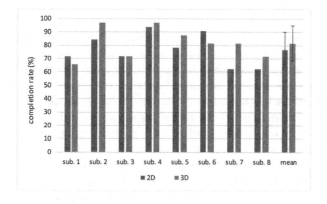

**Fig. 4.** TAC test task completion rate over subjects

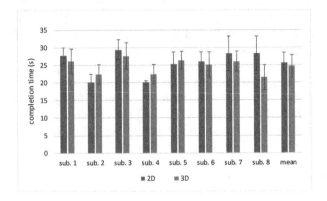

**Fig. 5.** TAC test task completion time over subjects

in the 3D world compared to a 2D image on the screen, but other subjects could take advantage of these extra information.

The results of the experiment and the questionnaire show the advantages and disadvantages of the PC and the HoloLens training environments, respectively: Virtual training environment shown on screen is easy to learn, while the virtual training environment based on HoloLens can provide a more immersive experience and more natural interaction, like spatial sound and voice input. However, the cost of a HoloLens is far higher, and it also requires more learning before one can control it. While these differences are agreed by all subjects, adaptation and preference can vary from person to person. As discussed before, some may experience negative influence from the 3-D training environment. It should also be reminded that all subjects in this experiment are able-bodied. It is worth researching whether the 3-D virtual training environment has the same effect on amputee.

## 4    Conclusion

The virtual training environment for prosthetic rehabilitation is the golden future of the market, due to its relative low cost and entertaining elements. And with the virtual training environment amputees are finally freed from the tedious classic physical therapies. For those advantages, the virtual training environment for prosthetic rehabilitation has already caught the eyes of the researchers, who have came up with some novel platforms. As a feature of these novel platforms, VR/AR technology are utilized and game elements are introduced, in order to build a more immersive environment and make the amputees more engaged. However, the performance of these novel methods still requires investigation, and most prototypes are still primitive, because the gap between lab researches and professional VR/AR/video game has not been bridged.

To bridge this gap, in this work, we present a common prosthetic rehabilitation platform based on modular design. The researchers can implement the central server of this platform to test new sensor, prosthetic control method and experiment scheme, and the professional developers can develop excellent applications and games as clients. The bridge based on mainstream communication modes can connect the contribution of researchers and developers and combine them to build a whole virtual training platform. This platform makes it possible to test and adopt the new research findings with a professional VR/AR applications and games for a brief time. To evaluate this platform, we implemented this platform to build both 2D and 3D versions of TAC test and conducted the experiment to research on the performance of 3D training environment. The setup process of this experiment proved the feasibility and expansibility of our platform and the results showed the trade-offs between the traditional 2D virtual training environment and the novel 3D one. This 3D virtual training environment based on HoloLens was more immersive and engaging for subjects but brought extra burden of learning and wearing to subjects, while the statistical analysis also shows individual variation on adaptation to these 2D and 3D training environments.

The next step of our work is to conduct our experiment on amputee, to find out their reaction to the 3-D virtual training environment and the vivid 3D virtual hand. Beside that, other up-to-date VR/AR devices will be tested in this experiment, to strike a balance between comfort, cost and immersiveness, such as the HTC VIVE Pro (with wireless kit) and VR/AR box with smart phone inside. Furthermore, games will also be adopted in this experiment to evaluate the performance of a game-based training environment [17].

**Acknowledgements.** This research was supported in part by National Natural Science Foundation of China under Grant 91748119 and 51620105002.

# References

1. Atzori, M., Müller, H.: Control capabilities of myoelectric robotic prostheses by hand amputees: a scientific research and market overview. Front. Syst. Neurosci. **9**, 162 (2015)
2. Geethanjali, P.: Myoelectric control of prosthetic hands: state-of-the-art review. Med. Devices (Auckland, NZ) **9**, 247 (2016)
3. Englehart, K., Hudgins, B.: A robust, real-time control scheme for multifunction myoelectric control. IEEE Trans. Biomed. Eng. **50**(7), 848–854 (2003)
4. Hahne, J.M., et al.: Linear and nonlinear regression techniques for simultaneous and proportional myoelectric control. IEEE Trans. Neural Syst. Rehabil. Eng. **22**(2), 269–279 (2014)
5. Farina, D., et al.: The extraction of neural information from the surface EMG for the control of upper-limb prostheses: emerging avenues and challenges. IEEE Trans. Neural Syst. Rehabil. Eng. **22**(4), 797–809 (2014)
6. Howard, M.C.: A meta-analysis and systematic literature review of virtual reality rehabilitation programs. Comput. Hum. Behav. **70**, 317–327 (2017)
7. Dunn, J., Yeo, E., Moghaddampour, P., Chau, B., Humbert, S.: Virtual and augmented reality in the treatment of phantom limb pain: a literature review. NeuroRehabilitation **40**(4), 595–601 (2017)

8. Ortiz-Catalan, M., et al.: Phantom motor execution facilitated by machine learning and augmented reality as treatment for phantom limb pain: a single group, clinical trial in patients with chronic intractable phantom limb pain. Lancet **388**(10062), 2885–2894 (2016)
9. Prahm, C., Kayali, F., Vujaklija, I., Sturma, A., Aszmann, O.: Increasing motivation, effort and performance through game-based rehabilitation for upper limb myoelectric prosthesis control. In: 2017 International Conference on Virtual Rehabilitation (ICVR), pp. 1–6. IEEE (2017)
10. Tabor, A., Bateman, S., Scheme, E., Flatla, D.R., Gerling, K.: Designing game-based myoelectric prosthesis training. In: Proceedings of the 2017 CHI Conference on Human Factors in Computing Systems, pp. 1352–1363. ACM (2017)
11. Simon, A.M., Hargrove, L.J., Lock, B.A., Kuiken, T.A.: The target achievement control test: evaluating real-time myoelectric pattern recognition control of a multifunctional upper-limb prosthesis. J. Rehabil. Res. Dev. **48**(6), 619 (2011)
12. Microsoft: HoloLens. https://www.microsoft.com/en-us/hololens. Accessed 22 Mar 2018
13. UnityTechnologies: Unity3D. https://unity3d.com/unity. Accessed 22 Mar 2018
14. Xu, K., Guo, W., Hua, L., Sheng, X., Zhu, X.: A prosthetic arm based on EMG pattern recognition. In: 2016 IEEE International Conference on Robotics and Biomimetics (ROBIO), pp. 1179–1184. IEEE (2016)
15. Scheme, E., Englehart, K.: Electromyogram pattern recognition for control of powered upper-limb prostheses: state of the art and challenges for clinical use. J. Rehabil. Res. Dev. **48**(6), 643 (2011)
16. Boschmann, A., Dosen, S., Werner, A., Raies, A., Farina, D.: A novel immersive augmented reality system for prosthesis training and assessment. In: 2016 IEEE-EMBS International Conference on Biomedical and Health Informatics (BHI), pp. 280–283. IEEE (2016)
17. Woodward, R.B., et al.: A virtual coach for upper-extremity myoelectric prosthetic rehabilitation. In: 2017 International Conference on Virtual Rehabilitation (ICVR), pp. 1–2. IEEE (2017)

# Comparative Analysis of Surface Electromyography Features on Bilateral Upper Limbs for Stroke Evaluation: A Preliminary Study

Hongze Jiang[1], Yang Li[2], Yu Zhou[1], and Honghai Liu[1(✉)]

[1] State Key Laboratory of Mechanical System and Vibration,
Shanghai Jiao Tong University, Shanghai, People's Republic of China
{1044053154,hnllyu,honghai.liu}@sjtu.edu.cn
[2] Shanghai Jing'an District Central Hospital, Shanghai, People's Republic of China

**Abstract.** The loss of upper limb functionality caused by stroke significantly influences patients daily living. Surface electromyography (sEMG) has been applied for study of stroke rehabilitation for tens of years. This paper is an attempt to evaluate stroke severity using sEMG. An experiment including four basic upper limb arm motions was carried out, with eleven able-bodied and six stroke patients being employed. Several sEMG features of bilateral upper limbs were compared for their relationship with stroke severity, and results showed that a new proposed feature named Envelope Correlation (EC) performed best. The experiment outcomes provided a prospect to evaluate stroke grade using sEMG.

**Keywords:** Surface electromyography (sEMG) · sEMG features
Stroke · Evaluation · Bilateral upper limbs

## 1 Introduction

Stroke is a medical condition in which poor blood flow to the brain results in cell death. There are about fifty million people suffering from stroke every year, and five million of them are permanently disabled [1]. Among the stroke patients, 55% to 75% survivors suffer from different levels of physical dysfunctions, more than 80% of which are on upper limbs [2]. It is reported that early assessment and cure will improve the functions of the stroke patient so as to improve their life qualities [3,4].

Nowadays, the assessments of the upper limb functions of the stroke patients are mainly divided into two types: subjective rating and objective rating. The subjective rating usually involves gauge charts and assessment done by the therapists. The common assessment includes Brunnstrom Assessment, Fugl-Meyer Assessment (FMA), Lindmark Assessment, and modified Ashworth Scale, MAS. These assessments are not able to capture the minor change of the muscles and involve subjective judgement, which varies between different therapists. Recently,

© Springer Nature Switzerland AG 2018
Z. Chen et al. (Eds.): ICIRA 2018, LNAI 10984, pp. 255–263, 2018.
https://doi.org/10.1007/978-3-319-97586-3_23

the sEMG sensing technology has been applied to the clinic applications, which provides new ways of the upper limb recovery [5]. Elena Dalla Toffola et al. studied the myoelectric fatigue in bilateral tibialis anterior muscles of stroke patients [6]. Han [7] and Cheng [8] also did researches on stroke combining sEMG.

Feature extraction is always essential in the research of sEMG. Normally, the sEMG features can be divided into three types: time domain, frequency domain, and time-frequency domain [9,10]. Time domain features yield high recognition accuracy and cost less computational efforts. The time domain features developed by Hudgins et al. [11,12] produced good performance for representing myoelectric patterns. Alkan et al. [13] used mean absolute value (MAV) as a key input to classification system, which achieved very high classification rate. Lin et al. [14] combined MAV and AR features to build a robust gesture recognition algorithm. In comparison with time domain features, frequency domain features have much worse performance in sEMG signal classification. Instead, they enjoy better stability. In 1970, Kwatny et al. [15] applied the mean frequency of the spectrum (MNF) to detect muscle fatigue. Later, median frequency (MF) was introduced to determine changes during muscle fatigue [16]. Time-frequency domain features give a super combination between time domain features and frequency domain features. Short time fourier transform (STFT) and wavelet transform (WT) are both commonly used in this area.

In this paper, we give priority to evaluating stroke severity using sEMG sensing technique. To evaluate the relationship between sEMG features and stroke level, an experiment including four basic upper limb arm motions was carried out, with eleven able-bodied and six stroke patients being employed. Since existing sEMG features have weak relationship between stoke level. A new feature was proposed and had good performance to reflect the similarity of the bilateral upper limbs. The experiment protocol, data process, results and discussions, and the future work will be shown in the following sections.

## 2 Methodology

### 2.1 Experiment Architecture

**Data Collection Device.** During the experiment, a 16-channel commercial device (Delsys Trigno) was used, which is shown in Fig. 1. Delsys Trigno is a wireless, high-performing device designed to make EMG signal detection reliable and easy. The sample rate is set to 1926 Hz. According to the official document, the sEMG signals range from $-5$ V to $5$ V, while the baseline noise is under $0.5$ mV RMS. To guarantee the signal quality, the acquired data was then processed through other filters in the software. As a result, the output sEMG data was 20 Hz to 500 Hz and 50 Hz powerline noise was removed. The sEMG data was further processed by Matlab on a Windows based PC. More details of data processing are stated later in this paper.

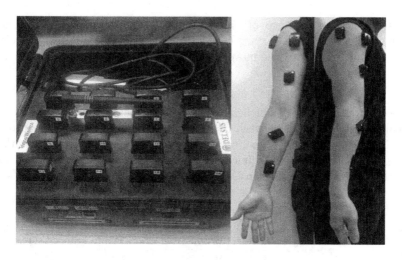

**Fig. 1.** Data collection device and electrode configuration.

### Experiment Protocol

*Electrode Configuration.* The sEMG electrodes were distributed over the subjects' upper limb to mainly cover muscle groups including biceps brachii, triceps brachii, anterior deltoid, lateral deltoid, flexor carpi radialis, and pronator teres, where 6 Delsys sensors were adopted. The detection site configuration is shown in Fig. 1.

*Motion Design.* Four primitive upper limb motions were performed in the experiment. The four motions included lifting arm forward, elbow flexion and extension, forearm pronation and supination, and lifting arm laterally. For the second and third motions, each motion primitive comprised a pair of basic rivalry movements.

*Data Collection.* Over the course of the experiment, the subjects were informed to sit up straight at their comfortable position, where hemiplegic subjects were assisted by their own therapists to maintain the stable trunk posture. One single trial of a motion primitive lasted for 15 s. In each trial of the first and last motion primitive, the subjects were asked to rest for the first 5 s and execute the motion for the last 10 s, While two rivalry movements were performed and held for 5 s respectively for the second and third motion primitive. In the meantime, 10 s rest was given between adjacent trials. For healthy subjects, 8 trials were performed and recorded for each motion primitive. However, due to the heavy burden of controlling the paralysed limb, 6 trials were performed and recorded instead.

All trials were conducted by the unilateral sides of the subjects respectively. The hemiplegic subjects were asked to conduct the trials with their healthy limb first to familiarise themselves with the paradigm of motions. Motions completed by the paralysed side of hemiplegic patients were assisted by the therapists throughout the session, more specifically by keeping the posture of joints or muscles fixed. The variance of paralysis degrees of patients contributed to the failure in completing certain tasks, which was mitigated through adopting assisted or limited muscle movement with fully voluntary motion intention instead of passive ones.

**Subject Information.** Eleven healthy subjects participated in this experiment. All healthy subjects were between the age of 20 and 30 and had no neuromuscular disease. At the same time, 6 stroke patients were also recruited. All six stroke patients had accepted clinical examinations in an authoritative medical institution. Detailed information is given in Table 1.

**Table 1.** Basic information of the stroke patients.

| Subject | Gender | Brunnstrom | Fugl-Meyer |
|---------|--------|------------|------------|
| 1 | *Male* | V-V | 45/66 |
| 2 | *Male* | III-IV | 18/66 |
| 3 | *Male* | V-VI | 49/66 |
| 4 | *Male* | III-II | 13/66 |
| 5 | *Male* | III-IV | 33/66 |
| 6 | *Female* | II-IV | 27/66 |

## 2.2   Data Processing

**Data Segmentation.** In general, two methods of EMG segmentation are widely acknowledged: disjoint segmentation and overlapped segmentation [17]. In disjoint segmentation, the EMG data is separated into several parts for further feature extraction. However, in overlapped segmentation, a sliding window with certain length slides over the data. Oskoei and Hu [17] proved that overlapped segmentation performed better during EMG classification.

In our study, data processing was performed in Matlab. As stated in the experiment protocol section, a single trial of a motion primitive lasted for 15 s. For each trial, the acquired sEMG data was saved in Excel format. In order to make data analysis easier, as shown in Fig. 2, the 15-s data was divided into three periods, 5 s for each period.

**Feature Extraction.** Following data segmentation, feature extraction was performed. Several classic sEMG features were extracted with the window length of 5 s, which have been shown in Table 2. For all subjects, under the same motion primitive, we focused on comparing these sEMG features on bilateral upper limbs within the same period of time.

Furthermore, a new feature was constructed to evaluate the similarity of bilateral upper limbs. The feature was named Envelope Correlation (EC). EC

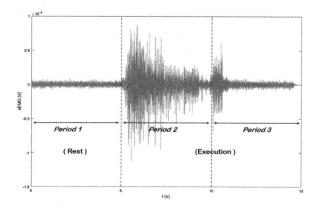

**Fig. 2.** Data segmentation for 15-s sEMG data.

**Table 2.** SEMG feature extracted in this study.

| Feature extraction | Mathematical equation |
|---|---|
| Root Mean Square (RMS) | $RMS = \sqrt{\frac{1}{N}\sum_{n=1}^{N} X_n^2}$ |
| Waveform Length (WL) | $WL = \sum_{n=1}^{N} |X_{n+1} - X_n|$ |
| Mean Absolute Value (MAV) | $MAV = \frac{1}{N}\sum_{n=1}^{N} |X_n|$ |
| Zero Crossing (ZC) | $ZC = \sum_{n=1}^{N-1}[sgn(X_n \times X_{n+1})) \cap |X_n - X_{n+1}| \geq threshold]\ sgn = \begin{cases} 1, & if\ X \geq threshold \\ 0, & otherwise \end{cases}$ |
| Median Frequency (MF) | $\int_0^{MF} P(f)df = \int_M^{\infty} P(f)df$ |

was built by the following steps: (a) The acquired 6-channel sEMG signals were linked into a single-channel signal. (b) Then we took root-mean-square envelopes of the single-channel data. (c) At last, we calculated the Pearson correlation coefficient of bilateral upper limbs by regarding the two set of enveloped signals as input.

## 3  Result

### 3.1  Classic sEMG Features

Table 3 shows the averaged performance of each extracted feature at the second period of the fourth motion. Two major muscle groups are mentioned in the table. Lateral deltoid is the primary muscle group for the fourth motion, while biceps brachii is the secondary muscle group. It is worth mentioning that Table 3 only shows the results of six stroke patients as the detailed results of healthy subjects shall be discussed in the future.

**Table 3.** SEMG Features of 4th motion at 2nd period of a trail

| | | | stroke 1 | stroke 2 | stroke 3 | stroke 4 | stroke 5 | stroke 6 |
|---|---|---|---|---|---|---|---|---|
| MAV ($10^{-6}V$) | Lateral deltoid | healthy | $66.03 \pm 5.53$ | $53.29 \pm 4.5$ | $26.05 \pm 1.07$ | $12.21 \pm 0.97$ | $24.66 \pm 2.01$ | $16.95 \pm 0.85$ |
| | | impaired | $17.96 \pm 0.44$ | $11.88 \pm 2.09$ | $12.62 \pm 1.33$ | $1.91 \pm 0.32$ | $3.85 \pm 0.78$ | $9.87 \pm 0.72$ |
| | Biceps brachii | healthy | $3.04 \pm 0.13$ | $2.11 \pm 0.05$ | $2.83 \pm 0.18$ | $2.73 \pm 0.22$ | $2.67 \pm 0.11$ | $2.25 \pm 0.07$ |
| | | impaired | $2.71 \pm 0.27$ | $12.98 \pm 2.04$ | $24.24 \pm 1.41$ | $8.73 \pm 4.81$ | $15.92 \pm 2.57$ | $7.19 \pm 4.13$ |
| RMS ($10^{-6}V$) | Lateral deltoid | healthy | $87.34 \pm 7.35$ | $71.72 \pm 4.42$ | $34.55 \pm 1.25$ | $16.29 \pm 1.26$ | $32.57 \pm 2.7$ | $22.51 \pm 1.13$ |
| | | impaired | $23.48 \pm 0.46$ | $16.26 \pm 3.39$ | $16.54 \pm 1.76$ | $2.5 \pm 0.43$ | $5.07 \pm 1.01$ | $12.98 \pm 0.95$ |
| | Biceps brachii | healthy | $4.02 \pm 0.17$ | $2.76 \pm 0.07$ | $3.72 \pm 0.24$ | $3.58 \pm 0.31$ | $3.51 \pm 0.15$ | $2.96 \pm 0.1$ |
| | | impaired | $3.58 \pm 0.37$ | $17.14 \pm 2.69$ | $31.89 \pm 1.76$ | $12.13 \pm 6.1$ | $21 \pm 3.24$ | $9.66 \pm 5.55$ |
| ZC | Lateral deltoid | healthy | $1002.33 \pm 31.44$ | $937 \pm 16.87$ | $1258.6 \pm 65.65$ | $1144 \pm 42.71$ | $1178 \pm 27.81$ | $1185.4 \pm 20.48$ |
| | | impaired | $1033.83 \pm 23.79$ | $704.17 \pm 32.43$ | $1032.63 \pm 36.65$ | $1403.6 \pm 141.84$ | $921 \pm 60.13$ | $889.57 \pm 18.44$ |
| | Biceps brachii | healthy | $1822.67 \pm 25.04$ | $1657.33 \pm 70.34$ | $1514.8 \pm 36.97$ | $1317.8 \pm 59.92$ | $1615.6 \pm 42.68$ | $1759 \pm 24.06$ |
| | | impaired | $1084.33 \pm 24.79$ | $927.5 \pm 147.73$ | $899.75 \pm 44.51$ | $1072 \pm 190.73$ | $722.6 \pm 5.82$ | $958.29 \pm 56.29$ |
| WL ($10^{-6}V$) | Lateral deltoid | healthy | $22.48 \pm 1.63$ | $16.91 \pm 1.75$ | $11.06 \pm 0.43$ | $4.73 \pm 0.39$ | $9.9 \pm 0.7$ | $6.8 \pm 0.29$ |
| | | impaired | $6.3 \pm 0.24$ | $2.78 \pm 0.46$ | $4.38 \pm 0.5$ | $0.89 \pm 0.06$ | $1.18 \pm 0.19$ | $2.99 \pm 0.21$ |
| | Biceps brachii | healthy | $1.85 \pm 0.06$ | $1.17 \pm 0.09$ | $1.45 \pm 0.07$ | $1.2 \pm 0.06$ | $1.45 \pm 0.06$ | $1.32 \pm 0.04$ |
| | | impaired | $1 \pm 0.12$ | $4.06 \pm 0.44$ | $7.42 \pm 0.26$ | $2.83 \pm 1.42$ | $3.91 \pm 0.58$ | $2.34 \pm 1.21$ |
| MF (Hz) | Lateral deltoid | healthy | $82.82 \pm 0.98$ | $74.06 \pm 1.04$ | $96.28 \pm 4.61$ | $79.25 \pm 1.55$ | $91.98 \pm 1.7$ | $92.34 \pm 2.16$ |
| | | impaired | $77.45 \pm 1.86$ | $49.85 \pm 2.88$ | $72.6 \pm 4.14$ | $66.02 \pm 5.3$ | $52.84 \pm 6.82$ | $62.02 \pm 1.44$ |
| | Biceps brachii | healthy | $130.13 \pm 4.78$ | $101.47 \pm 12.93$ | $102.06 \pm 1.89$ | $60.09 \pm 3.24$ | $106.57 \pm 3.04$ | $119.69 \pm 2.81$ |
| | | impaired | $60.15 \pm 3.55$ | $63.84 \pm 8.29$ | $69.05 \pm 3.53$ | $69.01 \pm 9.56$ | $56.79 \pm 3.16$ | $72.78 \pm 2.79$ |

For lateral deltoid, root mean square (RMS) of the impaired side is much smaller than the other one ($p < 0.05$). On the contrary, biceps brachii performs in the opposite way. As demonstrated by other authors [18], RMS of the impaired side is significantly different from the healthy side. This conclusion is verified in our results. Besides, the distribution of RMS among six channels reflects the pattern of strength within a particular period of time. After meticulous analysis, the results show that similar strength pattern resides in all the healthy subjects as well as bilateral upper limbs. But great differences arise in the results of stroke patients. In addition, numerically, waveform length (WL) and mean absolute value (MAV) perform similarly to RMS.

Zero crossing (ZC) also shows different performances between bilateral upper limbs. For biceps brachii, the value of the healthy side is much bigger than the other one ($p < 0.05$). However, there is no consistency in comparing these two values among the six stroke patients.

For median frequency (MF), the value of the impaired side is much lower the healthy side ($p < 0.05$) in either muscle group. Besides, the ratio of MF on the impaired side to MF on the healthy side is related to the severity of sickness.

## 3.2  Envelope Correlation (EC)

As shown in Fig. 3(a), EC reaches stable and high value on unilateral upper limb for each healthy subject. Under normal circumstances, for the same motion, each healthy subject behaves the same among different trials. We hope to construct a feature which may reflect the severity of sickness. So stability and high value on unilateral upper limb are just necessary conditions of a valid one.

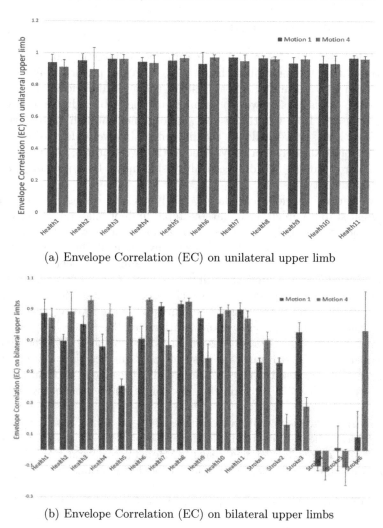

(a) Envelope Correlation (EC) on unilateral upper limb

(b) Envelope Correlation (EC) on bilateral upper limbs

**Fig. 3.** Envelope Correlation (EC) on unilateral upper limb and bilateral upper limbs

Figure 3(b) shows the performance of EC on bilateral upper limbs in healthy subjects and stroke patients. For the vast majority of healthy subjects, EC reaches 0.7. More than half of the healthy score over 0.8, and some even achieve 0.9. On the contrary, EC of the six stroke patients is much lower. Overall, the first and the third stroke patient obtain high value of EC. At the same time, they indeed have had better rehabilitation effect among the six patients.

## 4    Discussion and Conclusion

In this study, several classic sEMG features were investigated. The results demonstrated that the mentioned classic sEMG features performed much differently between bilateral upper limbs. Besides, the general performance of these patients was: (a) lack of power in primary muscles, (b) redundant power in secondary muscles. However, the results were obtained through qualitative comparison. That is to say, we could see the differences, but we could hardly evaluate the severity of sickness numerically. Among the five features, the ratio of MF on the impaired side to MF on the healthy side might be a possible index to evaluate the severity of sickness.

A new feature, envelope correlation (EC) was also discussed. The results showed great stability in either unilateral upper limb or bilateral upper limbs of healthy subjects. And EC was proved to be a valid feature which could roughly reflect the severity of sickness. The reasons why EC could become an effective feature were as follows: (a) By connecting six channels together, EC not only contained the valuable information of all the channels, but also intuitively showed the distribution of strength among the six channels. (b) Through the root-mean-square envelopes, some of the original "impurities" of the myoelectric data were removed. The data became smooth and relatively stable.

Meanwhile, four motion primitives were involved in this study. The first and the fourth motion showed much more stable performance than the other two motions, possibly because the myoelectric signal of the second and third motions were not strong enough. Therefore, the selection of motions is another essential topic.

In further work, more motion primitives shall be added based on a larger sample to verify effectiveness of envelope correlation (EC). Besides, we will continue to explore to construct new features. Hopefully, several features will be used to establish an assessment model for medical evaluations of stroke.

**Acknowledgments.** This work is supported by the National Natural Science Foundation of China (No. 51575338, 51575407, 51475427) and the Fundamental Research Funds for the Central Universities (No. 17JCYB03).

## References

1. Zhang, X., Zhou, P.: High-density myoelectric pattern recognition toward improved stroke rehabilitation. IEEE Trans. Biomed. Eng. **59**(6), 1649–1657 (2012)
2. Kim, M.S., et al.: The influence of laterality of pharyngeal bolus passage on dysphagia in hemiplegic stroke patients. Ann. Rehabil. Med. **36**(5), 696–701 (2012)
3. Kulishova, T.V., Shinkorenko, O.V.: The effectiveness of early rehabilitation of the patients presenting with ischemic stroke. Voprosy kurortologii, fizioterapii, i lechebnoi fizicheskoi kultury (6), 9–12 (2014). Vopr Kurortol Fizioter Lech Fiz Kult
4. Rasmussen, R.S., et al.: Stroke rehabilitation at home before and after discharge reduced disability and improved quality of life: a randomised controlled trial. Clin. Rehabil. **30**(3), 225–236 (2016)

5. Jiang, R.-r., Yan, C.H.E.N., Pan, C.-h.: Advance in assessment of upper limb and hand motor function in patients after stroke. Chin. J. Rehabil. Theor. Pract. **10**, 1173–1177 (2015)

6. Dalla Toffola, E.: Myoelectric manifestations of muscle changes in stroke patients. Arch. Phys. Med. Rehabil. **82**(5), 661–665 (2001)

7. Han, R., Ni, C.M.: Effect of electromygraphic biofeedback on upper extremity function in patients with hemiplegia after stroke. Zhongguo Kangfu Lilun yu Shijian **11**(3), 209–210 (2005)

8. Cheng, P.-T., et al.: Leg muscle activation patterns of sit-to-stand movement in stroke patients. Am. J. Phys. Med. Rehabil. **83**(1), 10–16 (2004)

9. Chowdhury, R.H., et al.: Surface electromyography signal processing and classification techniques. Sensors **13**(9), 12431–12466 (2013)

10. Ahsan, M.R., Ibrahimy, M.I., Khalifa, O.O.: EMG signal classification for human computer interaction: a review. Eur. J. Sci. Res. **33**(3), 480–501 (2009)

11. Hudgins, B., Parker, P., Scott, R.N.: A new strategy for multifunction myoelectric control. IEEE Trans. Biomed. Eng. **40**(1), 82–94 (1993)

12. Englehart, K., Hudgins, B.: A robust, real-time control scheme for multifunction myoelectric control. IEEE Trans. Biomed. Eng. **50**(7), 848–854 (2003)

13. Alkan, A., Gnay, M.: Identification of EMG signals using discriminant analysis and SVM classifier. Expert Syst. Appl. **39**(1), 44–47 (2012)

14. Lin, K., et al.: A robust gesture recognition algorithm based on surface EMG. In: Seventh International Conference on Advanced Computational Intelligence (ICACI). IEEE (2015)

15. Kwatny, E., Thomas, D.H., Kwatny, H.G.: An application of signal processing techniques to the study of myoelectric signals. IEEE Trans. Biomed. Eng. **4**, 303–313 (1970)

16. Sekulic, D., Medved, V., Rausavljevi, N.: EMG analysis of muscle load during simulation of characteristic postures in dinghy sailing. J. Sports Med. Phys. Fit. **46**(1), 20 (2006)

17. Oskoei, M.A., Huosheng, H.: Support vector machine-based classification scheme for myoelectric control applied to upper limb. IEEE Trans. Biomed. Eng. **55**(8), 1956–1965 (2008)

18. Ashby, P., Mailis, A., Hunter, J.: The evaluation of spasticity. Can. J. Neurol. Sci. **14**(S3), 497–500 (1987)

# Multi-length Windowed Feature Selection for Surface EMG Based Hand Motion Recognition

Dalin Zhou$^{(\boxtimes)}$, Yinfeng Fang, Zhaojie Ju, and Honghai Liu

Intelligent Systems and Biomedical Robotics Group, School of Computing,
University of Portsmouth, Portsmouth PO1 3HE, UK
{dalin.zhou,yinfeng.fang,zhaojie.ju,honghai.liu}@port.ac.uk

**Abstract.** Feature selection for surface electromyography based hand motion recognition has been seen to retrieve an optimal or quasi-optimal feature subset for classification. This work aims to consider the influence of channel, feature and window length simultaneously with an emphasis on the multiple segmentation. The bacterial memetic algorithm is applied to select the feature candidates from time domain and autoregressive coefficients, which is measured by the inter-day hand motion recognition accuracy. The evaluation is conducted on a case study of 3 able-bodied subjects performing 9 hand motions in consecutive 7 days with 4 different window lengths adopted for the electromyographic data segmentation. Classification in combination with the multi-length windowed feature selection achieved an improved recognition accuracy in comparison with using solely the single-length windowed features in inter-day scenarios and indicated that complementary information to full length segmentation resides in the sub-windows, thus providing feasible feature combinations for conventional pattern recognition based solutions to prosthetic control.

**Keywords:** Electromyography · Hand motion recognition
Window segmentation · Feature selection · Bacterial memetic algorithm

## 1 Introduction

Hand motions play a vital role in activity of daily livings and the recognition of hand motion is always a trending research topic. Besides the contribution towards a better recognition of hand motions for able-bodied people in the human-human and human-object interaction [1], active demands exist for subjects with limb impairment who would benefit more from improved hand motion recognition and execution. As a result, recent decades have witnessed the prompt development of the prosthetic hand design and its corresponding control strategies [2]. Among all the feasible noninvasive sensing modalities, surface electromyography (sEMG) has remained the mainly adopted access to upper

© Springer Nature Switzerland AG 2018
Z. Chen et al. (Eds.): ICIRA 2018, LNAI 10984, pp. 264–274, 2018.
https://doi.org/10.1007/978-3-319-97586-3_24

limb motion intention for its simplicity and robustness [3]. An accurate recognition of motion intention forms the basis of an intuitive control of the prosthesis which has been intensively investigated in this field. The most prominent recognition results in prosthetic control with a promising recognition accuracy rate over 90% are mostly confined to pattern recognition based solutions in laboratory environment. Despite the most recently growing research interest in deep learning approaches [4,5], conventional pattern recognition based methods with handcrafted features still play a dominant role in both academic and clinical scenarios. As a result, different stages of a typical pattern recognition problem have been continuously studied in the case of sEMG based hand motion recognition including signal preprocessing, feature extraction, classification and postprocessing. The success of a pattern recognition application empirically relies on the premise that distinguishable and consistent features are extracted. The importance of an effective feature set in sEMG based motion recognition has also been confirmed by Scheme et al. [6] with the observation of similar performance over most modern classifiers. Thus the sEMG feature extraction in domains of time, frequency and time-frequency has been elaborated on, as per the evaluation conducted by Phinyomark et al. [7].

Besides the numerous research of feature extraction strategies, feature selection is also addressed by researchers seeking for an optimised feature combination. Oskoei et al. [8] applied the genetic algorithm in the recognition of 6 motions to select the optimal subset for their 4-channel myoelectric system, while Khushaba et al. utilised the particle swarm optimisation to reduce the dimension of feature and channel combinations [9]. Al-Timemy et al. [10] looked into the finger movement classification and achieved comparable accuracy with subsets of selected channels instead of utilising the full set of all channels. Recent research has been seen on the feature selection against force and position variation. The classification accuracy under limb variation was improved by [11] with the selected subset out of 10 feature candidates. Adewuyi et al. [12] evaluated the optimal sEMG feature subset in varying wrist positions for subjects with partial hand amputation and proved its superiority over the ensemble TD or TDAR features. Let alone the emphasis on the feature selection, the importance of the data segmentation with various window length of sEMG stream is addressed by [13]. However to our knowledge, no studies have yet investigated the feature subset selection with fused multiple window lengths. The subset selection that simultaneously concerns feature type, detecting site and window segmentation is still missing. In our previous work, a memetic evolutionary method named bacterial memetic algorithm (BMA) has been applied in reducing the feature dimension while preserving a comparable recognition accuracy [14]. To further explore an optimised feature subset considering the three factors mentioned above, a BMA based feature selection of feature candidates extracted from multi-length windows is evaluated in this paper.

The rest of this paper is organised as follows. A brief introduction of the conventional routine of sEMG based hand motion recognition and the BMA based feature selection is outlined in Sect. 2. And the experiment setup and

results are presented and further discussed in Sect. 3. After the discussion, this
paper is finally concluded in Sect. 4.

## 2   EMG Feature Extraction and Selection

### 2.1   Handcrafted Feature Extraction

A typical conventional pattern recognition flowchart for sEMG based hand
motion recognition is illustrated in Fig. 1. Raw sEMG signals are first captured
by either dry or wet electrodes attached on the skin surface of upper limbs. Filters are then adopted to remove common mode noises, power line noises and
irrelevant components before feeding the signal stream to subsequent pattern
recognition system. The stream is later routinely segmented by windows with
either overlapping or non-overlapping increments. For each segment, its features are extracted and classified into predefined motion types, which forms the
sequence of recognition results. Postprocessing techniques such as majority vote
and velocity ramp are then used to concatenate the decision stream for prosthetic control. The assumption of this paper is that the combination of multiple
segmentation and feature extraction strategies can be optimised to improve the
classification results.

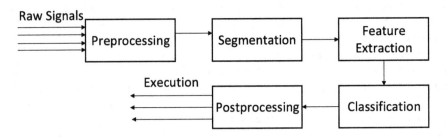

**Fig. 1.** A typical sEMG and pattern recognition based hand motion recognition

Despite the large amount of potential features, the Hudgins' time domain features [15] and the autoregressive coefficients (TDAR) have been mostly exploited
for their less time consumption and robust performance, and remained the state-
of-the-art for years. In this paper, mean absolute value (MAV), waveform length
(WL), zero crossings (ZC), slope sign changes (SSC) and 4-th order autoregressive coefficients (AR) of multiple window lengths are incorporated as candidate
features while segmentations of 250 ms, 200 ms, 150 ms and 100 ms are employed
as the window lengths.

### 2.2   Bacterial Memetic Algorithm

BMA is an evolutionary memetic algorithm inspired by the nature of microbial
evolution that combines both global and local optimisation [16]. Our previous

work of feature selection for dimension reduction with BMA can be seen in [14] and a brief introduction of the memetic evolutionary technique is given here. BMA comprises basic operators including mutation, transfer and the local search, where they take advantages of both evolutionary and memetic approaches. The optimisation problem of feature selection is to seek the optimal or quasi-optimal combinations of features, which can be encoded by the chromosomes. And the evaluation of each chromosome is based on the predefined fitness which measures the accuracy for a classification problem. After the initialisation of random chromosomes, mutation, transfer and local search operators will function on the chromosomes according to the calculated fitness. The loop of three operations will be carried out continuously till the termination criteria are met. The chromosome with the best fitness will then be retrieved as the final output.

In this work, the chromosomes are encoded by nonnegative integers representing feature candidates and the pattern recognition based inter-day hand motion accuracy is adopted as the fitness. Similarly to our previous work, the chromosome length is constrained by the regulariser as follows [14] to avoid overfitting, where $\sigma_i$ is the fitness of i-th chromosome/feature subset, $S$ is the number of hand motion samples for testing, $(x, t)$ is the sample from the testing domain $T$, $f(x)$ is the motion type classification which generates the labels, $L$ is the chromosome length, $U$ is the predefined upper boundary of chromosome length, $\lambda$ is the regularisation parameter and $\delta(k)$ is the Kronecker delta function.

$$\sigma_i = \frac{1}{S} \sum_{(x,t) \in T} \delta[f(x) - t] + \lambda \frac{L}{U}$$

$$\delta(k) = \begin{cases} 1, & k = 0 \\ 0, & k \neq 0 \end{cases} \tag{1}$$

The mutation operator functions on every chromosome in each loop to import new information. Multiple duplications of the original chromosome will be created with subsequent mutation of random positions. The muted chromosomes together with the unaltered one will be evaluated. The original chromosome is then replaced by the one with the best fitness. In our application, the chromosome length is variable in terms of including new feature candidates or excluding the current ones, and the length variation will only take place in the mutation stage. The parameters required for this operation include the probabilities of length variation and mutation. The transfer operation exchanges the segments of two chromosomes to acquire more information from the superior one. Chromosomes will be first sorted and split into two halves according to their fitness, where the superior set comprises those with better fitness and vice versa. A pair of chromosomes are then randomly selected from the two sets and the pointwise exchange will be conducted on their fragments with a fixed length. The changes of chromosomes which lead to an inferior fitness will be eliminated. The two sets will be sorted again with their new fitness when an valid exchange occurs. The exchange position and length are the parameters required for this operation. The local search operator acts on each chromosome based on the prior knowledge in the application scenario. The local search regions are firstly defined according to the neighbour candidates of each feature. The evaluation is then conducted within the region to find the local optimum in each loop. In our research, the criteria of determining the searching region is confined within the detection site distribution and feature type. For example, the local search of a channel will be carried out among its proximate channels while the searching candidates of a feature type will remain in the same domain, which exploits the most of the clinical context and boosts the convergence rate.

## 3    Experiment Setup and Analysis

### 3.1    Experiment Setup for Data Acquisition

The sEMG capturing system developed by the authors [17] was used for the data acquisition. The system consists of 16 channels covering the forearm muscles on both anterior and posterior sides with the bi-polar electrodes embedded in a customised sleeve connected by wired cables as shown in Fig. 2. The acquisition system is equipped with a sampling frequency of 1000 Hz, a gain of 3000 and an ADC resolution of 12 bits.

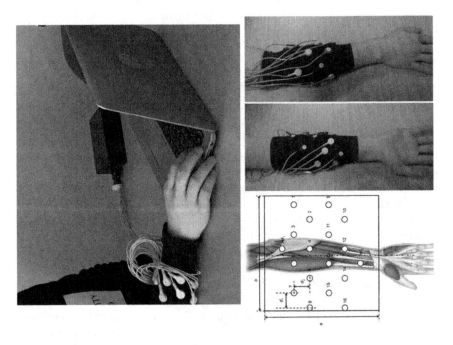

**Fig. 2.** A customised sEMG acquisition device [17]

Three able-bodied subjects participated in the experiment to conduct 9 hand motions including hand open/close, wrist extension/flexion, forearm supination/pronation, finger pinch/point and rest, as can be referred in Fig. 3. In this work, we evaluated the inter-day recognition and focused on the consistency of feature selection improvement. The training and validation data for feature selection came from the sEMG signals captured on one subject for 3 days while the test data was gathered from the other two subjects in consecutive 7 days to verify the feasibility. During the signal acquisition, subjects were asked to sit at their comfortable positions with elbows bent and forearm rest on the desk. Visual hints were displayed on the monitor to instruct the subjects of current motions to follow and each of them lasted for 10 s. A short rest was arranged between adjacent sessions to remedy the adverse effect of muscle fatigue.

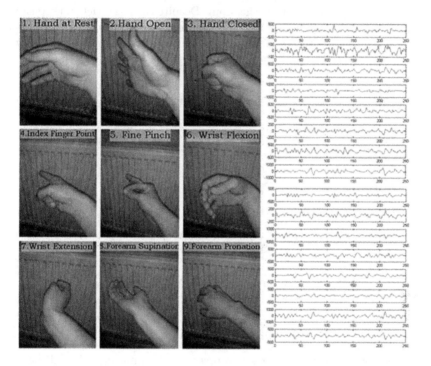

**Fig. 3.** Candidate motions and the 16-channel sEMG sample

The raw sEMG signals were firstly bandpass filtered by a Butterworth filter between 10 Hz and 500 Hz and a notch filter at the powerline frequency. Only the stationary phase of sEMG signals were kept for the recognition process the first and last 1 s of data between consecutive motions were discarded as the transient part. The filtered stationary stream was then segmented by the windows of 250 ms with an overlap of 200 ms. The segmentations were further segmented with the length/overlap combinations of 200 ms/190 ms, 150 ms/130 ms and 100 ms/50 ms as multi-length candidate sub-windows. The sub-windows were then concatenated and the feature extraction was performed on every one of them. For each window, a total of 128 dimensional features of 16 channels would be extracted which led to the final 2176 candidate features. Prior to the succeeding process, the features extracted on the same day were routinely normalised between 0 and 1. To exploit the most statistically invariant information, the 128 features of the 250 ms window were kept as the basis while the selection was conducted among those sub-windows. The feature selection was carried out following the BMA with the inter-day recognition accuracy as fitness. Based on our prior knowledge in feature selection for sEMG based hand motion recognition, the parameter configuration of BMA followed the setting in [14] with a similar convergence shown in Fig. 4. Linear discriminant analysis (LDA) was adopted as the classifier and a preceding principal components analysis (PCA) was used to collect the first 20 components as the input. Despite the less strict time consumption requirement for offline applications like feature selection, the convergence of BMA is achieved within a small amount of iterations, which coincides with our previous research.

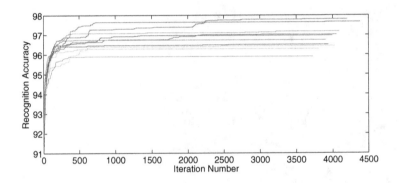

**Fig. 4.** Convergence for feature selection process

## 3.2  Experiment Results and Discussion

The feasibility of selected features is represented by inter-day recognition error respectively, on both testing and validating data from all three subjects in Fig. 5, on the testing data only captured from the two subjects shown in terms of training on 1 day and testing on the rest 6 days in Fig. 6. The trend can be seen that the average errors decrease with the adoption of selected features in comparison with solely using the single-length windowed ones. However, a discrepant case occurs for subject 2 when the data on day 2 is adopted for training and tested on the remaining days.

The enumeration of the channels, window lengths and feature types selected are demonstrated in Fig. 7 respectively and summarised in Table 1. An intuitive selection preference can be deducted by the results that windows with a longer segment length are more likely to be selected. The TD features especially the WL and SSC features and the first 2 components of the AR features are more favoured in the selection. The last 2 AR components are least preferred during

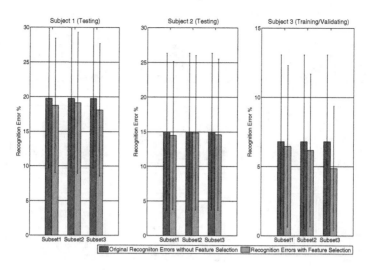

**Fig. 5.** Average recognition results with/without feature of 3 different subsets

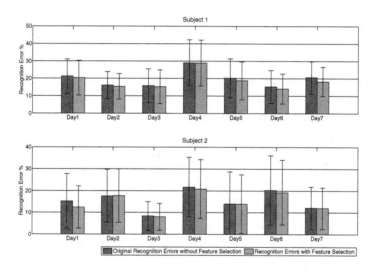

**Fig. 6.** Recognition with/without feature selection for inter-day use

the selection. The preference towards a larger segmentation is seen when using accuracy as the evaluation criteria, which coincides with the previous research [18] that an increased window length will provide features with less statistical variance, and results in a better recognition accuracy. However, due to the processing time, a tradeoff has to be considered between the time consumption and the classification accuracy. No explicit preference in the channel selection is possibly attributed to the preconditions of keeping all features extracted from all channels with 250 ms, where the baseline of an acceptable recognition accuracy can be achieved.

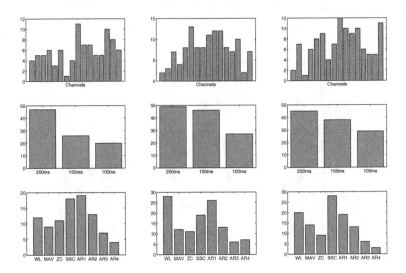

**Fig. 7.** Enumeration of the selected channels, window lengths, feature types

**Table 1.** Enumeration of window lengths and feature types

| Category | Candidate 1 | Candidate 2 | Candidate 3 |
|---|---|---|---|
| Window length 200 ms | 47 | 49 | 45 |
| Window length 150 ms | 26 | 46 | 38 |
| Window length 100 ms | 20 | 27 | 29 |
| Feature WL | 12 | 28 | 20 |
| Feature MAV | 9 | 12 | 14 |
| Feature ZC | 11 | 11 | 9 |
| Feature SSC | 18 | 19 | 28 |
| Feature AR1 | 19 | 26 | 19 |
| Feature AR2 | 13 | 13 | 13 |
| Feature AR3 | 7 | 6 | 6 |
| Feature AR4 | 4 | 7 | 3 |

## 4    Conclusion

In this paper, the feature selection among segmentation sub-windows with multiple lengths was investigated in an inter-day hand motion recognition scenario. Improvement in the average recognition accuracy has been seen on the two subjects as a case study. The potential of multi-length windowed feature fusion was preliminarily shown. The enumeration of multiple feature selection process showed support to the longer window lengths, and feature types of waveform length, slope sign changes and the first 2 components of autoregressive coefficients. Our next work is to further apply the feature selection results in scenarios with variant arm positions, exerted forces and various levels of muscle fatigue. A supplementary dataset captured from amputee subjects will be considered to validate its clinical usability. Besides, the inter-subject recognition accuracy will be checked upon the incorporation of more subjects.

**Acknowledgments.** The authors would like to acknowledge support from DREAM project of EU FP7-ICT (grant no. 611391), and National Natural Science Foundation of China (grant no. 51575338 and 51575412).

# References

1. Ji, X., Wang, C., Ju, Z.: A new framework of human interaction recognition based on multiple stage probability fusion. Appl. Sci. **7**(6), 567 (2017)
2. Fang, Y., Hettiarachchi, N., Zhou, D., Liu, H.: Multi-modal sensing techniques for interfacing hand prostheses: a review. IEEE Sens. J. **15**(11), 6065–6076 (2015)
3. Xue, Y., Ju, Z., Xiang, K., Chen, J., Liu, H.: Multimodal human hand motion sensing and analysis-a review. IEEE Trans. Cogn. Develop. Syst. (2018)
4. Atzori, M., Cognolato, M., Müller, H.: Deep learning with convolutional neural networks applied to electromyography data: a resource for the classification of movements for prosthetic hands. Front. Neurorobot. **10**, 9 (2016)
5. Zhai, X., Jelfs, B., Chan, R.H., Tin, C.: Self-recalibrating surface emg pattern recognition for neuroprosthesis control based on convolutional neural network. Front. Neurosci. **11**, 379 (2017)
6. Scheme, E., Englehart, K.: Electromyogram pattern recognition for control of powered upper-limb prostheses: State of the art and challenges for clinical use. J. Rehab. Res. Develop. **48**(6), 643 (2011)
7. Phinyomark, A., Quaine, F., Charbonnier, S., Serviere, C., Tarpin-Bernard, F., Laurillau, Y.: EMG feature evaluation for improving myoelectric pattern recognition robustness. Exper. Syst. Appl. **40**(12), 4832–4840 (2013)
8. Oskoei, M.A., Hu, H.: GA-based feature subset selection for myoelectric classification. In: 2006 IEEE International Conference on Robotics and Biomimetics, pp. 1465–1470. IEEE (2006)
9. Khushaba, R.N., Al-Jumaily, A.: Channel and feature selection in multifunction myoelectric control. In: 2007 IEEE 29th Annual International Conference of the Engineering in Medicine and Biology Society, EMBS 2007, pp. 5182–5185. IEEE (2007)
10. Al-Timemy, A.H., Bugmann, G., Escudero, J., Outram, N.: Classification of finger movements for the dexterous hand prosthesis control with surface electromyography. IEEE J. Biomed. Health Inform. **17**(3), 608–618 (2013)
11. Al-Angari, H.M., Kanitz, G., Tarantino, S., Cipriani, C.: Distance and mutual information methods for emg feature and channel subset selection for classification of hand movements. Biomed. Signal Process. Control **27**, 24–31 (2016)
12. Adewuyi, A.A., Hargrove, L.J., Kuiken, T.A.: Evaluating emg feature and classifier selection for application to partial-hand prosthesis control. Front. Neurorobot. **10**, 15 (2016)
13. Smith, L.H., Hargrove, L.J., Lock, B.A., Kuiken, T.A.: Determining the optimal window length for pattern recognition-based myoelectric control: balancing the competing effects of classification error and controller delay. IEEE Trans. Neural Syst. Rehab. Eng. **19**(2), 186–192 (2011)

14. Zhou, D., Fang, Y., Botzheim, J., Kubota, N., Liu, H.: Bacterial memetic algorithm based feature selection for surface emg based hand motion recognition in long-term use. In: 2016 IEEE Symposium Series on Computational Intelligence (SSCI), pp. 1–7. IEEE (2016)
15. Hudgins, B., Parker, P., Scott, R.N.: A new strategy for multifunction myoelectric control. IEEE Trans. Biomed. Eng. **40**(1), 82–94 (1993)
16. Botzheim, J., Cabrita, C., Kóczy, L.T., Ruano, A.: Fuzzy rule extraction by bacterial memetic algorithms. Int. J. Intell. Syst. **24**(3), 312–339 (2009)
17. Fang, Y., Liu, H., Li, G., Zhu, X.: A multichannel surface emg system for hand motion recognition. Int. J. Human. Robot. **12**(02), 1550011 (2015)
18. Englehart, K., Hudgins, B.: A robust, real-time control scheme for multifunction myoelectric control. IEEE Trans. Biomed. Eng. **50**(7), 848–854 (2003)

# Industrial Robot and Robot Manufacturing

# Posture Optimization Based on Both Joint Parameter Error and Stiffness for Robotic Milling

He Xie[⊠], Wenlong Li, and Zhouping Yin

State Key Laboratory of Digital Manufacturing Equipment and Technology,
Huazhong University of Science and Technology, Wuhan 430074, China
{xiehe,wlli,yinzhp}@hust.edu.cn

**Abstract.** Industrial robot has an increasingly wide application in automatic manufacturing of aerospace parts. However, the robot still faces challenges due to the hard processing material and high demand of machining quality. In this paper, a robot posture optimization method is proposed to improve the machining accuracy in aerospace skin milling. The basic idea is that the objective function is defined as absolute average machining accuracy to be minimized. Both joint parameter error and stiffness, the major two factors causing machining error, are considered in the objective function. The function relationship between machining error and joint parameter error is established using robot kinematic. Based on robot dynamics and force adjoint transformation, the machining error with respect to joint stiffness, milling force and robot posture is theoretically analyzed. For $n$ cutting location points, the optimization problem is formulated as $(n + 1)$ variables including rotation and translation redundant freedoms to be determined by genetic algorithm. Finally, the experiment of aerospace skin milling using ABB 6660 robot is provided.

**Keywords:** Posture optimization · Stiffness · Robotic milling

## 1 Introduction

Because of high flexibility and low cost, industrial robot is widely used in aerospace manufacturing, such as aircraft skin milling, blade grinding [1], and wing drilling [2]. However, compared with CNC machine tool, the problem of relative low positioning accuracy and stiffness still exists on the robot. Research [3] shows that both joint parameters and stiffness take a major influence on the positioning accuracy. As for aerospace parts, they usually require high machining accuracy. In addition, they are usually made of some hard processing materials such as stainless steel, titanium alloy and superalloy. When processing these materials, the robot may suffer serious deformation in the end-effector. Therefore, improving the stiffness and positioning accuracy becomes more important to extend the application of robotic aerospace manufacturing.

Many works have been developed around the positioning accuracy and stiffness. As for the positioning accuracy, it is mainly improved by error compensation and posture optimization. When the robot is in the empty load, the joint parameters errors are the main cause of positioning accuracy. Based on genetic algorithm, a compensation

© Springer Nature Switzerland AG 2018
Z. Chen et al. (Eds.): ICIRA 2018, LNAI 10984, pp. 277–286, 2018.
https://doi.org/10.1007/978-3-319-97586-3_25

method of positioning error is proposed by Dolinsky [4]. The position error in the end-effector is mapped and compensated to the joint parameters. Mitsi [5] optimizes the position of robot base frame by minimizing the objective function, which is defined as the sum of squared element difference between real and nominal $4 \times 4$ transformation matrix expressed in the robot end frame. Because there are differences in the dimension of each element (rotation matrix and translation vector) in the transformation matrix, this method cannot directly reflect the influence of robot positioning accuracy.

The research on stiffness improvement is mainly concentrated on three aspects: stiffness compensation [6, 7], posture optimization [8, 9] and robot structure optimization [10]. Zargarhsh [8] points out that there is an optimal posture with maximum stiffness for robotic drilling. An optimization index is also proposed to select the redundancy freedom. Based on ten DoF robot, Ming [11] applies the differential evolution algorithm to find the optimal posture configuration with optimal stiffness. Using KUKA KR270-2 robot, Colleoni [12] optimizes the workpiece posture by minimizing the tool average displacement, where the objective function is based on cutting force model and robot stiffness.

These methods above are mainly based on the individual joint parameter error or joint stiffness to improve the machining quality. For the aeronautical parts milling, both joint parameter error and stiffness are the major two factors causing positioning error, due to hard processing materials and high demand of machining quality. In this paper, an optimization method of robot posture is proposed by minimizing the absolute average machining error in the robotic milling system. The proposed method directly considers the effect of both joint parameter error and stiffness on machining error, which accords with the actual machining condition. Based on robot kinematics, the relationship between machining error and joint parameter error under the empty load is derived. Then the stiffness affecting the machining error is analyzed. Finally, the simulation experiment using ABB 6660 robot to mill the aircraft skin is executed to verify the availability of the proposed method.

## 2  Model of Milling Robot

The schematic of robotic milling is shown in Fig. 3. Before milling, the robot base frame $\{B\}$ can move straight on the guide rail, but it is fixed during the milling process. The guide rail is located on the global frame $\{G\}$ and is along the $x_G$ axis direction. The milling tool defined in the tool frame $\{T\}$ is fixed in the robot end frame $\{E\}$, and the milling axis is defined at the $z_T$ axis direction. The workpiece frame $\{W\}$ is attached to the edge of aircraft skin that is to be milled. The tool path consists of four segments $AB$, $BC$, $CD$ and $DA$, which are generated by offsetting the edge with tool radius. The tool path is dispersed into $n$ cutter location (CL) points $\{p_1, p_2, \cdots p_i, \cdots, p_n\}$ where a CL frame $\{p_i\}$ is attached to each point $p_i$. For point $p_i$, the allowance depth direction and feed direction are along the $x_{pi}$ and $y_{pi}$ axes, respectively. Note that the kinematic equation from global frame $\{G\}$ to CL frame $\{p_i\}$ is as follows

$$ {}_{W}^{G}T \, {}_{Pi}^{W}T = {}_{B}^{G}T \, {}_{E}^{B}T \, {}_{T}^{E}T \, {}_{Pi}^{T}T \tag{1} $$

where $ {}_{\psi_2}^{\psi_1}T $ define the 4 × 4 homogeneous transformation matrix for frame $\psi_1$ to $\psi_2$. Note the following symbols (Figs. 1 and 2).

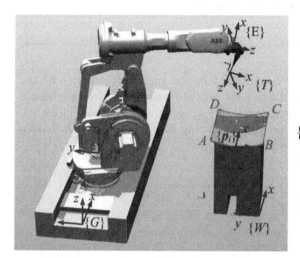 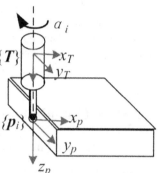

**Fig. 1.** Multicoordinate frames in robotic milling system    **Fig. 2.** Rotation $\alpha_i$ of tool around $z_T$ axis.

(1) Symbol $ {}_{B}^{G}T $ denotes the transformation from global frame $\{T\}$ to robot base frame $\{B\}$. Defining symbol $L$ as the moving distance of frame $\{B\}$ on the guide rail, the matrix $ {}_{B}^{G}T $ can be written as

$$ {}_{B}^{G}T(L) = \begin{bmatrix} 1 & 0 & 0 & L \\ 0 & 1 & 0 & 0 \\ 0 & 0 & 1 & 0 \\ 0 & 0 & 0 & 1 \end{bmatrix} \tag{2} $$

where distance $L$ is a variable to be determined for posture optimization.

(2) Symbol $ {}_{Pi}^{T}T $ denotes the transformation from tool frame $\{T\}$ to frame $\{p_i\}$. When frame $\{T\}$ and $\{p_i\}$ coincides with each other ($ {}_{Pi}^{T}T = I $), the point $p_i$ can be milled in correct position. However, the tool can still rotate around its $z_T$ axis without affecting the milling. Therefore, there is a kinematic redundancy called angle $\alpha_i$, which corresponds to the rotation angel of tool about the axis $z_T$ at point $p_i$. Then the matrix $ {}_{Pi}^{T}T $ can be written as

$$_{pi}^{T}\boldsymbol{T}(\alpha_i) = \begin{bmatrix} \cos \alpha_i & -\sin \alpha_i & 0 & 0 \\ \sin \alpha_i & \cos \alpha_i & 0 & 0 \\ 0 & 0 & 1 & 0 \\ 0 & 0 & 0 & 1 \end{bmatrix} \tag{3}$$

Then the robot posture $_{E}^{B}\boldsymbol{T}$ can be represented as

$$_{E}^{B}\boldsymbol{T} = \left(_{B}^{G}\boldsymbol{T}(L)\right)^{-1} {}_{W}^{G}\boldsymbol{T}_{pi}^{W}\boldsymbol{T}\left(_{T}^{E}\boldsymbol{T}_{pi}^{T}\boldsymbol{T}(\alpha_i)\right)^{-1} \tag{4}$$

Therefore, the robot posture $_{E}^{B}\boldsymbol{T}$ is decided by the angle $\alpha_i$ $(i = 1, 2, \cdots n)$ and distance $L$. Because both the joint stiffness and joint geometric parameter are related to the robot posture. They have a great impact on the machining accuracy. In the following, optimizing the robot posture and machining accuracy based on both joint stiffness and joint geometric parameters is presented.

# 3    Posture Optimization

## 3.1    The Objective Function

To optimize the robot posture for machining accuracy, the objective function is defined as

$$\min \quad F(\alpha_1, \alpha_2, \cdots, \alpha_n, L) = F_a(\alpha_1, \alpha_2, \cdots, \alpha_n, L) + F_k(\alpha_1, \alpha_2, \cdots, \alpha_n, L) \tag{5}$$

where symbols $F_a$ is the absolute average of the machining error $a_{qi}(L, \alpha_i)$ affected by the joint parameter error, and symbol $F_k$ is the absolute average of the machining error $a_{ki}(L, \alpha_i)$ affected by the stiffness. For $n$ CL points, there are $(n + 1)$ parameters $(\alpha_1, \alpha_2, \cdots \alpha_n, L)$ to be determined. The objective function $F$ reflects the overall machining quality. The smaller value $F$ means small machining error. The advantages of the objective function $F$ are as follows. (1) As the two major factors affecting machining error, the joint parameter error and joint stiffness are considered. (2) Error $a_{qi}$ and $a_{ki}$ have same geometric dimensions, and both of them directly reflect the relationship between machining error and joint parameter/stiffness.

## 3.2    Machining Error $a_{qi}$ with Respect to Joint Parameters

According to the Denavit Harvenberg (D-H) model, joint parameters can be represented $\boldsymbol{q} = \begin{bmatrix} \boldsymbol{a}^T & \boldsymbol{\gamma}^T & \boldsymbol{d}^T & \boldsymbol{\theta}^T \end{bmatrix}$ where $\boldsymbol{a}_{6 \times 1}$, $\boldsymbol{\gamma}_{6 \times 1}$, $\boldsymbol{d}_{6 \times 1}$ and $\boldsymbol{\theta}_{6 \times 1}$ are the vectors of link length, link angel, joint distance and joint angle, respectively. By calibrating the robot joint parameter [1], the joint parameter error $\Delta \boldsymbol{q}$ can be obtained. Using $\Delta \boldsymbol{q}$, the end posture error matrix in its end frame $\{E\}$ can be written as

$$^{E}\Delta = _{E}^{B}\boldsymbol{T}^{-1}\left(\Delta_{E}^{B}\boldsymbol{T}\right) = _{E}^{B}\boldsymbol{T}^{-1}\left(_{E}^{B}\boldsymbol{T}(\boldsymbol{q} + \Delta \boldsymbol{q}) - _{E}^{B}\boldsymbol{T}(\boldsymbol{q})\right) \tag{6}$$

When the error $\Delta q$ is small, the matrix ${}^E\Delta$ corresponds to a posture error vector ${}^ED_E = \begin{bmatrix} {}^Ed_E & {}^E\delta_E \end{bmatrix}^T$ where ${}^Ed_E$ and ${}^E\delta_E$ denote the $3 \times 1$ vectors of position error vector and orientation error vector, respectively. Define the transformation from the frame $\{p_i\}$ to end frame $\{E\}$ as ${}^{pi}_ET = \left({}^E_{pi}T^T T\right)^{-1} = \begin{bmatrix} {}^{pi}_ER & {}^{pi}p_{Eo} \\ 0 & 1 \end{bmatrix}$. Then according to the velocity adjoint transformation of differential motion, the posture error of point $p_i$ in its frame $\{p_i\}$ can be calculated as

$$
{}^{pi}D_{pi} = \begin{bmatrix} {}^{pi}_ER & [{}^{pi}p_{Eo}]{}^{pi}_ER \\ 0_{3\times3} & {}^{pi}_ER \end{bmatrix} {}^ED_E = \mathrm{Ad_V}\left({}^{pi}_ET\right){}^ED_E \tag{7}
$$

where symbol ${}^{pi}D_{pi}$ denotes a $6 \times 1$ vector, symbol $\mathrm{Ad_v}(\bullet)$ denotes the operator of velocity adjoint transformation, and symbol $[{}^{pi}p_{Eo}]_{3\times3}$ denotes the antisymmetric matrix of vector ${}^{pi}p_{Eo}$. The posture error of the workpiece origin in its frame $\{W\}$ can also be written as ${}^WD_W = \mathrm{Ad_V}\left({}^W_ET\right){}^ED_E$. Before milling, it is assumed that the origin of workpiece is selected to locate the tool. Therefore, it can be regarded as that the tool origin can reach the workpiece origin accurately. Based on the workpiece origin, the relative posture error of point $p_i$ is defined as $\Delta^{pi}D_{pi}$. Since the allowance depth direction is along the axis $x_{pi}$, the machining error affected by the joint parameter error can be written as

$$
a_{qi} = \left(\Delta^{pi}D_{pi}\right)^T D_x \tag{8}
$$

where $D_x = \begin{bmatrix} 1 & 0 & 0 & 0 & 0 & 0 \end{bmatrix}^T$ is a $6 \times 1$ matrix. Equations (6, 7 and 8) build the relationship between joint parameter errors to machining error.

### 3.3    Machining Error with Respect to Stiffness

In this section, the machining error caused by robot joint stiffness is derived. Define six-dimension milling wrench at point $p_i$ as ${}^{pi}F_i$. The wrench ${}^{pi}F_i$ is represented in the frame $\{p_i\}$. Assume that the tool is fully rigid without deformation. According to the force adjoint transformation, the force screw at the origin of end frame $\{E\}$ can be written as

$$
{}^EF_E = \begin{bmatrix} {}^E_{pi}R & 0 \\ [{}^Ep_{pio}]{}^E_{pi}R & {}^E_{pi}R \end{bmatrix} = \mathrm{Ad}\left({}^E_{pi}T\right){}^{pi}F_i \tag{9}
$$

where ${}^E_{pi}T = \begin{bmatrix} {}^E_{pi}R & {}^Ep_{pio} \\ 0 & 1 \end{bmatrix}$ is the transformation matrix, $\mathrm{Ad}(\bullet)$ is the operator of force adjoint transformation, and ${}^EF_E$ is expressed in its end frame $\{E\}$. Wrench ${}^EF_E$ can be converted to ${}^BF_E$ that is expressed in the base frame $\{B\}$. According to the robot dynamics, the deformation at the origin of robot end frame can be written as

$$^B\boldsymbol{D}_{Eo} = \left(\boldsymbol{J}\boldsymbol{K}_\theta^{-1}\boldsymbol{J}^{\mathrm{T}}\right)^B\boldsymbol{F}_E \tag{10}$$

where $^B\boldsymbol{D}_E$ is a $6 \times 1$ matrix and is expressed in the base frame $\{B\}$. Symbol $\boldsymbol{J} = \boldsymbol{J}(\boldsymbol{\theta})$ denotes the space Jacobian matrix and is related to the robot posture $\boldsymbol{\theta}$. Symbol $\boldsymbol{K}_\theta = \mathrm{diag}(k_1, k_2, \cdots, k_6)$ is the joint stiffness matrix where $k_j$ denotes the stiffness value of the $j$th joint. By transferring the error vector $^B\boldsymbol{D}_E$ to be expressed in the end frame $\{E\}$, the error at point $\boldsymbol{p}_i$ in its frame $\{\boldsymbol{p}_i\}$ can be written as $^{pi}\boldsymbol{D}_{pi} = \mathrm{Ad}_V\left(^{pi}_E\boldsymbol{T}\right)^E\boldsymbol{D}_E$. Then the machining error caused by the robot stiffness can be written as

$$a_{ki} = \left(^{pi}\boldsymbol{D}_{pi}\right)^T\boldsymbol{D}_x \tag{11}$$

where the machining error $a_{ki}$ is along the direction of $x_{pi}$ at point $\boldsymbol{p}_i$. Equations (9, 10 and 11) build the relationship between joint stiffness and machining error. It can be obtained that the machining error $a_{ki}$ is related to milling wrench $^{pi}\boldsymbol{F}_i$, robot posture $\boldsymbol{J}(\boldsymbol{\theta})$, and the robot stiffness $\boldsymbol{K}_\theta$. Unlike the machining error $a_{qi}$ with respect to joint parameter, machining error $a_{ki}$ is always positive due to the milling wrench. Based on both machining error $a_{qi}$ and $a_{ki}$, the minimization problem of objective function $F$ in (5) can be solved by genetic algorithm to calculate the variables $(\alpha_1, \alpha_2, \cdots, \alpha_n, L)$. Then the robot posture can be obtained using (4).

## 4    Experiment

In this simulation experiment, the robot joint parameter error is given in Table 1. And the milling wrench of each point $\boldsymbol{p}_i$ is $^{pi}\boldsymbol{F}_i = [500 \quad 0 \quad 0 \quad 0 \quad 0 \quad 0]^{\mathrm{T}}$ where the moment is neglected for small value. The aircraft skin has a size of $700\ mm \times 800\ mm$ There are 62 CL points on the tool path of the skin edge. The moving distance $L$ is at the range of $[-900\ mm, 400\ mm]$. For the angle variables $(\alpha_1, \alpha_2, \cdots, \alpha_n)$ where $0° \le \alpha_i \le 360°$, it can be simplified as the following two cases. (1) Each point $\boldsymbol{p}_i$ shares the same angle $(\alpha_1 = \alpha_2 = \cdots = \alpha_n)$, since the sudden change of angel $\alpha_i$ is improper for continuous and stable milling of four edges $(AB, BC, CD, DA)$. (2) Each edge shares the same rotation angel $(\alpha_1 = \cdots = \alpha_{15} = \alpha_{AB}, \alpha_{16} = \ldots = \alpha_{31} = \alpha_{BC}, \alpha_{32} = \ldots = \alpha_{46} = \alpha_{CD}, \alpha_{47} = \ldots = \alpha_{62} = \alpha_{DA})$, when the four edges $(AB, BC, CD$ and $DA)$ are milled separately.

**Table 1.** ABB 6660 joint parameters error.

| $i$ | $\Delta\gamma_{i-1}/°$ | $\Delta a_{i-1}/mm$ | $\Delta\theta_i/°$ | $\Delta d_i/mm$ |
|---|---|---|---|---|
| 1 | 0.0143 | 0.0510 | 0.1432 | 0.0220 |
| 2 | 0.0057 | 0.0100 | 0.0189 | 0.0330 |
| 3 | 0.0057 | 0.3000 | 0.0573 | 0.0120 |
| 4 | 0.0109 | 0.1200 | 0.0573 | 0.0170 |
| 5 | 0.0178 | 0.0340 | 0.1318 | 0.0170 |
| 6 | 0.1384 | 0.4000 | 0.0573 | 0.0300 |

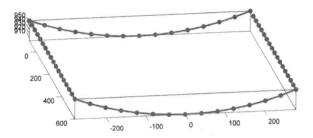

**Fig. 3.** 62 CL points at the tool path for aircraft skin milling

## 4.1 Each Point Shares the Same Rotation Angle $\alpha_i$

When each point $p_i$ share the same angle ($\alpha_1 = \cdots = \alpha_{62} = \alpha$), the variables to be determined in the objective function (5) are reduced to two variables $(\alpha, L)$ that can be calculated easily without using the genetic algorithm. Figure 4 shows the result of machining error $a_{ki}$ caused by joint stiffness. It reveals that error $a_{ki}$ is more sensitive to the distance $L$ than the angle $\alpha$, since the robot stiffness is more sensitive to the position than the orientation. It also can be know that machining error $a_{ki}$ is constant positive due to the milling wrench $^{pi}F_i$. The minimum value of the average machining error $a_{ki}$ is 0.831 $mm$ when $(\alpha, L) = (352.1°, -201.5\,mm)$. Figure 5 shows the machining error $\left(|a_{qi}| + |a_{ki}|\right)$, which corresponds to the objective function $F$ in (5). For $(\alpha, L) = (98.4°, -30.4\,mm)$, the machining error $\left(|a_{qi}| + |a_{ki}|\right)$ is minimized with 1.211 mm $\left(|a_{qi}| = 0.350\,mm \text{ and } |a_{ki}| = 0.861\,mm\right)$. Therefore, the stiffness has a major influence on stiffness. The joint angle at each CL point is given in Fig. 6. There is no sudden change of each joint during the milling process from the first point $p_1$ to the last point $p_{62}$, which corresponds to stable milling.

## 4.2 Each Edge Shares the Same Rotation Angle $\alpha_i$

When edge shares the same rotation angel $\alpha_i$, the solved variables is reduced to $S = \left(\alpha_{AB}, \alpha_{BC}, \alpha_{CD}, \alpha_{DA}, L\right)$. For the value $S^* = (101.7°, 11.3°, 90.3°, 0.4°, -30.5\,mm)$,

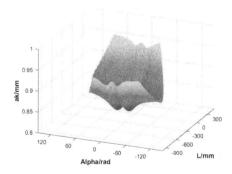

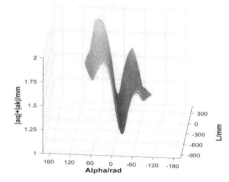

**Fig. 4.** Average machining error $a_{ki}$ with respect to variable $(\alpha, L)$

**Fig. 5.** Average machining error $\left(|a_{qi}| + |a_{ki}|\right)$ with respect to variable $(\alpha, L)$

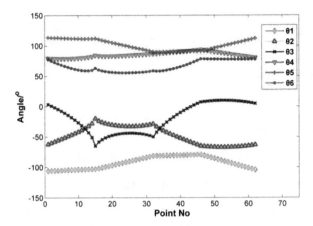

**Fig. 6.** The joint angle at each CL point when $(\alpha, L) = (98.4°, -30.4\,mm)$

the objective function $F$ is minimized with average machining error $0.958\ mm$ ($|a_{qi}| = 0.265\ mm$ and $|a_{ki}| = 0.693\ mm$). It is obvious that the machining error is reduced ($0.958\ mm$ VS $1.211\ mm$) when the variables are increased from $(\alpha, L)$ to $(\alpha_{AB}, \alpha_{BC}, \alpha_{CD}, \alpha_{DA}, L)$. However, if the machining error $a_{qi}$ is not considered, the $a_{ki}$ may not reflects the real machining error. When $L = -30.5\ mm$, the machining error $(|a_{qi}| + |a_{ki}|)$ with respect to CL point and rotation angel is shown in Fig. 7. The red line shows the best solution $S^*$. The four edges have different rotation angles, which contribute to the reduction of average machining error.

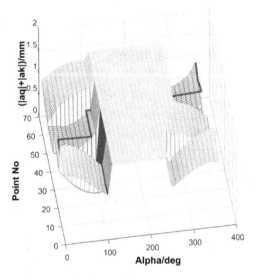

**Fig. 7.** Average machining error $(|a_{qi}| + |a_{ki}|)$ with respect to CL point and rotation angle (Color figure online)

# 5  Conclusion

To improve the machining accuracy for robotic milling in aerospace skin, this paper proposes a new optimization method for robot posture. The contributions of this paper are listed as follows. (1) Both joint parameter error and stiffness are considered to optimize the objective function. The position of robot base frame and the rotation angle of milling tool at each CL point are used as redundant freedoms to solve the optimization problem. It helps to expand the scope of optimal solution. (2) The function relationship between machining error and stiffness/joint parameters error is derived. The proposed method can also be applied into robotic grinding and drilling. Future work will add more additional constraint on the posture optimization problem for more rational optimization results. The constraints include collision avoidance, singularity avoidance, joint limit avoidance and specific machining requirement.

**Acknowledgement.** This work was by the supported by the National Natural Science Foundation of China (No. 51327801, 51535004), the National Basic Research Program of China (No. 2015CB057304), and the Outstanding Youth Foundation of Hubei Province (No. 2017CFA045).

# References

1. Li, W.L., Xie, H., Zhang, G., Yan, S.J., Yin, Z.P.: Hand–eye calibration in visually-guided robot grinding. IEEE Trans. Cybern. **46**(11), 2634–2642 (2016)
2. Devlieg, R., Sitton, K., Feikert, E., Inman, J.: Once (one-sided cell end effector) robotic drilling system (2002)
3. Judd, R.P., Knasinski, A.B.: A technique to calibrate industrial robots with experimental verification. IEEE Trans. Robot. Autom. **6**(1), 20–30 (1990)
4. Dolinsky, J.U., Jenkinson, I.D., Colquhoun, G.J.: Application of genetic programming to the calibration of industrial robots. Comput. Ind. **58**(3), 255–264 (2007)
5. Mitsi, S., Bouzakis, K.D., Sagris, D., Mansour, G.: Determination of optimum robot base location considering discrete end-effector positions by means of hybrid genetic algorithm. Robot. Comput.-Integr. Manuf. **24**(1), 50–59 (2008)
6. Schneider, U., Momeni-K, M., Ansaloni, M., Verl, A.: Stiffness modeling of industrial robots for deformation compensation in machining. In: IEEE/RSJ International Conference on Intelligent Robots and Systems, vol. 77, pp. 4464–4469. IEEE (2014)
7. Backer, J.D., Bolmsjö, G.: Deflection model for robotic friction stir welding. Ind. Robot **41** (4), 365–372 (2014)
8. Zargarbashi, S.H.H., Khan, W., Angeles, J.: The jacobian condition number as a dexterity index in 6r machining robots. Robot. Comput. Integr. Manuf. **28**(6), 694–699 (2012)
9. Bu, Y., Liao, W., Tian, W., Zhang, J., Zhang, L.: Stiffness analysis and optimization in robotic drilling application. Precis. Eng. **49**, 388–400 (2017)
10. Palpacelli, M.: Static performance improvement of an industrial robot by means of a cable-driven redundantly actuated system. Pergamon Press Inc, Oxford (2016)

11. Li, M., Wu, H., Handroos, H.: Stiffness-maximum trajectory planning of a hybrid kinematic-redundant robot machine. In: IECON 2011-37th Annual Conference on IEEE Industrial Electronics Society. vol. 6854, no. 5, pp. 283–288. IEEE (2011)
12. Colleoni, D., Miceli, G., Pasquarello, A.: Workpiece placement optimization for machining operations with a KUKA KR270-2 robot. In: IEEE International Conference on Robotics and Automation, vol. 147, pp. 2921–2926. IEEE (2013)

# A Robotic Belt Grinding Force Model to Characterize the Grinding Depth with Force Control Technology

Xiaohu Xu[1,2], Yifan Yang[1,2], Gaofeng Pan[3],
Dahu Zhu[4,5], and Sijie Yan[1,2,3(✉)]

[1] State Key Laboratory of Digital Manufacturing Equipment and Technology,
School of Mechanical Science and Engineering,
Huazhong University of Science and Technology, Wuhan 430074, China
sjyan@hust.edu.cn
[2] Blade Intelligent Manufacturing Division, HUST-Wuxi Research Institute,
Wuxi 214174, China
[3] Wuxi CRRC Times Intelligent Equipment Co., Ltd., Wuxi 214174, China
[4] Hubei Key Laboratory of Advanced Technology for Automotive Components,
Wuhan University of Technology, Wuhan 430070, China
[5] Hubei Collaborative Innovation Center for Automotive Components
Technology, Wuhan University of Technology, Wuhan 430070, China

**Abstract.** In the present paper, a new grinding force model is developed by analyzing and assessing the robotic abrasive belt grinding mechanism which is based on the fact that the chip formation during grinding process consists of three stages: ploughing, cutting and sliding. Then the grinding depth is predicted by the grinding force model to realize quantitative machining in the robotic belt grinding process. Next the grinding parameters optimization are implemented to further ensure the workpiece surface quality and profile accuracy with force control technology applied. Finally, a typical case on robotic abrasive belt grinding of test workpiece and aero-engine blade is conducted to validate the practicality and effectiveness of the grinding force model.

**Keywords:** Robotic belt grinding · Grinding depth · Force control

## 1 Introduction

With the improvement and advancement in performance of industrial robot, the robotic abrasive belt grinding has attracted worldwide attention in recent years owing to its advantages of low cost, excellent flexibility, intelligence, high efficiency, and large operating space, compared with the traditional multi-axis CNC machining [1, 2], especially in the machining of free-form geometries, such as turbine blades. The existing research on the robotic abrasive belt grinding and its application in parts with complex surface has been carried out for more than one decade. The research perspectives, however, are mainly focused on robot path planning [3, 4], material removal rate modeling [5, 6], robotic grinding force model [7, 8], abrasive grain property and wear [9, 10], etc.

© Springer Nature Switzerland AG 2018
Z. Chen et al. (Eds.): ICIRA 2018, LNAI 10984, pp. 287–298, 2018.
https://doi.org/10.1007/978-3-319-97586-3_26

During the process of robotic abrasive belt grinding, the material removal rate is regarded as one of the key evaluation criterion of machined surface contour and dimension accuracy. However, it is mainly decided by the grinding depth which is related to the grinding force, robot feed velocity and contact wheel speed, etc. Therefore, building and developing the grinding force model including the effects of grinding force, robot feed velocity and contact wheel speed is necessary to the robotic abrasive belt grinding of workpiece. Tang et al. [11] developed a grinding force model for surface grinding considering the friction and chip formation. Both the average contact pressure and the frictional coefficient are treated as variable parameters, unlike the previous research, where these parameters were considered as constant. Zhu et al. [12, 13] analyzed the grinding force based on the ploughing and chip formation without and with force control technology, then the effects of sliding, ploughing and cutting force on the surface quality are assessed in the robotic belt grinding system. Wang et al. [14] predicted the depth cut of robotic abrasive belt grinding of workpiece by modeling and simulating the contact wheel deformation, and then the FEM simulation is validated to calculate the material removal rate. Murtagian et al. [15] proposed a grinding depth model on grinding of gamma Titanium Aluminides considering the plastic deformation and then the PDD model was analyzed and measured in the abrasive belt grinding of workpiece.

The available literature mentioned above shows that there has been some progress in investigation of the belt grinding process and its cutting mechanisms, but little has been carried out on the machined surface quality at the level of grinding depth prediction, especially in the robotic belt grinding with force control technology. Moreover, a majority of grinding force model was built from the microscope force scale and then the three components of grinding force were analyzed and assessed to the surface quality in the abrasive belt grinding of workpiece. Therefore, a new force model from the macroscopic aspect is developed to predict the grinding depth and it could guide the robotic belt grinding of workpiece to realize the quantitative machining with force control applied. Next, a typical case on robotic abrasive belt grinding of test workpiece and aero-engine blade is conducted to validate the practicality and effectiveness of the grinding force model.

## 2   Mathematical Model of Abrasive Belt Grinding Force

The force control technology including PI/PD and hybrid force-position control [13, 16] is adopted to obtain the better surface quality in the robotic belt grinding of test workpiece. Most importantly, the process grinding force can be remained constant relatively and the depth of grinding (namely, the Z offset) also can be set as a fixed value in theory. In Fig. 1, the grinding force is changed into the reference force adaptively with force control strategy proposed by adjusting the real-time robot position. Particularly, $f1$ (F, Z), $f2$ (F, Z) and $f3$ (F, Z) are the curve about the relationship between grinding force and Z offset of robot with less-grinding, perfect grinding and over-grinding, respectively. When the robot position is located in Z1 or Z3, the actual force $f1$ (F1, Z1) or $f3$ (F3, Z3) would be increased or decreased to $f2$ (F2, Z2) in order to achieve the surface consistency of machined workpiece by employing the presented force control technology.

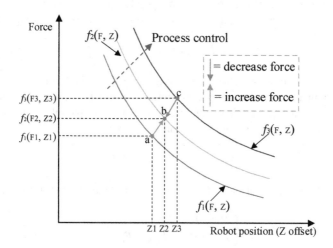

**Fig. 1.** The force control process in the robotic belt grinding.

The grinding process with force control technology which keep the normal grinding force constant can be divided into three stages: less-contact, in-contact and over-contact. Figure 2 illustrates the robotic belt grinding process with force control, in which the preset normal grinding force $F_n^0$ to be controlled is divided into three phrases: start phrase, process phrase and end phrase. In this way the sudden change in force can be reduced when the workpiece gets in touch with the abrasive belt at the beginning and it also could ensure the surface quality and machining consistency. Therefore, the normal grinding force $F_n^0$ would be treated as the basis of grinding force model owing to its controllability and adjustability.

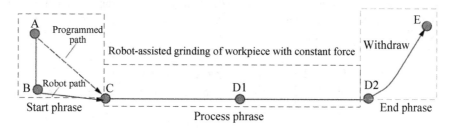

**Fig. 2.** The robotic belt grinding process with force control.

Many factors, including robot feed velocity, contact wheel speed, grinding force, depth of cut, abrasive belt granularity and type, etc., would affect the process of robotic abrasive belt grinding system [12]. The four uppermost factors (robot feed velocity $v_c$, contact wheel speed $v_w$, grinding force $F$, depth of cut $a_p$) are selected to demonstrate the grinding mechanism. Therefore, the corresponding force model is considered as an effective means to explore the cutting mechanism in robotic belt grinding, then it has:

$$F = f(v_c, v_w, a_p) \tag{1}$$

According to the process of wheel grinding forces [17], the forces generated during the belt grinding process also consist of cutting deformation force and sliding (frictional) force, and the former is separated into two parts, ploughing force and chip formation force. The belt grinding force can be expressed by:

$$F = F_{sliding} + F_{ploughing} + F_{chip} \tag{2}$$

Correspondingly, the normal grinding force and tangential belt grinding force is:

$$\begin{cases} F_n = F_{n \cdot sl} + F_{n \cdot pl} + F_{n \cdot ch} \\ F_t = F_{t \cdot sl} + F_{t \cdot pl} + F_{t \cdot ch} \end{cases} \tag{3}$$

In abrasive machining, the overall force ratio is defined as the ratio of tangential force to normal force [13]. Due to its similarity to a coefficient of friction, the symbol, $\mu$, is used:

$$\mu = \frac{F_t}{F_n} = \frac{F_{t \cdot sl} + F_{t \cdot pl} + F_{t \cdot ch}}{F_n} \tag{4}$$

Specifically, the coefficient of sliding friction is written as:

$$\mu_{sl} = \frac{F_{t \cdot sl}}{F_n} \tag{5}$$

The coefficient of ploughing friction is given by [18], in which the spherical asperity of grain is hypothesized. In Eq. (6), $R$ is termed as the average grain radius and $r$ is the distance between the axis of the spherical asperity and its point of contact with the surface of the workpiece.

$$\mu_{pl} = \frac{F_{t \cdot pl}}{F_n} = \frac{2}{\pi} \left\{ \left( \frac{2R}{r} \right)^2 \sin^{-1} \left( \frac{r}{2R} \right) - \left[ \left( \frac{2R}{r} \right)^2 - 1 \right]^{1/2} \right\} \tag{6}$$

According to the work [17], the cutting component of tangential force can be determined:

$$F_{t \cdot ch} = k \cdot \frac{v_r}{v_c} \cdot a'_p \tag{7}$$

where, $k$ is an experimental coefficient, $a'_p$ is the depth of cut, then it has:

$$\mu_{ch} = \frac{F_{t.ch}}{F_n} = k \cdot \frac{v_r}{v_c \cdot F_n} \cdot a'_p \tag{8}$$

Finally, Eq. (4) is finally rewritten as:

$$\mu = \mu_{sl} + \mu_{pl} + \mu_{ch} = \mu_{sl} + \mu_{pl} + k \cdot \frac{v_r}{v_c \cdot F_n} a'_p \tag{9}$$

Then the grinding force model can be calculated:

$$F_n = \frac{k \cdot v_r \cdot a'_p}{v_c \cdot \left( \mu - \mu_{sl} - \mu_{pl} \right)} \tag{10}$$

# 3 Experimental Procedures and Validation

## 3.1 Calculation of Grinding Force Model

Figure 3 illustrates the experimental setup of robotic abrasive belt grinding. The tests are conducted using a six degree-of-freedom ABB robot (IRB4400) with the load capacity of 60 kg, scope of work for 1.96 m, and repeat positioning accuracy of 0.19 mm. A F/T transducer (ATI Omega 160) with weight of 3.72 kg, force measurement range of 1500 N for X and Y axes and 3750 N for Z axis, torque scope of 240 N · m for X, Y and Z axes is mounted on the end of the robot to measure the three grinding force components once the constant force control strategy is implemented. The test workpiece made of Ti-6Al-4V alloy with the dimension of 100 mm × 200 mm × 20 mm is mounted on the transducer. For the belt grinding machine, the cylindrical contact wheel with the dimension of 180 mm in diameter and 25 mm in width is mainly composed of aluminium as the core material and elastic rubber as the out layer which has average hardness of 15–20 HRC, and the material of the abrasive grains on belt (Hermes RB590Y) is aluminum oxide with the average grain size of 61 μm. The belt grinding parameters, including contact wheel speed $V_c$, robot feed velocity $V_r$, and preset normal grinding force $F_n^0$ are selected to carry out the orthogonal experiments during the plunge belt grinding process, which is illustrated in Table 1. Among, $F_n$ and $F_t$ is the real measurement normal grinding and tangential grinding force, respectively, $a'_p$ is the real material grinding depth and $\mu$ is the force ratio.

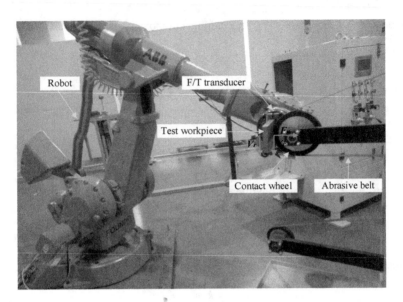

**Fig. 3.** Experimental setup of robotic abrasive belt grinding.

**Table 1.** Design of orthogonal experiments on robotic belt grinding of test workpiece.

| No. | $V_r$ (mm/s) | $V_c$ (m/s) | $F_n^0$ (N) | $F_n$ (N) | $F_t$ (N) | $a_p'$ (mm) | $\mu$ |
|---|---|---|---|---|---|---|---|
| 1 | 20 | 8.37 | 40 | 41 | 32 | 0.0296 | 0.78 |
| 2 | 20 | 8.37 | 60 | 62 | 44 | 0.0396 | 0.71 |
| 3 | 20 | 8.37 | 80 | 79 | 61 | 0.0482 | 0.77 |
| 4 | 40 | 8.37 | 60 | 61 | 43 | 0.0202 | 0.70 |
| 5 | 40 | 12.56 | 60 | 62 | 43 | 0.0425 | 0.69 |
| 6 | 40 | 16.75 | 60 | 61 | 44 | 0.024 | 0.72 |
| 7 | 20 | 12.56 | 80 | 79 | 61 | 0.0767 | 0.77 |
| 8 | 40 | 12.56 | 80 | 79 | 62 | 0.0449 | 0.78 |
| 9 | 60 | 12.56 | 80 | 78 | 59 | 0.0588 | 0.76 |

It can be seen from Eq. (9) that as $a_p'$ approaches zero, the overall force ratio $\mu$ can be considered as the sum of sliding and ploughing friction coefficients ($\mu_{sl} + \mu_{pl}$). Once $a_p'$ is measured, the ($\mu_{sl} + \mu_{pl}$) value by plotting $\mu$ versus $a_p'$ can be determined as $\mu_{sl} + \mu_{pl} = 0.702$, by extending the linear least squares fitting with Adj. R-Square 0.114 to the y-axis ($a_p' = 0$), as shown in Fig. 4.

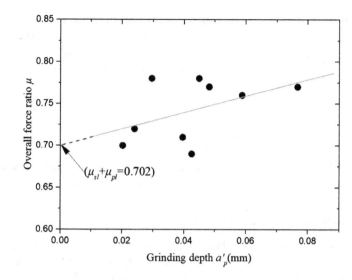

**Fig. 4.** Overall force ratio versus grinding depth in the robotic belt grinding of test workpiece.

Figure 5 shows the monitored contact force profiles during the process of robotic abrasive belt grinding of test workpiece with the Nos. 1–3 of Table 1. The real normal grinding force $F_n$ and tangential force $F_t$ are observed to be relatively stable during the machining process with the force control technology applied. The calculated force ratio $\mu$ ranges from 0.68 to 0.8 according to Table 1, and the average value $\mu = 0.742$ is employed to the grinding force modeling.

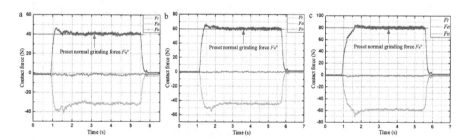

**Fig. 5.** The monitored contact force profiles during the process of robotic abrasive belt grinding of test workpiece.

According to Table 1, the experimental coefficient $k$ is calculated by the nine group tests and its average value is 24.47, then the force model Eq. (10) could be obtained:

$$F_n = \frac{k \cdot v_r \cdot a_p'}{v_c \cdot \left(\mu - \mu_{sl} - \mu_{pl}\right)} = \frac{611.75 \cdot v_r \cdot a_p'}{v_c} \tag{11}$$

The force model Eq. (11) is calculated and determined by the experiments and the related parameters such as $F_n$ and $a'_p$ are real measurement results. Therefore, the determined force model should be used to predict and improve the machining process before the robotic abrasive belt grinding with force control technology, then it can be get:

$$F_n^0 = \frac{k \cdot v_r \cdot a_p}{v_c \cdot \left(\mu - \mu_{sl} - \mu_{pl}\right)} = \frac{611.75 \cdot v_r \cdot a_p}{v_c} \tag{12}$$

Among, $F_n^0$ and $a_p$ are the preset normal grinding force and corresponding theoretical grinding depth.

## 3.2 Validation of Grinding Force Model

Experiments are carried out to validate the grinding force model in the process of robotic abrasive belt grinding of test workpiece. It is found that the predicted grinding depth $a_p$ is relative closest to the real grinding depth $a'_p$ under the variation of normal force $F_n^0$ from Fig. 6(a), next is robot feed velocity $v_r$ from Fig. 6(b) and the last is

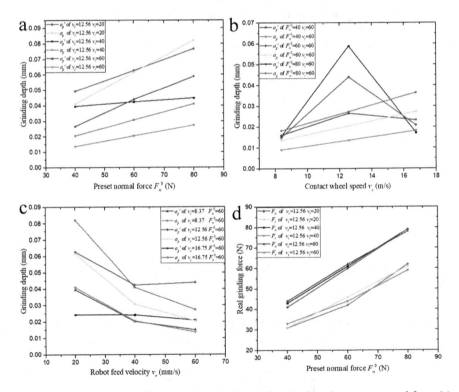

**Fig. 6.** Comparison of real ($a'_p$) and predicted ($a_p$) grinding depth under preset normal force (a), contact wheel speed (b) and robot feed velocity (c), of preset and real grinding force (d).

contact wheel speed $v_c$ from Fig. 6(c). Moreover, the real normal grinding force $F_n$ almost stabilizes at the preset normal force $F_n^0$ with the force control technology employed from Fig. 6(d).

The predicted grinding depth based on the proposed grinding force model doesn't exactly equal to the real grinding depth in the robotic abrasive belt grinding process owing to the effect of belt wear, elastic deformation and system rigidity. Therefore, the grinding force model proposed in the paper is mainly treated as an initial reference to the quantitative robotic abrasive belt grinding process, then the optimum grinding parameters, reducing the experimental time largely, would be determined to satisfy the surface quality of workpiece.

Figure 7 shows the overall evaluation of the effects of above three belt grinding process parameters on the grinding depth. In this chart, each process parameter has been equally weighted, and the points on the overlapping area constituted by the three areas represent the optimal combination of process parameters. A total of three points, including one from contact wheel speed and two from normal grinding force, are observed in Fig. 7, indicating that the optimal combination of process parameters to achieve the desired grinding depth is $\{v_c = 16.75\,\text{m/s}, F_n^0 = 40 - 60\,\text{N}\}$. In contrast, the robot feed velocity has little influence on the optimal solution, and a higher robot feed velocity value, however, is expected to further make the workpiece grinding depth desired.

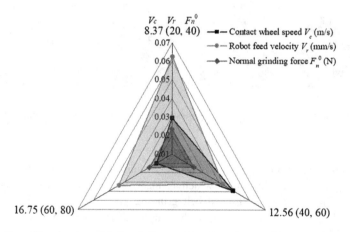

**Fig. 7.** Overall evaluation of belt grinding process parameters effects on grinding depth.

The presented grinding force model is adopted to predict the grinding depth under the optimized process parameters in the robotic abrasive belt grinding process which is illustrated in Table 2. It is observed that the predicted grinding depth $a_p$ is closer to the real grinding depth $a_p'$ neglecting the effect of belt wear, elastic deformation and system rigidity. It is also found that the robot feed velocity has the largest influence on the grinding depth, followed by the contact wheel speed and the normal grinding force. Therefore, it would meet the grinding needs of workpiece by further adjusting the process parameters based on the force model under the optical process parameters to realize the quantitative grinding. Meanwhile, the corresponding grinding effect is shown in Fig. 8 and the surface roughness also satisfy the requirements Ra < 0.4 μm.

**Table 2.** Comparison of predicted and real grinding depth under optical process parameters.

| No. | $V_r$ (mm/s) | $V_c$ (m/s) | $F_n^0$ (N) | $F_n$ (N) | $F_t$ (N) | $a_p'$ (mm) | $a_p$ (mm) |
|-----|-----|-----|-----|-----|-----|-----|-----|
| 1 | 40 | 16.75 | 40 | 42 | 29 | 0.0235 | 0.0274 |
| 2 | 40 | 16.75 | 60 | 61 | 41 | 0.0242 | 0.0410 |
| 3 | 60 | 16.75 | 40 | 42 | 26 | 0.0233 | 0.0183 |
| 4 | 60 | 16.75 | 60 | 61 | 41 | 0.0209 | 0.0273 |

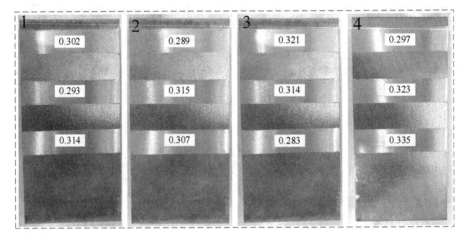

**Fig. 8.** The corresponding surface roughness of workpiece under optimized grinding parameters.

The proposed grinding force model would be also applied to the robotic abrasive belt grinding of aero-engine blade which is made of Ti-6Al-4V with force control technology. Then a set of optimized belt grinding parameters $v_c = 16.75$ m/s, $v_r = 40$ mm/s and $F_n^0 = 40$ N are selected to be employed in the robotic belt grinding of aero-engine blade and the grinding effect is shown in Fig. 9 which the average surface roughness Ra values are 0.343 μm in the concave surface and 0.313 μm in the convex surface. Moreover, the force-guided grinding effects have reached the surface roughness requirement for the aero-engine blade which should be below Ra 0.4 μm.

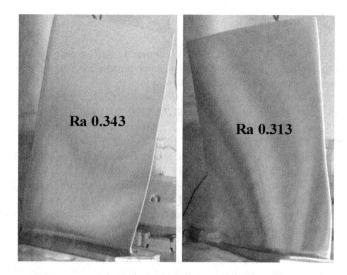

**Fig. 9.** Surface topographies of the aero-engine blade after robotic belt grinding.

## 4  Conclusion

Based on the force control technology, a new grinding force model is proposed to predict the grinding depth to realize the quantitative machining process, and then the predicted and real grinding depths under different grinding parameters are compared and analyzed. Following conclusions are achieved:

(1) The robot feed velocity has the largest influence on the grinding depth, followed by the contact wheel speed and the normal grinding force in the robotic abrasive belt grinding of test workpiece.

(2) The predicted grinding depth would approach the real grinding depth largely under the optimized grinding process parameters: $v_c = 16.75\,\text{m/s}$, $F_n^0 = 40 - 60\,\text{N}$ and $v_r = 40 - 60\,\text{mm/s}$ in the robotic belt grinding of test workpiece.

(3) With the combination of presented force model and force control technology under optical process parameters, desired grinding effects can be obtained and also meet the surface quality of test workpiece and aero-engine blade.

**Acknowledgements.** The authors would like to gratefully acknowledge the financial support from the National Nature Science Foundation of China (nos. 51675394, 51375196), the National Key Research and Development Program of China (nos. 2017YFB1303400), the State Key Laboratory of Digital Manufacturing Equipment and Technology (no. DMETKF2018018), and the Fundamental Research Funds for the Central Universities (no. 2017II33GX).

# References

1. Zhao, T., Shi, Y., Lin, X., Duan, J., Sun, P., Zhang, J.: Surface roughness prediction and parameters optimization in grinding and polishing process for IBR of aero-engine. Int. J. Adv. Manuf. Technol. **74**, 653–663 (2014)
2. Zhao, P., Shi, Y.: Posture adaptive control of the flexible grinding head for blisk manufacturing. Int. J. Adv. Manuf. Technol. **70**, 1989–2001 (2014)
3. Huang, H., Gong, Z.M., Chen, X.Q., Zhou, L.: Robotic grinding and polishing for turbine-vane overhaul. J. Mater. Process. Technol. **127**(2), 140–145 (2002)
4. Ren, X., Kuhlenkötter, B.: Real-time simulation and visualization of robotic belt grinding processes. Int. J. Adv. Manuf. Technol. **35**(11), 1090–1099 (2008)
5. Kountanya, R., Guo, C.: Specific material removal rate calculation in five-axis grinding. J. Manuf. Sci. Eng. **139**(12), 121010 (2017)
6. Ardashev, D.V., Dyakonov, A.A.: Mathematical model of the grinding force with account for blunting of abrasive grains of the grinding wheel. J. Manuf. Sci. Eng. **139**(12), 121005 (2017)
7. Zhang, X., Kuhlenkötter, B., Kneupner, K.: An efficient method for solving the Signorini problem in the simulation of free-form surfaces produced by belt grinding. Int. J. Mach. Tools Manuf. **45**(6), 641–648 (2005)
8. Wang, Y.J., Huang, Y., Chen, Y.X., Yang, Z.S.: Model of an abrasive belt grinding surface removal contour and its application. Int. J. Adv. Manuf. Technol. **82**(9–12), 2113–2122 (2016)
9. Dai, C., Ding, W., Xu, J., Fu, Y., Yu, T.: Influence of grain wear on material removal behavior during grinding nickel-based superalloy with a single diamond grain. Int. J. Mach. Tools Manuf. **113**, 49–58 (2017)
10. Jourani, A., Hagege, B., Bouvier, S., Bigerelle, M., Zahouani, H.: Influence of abrasive grain geometry on friction coefficient and wear rate in belt finishing. Tribol. Int. **59**, 30–37 (2013)
11. Tang, J., Du, J., Chen, Y.: Modelling and experimental study of grinding forces in surface grinding. Ann. CIRP, 1–8 (2008)
12. Zhu, D., Luo, S., Yang, L., Chen, W., Yan, S., Ding, H.: On energetic assessment of cutting mechanisms in robot-assisted belt grinding of titanium alloys. Tribol. Int. **90**, 55–59 (2015)
13. Zhu, D., Xu, X., Yang, Z., Zhuang, K., Yan, S., Ding, H.: Analysis and assessment of robotic belt grinding mechanisms by force modeling and force control experiments. Tribol. Int. **120**, 93–98 (2018)
14. Wang, W., Liu, F., Liu, Z., Yun, C.: Prediction of depth of cut for robotic belt grinding. Int. J. Adv. Manuf. Technol. **91**(1–4), 699–708 (2017)
15. Murtagian, G.R., Hecker, R.L., Liang, S.Y., Danyluk, S.: Plastic deformation depth modeling on grinding of gamma Titanium Aluminides. Int. J. Adv. Manuf. Technol. **49**(1–4), 89–95 (2010)
16. Xu, X., Zhu, D., Wang, J., Yan, S., Ding, H.: Calibration and accuracy analysis of robotic belt grinding system using the ruby probe and criteria sphere. Robot Comput. Integr. Manuf. **51**, 189–201 (2018)
17. Durgumahanti, U.S.P., Singh, V., Rao, P.V.: A new model for grinding force prediction and analysis. Int. J. Mach. Tools Manuf. **50**, 231–240 (2010)
18. Venkatachalam, S., Liang, S.Y.: Effects of ploughing forces and friction coefficient in microscale machining. J. Manuf. Sci. Eng. Trans. ASME **129**, 274–280 (2007)

# A Boundary Auto-Location Algorithm for the Prediction of Milling Stability Lobe Diagram

Mingkai Zhang[1], Xiaowei Tang[1], Rong Yan[1(✉)], Fangyu Peng[2], Chen Chen[1], Yuting Li[1], and Haohao Zeng[1]

[1] School of Mechanical Science and Engineering,
Huazhong University of Science and Technology, Wuhan 430074, China
yanrong@hust.edu.cn
[2] State Key Lab of Digital Manufacturing Equipment and Technology,
Huazhong University of Science and Technology, Wuhan 430074, China

**Abstract.** Chatter is known as a main factor that limits the machining quality and efficiency, and one universal solution is to predict occurrences of chatter via calculating the stability lobe diagram (SLD), such as time-domain methods, which are relatively time-consuming. Thus, based on time-domain methods, a boundary auto-location algorithm for the prediction of SLD in milling is proposed. In the proposed method, by setting a series of judgements based on the state of the transition matrix of the dynamic system, the calculation trajectory automatically surrounds the stability boundary line except isolated islands. Only the points on and around the stability boundary were calculated to draw SLD. The contrast simulations were conducted to verify the calculation efficiency of the given algorithm. And the results show that the computational time of the proposed method was cut down significantly than that of the traditional method.

**Keywords:** Milling process · Stability prediction · Boundary auto-location

## 1 Introduction

Prediction of stability lobe diagram is a vital method to avoid chatter that limits the quality and efficiency of manufacturing. However, the calculation of the stability prediction relate to plenty of parameters, such as modal parameters, cutting parameters and so on. It is time-consuming to calculate all of the points' stability of the parameters space, like the robotic milling. Therefore, a boundary auto-location algorithm is proposed to solve this issue in the prediction of SLD.

With respect to the prediction of cutting stability, so far there are two major methods: one is the frequency-domain method [1, 2], which is characterized by high-computational-efficiency and relatively low prediction accuracy [3]; another is time-domain method [4, 5], which is time-consuming and has higher prediction accuracy than the former. For the cutting system, the stability can be predicted by solving its feature equation in the former method and the time-periodic delayed-differential equation in the later method. Many efforts have been devoted to improving the prediction accuracy and efficiency in the previous literatures.

© Springer Nature Switzerland AG 2018
Z. Chen et al. (Eds.): ICIRA 2018, LNAI 10984, pp. 299–308, 2018.
https://doi.org/10.1007/978-3-319-97586-3_27

Altintas and Budak [1] proposed zero-order method that used the average of the Fourier series of the dynamic milling force to approximate the milling force variation, but is not capable in low radial immersion milling. Merdol and Altintas [2] solved this issue via developing multi-frequency method considering the harmonics of the tooth-passing frequencies. Insperger et al. [4] firstly proposed semi-discretization method (SDM), in which the time-delayed term is discretized. And it is widely adopted to predict the stability of linear-variant time-delayed milling system in time domain. Then, some novel and improved algorithms were proposed to improve the computational efficiency, like full-discretization method (FDM) presented by Ding et al. [5], numerical integration method (NIM) [7], linear approximation of acceleration method (LAAM) [8], Runge-Kutta-based complete discretization method(RKCDM) presented by Li et al. [9], and so on [10, 11]. Also, Li et al. [12] proposed a fast-straight forward method to plot SLD using modal parameters of milling system, but with a relatively low accuracy. In this paper, based on the time-domain methods, a fast algorithm of SLD prediction was developed without consideration of the special SLD contour that called stability isolated islands mentioned in Ref. [13].

All of the time-domain methods focus on the stability calculation of certain cutting parameters, like revolution speed and cut depth which are generally used to represent two-dimension SLD. Therefore, in the traditional method, SLD is given by calculating stabilities of all the points in the discretized cutting parameters space. It is a huge work and time-consuming. In this paper, by setting a series of judgements based on the state of the transition matrix of the dynamic system, only the points on and around the stability boundary were automatically calculated to draw SLD.

The remaining part of this paper is organized as follows: Sect. 2 proposes the boundary auto-location algorithm for prediction of the SLD; Sect. 3 gives the comparisons of computational efficiency between the proposed method and traditional methods. And in Sect. 3 this paper is concluded.

## 2 The Proposed Algorithm

### 2.1 The Stability Prediction Model in Milling Process

The milling system can be simplified as a two-DOF dynamic system, as shown in Fig. 1, the dynamic model is a time-periodic delayed-differential equation and can be expressed in the form of Eq. (1), as follow:

$$M\begin{bmatrix} \ddot{x}(t) \\ \ddot{y}(t) \end{bmatrix} + C\begin{bmatrix} \dot{x}(t) \\ \dot{y}(t) \end{bmatrix} + K\begin{bmatrix} x(t) \\ y(t) \end{bmatrix} = K_c(t)\begin{bmatrix} x(t) - x(t-T) \\ y(t) - y(t-T) \end{bmatrix} \tag{1}$$

where $M, C, K, K_c(t)$ is the mass, damping, stiffness matrices and the cutting force coefficient matrix of dynamic system respectively. $x(t) - x(t-T), y(t) - y(t-T)$ represent the dynamic displacement of the cutter between the previous and current revolution on the $X, Y$ direction. $T$ is the tooth-passing period.

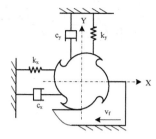

**Fig. 1.** The two-DOF dynamic model of milling system

By using one of the time-domain methods in the references, the Eq. (1) can be solved and the transition matrix $\Phi$ of the dynamic system can be obtained. And then according to the Floquet theory, the system is judged as stable when the maximum absolute value of the eigenvalues of the transition matrix $\Phi$ is less than 1. The following deduction process is developed based on this criterion.

## 2.2 The Algorithm for Locating the Original Point on the Stability Boundary

Firstly, it is necessary to evenly discretize the cutting parameters space of SLD. Here, the revolution speed and cut depth are used to configure the two-dimension SLD. The revolution speed and cut depth interval are divided into m and n discrete values as follows:

$$\begin{cases} S_i = \dfrac{i}{m}(S_{max} - S_{min}), i = 1, 2, \ldots, m \\ d_j = \dfrac{j}{n}(d_{max} - d_{min}), j = 1, 2, \ldots, n \end{cases} \tag{2}$$

where $S, d$ respectively represent the revolution speed and cut depth.

Then, along the revolution speed incremental direction $i = 1 \rightarrow m$, the limitation value of cut depth $d_{chatter}$ related to each discrete revolution speed $S_i$ can be sequentially obtained via the algorithm in this paper.

**Algorithm I.** Locate the original point on the stability boundary

```
1:   set the initial condition: i = 1, [a, b] = [1, n], j = [n/2]
2:   while 1 ≤ j ≤ n do
3:       if Eₚ(Sᵢ, dⱼ) = 1 then
4:           return d_chatter = dⱼ
5:       end algorithm
6:   else
7:       if j = a then
8:           if (Eₚ(Sᵢ, dₐ) − 1)·(Eₚ(Sᵢ, d_b) − 1) < 0 then
9:               return d_chatter = dₐ + (1 − Eₚ(Sᵢ, dₐ))·(d_b − dₐ)/(Eₚ(Sᵢ, d_b) − Eₚ(Sᵢ, dₐ))
10:              end algorithm
11:          else do the next iteration under S_{i+1}
12:          end if
13:      else
14:          if Eₚ(Sᵢ, dⱼ) > 1 then  b = j
15:          else  a = j
16:          end if
17:      end if
18:      j = [(a + b)/2]
19:   end if
20: end while
```

For the proposed boundary auto-location algorithm, there is an initial condition: the original point on the stability boundary should be determined, like the point A shown in the Fig. 2. To locate this original point, the Algorithm I is also developed based on the bisection method.

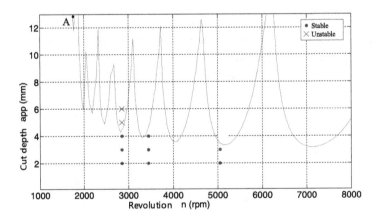

**Fig. 2.** The original point A on the stability boundary

As shown in the Fig. 2, SLD is divided into stable and unstable zone by the stability boundary, and this discrimination method could be formulated as the follow:

$$\begin{cases} E_\phi(S_i, d_j) \geq 1, & \text{system is unstable} \\ E_\phi(S_i, d_j) < 1, & \text{system is stable} \end{cases} \tag{3}$$

where $E_\Phi(S_i, d_j)$ refers to the maximum absolute value of the eigenvalues of the transition matrix $\Phi$ under the cutting parameters of $(S_i, d_j)$.

Note that $E_\Phi(S_i, d_j)$ of all the points on the stability boundary are equal to 1. Therefore, the boundary point $(S_i, d_{chatter})$ of current discrete revolution speed $S_i$ can be located rapidly based on the bisection method.

This bisection method iterates from $i = 1$ to $i = m$ until finding the first boundary point, which one would be identified as the original point A for the next algorithm. Each iteration start with an initial subscript interval of cut depth $[a, b] = [1, n]$, and then $a$ and $b$ would be updated corresponding to the value of $E_\Phi(S_i, d_j)$ as follow:

$$\begin{cases} b = j, & E_\phi(S_i, d_j) > 1 \\ a = j, & E_\phi(S_i, d_j) < 1 \end{cases} \tag{4}$$

Where $j = [(a + b)/2]$, and the bracket [ ] is a sign operator which means rounding down the number.

There exist the boundary point among interval $[a, b]$ when $(E_\Phi(S_i, d_a) - 1) \cdot (E_\Phi(S_i, d_b) - 1) \leq 0$. And then the cutting parameter $(S_i, d_{chatter})$ of the boundary point can be approximated by some interpolation methods, such as Lagrange Approximation. Here, the boundary point is given by Eq. (5) as follow:

$$d_{chatter} = d_a + (1 - E_\phi(S_i, d_a)) \cdot (d_b - d_a) / (E_\phi(S_i, d_b) - E_\phi(S_i, d_a)),$$
$$\text{if } (E_\phi(S_i, d_a) - 1) \cdot (E_\phi(S_i, d_b) - 1) \leq 0 \tag{5}$$

The detailed algorithm can be summarized as Algorithm I. Meanwhile, it is worthy to be mentioned, in general, if there exist one boundary point among subscript interval $[a, b]$, the relationship that $b = a + 1$ is satisfied. This attribute would be useful for the boundary auto-location algorithm.

## 2.3 The Boundary Auto-Location Algorithm

As can be seen in Fig. 2, generally, the stability boundary is a contour line, which is known for its continuity and equivalence. Therefore, once one boundary point has been founded, like the aforementioned point A $(S_i, d_{chatter})$, the next boundary point $(S_{i+1}, d_{chatter})$ can be determined in the neighborhood of it. Based on this it is unnecessary to calculate all the points among $[d_1, d_n]$ to determine the next $d_{chatter}$, only the points on and around the latest stability boundary point are needed.

Assume that, under the current discrete revolution speed $S_i$, the $d_{chatter}$ locate in the subscript interval $[a, b]$ with $b = a + 1$, such as point A. And the next iteration should start from $j = b$ or $j = a$ under the next discrete revolution speed $S_{i+1}$, in other words, directly calculating the value of the $E_\Phi \left( S_{i+1}, d_b \right)$ or $E_\Phi \left( S_{i+1}, d_a \right)$. Then same as the Algorithm I above, the Eq. (5) also can be used to determine the stability boundary point. But different from the update forms of the parameters $a, b$ and $j$ that used in Eq. (4), here, those parameters are updated in the form of Eq. (6).

$$\begin{cases} b = j, & E_\phi \left( S_i, d_j \right) > 1 \\ a = j, & E_\phi \left( S_i, d_j \right) < 1 \end{cases} \tag{6}$$

In this method, the calculation process would been proceeding along the incremental direction of cut depth when the current point $\left( S_i, d_j \right)$ is judged as stable and along the decremental direction in the unstable zone. Only a few steps are performed for this process to locate the new stability point, because the attributes of continuity and equivalence of the contour line. The whole trajectory of the calculation process closely surround the stability boundary, just like automatically keep following the tangential direction of the boundary line as can be seen in Fig. 3. The method of this part can be summarized as Algorithm II. Note that, in a special case where the system is stable in whole given cut depth interval $\left( d_{min}, d_{max} \right)$ under current revolution speed, the algorithm would proceed along $d_{max}$ until entering into the unstable zone.

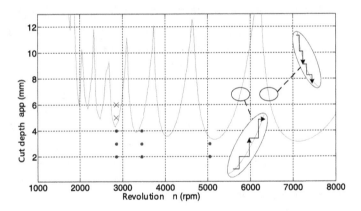

**Fig. 3.** Diagram of calculation process of the boundary auto-location algorithm

---

**Algorithm II.** Boundary auto-location algorithm for the SLD prediction

1:  input the parameters of the latest boundary point: $S_{i-1}, d_j, [a, b]$
2:  while $1 \leq j \leq n$ do
3:      if $\left(E_\phi(S_i, d_a) - 1\right) \cdot \left(E_\phi(S_i, d_b) - 1\right) \leq 0$ then
4:          return $d_{chatter} = d_a + \left(1 - E_\phi(S_i, d_a)\right) \cdot (d_b - d_a) / \left(E_\phi(S_i, d_b) - E_\phi(S_i, d_a)\right)$
5:          do the next iteration under $S_{i+1}$
6:      else
7:          if $E_\phi(S_i, d_j) > 1$ then $b = a, \; a = a - 1$
8:          else $a = b, \; b = b + 1$
9:          end if
10:          $j = b$
11:      end if
12:  end while
13:  do the next iteration under $S_{i+1}$

---

## 3   Simulations

The experimental platform is Intel(R) Core(TM) i5-4460, CPU@3.20 GHz, RAM 8 GB, and all of the algorithms are written and executed in MATLAB R2012a. The transition matrix $\Phi$ of the dynamic system is calculated via using the 2nd-FDM, NIM, and LAAM in the literatures [6], [7], and [8], respectively. The computational efficiency of the proposed method is studied and compared with that of the traditional method. Besides, the SDM in literature [4] is used as a criterion-reference to validate the accuracy of the calculation results.

Here, SLD of single-DOF milling model is represented by parameters space with cut depth and revolution speed under $0\,mm \leq d \leq 4\,mm$ and $5000\,rpm \leq S \leq 10000$ rpm. The quantity of discretization are $m = 160$ and $n = 80$. Besides, the tooth-passing period $T$, which is also an important parameter and proportional to the prediction accuracy in time-domain method to solve Eq. (1), is divided into 40 time intervals in the first three methods. And the tooth-passing period $T$ in the last method is divided into 200 time intervals as a criterion-reference.

The other cutting parameters, including tool, cutting force coefficients, etc., and dynamic parameters, including modal mass, natural frequency, etc., are same as the benchmark data adopted in the references [4, 6–8], as can be seen in Table 1.

**Table 1.** The cutting and dynamic parameters for single-DOF milling model.

| Radial immersion ratio | Tangential cutting force coefficient | Normal cutting force coefficient | Natural frequency | Relative damping | Modal mass |
|---|---|---|---|---|---|
| 1 | $6 \times 10^8$ N/m$^2$ | $2 \times 10^8$ N/m$^2$ | 922 Hz | 0.011 | 0.03993 kg |

The calculation results and comparisons of SLD for single-DOF milling model based on different time-domain methods are shown as Fig. 4. Because the transition matrix $\Phi$ of every point of SLD is same in each case, there is minimal difference of the stability boundary between the proposed fast algorithm and the traditional algorithm. And the tiny difference here is caused by the calculation method of contour line in MATLAB, which is used in later. The boundary point of the former is determined by Eq. (5), while the interpolation method, like Eq. (5), is also used in revolution speed interval to determine the boundary point of the later.

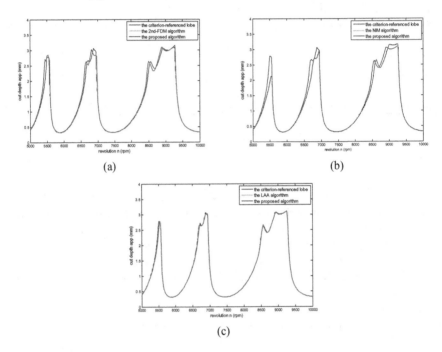

**Fig. 4.** SLD of the fast algorithm based on: (a) 2nd-FDM (b) NIM (c) LAAM

Meanwhile, in the terms of prediction accuracy the LAAM has the highest agreement with the criterion-referenced lobe in Fig. 4(c). The 2nd FDM in Fig. 4(a) is inferior to the former, and the NIM in Fig. 4(b) is lowest.

The comparisons of the computational time of the fast algorithm based on different method are shown in Table 2. There are three results of each case are given on account of the fluctuations of computer performance. And the computational efficiency is represented by the average of those three times. Compared with the traditional method that calculates all the points of SLD, the proposed algorithm enhance the computational time by 18.7 times in the 2nd-FDM-based case, 17.4 times in the NIM-based case and 19.5 times in the NIM-based case.

Table 2. The comparisons of the computational time.

| Algorithm | | 2nd-FDM-based | | NIM-based | | LAA-based | |
|---|---|---|---|---|---|---|---|
| | | Convention | This paper | Convention | This paper | Convention | This paper |
| Time(s) | 1 | 28.14648 | 1.49503 | 13.28893 | 0.75696 | 6.98059 | 0.35498 |
| | 2 | 28.19411 | 1.50136 | 13.29482 | 0.76778 | 6.97950 | 0.35864 |
| | 3 | 28.26346 | 1.51737 | 13.31026 | 0.77051 | 7.00041 | 0.35882 |
| Average | | 28.20135 | 1.50459 | 13.29802 | 0.76508 | 6.98833 | 0.35748 |

Therefore, the proposed boundary auto-location algorithm significantly improved the computational efficiency of the SLD prediction with same accuracy as the time-domain methods.

# 4   Conclusions

This paper proposes a boundary auto-location algorithm for the prediction of milling stability lobe diagram without consideration of stability isolated islands. In the proposed method, the calculation trajectory automatically follows the tangential direction of the stability boundary line via setting the judgements based on the state of the transition matrix of the dynamic system and Floquet theory. Only the points on and around the stability boundary were calculated, thereby avoiding to calculate all of the points on SLD. The high-computational-efficiency of the proposed algorithm is validated by the simulations, in which the computational time is enhanced significantly than that of the traditional method.

**Acknowledgements.** This research is supported by National Natural Science Foundation of China under Grant no. 51625502, Innovation Group of Hubei under Grant no. 2017CFA003. Natural Science Foundation of Jiangsu under Grant no. BK20161473.

# References

1. Altintas, Y., Budak, E.: Analytical prediction of stability lobes in milling. CIRP **44**(1), 357–362 (1995)
2. Merdol, S., Altintas, Y.: Multi frequency solution of chatter stability for low immersion milling. ASME J. Manuf. Sci. Eng. **126**(3), 459–466 (2004)
3. Bayly, P., Mann, B., Schmitz, T., et al.: Effects of radial immersion and cutting direction on chatter instability in end-milling. In: 2002 ASME International Mechanical Engineering Congress and Exposition, pp. 351–363. American Society of Mechanical Engineers, New Orleans (2002)
4. Insperger, T., Stépán, G.: Updated semi-discretization method for periodic delay-differential equations with discrete delay. Int. J. Numer. Methods Eng. **61**(1), 117–141 (2004)
5. Ding, Y., Zhu, L.M., Zhang, X.J., et al.: A full-discretization method for prediction of milling stability. Int. J. Mach. Tools Manuf. **50**(5), 502–509 (2010)
6. Ding, Y., Zhu, L.M., Zhang, X.J., et al.: Second-order full-discretization method for milling stability prediction. Int. J. Mach. Tools Manuf. **50**, 926–932 (2010)

7. Ding, Y., Zhu, L.M., Zhang, X.J., et al.: Numerical integration method for prediction of milling stability. ASME J. Manuf. Sci. Eng. **133**(3), 031005 (2011)
8. Huang, T., Zhang, X.M., Zhang, X.J., Ding, H.: An efficient linear approximation of acceleration method for milling stability prediction. Int. J. Mach. Tools Manuf. **74**, 56–64 (2013)
9. Li, Z., Yang, Z., Peng, Y., et al.: Prediction of chatter stability for milling process using Runge-Kutta-based complete discretization method. Int. J. Adv. Manuf. Technol. **1**, 1–10 (2015)
10. Zhang, Z., Li, H., Meng, G., et al.: A novel approach for the prediction of the milling stability based on the Simpson method. Int. J. Mach. Tools Manuf. **99**, 43–47 (2015)
11. Tang, X., Peng, F., Rong, Y., et al.: Accurate and efficient prediction of milling stability with updated full-discretization method. Int. J. Adv. Manuf. Technol. **88**(9–12), 2357–2368 (2017)
12. Li, Z., Wang, Z., Shi, X.: Fast prediction of chatter stability lobe diagram for milling process using frequency response function or modal parameters. Int. J. Adv. Manuf. Technol. **89**(9–12), 2603–2612 (2017)
13. Patel, B.R., Mann, B.P., Young, K.A.: Uncharted islands of chatter instability in milling. Int. J. Mach. Tools Manuf. **48**(1), 124–134 (2008)

# Deformation Error Prediction and Compensation for Robot Multi-axis Milling

Xiaowei Tang[1], Rong Yan[1], Fangyu Peng[1,2(✉)], Guangyu Liu[1],
Hua Li[1], Dequan Wei[1], and Zheng Fan[1]

[1] School of Mechanical Science and Engineering,
Huazhong University of Science and Technology,
No. 1037 LuoYu Road, Wuhan, China
txwysxf@126.com, {yanrong,pengfy,M201570333,
quan_mech}@hust.edu.cn, 13296614195@163.com,
346332909@qq.com
[2] State Key Lab of Digital Manufacturing Equipment and Technology,
Huazhong University of Science and Technology,
No. 1037 LuoYu Road, Wuhan, China

**Abstract.** Robot milling is an alternative to the expensive multi-axis NC for the large components, such as large scale marine propeller, with the advantage of less expensive and more flexible. However, the machining error mainly caused by poor stiffness of the joint is a critical obstacle for robot milling application. Especially for the robot multi-axis milling, the machining error is more difficult to predict and compensate due to the complicated coordinate transformation between the deformation of tool point and the cutting force feedback. The static stiffness model of robot milling system is established based on the joint stiffness matrix and Jacobian matrix. Using the static stiffness model and the cutting force model, an equilibrium equation with the variable of tool point deformation is established based on the constructed coordinate transformation for the theoretical cutting position and the actual cutting position in different coordinate systems. The analyzed results indicate that the milling error is mainly influenced by the translation of tool point. The proposed error prediction and compensation method is validated by the cutting experiments.

**Keywords:** Robot milling · Multi-axis · Static stiffness · Error
Prediction and compensation

## 1 Introduction

The high demand of aerospace and navigation industry result a great number of applications of large components. Multi-axis NC is the main equipment to machining the large components, however it has the disadvantage of operating time and costs. Machining robot is an alternative strategy since it's less expensive and more flexible, as shown in Fig. 1. The main disadvantage of industrial robots used for machining processes is their poor stiffness which is less than 1 N/m, while a standard CNC machine

© Springer Nature Switzerland AG 2018
Z. Chen et al. (Eds.): ICIRA 2018, LNAI 10984, pp. 309–318, 2018.
https://doi.org/10.1007/978-3-319-97586-3_28

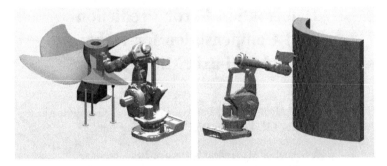

**Fig. 1.** Robot milling for large components (left: marine propeller; right: rocket tank).

very often has stiffness greater than 50 N/m [1]. Thus, the machining error caused by poor stiffness is a critical obstacle for robot milling application.

There have been many researches on the identification of robot joint stiffness and machining error compensation of robot milling. In order to develop the milling application of robot, Zaeh and Roesch [2] identified the stiffness of robot joints and modified the static deflection along the cutting path based on fuzzy control theory. Wu et al. [3] identified the stiffness of robot joints by hanging the mass block on the end of KUKA KR-270 robot. Tyapin and Kaldestad et al. [4, 5] compensated the machining path using the stiffness of robot joints and the static cutting force, and verified the compensation result by utilizing laser tracker. Kamali et al. [6] proposed an elastic geometric calibration method with multiple external loads for the ABB IRB 1600 robot, and calibrated geometrical errors and joint stiffness parameters of various robot poses based on the position measurement of the robot end-effector. Möller et al. [7] measured the spindle location of the robot milling system by using the stereo camera system, and made it as the feedback to adjust the robot movement.

The existing researches have mainly focused on the identification of robot joint stiffness and error compensation for three-axis milling along the cutting path. For the large curved surface parts, such as marine propeller, the multi-axis milling is a necessary cutting mode for the robot machining system. Thus, this paper studies the static stiffness error prediction and compensation of robot multi-axis milling. The static stiffness model of robot milling system is established, and the relationship of the theoretical and actual cutting location based on the tool point deformation is deduced. Based on the static stiffness model, cutting force model and the relationship of cutting location, an equilibrium equation with the variable of tool point deformation is established to predict and compensate the milling error.

## 2   Error Compensation Model

The robot multi-axis milling system is constructed by the ABB IRB 6660-205/1.9 robot, IBAG high speed spindle and a bull-nose end cutter, as shown in Fig. 2. The tool point of the robot milling system deforms under the action of milling force due to the

**Fig. 2.** Robot multi-axis milling system.

compliance manipulator base, gearboxes, motors, links and the cutter. For many industrial robots, the compliance of the joints is regarded as the dominant source of the robot deformation [8], and the cutter used in this paper has well rigidity with diameter 25 mm. Thus, the links and cutter are assumed as infinitely stiff, therefore the error prediction and compensation is based on the static rigidity analysis of the joints.

## 2.1   Static Stiffness Model of Robot Milling System

In order to establish the static stiffness model of robot milling system, the kinematic relationship of the tool point and the robot joints movement should be determined firstly. The Denavit Hartenberg (D-H) notion is a classic model to describe the kinematic movement of robot and the coordinate frames assignment for the ABB IRB 6660-205/1.9 robot is shown in Fig. 3. The D-H parameters are provided by the manufacturer and listed in Table 1.

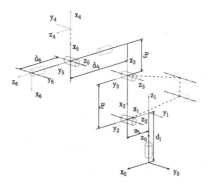

**Fig. 3.** Kinematic model of ABB IRB 6660-205.

The Jacobian matrix $J$ of the robot (the Link 6 contains the structure of spindle-tool) is obtained by using the D-H parameters, and the movement of the tool point can be expressed by the angle values of robot joints.

**Table 1.** D-H kinematic parameters.

| Link $i$ | $d_i$ (mm) | $a_{i-1}$ (mm) | $\alpha_{i-1}$ (rad) | $\theta_i$ (rad) |
|---|---|---|---|---|
| 1 | 814.5 | 0 | 0 | $\theta_1$ |
| 2 | 0 | 300 | $-\pi/2$ | $\theta_2 - \pi/2$ |
| 3 | 0 | 700 | 0 | $\theta_3 - \theta_2$ |
| 4 | 893 | 280 | $-\pi/2$ | $\theta_4$ |
| 5 | 0 | 0 | $\pi/2$ | $\theta_5$ |
| 6 | 200 | 0 | $-\pi/2$ | $\theta_6 + \pi/2$ |

$$D = J\Delta q \tag{1}$$

where $D$ denotes the translation and rotation of tool point, $\Delta q$ represents the rotation of the six joint angles.

$$D = \begin{bmatrix} \Delta x & \Delta y & \Delta z & \Delta\theta_x & \Delta\theta_y & \Delta\theta_z \end{bmatrix}^T$$
$$\Delta q = \begin{bmatrix} \Delta\theta_1 & \Delta\theta_2 & \cdots & \Delta\theta_6 \end{bmatrix} \tag{2}$$

The relationship of torque applied on joint and joint angle rotation can be expressed as

$$\tau = K\Delta q \tag{3}$$

where $\tau$ is the joint moment vector, and diagonal matrix $K = \text{diag}(k_1, k_2, \ldots, k_6)$ denotes the joint stiffness which should be calibrated by experiments.

Based on the Jacobian matrix $J$, the torque applied on joint caused by the load on tool point can be expressed as

$$\tau = J^T F \tag{4}$$

Substituting Eqs. (1) and (3) into Eq. (4), the static stiffness model for tool point of robot milling system is described as

$$D = JK^{-1}J^T F \tag{5}$$

The load $F$ in Eq. (5) is result from the multi-axis cutting force which is described in Cutter Coordinate System (CCS) as follow referring to Ref. [10].

$$\begin{bmatrix} f_x \\ f_y \\ f_z \end{bmatrix} = \begin{bmatrix} -\cos\varphi_j(z) & -\sin\kappa(z)\sin\varphi_j(z) & -\cos\kappa(z)\sin\varphi_j(z) \\ \sin\varphi_j(z) & -\sin\kappa(z)\cos\varphi_j(z) & -\cos\kappa(z)\cos\varphi_j(z) \\ 0 & \cos\kappa(z) & -\sin\kappa(z) \end{bmatrix}$$

$$\begin{bmatrix} \sum_{j=1}^{N}\int_0^{a_p} g(\varphi_j(z))K_t \\ \sum_{j=1}^{N}\int_0^{a_p} g(\varphi_j(z))K_r \\ \sum_{j=1}^{N}\int_0^{a_p} g(\varphi_j(z))K_a \end{bmatrix} h_d(\varphi_j(z))\frac{\mathrm{d}z}{\sin\kappa} \tag{6}$$

In this paper, the torque in $F$ is ignored, thus the load applied on tool point is expressed as

$$F = \begin{bmatrix} f_x & f_y & f_z & 0 & 0 & 0 \end{bmatrix} \tag{7}$$

## 2.2  Error Prediction and Compensation

Different from the multi-axis NC milling only considering the radial translational deformation of tool point, the machining error of robot milling system is influenced by the translational and rotational deformation under X, Y and Z direction simultaneously.

In the multi-axis milling process shown in Fig. 4, the theoretical cutter location and orientation only considering kinematics are denoted by $CL$ and $OZ$ in Workpiece Coordinate System (WCS). The actual cutter location and orientation transfer to $CL'$ and $OZ'$ when the joints appear deformation under the action of milling force. However, the milling force is influenced by the variation of the cutter location and orientation again, which produce a closed loop feedback system. Based on the cutter geometry, cutter-contact point (CCP) can be calculated according to the cutter location and orientation. Consequently, the difference between theoretical $CCP$ and actual $CCP'$ coordinate value in WCS is the milling error. Thus, the key issue is to identify the actual cutter location $CL'$ and orientation $OZ'$ according to the theoretical $CL$ and $OZ$ based on static stiffness model of robot milling system in Eq. (5).

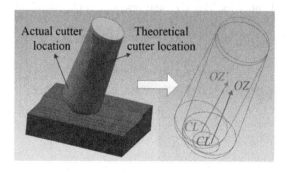

**Fig. 4.** The theoretical and actual cutter location and orientation.

Firstly, the $CL'$ and its relationship with $CL$ should be predicted. The translation of the tool point in Eq. (2) describes the difference between $CL$ and $CL'$ in X, Y and Z direction, but it is calculated in the Base Coordinate System (BCS) of robot milling system. The homogeneous transformation matrix ${}^{W}_{B}T$ of BCS to WCS can be calculated by the Jacobian matrix $J$, the cutter location and orientation.

$$
{}^{W}_{B}T = \begin{bmatrix} {}^{W}_{B}R & CL_W \\ 0 & 1 \end{bmatrix} \tag{8}
$$

Transforming the deformation of tool point $\Delta x$, $\Delta y$ and $\Delta z$ in Eq. (2) from BCS to the WCS, and the geometrical relationship between $CL$ and $CL'$ can be expressed as

$$
\begin{bmatrix} CL' \\ 1 \end{bmatrix} = [{}^{W}_{B}T] \begin{bmatrix} \Delta x \\ \Delta y \\ \Delta z \\ 1 \end{bmatrix} + \begin{bmatrix} CL \\ 1 \end{bmatrix} \tag{9}
$$

Secondly, the relationship of $OZ'$ and $OZ$ should be calculated. Similarly, the rotation of tool point $\Delta\theta_x$, $\Delta\theta_y$ and $\Delta\theta_z$ in Eq. (2) describes the difference between $OZ$ and $OZ'$ in BCS. Using the rotational matrix ${}^{W}_{B}R$, the $OZ'$ in WCS is expressed as

$$
OZ' = \left[ R_{zyx}\left( {}^{W}_{B}R[ \Delta\theta_x \quad \Delta\theta_y \quad \Delta\theta_z ]^{\mathrm{T}} \right) \right] OZ \tag{10}
$$

where $R_{zyx}$ is the z-y-x Euler angle rotation transformation [9].

According to $CL'$ and $OZ'$, the cutting parameters required in the cutting force prediction in Eq. (6) can be determined, thus the cutting force can be expressed as the function of $CL'$ and $OZ'$.

$$
F = f(CL', OZ') \tag{11}
$$

Combining Eqs. (5), (9), (10) and (11), an equilibrium equation with the variable $\Delta x$, $\Delta y$, $\Delta z$, $\Delta\theta_x$, $\Delta\theta_y$ and $\Delta\theta_z$ can be established. After solving the equilibrium equation, the actual cutter location $CL'$ and orientation $OZ'$ is calculated to determine the actual CCP.

Referring to Ref. [10], the CCP in CCS is expressed as

$$
CCP_{CCS} = \begin{bmatrix} (D/2 - r + r\,\sin(LEAD))\,\sin(Ficc) \\ (D/2 - r + r\,\sin(LEAD))\,\cos(Ficc) \\ r - r\,\cos(LEAD) \end{bmatrix} \tag{12}
$$

where

$$\begin{cases} LEAD = \arccos(\cos(lead)\cos(tilt)) \\ Ficc = \pi/2 + \arcsin\left(\dfrac{\sin\left(\arctan\left(\frac{\sin(tilt)}{\tan(lead)}\right)\right)}{\cos(lead)}\right) \end{cases} \tag{13}$$

*lead* and *tilt* in Eq. (13) are the lead and tilt angle respectively which are identified by the cutter orientation.

The homogeneous transformation matrix $_C^W T$ of CCS to WCS can be constructed utilizing the cutter location and orientation. Thus the value of the *CCP* in WCS is calculated as

$$\begin{bmatrix} CCP_{WCS} \\ 1 \end{bmatrix} = [_C^W T] \begin{bmatrix} CCP_{CCS} \\ 1 \end{bmatrix} \tag{14}$$

Using Eq. (14), the theoretical and actual *CCP* in WCS are calculated as $CCP_{WCS}$ and $CCP'_{WCS}$. The milling error is expressed as

$$Error = \begin{bmatrix} CCP'_{WCS} - CCP_{WCS} \end{bmatrix} \tag{15}$$

In order to compensate the milling error, the actual cutter location $CL'$ and orientation $OZ'$ in the equilibrium equation are set as the required value and regarded as the input parameters; the theoretical cutter location $CL$ and orientation $OZ$ are regarded as the output parameters. And, the solved $CL$ and $OZ$ are set as the cutting parameters of the robot milling system.

## 3  Experimental Verification

The verification experiment is carried out on the robot milling system shown in Fig. 2. The calibrated joint stiffness of the ABB IRB 6660 robot is listed in Table 2 referring to Ref. [11]. The theoretical cutting depth is 1.5 mm and a feed per tooth is 0.07 mm. The used cutting style is slotting and the lead and tilt angle of the tool are set as 5° and 10° which determined the cutter orientation. The calculated theoretical $CL$ in WCS is $x = 0$, $y = 0$, $z = 0.048$.

**Table 2.** Joint stiffness of the ABB IRB 6660 robot.

| Joint no. | 1 | 2 | 3 | 4 | 5 | 6 |
|---|---|---|---|---|---|---|
| Stiffness ($10^8$ N·m/rad) | 88.6 | 55 | 185 | 33 | 21 | 8.81 |

First, using the numerical iterative method, the deformation of tool point in WCS is calculated by the proposed method under the given theoretical cutting parameters and listed in Table 3. As shown in Table 3, the maximal error occurs along the $z$ direction, and the rotation of tool point can be ignored compared to the translation.

**Table 3.** The deformation of tool point in WCS.

| $x$ (mm) | $y$ (mm) | $z$ (mm) | $\theta_x$ (°) | $\theta_y$ (°) | $\theta_z$ (°) |
|---|---|---|---|---|---|
| −0.022 | −0.058 | 0.127 | −0.0001 | −0.0001 | −0.00001 |

The verification experiment is show in Fig. 5, the machining error is predicted by the proposed method, and the predicted and experimental errors are listed in Table 4.

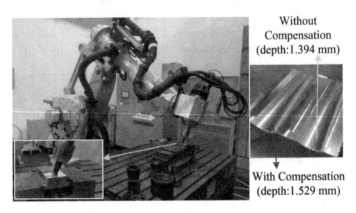

Without
Compensation
(depth:1.394 mm)

With Compensation
(depth:1.529 mm)

**Fig. 5.** Cutting experiment with set cutting depth 1.5 mm.

**Table 4.** Machining error and compensation value.

| Theoretical cutting depth (mm) | Predicted error (mm) | Actual cutting depth (mm) | Actual error (mm) |
|---|---|---|---|
| 1.5 | 0.127 | 1.394 | 0.106 |

In order to obtained the required cutting depth (1.5 mm), the cutter location **CL** set in the machining program is calculated by using the error compensation method proposed in this paper, and the set **CL** value and the actual cutting depth are listed in Table 5.

As seen in Tables 4 and 5, the actual cutting depth without compensation is 1.394 mm, however it increase to 1.529 mm (the required cutting depth is 1.5 mm) after compensation. Thus, the compensation method can improve the machining accuracy observably.

**Table 5.** The set **CL** value and actual cutting depth.

| Set CL (mm) | | | Theoretical cutting depth (mm) | Predicted cutting depth (mm) | Actual cutting depth (mm) |
|---|---|---|---|---|---|
| $x$ | $y$ | $z$ | | | |
| 0.035 | 0.069 | −0.133 | 1.680 | 1.491 | 1.529 |

Because of only the static stiffness and the average cutting force is considered in the compensation method proposed by this paper, the dynamic error still influence the machining accuracy of the robot milling. To further improve the robot milling accuracy, the research on the dynamic error prediction and compensation will be carried out in our future work.

# 4    Conclusions

The machining error caused by poor stiffness of the joint is a critical obstacle for robot milling application. For the multi-axis milling based on industrial robot, the machining error is more difficult to predict and compensate due to the complicated coordinate transformation between the deformation of tool point and the cutting force feedback. The static stiffness error prediction and compensation of robot multi-axis milling is studied in this paper. In order to predict the machining error, the static stiffness model of robot milling system is established based on the joint stiffness matrix and Jacobian matrix. Based on the static stiffness model and the cutting force model, an equilibrium equation with the variable of tool point deformation is established by using the relationship of theoretical cutting location and actual cutting location. And the milling error is defined by the solved actual cutting location. The predicted results show that the rotational deformation under the cutting force can be ignored compared to the translational deformation, thus the milling error is mainly influenced by the translation of tool point. The compensation result and the experiment verification demonstrate that the proposed compensation method can improve the machining accuracy observably.

However for actual robot milling of large components, such as large scale marine propeller, the milling error is influenced by the translation of tool point and workpiece simultaneously. The stiffness of the workpiece should be considered in the deformation compensation model. The next steps to achieve that goal are already planned at our research group.

**Acknowledgment.** This work was partially supported by the National Science Fund for Distinguished Young Scholars under Grant No. 51625502, Innovative Group Project of Hubei Province under Grant No. 2017CFA003 and the China Postdoctoral Science Foundation Funded Project (Project No. 2017M622412).

# References

1. Pan, Z.X., Zhang, H., Zhu, Z.Q., Wang, J.J.: Chatter analysis of robotic machining process. J. Mater. Process. Technol. **173**, 301–309 (2006)
2. Zaeh, M.F., Roesch, O.: Improvement of the machining accuracy of milling robots. Prod. Eng. **8**(6), 737–744 (2014)
3. Wu, Y., Klimchik, A., Caro, S., Boutolleau, C., Furet, B., Pashkevich, A.: Experimental study on geometric and elastostatic calibration of industrial robot for milling application. In: IEEE/ASME International Conference on Advanced Intelligent Mechatronics, pp. 1689–1696. Institute of Electrical and Electronics Engineers Inc., Besacon (2014)

4. Tyapin, I., Kaldestad, K.B., Hovland, G.: Off-line path correction of robotic face milling using static tool force and robot stiffness. In: 2015 IEEE/RSJ International Conference on Intelligent Robots and Systems, pp. 5506–5511. Institute of Electrical and Electronics Engineers Inc., Hamburg (2015)

5. Kaldestad, K.B, Tyapin, I., Hovland, G.: Robotic face milling path correction and vibration reduction. In: 2015 IEEE International Conference on Advanced Intelligent Mechatronics, pp. 543–548. Institute of Electrical and Electronics Engineers Inc., Busan (2015)

6. Kamali, K., Joubair, A., Bonev, I.A., Bigras, P.: Elasto-geometrical calibration of an industrial robot under multidirectional external loads using a laser tracker. In: IEEE International Conference on Robotics and Automation, pp. 4320–4327. Institute of Electrical and Electronics Engineers Inc., Stockholm (2016)

7. Möller, C., Schmidt, H.C., Shah, N.H., Wollnack, D.J.: Enhanced absolute accuracy of an industrial milling robot using stereo camera system. Procedia Technol. **26**, 389–398 (2016)

8. Wang, J.J., Zhang, H., Fuhlbrigge, T.: Improving machining accuracy with robot deformation compensation. In: 2009 IEEE/RSJ International Conference on Intelligent Robots and Systems, pp. 3826–3831. IEEE Computer Society, St. Louis (2009)

9. Zhu, Z.R., Yan, R., Peng, F.Y., Duan, X.Y., Zhou, L., Song, K., Guo, C.Y.: Parametric chip thickness model based cutting forces estimation considering cutter runout of five-axis general end milling. Int. J. Mach. Tools Manuf. **101**, 35–51 (2016)

10. Bruno, S., Lorenzo, S., Luigi, V., Giuseppe, O.: Robotic: Modeling, Planning and Control. Springer, London (2009). https://doi.org/10.1007/978-1-84628-642-1

11. Cordes, M., Hintze, W.: Offline simulation of path deviation due to joint compliance and hysteresis for robot machining. Int. J. Adv. Manuf. Technol. **90**(1–4), 1075–1083 (2017)

# Chatter Detection Based on ARMAX Model-Based Monitoring Method in Thin Wall Turning Operation

Yang Liu and Zhenhua Xiong[✉]

School of Mechanical Engineering, Shanghai Jiao Tong University, Shanghai 200240, China
mexiong@sjtu.edu.cn

**Abstract.** The ARMAX model-based monitoring method is proposed for chatter detection in thin wall turning operation. ARMAX modelling is deduced to fit the cutting force for the time varying dynamic process caused by the stiffness variation of thin wall workpiece. Residuals closely connected with chatter are extracted and monitored by control charts in real time. Cutting experiments for two different depth of cuts are performed for verification, from which it is found that the ARMAX model-based monitoring process has lower false alarm rate than the typical ARMA modelling. In addition, the model parameters affected by the varying stiffness is also time dependent, so the RELS parameter estimation algorithm is employed. The forgetting factor in the RELS algorithm is optimized through experiments to further reduce the false alarm rate. It is observed that the RELS ARMAX model-based algorithm with forgetting factors between 0.85 and 0.9 has the best monitoring performance with zero false alarms.

**Keywords:** Chatter · Thin wall turning · Stiffness variation · ARMAX · RELS

## 1 Introduction

Three different types of mechanical vibrations, known as free vibrations, forced vibrations and self-excited vibrations, are generally found in metal cutting processes. Researchers have developed several effective techniques to reduce or eliminate the occurrence of the free and forced vibrations. The self-excited vibration (i.e. chatter) is mainly caused by the regenerative effect between the cutting tool and the workpiece and brings the machining process to instability, which is the most threatening and uncontrollable type of vibration. The external reason for regenerative chatter is the lack of stiffness or damping of the machine tool, the tool holder, the cutting tool and the workpiece material, in case of the machining of thin wall workpieces, deep holes and long overhang workpieces. For this reason, the mechanism analysis, detection and suppression techniques of the regenerative chatter have been a popular topic for academic and industrial research.

The automatic detection of regenerative chatter is of vital significance to avoid damage to workpieces or machine tools. A number of studies have been reported about the development of tool condition monitoring (TCM) techniques, which are focused on

© Springer Nature Switzerland AG 2018
Z. Chen et al. (Eds.): ICIRA 2018, LNAI 10984, pp. 319–330, 2018.
https://doi.org/10.1007/978-3-319-97586-3_29

the abnormal state monitoring and diagnosis caused by tool wear, tool breakage as well as regenerative chatter [1–8]. One of the TCM approaches is time series analysis, in which a specific time series modelling is employed to the measured signals, such as cutting forces, vibration, or acoustic emission. A popular solution for linear time series monitoring involves autoregressive (AR) or autoregressive moving average (ARMA) modelling.

The AR(1) modelling was used to the acoustic emission signal for online tool wear monitoring. The AR parameters and power of the AR residual signals were selected as features and found to be effective in tool condition monitoring [9]. A time series based tooth period modelling technique was proposed for detecting tool breakage by monitoring a cutting force or torque signal, which can also be used for chatter detection or tool wear evaluation as stated [10]. A tool breakage detection algorithm was developed after removing the transient process by AR(1) filtering [11]. The AR(5) model was fitted to the cutting forces and it was found that the power of the residual signal was most effective and consistent. The AR parameters classified by pattern recognition technique had showed distinct response towards tool wear [12]. The AR(1) model was deduced to the amplitudes of the relevant frequencies of the drilling torque signals, while chatter vibration was detected with respect to the residual signals based on a ranked EWMA control chart [13]. The relationship between the mechanical model of cutting process and its corresponding time series model was discussed, in which the ARMA(4,3) model was derived to deal with the acceleration vibration signals [14]. The ARMA modelling was employed to the tool vibration signals in a turning process. The ARMA distance calculated based on eigenvalues of the tool/holder system, was applied in an online tool wear estimation algorithm [15].

The time series modelling had been widely investigated for TCM, but few researches have focused the time dependent property of the model used in case of flexible cutting tool or workpiece. In this work, the ARMAX modelling is developed based on the time varying depth of cut caused by stiffness variation and dynamic vibration of the thin wall workpiece. This model is more appropriate and more accurate to describe the cutting process after considering the exogenous input of stiffness. In addition, the RELS parameter estimation algorithm is employed to further increase the accuracy for online model identification. At last, the ARMAX modelling with residual control chart is performed for thresholding or decision-making online.

## 2　Time Series Modelling of Thin Wall Turning Operation

### 2.1　Stiffness Variation

The experimental setup is shown in Fig. 1, where the turning experiments are performed on a CK 6150A lathe and the cutting force signal is recorded with a Kistler 9257B dynamometer sampled at 20 kHz. The thin wall disc workpiece with two-stepped shape is shown in Fig. 2. As the cutting tool moves along the feed direction from inner to outer of the free-end, the axial stiffness is debilitated due to the cantilever structure of thin wall workpiece. In addition, the removal of workpiece material causes the stiffness variation slightly.

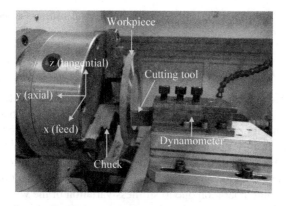

**Fig. 1.** Turning experimental setup.

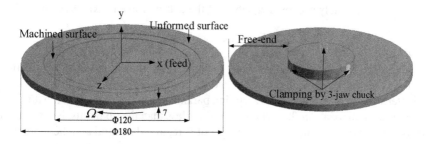

**Fig. 2.** Thin wall disc workpiece (All dimensions are in mm).

In this work, the stiffness variation of workpiece is obtained offline by finite element simulation before machining. An axial moving unit static force is imposed at the tool nose as the cutting goes on. The axial static deformation is simulated, then the axial stiffness is calculated. However, only integral locations on the machined surface are considered due to the time-consuming process of simulation, and the whole stiffness curve is obtained by spline interpolation. The axial stiffness for different depth of cuts

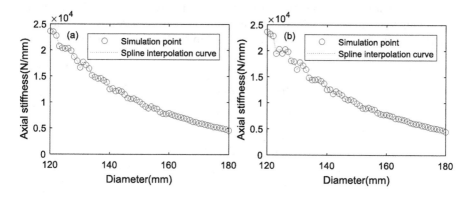

**Fig. 3.** Axial stiffness for different depth of cuts (a) 0.05 mm; (b) 0.1 mm.

is shown in Fig. 3, from which it is observed that the axial stiffness is reduced severely as the cutting tool moves along the feed direction.

## 2.2 ARMAX Modelling of the Cutting Force

It was reviewed that the cutting force was the best signal for chatter detection compared with acoustic emission and acceleration signals, and the characteristic patterns of cutting force variation make it possible to clearly distinguish chatter [16]. Apart from chatter, the cutting forces are also sensitive to other cutting conditions such as cutting parameters, temperature, hardness of material and dynamics behavior of cutting tool, making correlation with chatter more complicated. In this work, time varying process dynamics of the thin wall turning operation due to stiffness variation of the workpiece is mainly considered, in order to obtain the key information which is closely connected with chatter.

A simple and widely accepted equation of the cutting force is expressed as [17]

$$F_y = K_y ah \tag{1}$$

where $F_y$ and $K_y$ are the cutting force and the cutting coefficient in the axial direction, respectively. $a$ and $h$ are the depth of cut and the feed rate, respectively.

Chatter will occur even under low cutting parameters at the free-end of the thin wall disc. Thus, a small axial depth of cut, which is far less than the tip radius, is assumed to analysis the tool displacement. The cutting tool is regarded as rigid and the workpiece is flexible in the axial direction. It is noticed that the actual depth of cut is varying during cutting process. When the second revolution starts, there are undulations, caused by chatter vibration and weak stiffness, both in the inner machined surface of current revolution and the outside surface of previous revolution. As shown in Fig. 4, the cutting area $ah$ is changed to the gray part, which can be expressed as

$$ah \approx \sum \Delta S_i = h_0 \left[ a_0 - \frac{(y_{t-\tau} + 2y_{t-\tau+1} + \cdots + 2y_{t-\tau+1} + y_t)}{2n} \right] \tag{2}$$

where $a_0$ and $h_0$ are the selected depth of cut and feed rate, respectively. $n$ is the number of sampling points in one revolution period $\tau$. $y_t$ is the feed displacement at time $t$.

The axial workpiece displacement is assumed to be divided into two parts, i.e. the static deflection and the dynamic vibration.

$$y_t = y_t^d + y_t^s \tag{3}$$

here

$$y_t^s = \frac{F_t}{k_t} \tag{4}$$

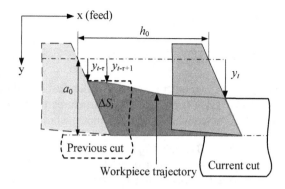

**Fig. 4.** Dynamic cutting process of thin wall turning in one revolution period.

where the superscripts, s and d represent the static deflection and the dynamic vibration. $k_t$ is the axial workpiece stiffness at current time.

The workpiece trajectory from A to B is nonlinear due to the time varying dynamics of thin wall turning operation. For simplification, the sum of tool displacements in Eq. (2) may be represented only by the displacements at present time and one spindle revolution period before, which can be expressed as

$$ah \approx h_0 \left[ a_0 - \frac{\left( p_1 y^s_{t-\tau} + p_2 y^d_{t-\tau} + p_3 y^s_t + p_4 y^d_t \right)}{2} \right] \tag{5}$$

where $p_i$ are the unsolved parameters. Substituting Eqs. (3–5) into Eq. (1) and rearranging the result as

$$\left( \frac{2}{K_c h_0} + \frac{p_3}{k_t} \right) F_t + \left( \frac{p_1}{k_{t-\tau}} \right) F_{t-\tau} = 2a_0 - p_2 y^d_{t-\tau} - p_4 y^d_t \tag{6}$$

As mentioned, the cutting force signals usually contain some detrimental components due to external cutting conditions such as stiffness variation in this case. It can surely influence the performance of common used chatter detection methods. In contrast, the model in Eq. (6) can extract the significative information hiding in the complex cutting force signals. In the stable state, the dynamic vibration is assumed to be normally distributed as white noise, thus time series modelling can be performed [18]. Previous observations are substituted by current data using the shaft operator $q$ as

$$F_{t-\tau} = q^{-1} F_t, \ y^d_{t-\tau} = q^{-1} y^d_t \tag{7}$$

Substitution Eq. (7) into Eq. (6) yields

$$\left[ 1 + \frac{p_1}{P_t k_{t-\tau}} q^{-1} \right] F_t = \frac{2a_0 K_c h_0 k_t}{2k_t + p_3 K_c h_0} + \left( -\frac{p_2}{P_t} q^{-1} - \frac{p_4}{P_t} \right) e_t \tag{8}$$

here

$$P_t = \frac{2}{K_c h_0} + \frac{p_3}{k_t}, \; e_t = y_t^d \qquad (9)$$

where $-p_4/P_t$ is assumed to be 1 and other parameters will be estimated by model fitting.

Then it is observed that Eq. (8) has the same structures to the ARMAX model (AutoRegressive Moving Average with eXogenous input) whose general form is

$$\begin{aligned}
A(q)y_t &= B(q)u_t + C(q)e_t \\
A(q) &= 1 + a_1 q^{-1} \cdots + a_{na} q^{-na} \\
B(q) &= b_1 + b_2 q^{-1} \cdots + b_{nb} q^{-nb+1} \\
C(q) &= 1 + c_1 q^{-1} \cdots + c_{nc} q^{-nc}
\end{aligned} \qquad (10)$$

The output $y_t$ is the measured axial cutting force, while the exogenous input $u_t$ is proportional to the cutting coefficient, the depth of cut, the feed rate and the current axial stiffness as follows.

$$u_t = 2K_c a_0 h_0 k_t \qquad (11)$$

In addition, the orders for polynomials $A(q)$, $B(q)$ and $C(q)$ are 1, 0 and 1, respectively. The model parameters are also time varying with respect to the stiffness variation as follows.

$$a_1 = \frac{p_1}{P_t k_{t-\tau}}, \; b_0 = \frac{1}{2k_t + p_3 K_c h_0}, \; c_1 = -\frac{p_2}{P_t} \qquad (12)$$

## 3   Chatter Detection

### 3.1   Online Model Identification and Forecast

For the ARMAX model, all parameters to be estimated are arranged in a vector $\vartheta$, while the relevant input-output data are collected into the regressor vector $\varphi$ as

$$\vartheta_t = [a_1 \cdots a_{na} \, b_1 \cdots b_{nb-1} \, c_1 \cdots c_{nc}]^T, \; \varphi_t = [-y_{t-1} \cdots - y_{t-na} \, u_t \cdots u_{t-nb+1} \, \bar{\varepsilon}_{t-1} \cdots \bar{\varepsilon}_{t-nc}]^T \qquad (13)$$

where $\bar{\varepsilon}_t = y_t - \varphi_t^T \hat{\vartheta}_t$ represents the residual.

The parameters need to be identified using an appropriate estimation algorithm. Due to the time varying properties of the process, the adaptive recursive extended least square (RELS) algorithm may be applied to perform online model identification [19]. The RELS approach allows the estimate to be updated at each step, by correcting it by a term proportional to the innovation brought about by the last collected observation. In other words, the estimate at time $t$ can be obtained from the estimate at time $t-1$, without the need to explicitly reconsider all of the past data. The RELS algorithm is expressed as

$$\hat{\vartheta}_t = \hat{\vartheta}_{t-1} + \gamma_t R_t^{-1} \varphi_t \varepsilon_t$$
$$R_t = R_{t-1} + \gamma_t \left( \varphi_t \varphi_t^T - R_{t-1} \right) \tag{14}$$
$$\varepsilon_t = y_t - \varphi_t^T \hat{\vartheta}_{t-1}$$

where $\gamma_t$ is the gain which has a constant value $\gamma$ in the following algorithm. $\varepsilon_t$ is the prediction error.

For a time dependent system that changes gradually, the recent measurements are usually considered more reliable and the older ones are no longer representative of the process dynamics [20]. In order for the estimation algorithm to progressively "forget" older data, the most common choice is to take a constant forgetting factor $\lambda$. In this way, the most recent data are given unitary weight in the minimization process, while an exponentially time decreasing weight is assigned to past measurements. The parameters will be estimated by minimizing the sum of the exponential weighted least squares of the prediction errors as follows.

$$\vartheta_t = \arg \min_{\vartheta} \sum_{k=1}^{t} \lambda^{t-k} \varepsilon_t^2 \tag{15}$$

The relationship between the loss factor and the constant gain is $\gamma + \lambda = 1$. With a predefined forgetting factor, all of the autoregressive, moving average and exogenous parameters can be estimated with the measured cutting force and simulated stiffness. Then the forecast for residuals is calculated by

$$\bar{\varepsilon}_t = F_t + \hat{a}_1 F_{t-1} - \hat{b}_1 u_t - \hat{c}_1 \bar{\varepsilon}_{t-1} \tag{16}$$

### 3.2 Thresholding Based on Residual Control Chart

It is crucial to set up an appropriate threshold for decision-making of chatter detection, while the traditional determination of thresholds is to design preliminary cutting experiments and find a constant threshold between stable state and chatter state empirically. However, this method has limitation in choosing the threshold and is less reliable and less adaptive for variable cutting conditions.

Statistical process control or control charts, which is focused on the inherent characteristic of data series, are usually used to monitor the statistical state and alarm out-of-control in quality engineering. The advantage of the statistical process control is to determine the threshold non-empirically and update it automatically. It should also be noted that control charts are built based on the basic assumption that the process is independently and normally distributed. However, this assumption rarely holds in practice. In thin wall tuning, any sensor signal acquired during one spindle revolution is expected to be characterized by some level of autocorrelation due to the time varying process dynamics. The autocorrelation in axial cutting forces is mainly caused by stiffness variation, vibration and other reasons include cutting impact, tool wear and temperature. Control charts should not be used directly, because more false alarms will occur for data series with high level autocorrelation. An approach that has proved useful in dealing with autocorrelated data is residual control chart [21]. Firstly, an appropriate time series model is fitted to the data series to remove the autocorrelation from the data. Then the control charts like Shewhart charts can be applied to the residuals, where the control limits or thresholds are calculated as

$$
\left.
\begin{array}{l}
\text{Upper control limit: } \mu_t + L_t\sigma_t \\
\text{Center line: } \mu_t \\
\text{Lower control limit: } \mu_t - L_t\sigma_t
\end{array}
\right\}
\tag{17}
$$

where $\mu_t$ and $\sigma_t$ are the mean and standard deviation of the residuals. The width of control limits $L_t$ is generally constant as 3, which is also well known as the 'three-sigma' criterion.

## 4 Experimental Validation

The measured axial cutting forces for different depth of cuts are shown in Fig. 5. Larger forces are noted at the entry point probably due to the impact when the tool cuts into the workpiece, then in the stable state it appears to be distributed as a white noise. It has a considerable oscillation around zero when the cutting state turns into chatter. The cutting forces change dramatically between stable and chatter states. This in-between state is generally called the onset state when the cutting forces have slight fluctuations and no visible chatter marks are generated on the workpiece before chatter has fully developed. These three cutting states can be identified artificially from experiments based on the noise and the surface appearance. It is observed that chatter occurs about 10 s earlier at 0.1 mm than that at 0.5 mm.

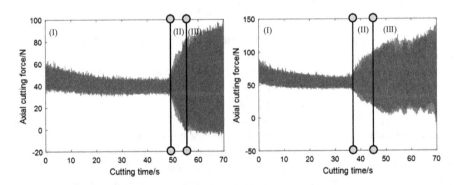

**Fig. 5.** Axial cutting forces in real cutting experiments under spindle speed of 220 rpm, feed rate of 0.1 mm/r, depth of cuts of 0.05 mm (left) and 0.1 mm (right) (I) Stable state; (II) Onset state; (III) Chatter state.

### 4.1 ARMAX Model-Based Monitoring

The outputs of the ARMAX model, i.e. measured cutting forces, are extracted in a period of one spindle revolution about 0.27 s, which means from all sampling points there are

5455 available groups of data. Different groups make little influence, so the first group is chosen uniformly for the following algorithm. In contrast to the previous research, the ARMA model-based monitoring is also performed.

Both the ARMA model and the ARMAX model are identified online by least squares (LS) algorithm every time a new data is collected. Then the residuals are obtained and control limits are calculated by the Shewhart chart with the time histories of residuals from the beginning to the current time. Without other criterions, the process is regarded as in-control (i.e. stable state) when the observations remain inside the control limits and out-of-control (i.e. onset state or chatter state) when they fall outside the upper or lower control limits. The numbers of out-of-control for the Shewhart charts with ARMAX or ARMA modelling and without modelling are shown in Fig. 6.

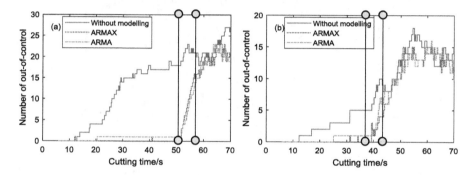

**Fig. 6.** Monitoring processes with and without modelling (a) 0.05 mm; (b) 0.1 mm.

It is found that there are too many false alarms for monitoring process without modelling, so direct monitoring of the cutting force signal is infeasible in this case. Although a steep increase of the out-of-control number occurs in the onset state both with ARMA and ARMAX modelling, the least out-of-control number is observed in the stable state for the ARMAX model-based monitoring processes.

The false alarm rates are calculated as ratios between the times of out-of-control number larger than 0 in the stable state and the sampling times of the whole stable state. The results are shown in Table 1, where the false alarm rates of ARMAX model-based monitoring processes are reduced substantially to small values under 2%.

**Table 1.** False alarm rate (%) with and without model-based monitoring processes.

| Time series modelling | Depth of cut/mm | |
|---|---|---|
| | 0.05 | 0.1 |
| No model fitting | 84.34 | 80.33 |
| ARMAX | 1.87 | 1.64 |
| ARMA | 67.47 | 41.8 |

## 4.2 RELS ARMAX Model-Based Monitoring

As mentioned in Eq. (12), the ARMAX model parameters are also time dependent due to the stiffness variation, so the RELS algorithm with 'forget' capacity may be more appropriate for parameter estimation. The key coefficient of the RELS algorithm need to be defined is the forgetting factor. The residuals will be forecasted recursively by the RELS ARMAX modelling with different forgetting factors, then the cutting process is monitored by the residual control chart. The compassion of false alarm rates between RELS ARMAX model-based monitoring processes with different forgetting factors is shown in Fig. 7.

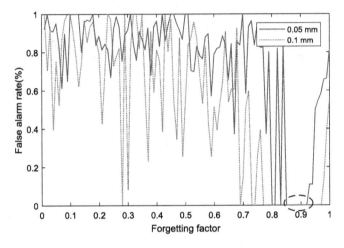

**Fig. 7.** False alarm rate (%) of RELS ARMAX model-based monitoring processes with different forgetting factors.

Enormous differences between the false alarm rates with different forgetting factors are observed. Zero false alarm rate can be achieved, while large false alarm rates are also noticed, which means the forgetting factor should be carefully selected. A zone of optimized forgetting factor, i.e. 0.85 to 0.9, is found through the experimental results of two depth of cuts. The monitoring performance of the RELS ARMAX model-based method with optimized forgetting factor is further improved than that of the LS ARMAX modelling.

## 5   Conclusion

In the case of thin wall turning operation, an ARMAX model for the cutting force is deduced by analyzing the dynamic cutting process caused by stiffness variation and dynamic vibration. The model residuals are monitored and the chatter state is identified by the residual control charts. The main contribution of this work is summarized as

1. The residuals from ARMAX modelling of the cutting force are determinant information closely connected with chatter, which are monitored by the residual control charts. This ARMAX model-based monitoring process shows better performance than the typical ARMA modelling in real cutting experiments.
2. Base on the time varying properties of the model, time dependent parameters of ARMAX model are estimated online by the RELS parameter estimation algorithm. The false alarm rate of the RELS ARMAX model-based monitoring process with an experimental optimized forgetting factor is further decreased to zero, which may be an effective chatter detection method for engineering application.

# References

1. Tansel, I.N., Wagiman, A., Tziranis, A.: Recognition of chatter with neural networks. Int. J. Mach. Tools Manuf **31**(4), 539–552 (1991)
2. Wu, Y., Du, R.: Feature extraction and assessment using wavelet packets for monitoring of machining processes. Mech. Syst. Signal Process. **10**(1), 29–53 (1996)
3. Govekar, E., et al.: A new method for chatter detection in grinding. CIRP Ann. Manuf. Technol. **51**(1), 267–270 (2002)
4. Tangjitsitcharoen, S., Pongsathornwiwat, N.: Development of chatter detection in milling processes. Int. J. Adv. Manuf. Technol. **65**(5–8), 919–927 (2013)
5. Sun, Y., Zhuang, C., Xiong, Z.: Real-time chatter detection using the weighted wavelet packet entropy. In: IEEE/ASME International Conference on IEEE, pp. 1652–1657 (2014)
6. Nouri, M., et al.: Real-time tool wear monitoring in milling using a cutting condition independent method. Int. J. Mach. Tools Manuf **89**, 1–13 (2015)
7. Sun, Y., Xiong, Z.: An optimal weighted wavelet packet entropy method with application to real-time chatter detection. IEEE/ASME Trans. Mechatron. **21**(4), 2004–2014 (2016)
8. Fu, Y., et al.: Timely online chatter detection in end milling process. Mech. Syst. Signal Process. **75**, 668–688 (2016)
9. Ravindra, H.V., Srinivasa, Y.G., Krishnamurthy, R.: Acoustic emission for tool condition monitoring in metal cutting. Wear **212**(1), 78–84 (1997)
10. Tansel, I.N., McLaughlin, C.: Detection of tool breakage in milling operations—I. The time series analysis approach. Int. J. Mach. Tools Manuf **33**(4), 531–544 (1993)
11. Altintas, Y.: In-process detection of tool breakages using time series monitoring of cutting forces. Int. J. Mach. Tools Manuf **28**(2), 157–172 (1988)
12. Kumar, S.A., Ravindra, H.V., Srinivasa, Y.G.: In-process tool wear monitoring through time series modelling and pattern recognition. Int. J. Produc. Res. **35**(3), 739–751 (1997)
13. Messaoud, A., Weihs, C., Hering, F.: Detection of chatter vibration in a drilling process using multivariate control charts. Comput. Stat. Data Anal. **52**(6), 3208–3219 (2008)
14. Song, D.Y., et al.: A new approach to cutting state monitoring in end-mill machining. Int. J. Mach. Tools Manuf **45**(7–8), 909–921 (2005)
15. Aghdam, B.H., Vahdati, M., Sadeghi, M.H.: Vibration-based estimation of tool major flank wear in a turning process using ARMA models. Int. J. Adv. Manuf. Technol. **76**(9–12), 1631–1642 (2015)
16. Tlusty, J., Andrews, G.C.: A critical review of sensors for unmanned machining. CIRP Ann. Manuf. Technol. **32**(2), 563–572 (1983)
17. Altintas, Y.: Manufacturing Automation: Metal Cutting Mechanics, Machine Tool Vibrations, and CNC Design, 2nd edn. Cambridge University Press, New York (2012)

18. Brockwell, P.J., Davis, R.A.: Introduction to Time Series and Forecasting. Springer, New York (2016). https://doi.org/10.1007/b97391
19. Paleologu, C., Benesty, J., Ciochina, S.: A robust variable forgetting factor recursive least-squares algorithm for system identification. IEEE Signal Process. Lett. **15**, 597–600 (2008)
20. Fung, E.H.K., et al.: Modelling and prediction of machining errors using ARMAX and NARMAX structures. Appl. Math. Modell. **27**(8), 611–627 (2003)
21. Messaoud, A., Weihs, C.: Monitoring a deep hole drilling process by nonlinear time series modeling. J. Sound Vib. **321**(3–5), 620–630 (2009)

# Out-of-Plane Vibration Frequency Estimation for Flexible Substrate in Roll-to-Roll Processing

Jiankui Chen$^{(\boxtimes)}$, Yufei Zhu, Hua Yang,
Yongan Huang, and Zhouping Yin

State Key Laboratory of Digital Manufacturing Equipment and Technology,
Flexible Electronics Research Center,
Huazhong University of Science and Technology, Wuhan 430074, China
chenjk@hust.edu.cn

**Abstract.** Out-of-plane displacements of a moving web is well known to be a main limiting factor of roll-to-roll (R2R) manufacturing for flexible electronics. To tackle this problem, in this paper, a new contactless approach for the measurement of membrane vibration is addressed. This technique, which allows catching out-of-plane membrane vibration involves a vision system composed of a CCD camera and a laser stripe device. With gray centroid method and principal component analysis, out-of-plane vibration distribution is estimated in real time. Experiments are conducted on the RFID inlay in R2R experimental platform to confirm the effectiveness and accuracy of the proposed method.

**Keywords:** Out-of-plane vibration · Flexible electronics · Roll-to-roll
Frequency estimation · Structured light

## 1 Introduction

Flexible electronics, also known as printed or organic electronics, are depositing organic/inorganic electronic devices on a flexible web [1]. Its production process can be integrated by roll-to-roll (R2R) manufacturing, and the main objective is to increase the web velocity while controlling the tension of the web. This can improve the production efficiency of flexible electronics. However, some disturbances limit the velocity, like the non-circularity of the roll, roll misalignment and web sliding, etc. What's more, since there exists a coupling introduced by the elastic property of the flexible web, disturbances are transmitted to the web tension, resulting in web out-of-plane vibrations. Therefore, the out-of-plane vibration measurement is meaningful for perceiving the web tension change and improving the production efficiency of flexible electronics.

The out-of-plane vibration can be measured by several methods in the R2R system. Traditionally, laser sensor set vertical to the web is used to measure out-of-plane displacements [2–4], which is fast and high-precise. However, this method is affected by the accuracy of laser sensors and can only achieve the vibration information of one point in the web surface. With the development of machine vision and image processing technology, vision-based measuring methods are used to measure vibration

© Springer Nature Switzerland AG 2018
Z. Chen et al. (Eds.): ICIRA 2018, LNAI 10984, pp. 331–341, 2018.
https://doi.org/10.1007/978-3-319-97586-3_30

gradually. Maurice [5] proposed a new framework for the design of patterns driven by the Hamming distance with coded structured light, and, the 3-D information can be obtained accurately. Ishii [6] investigated a high-frame-rate laryngoscope including a high-speed camera at 4000 fps to measure the vibration distribution of a human vocal fold. Zhai [7] studied a method by projecting $\pi$ phase shifting bi-color sinusoidal fringe to analyze the vibration of membrane. Doignon [8] proposed a new technique to estimate out-of-plane web vibration properties with a camera and a laser dot pattern device and experiments was performed. Doignon [9] also presented a faster stereovision system in conjunction with a more appropriate laser stripe line pattern. This method can be used for non-contact measurement of the whole web surface.

In this paper, a measuring system composed of a CCD camera and a laser stripe device is built to measure the out-of-plane displacement of the horizontal moving web between two rolls. Firstly, with stripe structured light principle, the geometric model of vibration measurement system is built to get the relationship between the out-of-plane displacements of moving web and the coordinates of the laser stripe in the images. Then, gray centroid method is used to derive initial centerline coordinates of laser stripe. Principal component analysis is used to derive initial centerline normal vectors to get precise coordinates. Finally, coordinates are transformed into out-of-plane displacements and the natural frequencies are also calculated based on the Fourier transform. Experiments are carried out to confirm the proposed method.

## 2    Geometrical Model

The geometrical model of the whole vision system is briefly described in this section. To get actual coordinates of points in the web surface, the relationship between image plane coordinates and world coordinates is built. The stationary web plane is on the base plane $O_w X_w Y_w$. The well-known pinhole camera model is used to relate a 3-D point $P$ in the web surface, with homogeneous coordinates $(x^W, y^W, z^W, 1)^T$ expressed in the world coordinate system and its corresponding projection $(m, n, 1)^T$ in the image plane coordinate system. That is:

$$z^C \begin{bmatrix} m \\ n \\ 1 \end{bmatrix} = \begin{bmatrix} 1/m_x & 0 & 0 \\ 0 & 1/n_x & 0 \\ 0 & 0 & 1 \end{bmatrix} \begin{bmatrix} f & 0 & 0 & 0 \\ 0 & f & 0 & 0 \\ 0 & 0 & 1 & 0 \end{bmatrix} \begin{bmatrix} R & T \\ 0 & 1 \end{bmatrix} \begin{bmatrix} x^W \\ y^W \\ z^W \\ 1 \end{bmatrix} \tag{1}$$

where $m_x$, $n_x$ are the sizes of CCD camera cells and $f$ is focal length. Rotation matrix $R$ and translation vector $T$ are used to describe the conversion relation between world coordinate system and camera coordinate system.

Also, in order to get the out-of-plane displacements, the relationship between the displacements at any point of the projected stripe and the coordinates of the point in the image is established. The geometrical model of the web vibration measuring system is shown in Fig. 1. Based on the traditional 3-D measurement system of structured light, we optimize the system geometry and weaken the geometric constraints (see Fig. 1(a)). The optical axis of the camera is not intersected with the light axis of the laser device on the base plane. The optical axis of the camera and the light axis of the laser device don't need to be perpendicular to the base plane. Also, the line of the camera optical center and laser optical center don't need to be parallel. In general, we assume that laser stripe projected on the base plane is not perpendicular to the $Y^W$ axis. Instead, it is perpendicular to the auxiliary line $L$, and the angle between the line $L$ and the $X^W$ axis is $\theta$ (see Fig. 1(b)).

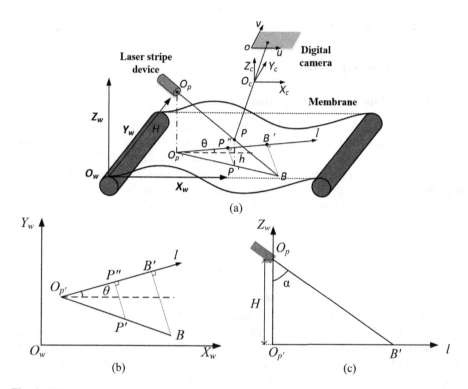

**Fig. 1.** The geometrical model of the web vibration measuring system (a) Optical path of measuring system; (b) Geometric distribution of base plane; (c) 2-D structure of projection system.

According to the similar triangle and geometric relationship, the formula between the world coordinates on the web surface and the vibration displacements of the point $P$ can be established as follow:

$$h = H - \frac{(x^W - x_0)\cos\theta - (y^W - y_0)\sin\theta}{H\tan\alpha} \tag{2}$$

where $H$ is the distance of laser optical center $O_p$ to the base plane $O^W X^W Y^W$, $(x^W, y^W, 0)$ is the world coordinates of the projected point $P'$ and $(x_0, y_0, 0)$ is the world coordinates of the projected point $O'_p$.

Finally, the approximate fitting formula of the web vibration measuring system model can be described as

$$h = (A_1 + A_2 m + A_3 n)/(B_1 + B_2 m + B_3 n) \tag{3}$$

where the coefficients $A_1$, $A_2$, $A_3$, $B_1$, $B_2$ and $B_3$ are dependent on the geometric parameters and camera intrinsic parameters.

Therefore, once the image plane coordinates of the points are obtained, the out-of-plane displacements can be computed from Eq. (3) immediately.

## 3    Theoretical Analysis

### 3.1    Out-of-Plane Vibration Equation

The equations of an axially moving a string or beam can be derived from the extended Hamilton principle and consist of two partial different equations [10]. A numerical study shows that, in our experimental conditions, the coupling in the longitudinal and out-of-plane vibrations through the web tension is not significant, and this was experimentally observed [11]. Therefore, these equations can be decoupled and simplified under the realistic assumption that the web velocity and tension remain constant. The equation is given as

$$(V^2 - \frac{T}{\rho})\frac{\partial^2 y}{\partial x^2} + 2V\frac{\partial^2 y}{\partial t \partial x} + \frac{\partial^2 y}{\partial t^2} = 0 \tag{4}$$

where $y$, $\rho$, $V$, $T$ represent the out-of-plane displacement, the linear web density, the web speed and the web tension. So, the natural frequencies for a stationary web model without stiffness can be given as $f_1 = c/2L$, and the natural frequencies for a moving web model without stiffness can be given as $f_2 = c(1 - V^2/c^2)/2L$. The natural frequencies will be used to compare with the experimental results. It is important to emphasize that the proposed method in this paper is to measure the vibration distribution of the web surface and its working condition is different from the realistic condition of this string model. Therefore, the contrast can only be used as a general reference.

### 3.2    Centerline Extraction of Structured Light Strip

In this section, we propose a new method to estimate the accurate coordinates of the laser stripe centerline, which is based on principal component analysis (PCA). The measuring procedure includes grabbing the laser strip images, extracting the region of interest, computing the rough coordinates of strip, calculating the normal vectors of the rough coordinates and the accurate coordinates.

**Extraction of Rough Coordinates.** In order to extract the centerline rough coordinates, the region of interest (ROI) should be extracted firstly. The grayscale distribution of the laser stripe image modulated by the amplitude of web vibration is very obvious, which is divided into two parts: laser stripe (foreground) and web (background). As the high energy characteristic of the laser stripe, its gray value is mostly distributed near the high value. On the contrary, the web is in the dark environment and its gray value is distributed near low value. Therefore, a suitable threshold $T$ needs to be found to separate the foreground and background. Adaptive threshold segmentation called Otus's method [12] is used in this paper to achieve this threshold $T$.

The ratio of the laser stripe pixels to the image is $\mu_0$ and the average gray value of the laser stripe is $\omega_0$. The ratio of the web pixels to the image is $\mu_1$ and the average gray value of the web is $\omega_1$. The average gray value of the whole image is $\mu$. So, the variance $g$ between the laser stripe and the web is given as $g = \omega_0(\mu_0 - \mu)^2/(1 - \omega_0)$. Therefore, the best threshold $T$ for separating the laser stripe and the web is the threshold which makes the variance $g$ reaches the maximum.

Since ROI has been found, the rough coordinates of centerline can be obtained next step. Considering that the noise is small in the ROI of laser stripe, and the amplitude of the web vibration is also small which means that the amount of the laser stripe deformation is small, gray centroid method is utilized to achieve the rough coordinates.

In the ROI of the stripe, $(x_m, y_i)$ is used to denote the coordinates of the $m$ column in the stripe section and the corresponding gray value is $I(x_m, y_i)(i = 1, 2, ..., N)$. $N$ is used to denote the number of points in the stripe section, which is usually taken as an odd number. Then, the column coordinate $y_m$ is given as

$$y_m = \sum_{i=1}^{N} I(x_m, y_i) \cdot y_i / \sum_{i=1}^{N} I(x_m, y_i) \tag{5}$$

The results obtained by the Eq. (5) are shown in Fig. 2.

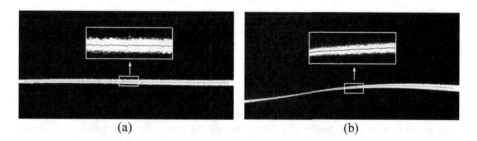

**Fig. 2.** The rough coordinates obtained by gray centroid method (a) linear laser stripe; (b) curvilinear laser stripe.

**Extraction of Normal Vectors.** In this section, the method for obtaining the normal vectors of laser strip will be discussed. In order to obtain the normal vectors, gray images need to be converted from gray space to the gradient vector space. $I(x, y)$ is used to denote the gray value of the rough point $P(x, y)$. Then using the Sobel operator, the gradient vector at point $P$ is given as

$$[G_x, G_y]^T = [\frac{\partial I(x,y)}{\partial x}, \frac{\partial I(x,y)}{\partial y}]^T \tag{6}$$

Next, $v$ is used to denote the unit normal vector at the point $P$. The sum of the projection of gradient vector dataset on the unit normal vector $v$ can be defined as

$$G^T v = \frac{1}{N}\sum_{i=1}^{N}(G^{i^T} \cdot v) = v^T(\frac{1}{N}\sum_{i=1}^{N} G^i \cdot G^{i^T})v \tag{7}$$

in a window $W$. Since $v$ is a unit normal vector, solving the maximum value of the formula (7) is to obtain the main eigenvector of the auto covariance matrix $C$, i.e.,

$$C = \begin{bmatrix} Cov(G_x, G_x) & Cov(G_x, G_y) \\ Cov(G_y, G_x) & Cov(G_y, G_y) \end{bmatrix}$$
$$= \begin{bmatrix} E[G_x^2] - (E[G_x])^2 & E[G_xG_y] - E[G_x]E[G_y] \\ E[G_xG_y] - E[G_x]E[G_y] & E[G_y^2] - (E[G_y])^2 \end{bmatrix} \tag{8}$$

In this estimate of the auto covariance matrix $C$, the assumption is made that the gradient vectors are zero-mean, i.e., $E[G_x] = E[G_y] = 0$, in the window $W$. Therefore, the eigenvalues and eigenvectors of the auto covariance matrix are derived as

$$\lambda_1 = \frac{E[G_x^2] + E[G_y^2]}{2} + \frac{\sqrt{(E[G_x^2] - E[G_y^2])^2 + 4E[G_xG_y]^2}}{2}$$

$$\lambda_2 = \frac{E[G_x^2] + E[G_y^2]}{2} - \frac{\sqrt{(E[G_x^2] - E[G_y^2])^2 + 4E[G_xG_y]^2}}{2}$$

$$v_1 = \left[\frac{E[G_x^2] - E[G_y^2]}{2} + \frac{\sqrt{(E[G_x^2] - E[G_y^2])^2 + 4E[G_xG_y]^2}}{2} \quad E[G_xG_y]\right]^T$$

$$v_2 = \left[\frac{E[G_x^2] - E[G_y^2]}{2} - \frac{\sqrt{(E[G_x^2] - E[G_y^2])^2 + 4E[G_xG_y]^2}}{2} \quad E[G_xG_y]\right]^T \tag{9}$$

By the definition of the gradient vectors auto covariance matrix, the eigenvector corresponding to the maximum absolute value of the eigenvalues is the normal vector of stripe. The results obtained by the Eq. (9) are shown in Fig. 3.

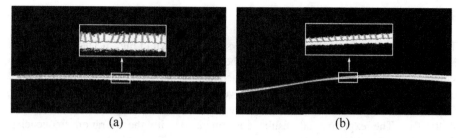

(a)                                          (b)

**Fig. 3.** The normal vectors obtained based on principal component analysis (a) linear laser stripe; (b) curvilinear laser stripe.

Finally, along the normal direction of each rough coordinate, the accurate coordinate of point $P(x_a, y_a)$ can be given as

$$
x_a = \begin{cases} \dfrac{\sum\limits_{i=-W/2}^{i=W/2} I(x_0+i,y_0+ki)\cdot(x_0+i)}{\sum\limits_{i=-W/2}^{i=W/2} I(x_0+i,y_0+ki)}, & \theta \in \left(-\frac{\pi}{2},-\frac{\pi}{4}\right] \cup \left[\frac{\pi}{4},\frac{\pi}{2}\right) \\[4ex] \dfrac{\sum\limits_{i=-W/2}^{i=W/2} I(x_0-\frac{1}{k}i,y_0+i)\cdot(x_0-\frac{1}{k}i)}{\sum\limits_{i=-W/2}^{i=W/2} I(x_0-\frac{1}{k}i,y_0+i)}, & \theta \in \left(-\frac{\pi}{4},\frac{\pi}{4}\right) \end{cases}
$$

$$
y_a = \begin{cases} \dfrac{\sum\limits_{i=-W/2}^{i=W/2} I(x_0+i,y_0+ki)\cdot(y_0+ki)}{\sum\limits_{i=-W/2}^{i=W/2} I(x_0+i,y_0+ki)}, & \theta \in \left(-\frac{\pi}{2},-\frac{\pi}{4}\right] \cup \left[\frac{\pi}{4},\frac{\pi}{2}\right) \\[4ex] \dfrac{\sum\limits_{i=-W/2}^{i=W/2} I(x_0-\frac{1}{k}i,y_0+i)\cdot(y_0+i)}{\sum\limits_{i=-W/2}^{i=W/2} I(x_0-\frac{1}{k}i,y_0+i)}, & \theta \in \left(-\frac{\pi}{4},\frac{\pi}{4}\right) \end{cases}
$$

(10)

where $k$ denotes the normal slope at the rough coordinate of the stripe and $\theta$ denotes the tilt angle of normal slope at the rough coordinate. The results obtained by the Eq. (10) are shown in Fig. 4.

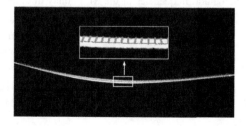

**Fig. 4.** The accurate coordinates of centerline laser stripe.

## 4  Experiments

The experiments are performed on a R2R system, as shown in Fig. 5. The CCD camera is vertical to the web surface to measure the out-of-plane displacements, which has a sampling frequency of 50 Hz. The data acquisition, A/D conversion and data processing are all carried out using the Universal Motion and Automation Controller (UMAC) and Visual Studio 2015 software. The UMAC is a systemic controller of Delta Tau. The experimental results are compared with the vibration frequencies derived from the string model.

(a)                                              (b)

**Fig. 5.** The R2R experimental environment (a) The R2R platform; (b) The web used in experiments.

The RFID inlay (see in Fig. 5(b)), one PET substrate with patterning antenna, is used as the travelling web in this experiment. The dimensions of the web are $L = 0.380\,$m, $b = 0.084\,$m, $h = 5 \times 10^{-5}\,$m, $\rho = 0.0580\,$kg/m$^3$. Because of large amount of experimental data, three feature points in the web are selected at the same interval to explain. The natural frequencies in different states calculated by the string model is shown in Table 1.

**Table 1.** The natural frequencies of the RFID inlay in different states.

| Tension (N) | Stationary web $f_1$ (HZ) | Moving web ($V = 10$) $f_2$ (HZ) | Moving web ($V = 20$) $f_3$ (HZ) |
|---|---|---|---|
| 4 | 10.927 | 10.923 | 10.909 |
| 6 | 13.383 | 13.378 | 13.361 |
| 8 | 15.453 | 15.447 | 15.428 |
| 10 | 17.277 | 17.270 | 17.249 |

## 4.1 Static Experiments

For the stationary web, the tension is set 4 N. From the centerline extraction algorithm proposed in this paper, the vibration displacements are calculated. From the Fourier transform, the natural frequencies are also calculated in the range from 0 Hz to 30 Hz. The displacement-time curves and amplitude-frequency diagram are shown in Fig. 6. It is seen that the detected first natural vibration frequencies of the three points are 11.52 Hz, 10.55 Hz, 11.23 Hz, respectively. The experimental results show good agreement with the theoretical results predicted with the simple string model.

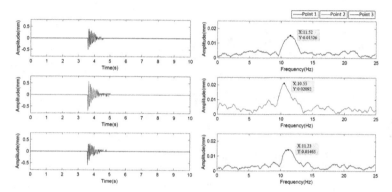

**Fig. 6.** The displacement-time curve are amplitude-frequency curve for the stationary web when $T = 4$ N.

## 4.2 Dynamic Experiments

For the web in dynamic state, the effect of the travelling speed on the natural frequencies should be taken into consideration [13, 14]. The web tension is set 4 N while the web velocity is set 10 m/min and the web tension is set 6 N while the web velocity is set 20 m/min. The displacement-time curves and amplitude-frequency diagram are shown in Figs. 7 and 8, respectively.

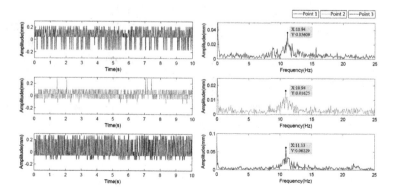

**Fig. 7.** The displacement-time curve are amplitude-frequency curve for the web when $T = 4$ N and $V = 10$ m/min.

The comparisons of the web vibration frequencies estimated by the proposed method and the string model are shown in Table 2. As a result, the frequencies of the different position are close to the frequencies calculated by the string model. The error of the web vibration frequency estimated by the proposed method is within 6%. The proposed method based on the vision technique can conveniently measure the web vibration distribution. It is noted that, for the web vibration measurement, discrepancies of results may be caused by following reasons: (a) the fitting error in the structured light stereovision geometrical model; (b) the material parameters of web; (c) the resolution of measured amplitudes using CCD camera.

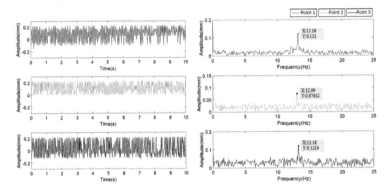

**Fig. 8.** The displacement-time curve are amplitude-frequency curve for the web when $T = 6$ N and $V = 20$ m/min.

**Table 2.** The comparison of the web vibration frequency estimated by the proposed method and the frequency by the string model.

| Web speed V (m/min) | Tension T (N) | Point | Measured frequency $f_m$ (Hz) | Theoretical frequency $f_t$ (Hz) | Error $|f_m - f_t|/f_t$ |
|---|---|---|---|---|---|
| 0 | 4 | 1 | 11.52 | 10.927 | 5.42% |
| | | 2 | 10.55 | 10.927 | 3.45% |
| | | 3 | 11.23 | 10.927 | 2.77% |
| 10 | 4 | 1 | 10.94 | 10.923 | 0.16% |
| | | 2 | 10.94 | 10.923 | 0.16% |
| | | 3 | 11.13 | 10.923 | 1.90% |
| 20 | 6 | 1 | 13.18 | 13.361 | 1.35% |
| | | 2 | 12.99 | 13.361 | 2.78% |
| | | 3 | 13.18 | 13.361 | 1.35% |

# 5 Conclusions

In this paper, we have presented a new measurement technique for the out-of-plane vibration distribution of a moving web. A structured light stereovision geometrical model is used to achieve the precise amplitudes of web vibration. Also, the centerline extraction of laser stripe based on the principal component analysis is proposed. The vibration frequencies distributed along the stripe is successfully estimated and the error is generally less than 6%. The main advantage of this measurement is its non-contact with web, which has no additional influence on the moving web. Also, there is no need to change or modify the existing layout of the web handling machine.

# References

1. Rogers, J., Someya, T., Huang, Y.: Materials and mechanics for stretchable electronics. Science **327**, 1603–1607 (2010)
2. Ma, L., Chen, J., Tang, W., Yin, Z.: Vibration-based estimation of tension for an axially travelling web in roll-to-roll manufacturing. Measur. Sci. Technol. **29**, 015102 (2018)
3. Guo, J., Zheng, Z., Wu, S.: An impact vibration experimental research on the pretension rectangular membrane structure. Adv. Mater. Sci. Eng. **2015**, 1–8 (2015)
4. Prakash, S., Upadhyay, S., Shakher, C.: Real time out-of-plane vibration measurement/monitoring using Talbot interferometry. Optics Lasers Eng. **33**, 251–259 (2000)
5. Maurice, X., Graebling, P., Doignon, C.: A pattern framework driven by the hamming distance for structured light-based reconstruction with a single image. In: Computer Vision and Pattern Recognition, pp. 2497–2504 (2011)
6. Ishii, I., Takemoto, S., Takaki, T., et al.: Real-time laryngoscopic measurements of vocal-fold vibrations. In: International Conference of the IEEE Engineering in Medicine & Biology Society, p. 6623 (2011)
7. Zhai, S.L.: Vibration mode analysis of membrane based on bi-color fringe projection. Optics Optoelectron. Technol. **2**, 015 (2013)
8. Doignon, C., Knittel, D.: A structured light vision system for out-of-plane vibration frequencies location of a moving web. Mach. Vis. Appl. **16**, 289 (2005)
9. Doignon, C., Knittel, D., Maurice, X.: A vision-based technique for edge displacement and vibration estimations of a moving flexible web. IEEE Trans. Instr. Measur. **57**, 1605–1613 (2008)
10. Shin, C., Kim, W., Chung, J.: Free in-plane vibration of an axially moving membrane. J. Sound Vibr. **272**, 137–154 (2004)
11. Vedrines, M., Gassmann, V., Knittel, D.: Moving web-tension determination by out-of-plane vibration measurements Using a Laser. IEEE Trans. Instr. Measur. **58**(1), 207–213 (2009)
12. Otsu, N.: A threshold selection method from gray-level histograms. IEEE Trans. Syst. Man Cybern. **9**(1), 62–66 (1979)
13. Jaberzadeh, E., Azhari, M., Boroomand, B.: Free vibration of moving laminated composite plates with and without skew roller using the element-free Galerkin method. Iran. J. Sci. Technol. Trans. Civ. Eng. **38**, 377–393 (2014)
14. Tang, Y.Q., Chen, L.Q., Zhang, H.J., Yang, S.P.: Stability of axially accelerating viscoelastic Timoshenko beams: recognition of longitudinally varying tensions. Mech. Mach. Theor. **62**, 31–50 (2013)

# Digital Template System for Measuring Turbine Blade Forging and Its Calibration Method

Yongkai Cai[1,2], Xu Zhang[1,2(✉)], Zelong Zheng[2], Limin Zhu[3], and Yuke Zhu[2,3]

[1] School of Mechatronic Engineering and Automation, Shanghai University,
No. 99 Shangda Road, BaoShan District, Shanghai 200444, China
xuzhang@shu.edu.cn
[2] HUST-Wuxi Research Institute,
No. 329 YanXin Road, Huishan District, Wuxi 214100, China
[3] Robotics Institute, Shanghai Jiao Tong University,
No. 800 Dongchuan Road, Minhang District, Shanghai 200240, China

**Abstract.** Because of complex profile and large size, the measurement of turbine blade is complicated and difficult. A digital template measurement system is developed, which is composed of two laser scanners, two linear motion modules, one turntable and the developed blade measure & analysis software. In order to achieve high precision, the specific calibration method is proposed for each part, such as the laser scanner, the motion direction of scanners, the relative pose between scanners, and the axis of the rotation module. In the first and second calibration stage, only the calibration board is adopted. After the camera is calibrated, the laser plane and the motion direction of scanner are determined with PnP method. In the third and fourth calibration state, the sphere is adopted as the feature to determine the parameters. In the experimental section, we analyze the precision of calibrated system and introduce the application of blade measurement.

**Keywords:** Calibration · Laser scanner · Turbine blade measurement

## 1 Introduction

As one of the most important parts of turbine machinery, turbine blade has complex profile, large size span and numerous description parameters. Its profile quality directly affects the energy conversion efficiency, so blade detection becomes a significant link in blade manufacturing. In addition, because of the large number and size of blade, high demands are placed on measurement efficiency and accuracy.

The contact measurement method includes standard template method and coordinate measurement machine(CMM). The accuracy of the standard template method is low because it depends too much on operators experience. Although CMM is widely used to measure complex surface, it not only has

© Springer Nature Switzerland AG 2018
Z. Chen et al. (Eds.): ICIRA 2018, LNAI 10984, pp. 342–353, 2018.
https://doi.org/10.1007/978-3-319-97586-3_31

low efficiency for the point-by-point measurement of blade profile and pattern, but also requires complex path planning for the object and probe radius compensation.

Non-contact measurement methods include optical theodolite, stereo photography, laser interferometry and laser triangulation method [1]. Fu et al. [2] proposed a method to measure the blade profile by structured light which expands the blade phase rapidly with a new coding method. But the stripe pattern is not suitable for low exposure situations. Sun et al. [3–5] designed a non-contact measurement system with four axis which can only measure the aero-engine blades with height less than 0.36 m.

Aiming at the efficient measurement and result analysis of turbine blades with height less than 1 m, we design the digital template measurement system based on line laser triangulation. In this system, one-dimensional motion module is used to make the laser scanner quickly reconstruct the blade profile, the complete point cloud of the blade is obtained by the rotation of turntable, and customized measurement report is exported automatically. According to the system model, it requires to calibrate the laser scanner, motion direction, relative pose between scanners and turntable coordinate system. The calibration of laser scanner is to solve the relative pose of laser plane and camera, and the main methods are mechanical adjustment method [6,7], drawing calibration method [8,9], invariance of cross-ratio method [10–12] and so on. The calibration of rotation axis of turntable is to obtain the rotation center and the direction of rotation axis, mainly divided into calibration object method [13–16] and no calibration object method [17].

In the above-mentioned measurement methods, contact measurement is inefficient, while non-contact measurement is designed for small blades. Each of the mentioned calibration methods is only applicable to specific conditions, and the calibration objects are different. According to the characteristics of digital template system, our calibration method unifies the calibration objects of each calibration link, simplifies the overall calibration process and improve the efficiency of system calibration.

This paper presents a complete calibration method of digital template measurement system for turbine blade forging. We establish the system coordinate frame and provide the transformation relationship and calibration method between each coordinate system. In the experimental part, the reconstruction accuracy of the calibrated measurement system is analyzed in detail. Finally, we also show the applications of the calibrated system in the actual blade measurement tasks.

## 2    Digital Template Measurement System

The measurement system consists of two vision scanners, two high-precision linear modules, turntable, working platform and control cabinet. The turbine blade should be fixed with a fixture vertically to the turntable before the measurement. Each linear module loads a vision scanner for vertical motion, and turntable rotates the blade.

As illustrated in Fig. 1, the blade measurement system requires to establish the coordinate system of camera, laser plane, measurement and turntable. The origin of the camera coordinate system is the optical center of camera, the direction of the optical axis of camera is taken as the direction of $Z_c$ axis, the $Y_c$ axis is on the plane where laser projector and camera are located and perpendicular to the $Z_c$ axis, and the direction of $X_c$ axis is the direction of the multiplication cross result of their unit vectors. The origin of the laser plane coordinate system is the vertical foot from the camera optical center to the laser plane, the direction of $Z_l$ axis is the projection direction of the camera optical axis on the laser plane, the direction perpendicular to $Z_l$ axis and passing through the origin is taken as $Y_l$ axis, and the direction of $X_l$ axis is the direction of multiplication cross of both unit vectors. The origin of the measurement coordinate system is the pulse zero point of the linear module, the direction of $Y_m$ axis is the actual movement direction of scanner, the direction of $Z_m$ axis is the projection direction of $Z_l$ axis perpendicular to $Y_m$ axis in the laser plane coordinate system, and $X_m$ axis is created by right hand coordinate system. The origin of the turntable coordinate system is established in the center of turntable, the plane normal direction of turntable is taken as the direction of $Z_t$ axis, the direction of $X_t$ axis is the direction of multiplication cross of unit vectors of camera optical axis and $Z_t$ axis, and $Y_t$ is also established by right hand coordinate system.

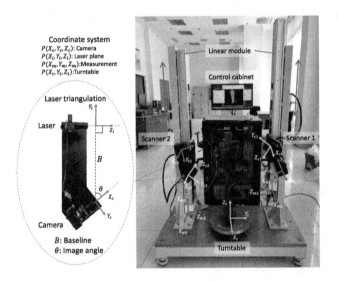

**Fig. 1.** System structure and coordinate system model

After taking laser images by camera, we convert the coordinates of the extracted laser stripe points from the camera coordinate system to the laser plane coordinate system. Due to the calibration of motion direction, these points can be transformed into the measurement coordinate system. Finishing the transformation of point cloud reconstructed by Scanner 2 to the measurement coordinate system of Scanner 1, we can unify all measuring results into the turntable coordinate system so that the complete point cloud can be obtained.

# 3   Calibration Method

The whole system calibration process is divided into four calibrations of laser scanner, motion direction, relative pose between scanners and turntable coordinate system. The first two parts only needs a general planar calibration board, and the other two will use a criterion sphere.

## 3.1   Calibration of Laser Scanner

We adopt an iterative method based on ring characteristic pattern to calibrate the camera. After taking pictures of the calibration board under different poses in the view field of the camera, we can obtain the interior and exterior parameters of the camera according to the calibration method of Zhang [18]. Then we convert the original image into a view parallel to the calibration plate to extract the center of the ring and map the ring center points to three-dimensional planar points. The last step is mapping three-dimensional points of the ring centers to the original image to get corresponding information. Repeat the above process and the convergence is judged by whether the direction projection error is reduced or the number of iterations is maximum.

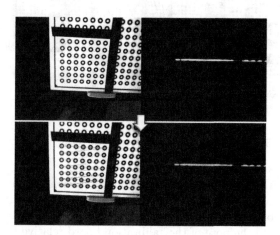

**Fig. 2.** Ring detection and laser stripe center extraction

As is shown in Fig. 2, laser plane is calibrated by taking multiple sets of calibration board images with and without laser. The specific process comprises the following steps: firstly, take a calibration board without laser and keep the calibration board still. Then turn on the laser and reduce the exposure. Finally, take a calibration board image with laser in the same pose. The poses of the calibration board in the camera coordinate system are obtained from the images without laser by PnP method [19], and the laser stripe center is extracted from the pictures with laser. Since $y_l$ of the laser stripe center is 0, the transformation of the laser stripe center between the image coordinate system and the laser plane coordinate system can be expressed as:

$$
\begin{bmatrix} u \\ v \\ 1 \end{bmatrix} = \begin{bmatrix} \alpha_u & 0 & u_o \\ 0 & \alpha_v & v_o \\ 0 & 0 & 1 \end{bmatrix} \begin{bmatrix} r_1 & r_3 & t \end{bmatrix} \begin{bmatrix} x_l \\ z_l \\ 1 \end{bmatrix} = H \begin{bmatrix} x_l \\ z_l \\ 1 \end{bmatrix} \tag{1}
$$

where $r_i$ is the $i$ column vector of rotation matrix, $\alpha_u$ and $\alpha_v$ are the scale factors on the axis $u$ and axis $v$ of the image coordinate system, $(u_o, v_o)$ is the coordinate of principal point, $H$ is the scanner parameter matrix.

Finally, we fit laser plane in the camera coordinate system by least square method.

### 3.2  Motion Direction

Due to the error of installation and mechanical structure, we must calibrate the vector of motion direction in the camera coordinate system, which is namely the $Y$ axis of the measurement coordinate system.

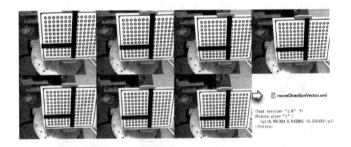

**Fig. 3.** Calibration images of motion direction and vector

As is shown in Fig. 3, the motion direction calibration is to fix the calibration board in the measurement space and make the linear module carry the scanner to move a plurality of positions to take photos of the calibration plate. We use PnP method to calculate the pose of the calibration board relative to the camera, fit the straight line of each pose, and obtain the vector of the motion direction in the camera coordinate system. The conversion of the laser stripe center coordinate from the laser plane coordinate system to the measurement coordinate system can be described as:

$$
\begin{bmatrix} x_m \\ y_m \\ z_m \end{bmatrix} = {}^m R_l \begin{bmatrix} x_l \\ y_l \\ z_l \end{bmatrix} + Y_m = {}^m R_l \begin{bmatrix} x_l \\ 0 \\ z_l \end{bmatrix} + \begin{bmatrix} 0 \\ P_n \cdot P_s \\ 0 \end{bmatrix} \tag{2}
$$

where $^{m}R_{l}$ is rotation matrix from the laser plane coordinate system to the measurement coordinate system, $P_{n}$ and $P_{s}$ are the number of input pulses and the physical distance corresponding to a unit pulse for linear module motion.

So when the scanner moves $H_{c}$ in the camera coordinate system, the transformation of laser plane coordinate system and measurement coordinate system can be expressed as:

$$^{m}T_{c} = \begin{bmatrix} E & Y_{m}H_{c} \\ 0 & 1 \end{bmatrix} {}^{l}T_{c} \tag{3}$$

where $m$ and $l$ represent the coordinate system of measurement and laser plane.

### 3.3   Relative Pose Between Scanners

As is shown in Fig. 4, the relative pose between scanners calibration is to reconstruct the criterion sphere in multiple positions with two scanners in the same time. After fitting point cloud of both scanners into spheres, we obtain the spherical center coordinates and the correspondence between two sets of spatial points. Firstly, calculate the centers of gravity of two sets of points:

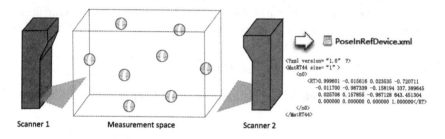

**Fig. 4.** Calibration of relative pose between scanners and transformation matrix

$$\begin{bmatrix} x_{ag} & y_{ag} & z_{ag} \end{bmatrix}^{T} = \frac{1}{n} \begin{bmatrix} \sum_{i=1}^{n} x_{ai} & \sum_{i=1}^{n} y_{ai} & \sum_{i=1}^{n} z_{ai} \end{bmatrix}^{T}$$
$$\begin{bmatrix} x_{bg} & y_{bg} & z_{bg} \end{bmatrix}^{T} = \frac{1}{n} \begin{bmatrix} \sum_{i=1}^{n} x_{bi} & \sum_{i=1}^{n} y_{bi} & \sum_{i=1}^{n} z_{bi} \end{bmatrix}^{T} \tag{4}$$

where $a$, $b$ represent Scanner 1 and Scanner 2, $n(n \geq 3)$ is the number of points, $(x_{ai}\ y_{ai}\ z_{ai})(i = 1, 2, \ldots, n)$ and $(x_{bi}\ y_{bi}\ z_{bi})(i = 1, 2, \ldots, n)$ are points from Scanner 1 and Scanner 2.

Then rotation matrix $^{a}R_{b}$ of Scanner 2 relative to Scanner 1 can be calculated by

$$\begin{bmatrix} x_{a1}\ y_{a1}\ z_{a1} \\ x_{a2}\ y_{a2}\ z_{a2} \\ \vdots \\ x_{an}\ y_{an}\ z_{an} \end{bmatrix} = {}^{a}R_{b} \begin{bmatrix} x_{b1}\ y_{b1}\ z_{b1} \\ x_{b2}\ y_{b2}\ z_{b2} \\ \vdots \\ x_{bn}\ y_{bn}\ z_{bn} \end{bmatrix} \tag{5}$$

Translation vector $^{a}t_b$ is described as

$$^{a}t_b = \begin{bmatrix} x_{ag} \\ y_{ag} \\ z_{ag} \end{bmatrix} - {}^{a}R_b \begin{bmatrix} x_{bg} \\ y_{bg} \\ z_{bg} \end{bmatrix} \tag{6}$$

Taking advantage of the calibrated relationship between the two camera coordinate systems, we can directly transform the point cloud reconstructed by Scanner 2 to the camera coordinate system of Scanner 1.

### 3.4  Turntable Coordinate System

All point cloud is finally unified into the turntable coordinate system after the transformation of each coordinate system. The calibration of the turntable coordinate system is to obtain the pose of the measurement coordinate system of the Scanner 1 under the turntable coordinate system.

As is shown in Fig. 5, our approach is to make the Scanner 1 construct the criterion sphere placed on the edge of turntable at different angles by rotating turntable. Then fit the spherical centers of point clouds at each pose. Finally all the spherical centers are fitted into a spatial circle and the obtained center coordinate is the origin of the turntable coordinate system [13]. The transformation matrix of Scanner 1 in the measurement coordinate system relative to the turntable coordinate system is

$$^{t}T_m = (^{m}T_c\,^{c}T_t)^{-1} \tag{7}$$

where $c$, $m$ and $t$ represent the coordinate system of camera, measurement and turntable.

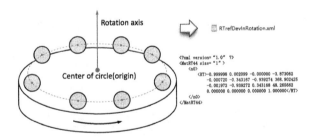

**Fig. 5.** Rotation axis and transformation matrix

## 4   Experiments

In this section, we first provide several experiments to show the accuracy of our calibration method and then show the application of blade measurement. The experimental object is a white matte ceramic bat. Sphere 1 is 49.975 mm, sphere 2 is 49.971 mm and distance between centers of sphere is 500.310 mm. And the calibration board is $7 \times 7$ with ring pattern.

## 4.1   Results and Analysis

Since the $X$ coordinate of each center of light stripe in the image is fixed, we extract the light stripe 200 times from a single image of criterion sphere at the same position to test the repeatability of laser stripe extraction. As illustrated in Fig. 6, the random error is analyzed by selecting the same $X$ coordinate interval and comparing the corresponding $Y$ coordinate. The maximum of maximum error is 0.452 pixel and the maximum of standard deviation is 0.188 pixel, which indicates that the repeatability error has little influence on the measurement system.

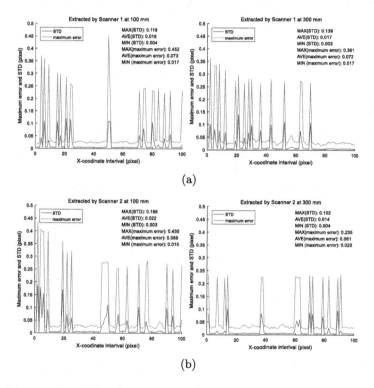

(a)

(b)

**Fig. 6.** (a) and (b) are change of $Y$ coordinate corresponding to each $X$ coordinate obtained by extracting the center of laser stripe 200 times from the same position at 100 mm and 300 mm of view field.

As illustrated in Fig. 7, we repeated 50 scans at the same position using a scanner and fitted sphere 1 in order to compare center coordinates and diameters. Table 1 shows the reconstructed repeatability of each scanner. The results indicates the accuracy is within ±0.02 mm in the view field near the scanner, the maximum standard deviation of center coordinates is 0.011 mm, and that of sphere diameter is 0.005 mm. As shown in Fig. 8, we also tested the accuracy of each registration link. Figure 8(a) analyses the accuracy of motion direction

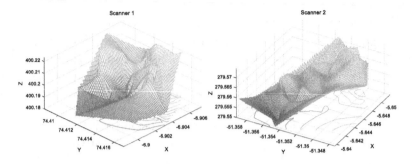

**Fig. 7.** Spherical center coordinate distribution after 50 times fitting repeatedly

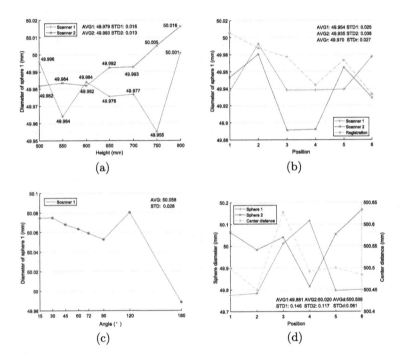

**Fig. 8.** (*a*) Accuracy of single scanner at different heights (*b*) Registration accuracy between scanners (*c*) Registration accuracy of rotation (*d*) Results of reconstructing bat

calibration. The accuracy decreased with the height increasing, and the maximum deviations of two scanners are 0.026 mm and 0.041 mm. From position 1 to 6 in Fig. 8(*b*), the accuracy of single scanner changes as the sphere is placed from Scanner 1 to Scanner 2. And the maximum error after registration is −0.042 mm. The rotation error analysis of turntable is to rotate sphere 1 around by different angles, fit the point cloud after rotation, and analyze the influence of different angles on the measurement accuracy. The experimental results are shown in

**Table 1.** Repeatability and accuracy of 50 times scanning by single scanner $(mm)$

| Equipment | $STD_X$ | $STD_Y$ | $STD_Z$ | $AVE_{diameter}$ | $STD_{diameter}$ | $MAX_{error}$ |
|-----------|---------|---------|---------|------------------|------------------|---------------|
| Scanner 1 | 0.002 | 0.002 | 0.011 | 49.977 | 0.005 | 0.012 |
| Scanner 2 | 0.004 | 0.002 | 0.006 | 49.973 | 0.004 | 0.011 |

Fig. 8($c$). Obviously, the smaller the rotation angle, the more rotations, and the lower the accuracy. The maximum error of rotation registration is 0.100 mm. As shown in Fig. 8($d$), we scanned the bat completely and observed the combined deviation of the sphere diameter and center distance at different positions. All of the rotations in this experiment were performed four times at 90 degrees each time. Due to the error accumulation of mechanical installation, motion mechanism and multi-calibration, the maximum deviation of sphere diameter is $-0.200$ mm and that of distance between spherical centers is 0.318 mm.

**Table 2.** Comparison of different methods of blade measurement

| Method | Accuracy($mm$) | Workplace | Time($min$) | Range($mm$) |
|--------|----------------|-----------|-------------|-------------|
| Standard template | 0.400 | Workshop | 10 | Customized |
| CMM | 0.002 | Lab | 25 | Max height of 3000 |
| LDS | 0.010 | Lab | 11 | $220 \times 180 \times 360$ |
| Our system | 0.350 | Workshop | 3 | $400 \times 500 \times 1000$ |

The comparison among Standard template, CMM, LDS [3] and our system is shown in Table 2. Obviously, our measuring system has high measuring efficiency and large measuring range. And the measurement accuracy also meets the measurement requirements of turbine blade forging.

## 4.2  Application of Blade Measurement

The specified point & cross section measurement is to measure a number of known coordinate points on the important cross sections of blade. As shown in Fig. 9, we make the template with CAD of blade, point & cross section characteristics and report form. Then the real-time point cloud is automatically imported and registered with CAD. The software calculates the normal deviation of each point and section parameters. For the blade with height of 0.7 m, it takes approximately 3 min from the start of scan to the end of exporting report, which greatly improves the measurement efficiency.

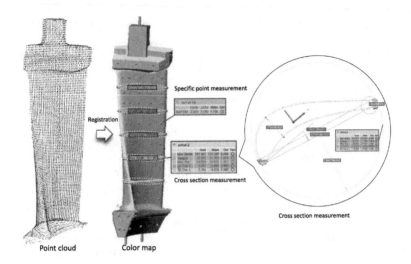

Point cloud          Color map

**Fig. 9.** Measurement process for specified points

## 5    Conclusion

At present, most of the multi-axis vision measuring devices are designed for small turbine blades. For the profile measurement of large turbine blade forgings, our calibrated system can greatly increase the measurement efficiency and reduce the labor cost. In this paper, we establish the coordinate system model of digital template measurement system, provide the complete calibration process and method, and analyze the error of each calibration link. It is simple and convenient to finish the calibration, and the experimental results show that the accuracy of our calibration method can meet the requirements of turbine blade forging measurement.

**Acknowledgments.** This research was partially supported by The key research project of Ministry of science and technology (Grant no. 2017YFB1301503) and the National Nature Science Foundation of China (Grant No. 51575332).

## References

1. Junhui, H., et al.: Overview on the profile measurement of turbine blade and its development. In: Proceedings of SPIE - The International Society for Optical Engineering, vol. 7656, no. 2 (2010)
2. Fu, Y., et al.: Three-dimensional profile measurement of the blade based on surface structured light. Optik - Int. J. Light Electron Opt. **124**(18), 3225–3229 (2013)
3. Sun, B., Li, B.: A rapid method to achieve aero-engine blade form detection. Sensors **15**(6), 12782–12801 (2015)
4. Sun, B., Li, B.: Laser displacement sensor in the application of aero-engine blade measurement. IEEE Sens. J. **16**(5), 1377–1384 (2016)

5. Zhang, F., et al.: Error modeling and compensation for high-precision non-contact four coordinate measuring system. Key Eng. Mater. **437**, 232–236 (2010)
6. Niola, V., et al.: A method for the calibration of a 3-D laser scanner. Robot. Comput. Integr. Manuf. **27**(2), 479–484 (2011)
7. Vilaca, J.L., Fonseca, J.C., Pinho, A.M.: Calibration procedure for 3D measurement systems using two cameras and a laser line. Opt. Laser Technol. **41**(2), 112–119 (2009)
8. Niu, M., et al.: Research on the method to calibrate structure parameters of line structured light vision sensor. In: International Conference on Mechanics, materials and Structural Engineering 2016 (2016)
9. Dewar, R.: Self-generated targets for spatial calibration of structed-light optical sectioning sensors with respect to an external coordinate system. In: Proceedings of Robots and Vision 1988 Conference, Detroit, pp. 5–13 (1988)
10. Wei, J., Liu, Z., Cheng, F.: Vision sensor calibration method based on flexible 3d target and invariance of cross ratio. In: International Conference on Information Sciences, Machinery, Materials and Energy 2015 (2015)
11. Xu, G., et al.: Global calibration and equation reconstruction methods of a three dimensional curve generated from a laser plane in vision measurement. Opt. Express **22**(18), 22043 (2014)
12. Sun, J., et al.: A vision measurement model of laser displacement sensor and its calibration method. Opt. Lasers Eng. **51**(12), 1344–1352 (2013)
13. Yong-An, X.U., Qin, Y., Jin-Peng, H.: Calibration of the axis of the turntable in 4-axis laser measuring system and registration of multi-view. Chin. J. Lasers **32**(5), 659–662 (2005)
14. Park, S.Y., Subbarao, M.: A multiview 3D modeling system based on stereo vision techniques. Mach. Vis. Appl. **16**(3), 148–156 (2005)
15. Li, J., et al.: Calibration of a multiple axes 3-D laser scanning system consisting of robot, portable laser scanner and turntable. Opt. Int. J. Light Electron Opt. **122**(4), 324–329 (2011)
16. Kurnianggoro, L., Hoang, V.D., Jo, K.H.: Calibration of a 2D laser scanner system and rotating platform using a point-plane constraint. Comput. Sci. Inf. Syst. **12**(1), 307–322 (2015)
17. Pang, X., et al.: A tool-free calibration method for turntable-based 3D scanning systems. IEEE Comput. Graph Appl. **36**(1), 52–61 (2016)
18. Zhang, Z.: A flexible new technique for camera calibration. TPAMI **22**(11), 1330–1334 (2000)
19. Wu, Y., Hu, Z.: PnP problem revisited. J. Math. Imaging Vis. **24**(1), 131–141 (2006)

# A Teaching System for Serial Robots Under ROS-I

Tengfei Shan, Cheng Ding, Chao Liu, ChunGang Zhuang,
Jianhua Wu$^{(\boxtimes)}$, and Zhenhua Xiong

Shanghai Jiao Tong University,
800 Dongchuan Road, Shanghai 200240, People's Republic of China
wujh@sjtu.edu.cn

**Abstract.** The teaching pendant, aiming to teach robots and control it, is an essential subsystem of industrial robots. However, robots from different manufacturers have different programming languages and interactive modes. If a new robot is built, robot's developers have to rebuild the teach pendant meanwhile robot's users have to restart learning how to use it. In addition, all teaching boxes can not work without connecting with robots thus it is inefficient to teach robots by keeping the teach pendant in hand. In this paper, a teaching system is proposed based on Robot Operating System Industrial (ROS-I). There are some improvements compared with other teach pendants. Firstly, it packages functions into different plug-ins, improving the flexibility for both robot users and developers. Secondly, we build the simulation environment using Gazebo. Thirdly, voice control module allows robot users to use robots easily without teaching pendant.

**Keywords:** ROS-I · Moveit · Plug-ins · Voice control · Simulation

## 1 Introduction

With the development of the manufacturing requirements, robots are increasingly used in many factories to complete onerous, duplicating work typically done by people. Considering that robots do individual work in different manufacturing process, teaching system plays an extremely essential role. Since higher teaching efficiency makes it easier to change robots' assignments conveniently in flexible manufacture, how to improve it is proposed in recent research [8]. The research proposes that researchers can focus on human-robot interactive efficiency and scalable software architectures for interactive robotics. As can be seen, there are researches on these directions. Research [13] proposes an universal teach pendant for serial robots. Based on the unified pendant, developers can pay much attention to connecting with robots and configuring the parameters, thus they can develop robots fast. Another research [9] develops a virtual teaching software which is run in computer. Since the software does not depend on pendant, it is flexible to connect with robots as well. Research [12] uses

© Springer Nature Switzerland AG 2018
Z. Chen et al. (Eds.): ICIRA 2018, LNAI 10984, pp. 354–365, 2018.
https://doi.org/10.1007/978-3-319-97586-3_32

Rviz to visualize robots' motion and Moveit to do motion planning. By using Moveit, users can do motion planning. A good simulation environment [17,21] for robots manipulation is built. Based on Gazebo and ROS, users can do quick simulation without robots. Research [19] presents means of adapting trajectories by transforming them into sequences of geometric primitives to improve the efficiency in programming. In addition, some new functions like "Linear Navigation" and "Joint Interpolation Navigation" are proposed [15], which improves efficiency in some work environments. Although there are a series of improvements in these researches, there are several drawbacks which have negative effects on robot's efficiency, flexibility and extensibility of functions. Firstly, it is necessary for robots to have an universal teaching system. Different robots have their unique teach pendants as shown in Fig. 1 due to different sizes, programming languages and interactive logic. It means that robot manufacturers have to pay much attention to rebuild the teach pendant once a new robot is created. Although research [13] proposes an universal teach pendant, it is unfriendly for developers to rebuild functions due to restriction of unified hardware. Another software [9] is built as a whole unit, which is inflexible considering future extension. If robot developers want to add more functions, they should modify whole software. Secondly, researches [9,12,13] do not provide a good simulation environment. As can be seen, researches [12,13] use Rviz to visualize robots' motion. Lacking robot's physical properties makes it hard to show the motion like real robots. So it is hard to know how much manufacturing efficiency raises or how many workers can be replaced exactly before robots are installed in the production lines without simulation function. Other researches [17,21] build a good simulation for robots manipulation, but they don't provide efficient human-robots interfaces and cannot teach robots without using motion planning tools like Moveit. Thirdly, it is not convenient and effective since the pendant device will be in hand and out of hand frequently when users teach the robots. Researches [15,19] completed the improvements based on teach pendants which has potential room to improve the teaching efficiency.

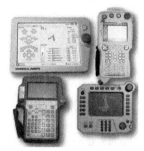

**Fig. 1.** Different pendants of robots.

In this paper, we proposed a teaching system for common serial robots based on ROS-I [5]. This system is executed in computer instead of unique teaching pendant in hand, making it more adaptable and universal. We packaged main functions into different plug-ins using QT creator, giving users more options to choose functions they want to use. It is also more flexible for developers to add more functions or modify them. What is more, it integrates with Gazebo [2], allowing robots' users to complete the teaching process or estimation without connecting robots. If users are satisfied with simulation data, they can save them in the offline library. If the position data cannot be executed perfectly in real robots, users can do necessary adjustments quickly. Moreover, the voice control mode is built to improve the teaching performance. By using voice control, users can complete the teaching process without pendant box.

The rest of the paper consists of these sections: The technical background is in Sect. 2, followed by the software's design and implementation. The demonstrations is in Sect. 4. The conclusions and future work is in the last section.

## 2    Technical Background

### 2.1    ROS

Robot Operating System (ROS) [4,18] is an open software architecture used to build robots fast. As shown in Fig. 2, there is a core node called master, responsible for naming and locating other nodes, tracking the topics and services that other nodes publish or subscribe. All nodes communicate with each other by publishing or subscribe their message to topics and services.

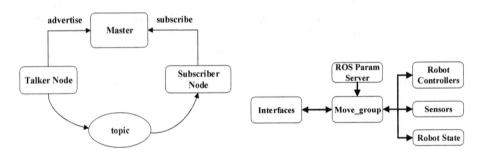

**Fig. 2.** Architecture of ROS.                 **Fig. 3.** Architecture of Moveit

**ROS-I.** ROS-I integrates some powerful tools like Moveit, TF, ros_contol etc. It has excellent program interfaces for many sensors and actuators. ROS-I is divided into several typically layers shown in Fig. 4 [6].

**Moveit.** Moveit [3,11] is the core tool of ROS-I for robots' manipulation. It [20] provides essential functions like collision checking [16], kinematics, motion planning and control. As shown in Fig. 3, the move_group node is the core of Moveit.

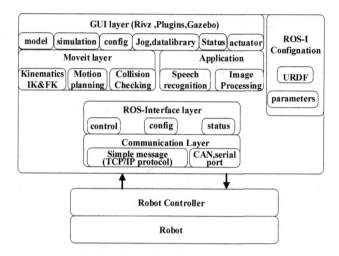

**Fig. 4.** Architecture of ROS-I.

**Ros_control.** Ros_control [1,10] is the Bridge between high layers and low layers. It contains a series of interfaces like controller interfaces, transmission interfaces, hardware interfaces and the controller toolbox.

### 2.2 Voice Interactive System

Speech interactive system [7] consists of automatic speech recognition (ASR) and text to speech (TTS). ASR collects the user's voice and converts the voice into words that can be put into other functions or TTS. TTS is responsible for converting the text words into voice and outputting it by audio.

## 3 Design and Implementation of the Software

Based on the architecture of ROS-I shown in Fig. 4, there are five modules including modelling module, visualization module, planning module, controlling module, voice control module as shown in Fig. 5. The architecture is in Fig. 10 and the GUI is shown in Fig. 11.

### 3.1 Design of the Software

**Modelling and Visualization.** As shown in the Fig. 4, before using the software, the robot description and parameters should be given using URDF. URDF is built according to the DH convention of the robot. Moreover, URDF [14] records the whole physical information like joints, links, transmission, properties, inertial and collision etc. Based on robot's URDF, robots with it's actuators and sensors can be visualized in the Rviz and Gazebo to show the robots' motion and simulation.

**Fig. 5.** Modules of system            **Fig. 6.** Flow of position data

As shown in Fig. 6, when robots or the virtual robots in Gazebo execute the motion, the joints data information will be published by node named joint_state_publisher and finally published to the topic TF. Then the motion will be visualized in Rviz and Gazebo Synchronously.

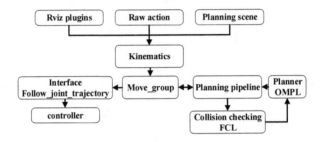

**Fig. 7.** Flow of Moveit

**Planning.** Due to the powerful functions in robot's manipulation, we use Moveit to achieve the manipulation planning tasks. As shown in Fig. 7, when getting the input, we firstly compute that whether the goal is out of the robot's workspace. Once the goal is not in the workspace, we give up doing motion planning. After planning and collision checking, the planning pipeline gives the result back to move_group and move_group publishes the data to an action named follow_joint_trajectory. The data is transported to robot controllers eventually.

**Control.** The high layer does not communicate with hardware directly because we use ros_control as the bridge between application layers and robots.

As can be seen in Fig. 8, the data from follow_joint_trajectory is put into the controller manager and got by Gazebo or robot from joint_path_command. Before transferred to robots or Gazebo, the way points data will be interpolated which can be fed into position controllers. By using simple message protocol, all position data can be transferred to robot controllers. During the execution, the joint state and robot status will be updated to joint_states and robot_status continuously.

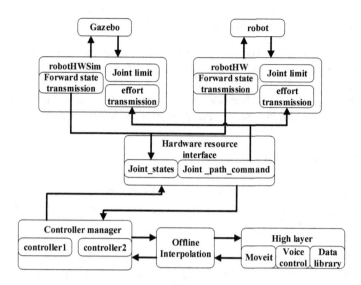

**Fig. 8.** Controller layer

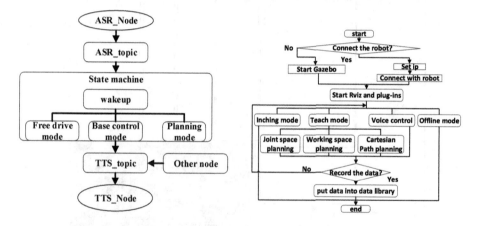

**Fig. 9.** Flow of voice control          **Fig. 10.** Architecture of software

**Voice Control.** The voice control function is built to improve teaching efficiency. As shown in Fig. 9, there are two main parts including ASR and TTS. We packaged the speech recognition into node named ASR_Node to get the user's commands and convert it into text words which are put into a state machine node immediately. To improve the flexibility and accuracy of command recognition, many synonyms for each command are used in state machine. The state machine jumps into different states to execute different motions and puts the results into TTS_topic which is subscribed by TTS_node. The TTS_node can output any other results once other node publishes the message into TTS_topic. All messages in this topic will be output by audio.

### 3.2    Architecture of the Software

As can be seen in Fig. 10, there are two big parts. The first part is starting the software and deciding that whether connecting with a real robot while the second part is choosing functions' modes. There are four modes that can be executed including inching mode, teach mode, voice control mode and offline mode. After initialization, users can choose or change to the mode they want to use.

## 4    Implementation

The system is developed using C++ and QT based on plug-ins mechanism. As can be seen in Fig. 11, there are several plug-ins responsible for individual functions separately.

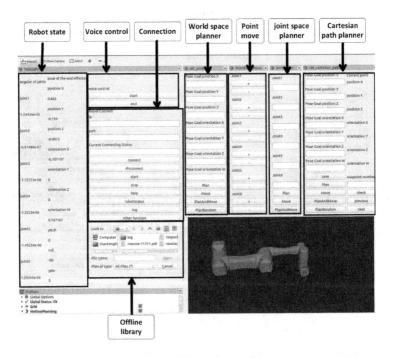

**Fig. 11.** GUI of the system

### 4.1    Connection

Connection module decides that whether the system connects with robots or with Gazebo. Since using TCP/IP protocol, the software connects with robots with correct IP address and port. It shows connecting status as well.

## 4.2   Robot State

Robot states module monitors all joints angles and the robot's pose. As can be seen in Fig. 6, the data is subscribed from the topic joint_states after transformed by TF package.

## 4.3   Planners

There are three planners including the world space planner, the joint space planner and Cartesian path planner. World space planner does the motion planning in world space. Users give the pose of destination in world space, the planner will compute the way points. Joint space planner needs all axis angles to do the planning. Cartesian path planner focuses on Cartesian path's computation. Users give the destination's pose then this plug-ins will compute the path taking the obstacle in the environment into account. If the destination is out of the robot's workspace, the Cartesian computing will be incomplete. All the data of the planning process can be saved as a file and put into the library once users are satisfied with the planning which can be reused if users prefer.

## 4.4   Voice Control

Voice control module only has the start and end buttons. This module has excellent performance combined with free drive mode of robots which substitutes the teach pendant system completely. By using it, the command will be recognized by ASR node and put into a state machine. Corresponding motion will be started and the execution process will be published by TTS continuously. Based on the high recognition, the voice control module improve teaching efficiency without using the GUI.

**Fig. 12.** UR5 and ZU.

# 5   Demonstrations

In this paper, the system is tested on two robots including UR5 and ZU shown in Fig. 12. The GUI part is tested on ZU and voice control part is tested on UR5 to demonstrate the functionality and university.

## 5.1  Preparation

Before using the system, we complete the preparations as follows.

(1) We described the robot's DH convention shown in Fig. 13 by building the robot's URDF. Other parameters like transmissions, inertial, Gazebo parameters are configured as well. After configuration, Zu can be visualized.
(2) We Used Moveit set_up_assistant to set the planning group, self-collision matrix etc. Then robots described in URDF can be visualized in Rviz.
(3) We Configured the ros_control package including a series of controllers which are responsible for connecting with real robots and Gazebo.
(4) We built the connection using Ethernet with ZU. Any kinds of robot controller which supports TCP/IP protocol can connects with this system.

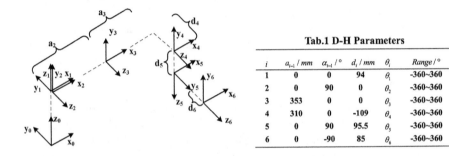

**Tab.1 D-H Parameters**

| $i$ | $a_{i-1}$ / mm | $\alpha_{i-1}$ / ° | $d_i$ / mm | $\theta_i$ | Range / ° |
|---|---|---|---|---|---|
| 1 | 0 | 0 | 94 | $\theta_1$ | -360~360 |
| 2 | 0 | 90 | 0 | $\theta_2$ | -360~360 |
| 3 | 353 | 0 | 0 | $\theta_3$ | -360~360 |
| 4 | 310 | 0 | -109 | $\theta_4$ | -360~360 |
| 5 | 0 | 90 | 95.5 | $\theta_5$ | -360~360 |
| 6 | 0 | -90 | 85 | $\theta_6$ | -360~360 |

**Fig. 13.** DH convention of ZU.

## 5.2  Functional Test

We tested the system in two robots. Motion planning and simulation were tested on ZU and voice control was tested on UR5.

**Simulation and Motion Planning Test.** Users can built the work environment fast by adding models. Then they can do simulation in Gazebo. In Fig. 14, we built a work environment and do planning simulation. Due to physical properties in URDF, ZU can move like a real robot. We used joint-space planner by putting all joint position. The result is shown in Fig. 15. Other work space like world space planning and Cartesian path computation are similar with it.

During motion planning, ZU can do collision checking real-time. In URDF, we made the collision model slightly bigger than real robots which makes ZU stops before colliding. As can be seen in Fig. 16, when ZU collides with itself or with other objects, the colliding part's color will change to red and ZU stops do any motion.

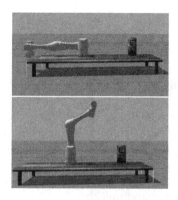
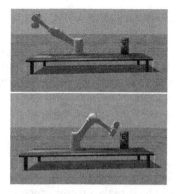

**Fig. 14.** Simulation in Gazebo

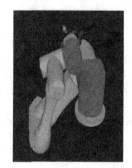

**Fig. 15.** Joint-space planning          **Fig. 16.** Collision checking

**Voice Control Test.** Voice control module is tested on UR5. By using voice control, we teach UR5 to a series of way points in free drive mode. After recording the way points, UR5 can repeat the way we teach it as shown in Fig. 17. If we are satisfied with this motion, we can save it in offline library.

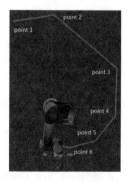
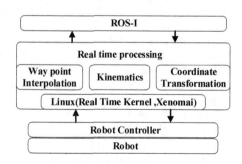

**Fig. 17.** Voice control.              **Fig. 18.** Future work.

# 6    Conclusions and Future Work

In this paper, we proposed a new teaching system for serial robots. Based on Moveit, it can complete motion planning, collision checking and teaching. Since we packaged main functions into different plug-ins, users can choose the function plug-ins they need and developers can build new function plug-ins easily. The modular designing method gives them more flexibility. In addition, Gazebo allows users to do simulation without robots. Users can complete teaching process and do research just like using a real robot. After simulation, satisfied data put into offline library, which then be fed into robots drivers directly. Moreover, by using voice control, users can use robots avoid keeping teaching pendant in hands, which makes using process more efficiency and quickly.

The system still has much work to improve. It is commonly known that all the command to robots should be received by robots in real time [22]. However, ROS and Linux do not support real time task. In order to have a better performance in real-time assignments, the architecture of this system is being rebuilt recently. As can be seen in Fig. 18, some essential work such as interpolation, kinematics (FK, IK) and coordination transform which require real time performance will be put into real time process module of the computer based on Linux and Xenomai. By doing this, the system can promise the command execution with a little delay.

**Acknowledgments.** This research was supported in part by National Natural Science Foundation of China under Grant 51575355 and National Science and Technology Major Project under Grant 2015ZX04005006.

# References

1. control architecture. http://wiki.ros.org/ros_control
2. Gazebo advantage. http://www.gazebosim.org/
3. Moveit architecture. http://moveit.ros.org/
4. Ros advantage. http://www.ros.org/core-components/
5. Ros industrial framework. http://wiki.ros.org/Industrial
6. rosi architecture. http://www.guyuehome.com/1444
7. voice architecture. http://www.xfyun.cn/?ch=bdtg
8. Brogårdh, T.: Present and future robot control development an industrial perspective. Annu. Rev. Control **31**(1), 69–79 (2007)
9. Chen, L., Wei, Z., Zhao, F., Tao, T.: Development of a virtual teaching pendant system for serial robots based on ROS-I. In: 2017 IEEE International Conference on Cybernetics and Intelligent Systems (CIS) and IEEE Conference on Robotics, Automation and Mechatronics (RAM), pp. 720–724. IEEE (2017)
10. Chitta, S., et al.: ros_control: a generic and simple control framework for ROS. J. Open Source Softw. **2**, 456 (2017)
11. Chitta, S., Sucan, I., Cousins, S.: Moveit! [ROS topics]. IEEE Robot. Autom. Mag. **19**(1), 18–19 (2012)
12. Deng, H., Xiong, J., Xia, Z.: Mobile manipulation task simulation using ROS with Moveit. In: 2017 IEEE International Conference on Real-time Computing and Robotics (RCAR), pp. 612–616. IEEE (2017)

13. Gao, Y., et al.: U-Pendant: a universal teach pendant for serial robots based on ROS. In: 2014 IEEE International Conference on Robotics and Biomimetics (ROBIO), pp. 2529–2534. IEEE (2014)
14. Joseph, L.: Mastering ROS for Robotics Programming. Packt Publishing Ltd., Birmingham (2015)
15. Kaluarachchi, M.M., Annaz, F.Y.: GUI teaching pendant development for a 6 axis articulated robot. In: Ponnambalam, S.G., Parkkinen, J., Ramanathan, K.C. (eds.) IRAM 2012. CCIS, vol. 330, pp. 111–118. Springer, Heidelberg (2012). https://doi.org/10.1007/978-3-642-35197-6_12
16. Pan, J., Chitta, S., Manocha, D.: FCL: a general purpose library for collision and proximity queries. In: 2012 IEEE International Conference on Robotics and Automation (ICRA), pp. 3859–3866. IEEE (2012)
17. Qian, W., et al.: Manipulation task simulation using ROS and Gazebo. In: 2014 IEEE International Conference on Robotics and Biomimetics (ROBIO), pp. 2594–2598. IEEE (2014)
18. Quigley, M., et al.: ROS: an open-source robot operating system. In: ICRA Workshop on Open Source Software, Kobe, Japan, vol. 3, p. 5 (2009)
19. Schraft, R.D., Meyer, C.: The need for an intuitive teaching method for small and medium enterprises. VDI Berichte **1956**, 95 (2006)
20. Sucan, I.A., Moll, M., Kavraki, L.E.: The open motion planning library. IEEE Robot. Autom. Mag. **19**(4), 72–82 (2012)
21. Torres-Torriti, M., Arredondo, T., Castillo-Pizarro, P.: Survey and comparative study of free simulation software for mobile robots. Robotica **34**(4), 791–822 (2016)
22. Wei, H., et al.: RT-ROS: a real-time ros architecture on multi-core processors. Future Gener. Comput. Syst. **56**, 171–178 (2016)

# Real-Time HALCON-Based Pose Measurement System for an Astronaut Assistant Robot

Lihong Dai[1,2,3], Jinguo Liu[1(✉)], Zhaojie Ju[1,4], and Yuwang Liu[1]

[1] State Key Laboratory of Robotics, Shenyang Institute of Automation,
Chinese Academy of Sciences, Shenyang 110016, China
`dailihong2004@163.com, {liujinguo,liuyuwang}@sia.cn`
[2] School of Electronic and Information Engineering,
University of Science and Technology Liaoning, Anshan 114051, China
[3] University of the Chinese Academy of Sciences, Beijing 100049, China
[4] Intelligent Systems and Biomedical Robotics,
School of Creative Technologies, University of Portsmouth, Portsmouth, UK
`zhaojie.ju@port.ac.uk`

**Abstract.** The manned space program in China has entered the space station stage. Space astronauts normally have very limited time for certain tasks, so in order to free space astronauts from some repetitive routine tasks, various robots in space appear. Free-flying robots in space can also provide conditions for various scientific experiments with their unique microgravity advantages. The measurement of their positions and attitudes is the premise of their autonomous flight and remote operations. So, the pose measurements system of the Astronaut Assistant Robot is designed in this paper. With the camera calibrated, the image is acquired and processed in real time, and the robot is tracked in real time. In the end poses are estimated by virtue of PnP algorithm. The pose measurements system is modeled and simulated, with an automatic global threshold method to improve the reliability of the system. In addition, a series of measures are adopted to improve its efficiency with a satisfactory accuracy. The experimental results show that the measurement system meets the requirements of both effectiveness and efficiency.

**Keywords:** Astronaut Assistant Robot · Pose measurement system
PnP algorithm · Real time

## 1 Introduction

China's manned space program has entered the space station phase [1]. Compared with the ground laboratory, the space station can provide unique microgravity environment for science research to obtain the results which are difficult to get from the ground station. Astronauts in the space shuttle normally have limited time for certain tasks. Robots in space can help astronauts to do some repetitive work and provide support for astronauts. And free-flying robots in space normally do not touch the capsule, so there is no mechanical vibration, which can provide a kind of microgravity environment.

© Springer Nature Switzerland AG 2018
Z. Chen et al. (Eds.): ICIRA 2018, LNAI 10984, pp. 366–378, 2018.
https://doi.org/10.1007/978-3-319-97586-3_33

On the one hand, only small power is needed to control their operation. On the other hand, microgravity environment needed for the scientific experiment is provided. It can be seen that the free-flying robots in space play an important role in the space cabin. There are various free-flying in-cabin robots, such as SPHERE [2] developed by the university of MIT; PSA, Smart SPHERE [3] and Astrobee [4, 5] developed by the NASA AMS research center; SHB developed in Japan; SCAMP developed by the university of Maryland in the United States; Mini AERCam developed at NASA Johnson space center; and AAR [6] (The first generation in-cabin robot) and AAR-2 [7] (The second generation in-cabin robot) developed by Shenyang institute of automation, Chinese academy of sciences.

SPHERE [2] is mainly used for formation flight and docking, where dynamic programming method is adopted. However, its beacon-based localization limit the operating space. Based on SPHERE, Smart SPHERE is installed a commercial smartphone as an embedded controller to improve performance, which can capture images using embedded camera in smartphone. Astrobee [4, 5] is developed recently. Compared with SPHERE, Astrobee extends the space largely, by using the tracking and positioning method of monocular vision. But in measurement system, SURF feature detection and BRISK feature matching algorithms are used, which are more complex, limiting its efficiency. AAR [6] is controlled by PID Network algorithm, and is positioned by binocular visual system. Based on AAR of the first generation in-cabin robot, AAR-2 [7] is designed and developed, which is controlled by using the fuzzy sliding mode control algorithm. There are few detailed descriptions of the pose measurement system on robots in space in literatures. However the measurement of their position and attitude is the premise of their autonomous flight and remote operation. So, the pose measurement system on AAR-2 is designed in the paper. The picture of AAR-2 is shown in Fig. 1.

**Fig. 1.** The picture of AAR-2

Monocular vision measurement system has the advantages of simple structure, convenient operation, easy calibration, and fast calculation speed. So it is suitable for the real-time application. In this paper, the method based on monocular vision is used in position and attitude measurement system.

## 2  Pose Measurement System

A camera is mounted on an AAR-2, so it moves with AAR-2. The target is installed in the inner wall of the cabinet or sealed cabin, with known feature point distribution patterns. The image of target is acquired in real-time by the camera. A series of measurement is taken in order to improve the performance of real-time. Furthermore, the system is simulated by the powerful HALCON visual software.

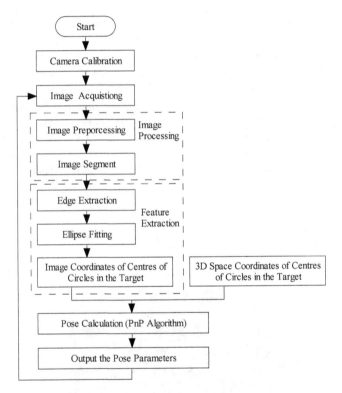

**Fig. 2.** Schematic diagram of the pose measurement system

The pose measurement system mainly can be divided into five sections: camera calibration, image acquisition, image processing, feature extraction and pose calculation, as shown in Fig. 2.

The purpose of the camera calibration is to calculate the interior camera parameters which provide the basic data for pose calculation of camera. After image acquisition, image processing and feature extraction are carried out. Image processing mainly includes image preprocessing and image segmentation. And feature extraction mainly includes edge extraction, elliptical fitting and coordinate extraction of target center. On the basis of 2D image coordinates of target circles centers and their 3D space coordinates, the position and attitude parameters of the camera relative to cooperative target are calculate with PnP algorithm, which are saved to a file, and provide the foundation for better control and operation of the AAR-2.

## 2.1  Camera Calibration

In order to calculate the position and attitude more accurately, it is necessary to calibrate the camera first. The camera calibration model is set up, the method of camera calibration proposed by Zhengyou Zhang is employed, and the process of the camera calibration is realized by virtue of the HALCON software.

Firstly, when the distortion of camera lens is not considered, the camera model is the ideal small pinhole imaging model. The conversion process of 3D world coordinates to 2D image pixel coordinates is obtained through the camera model, in which the parameters of the camera is involved. The camera calibration process is that of determining the camera parameters.

Then, in the plane template calibration method proposed by Zhang [8], the world coordinate system is set on the plane of the calibration plate. The interior camera parameters and the distortion coefficient obtained are computed as the initial values. And the parameters with high final accuracy are obtained by the method of levenberg-marquarat nonlinear optimization.

HALCON has been recognized as the most effective machine vision software in industry in Europe and Japan. It provides some assistants and visual tools, as well as programming hints, which make programming and modification easy, development cycles short, and development costs low. It is widely used to develop visual system, such as in the literatures [9, 10]. The process of calibrating camera is as follows. Calibration data model is created at first. Initial camera parameters are set in the calibration data model. Then calibration object is set. The image of calibration plate is read to find the coordinate of marks and pose of target, which are saved in data model. Next, camera is calibrated to get the interior camera parameters. Finally the interior camera parameters is written in files.

## 2.2  Image Acquisition

The image acquisition assistant in HALCON can be used for real-time image acquisition. If you connect the camera to the computer via the USB port, and then click "the image acquisition assistant" in the assistant menu, an image acquisition window will pop up. Under the resource tab, the camera can be detected by clicking on "the image acquisition interface" radio button. And under the connection tab, the connected camera device can be seen. Furthermore, under the code generation tab, if you click "insert code" button, image acquisition code can be generated automatically, which makes programming easy.

## 2.3  Image Processing

(1)  Image Preprocessing.

In order to improve the anti-interference ability of the image, and decrease the noise in the image, image often is preprocessed with filter.

Filtering is classified into time domain filtering and frequency domain filtering. Image filtering in time domain includes linear filtering and nonlinear filtering. The common linear filtering mainly includes mean filtering and Gaussian filtering. The nonlinear filtering mainly includes median filtering, bilateral filtering and anisotropic diffusion filtering. Because the frequency domain filtering is processed by the Fourier transform and inverse transform, and the nonlinear filtering is often complicated, which are not conducive to the real-time processing of the image, the linear filter is selected in the paper. In the linear filter, the mean filter is the average gray value of the pixel in the template, whose filtering effect is often not very ideal. The Gaussian filter is a Gaussian weighted average gray value of pixel in the template. Most of image signal or energy concentrated in low frequency and medium frequency amplitude spectrum, while at higher frequencies, useful information is often submerged by noise, and so one filter which can reduce high frequency amplitude can reduce noise. Gaussian filtering is essentially a low-pass filter, which is widely used in image denoising and is very effective in suppressing the noise that obeys normal distribution. The image was pre-treated with function named gauss_image in HALCON.

(2)  Image Segmentation

The gray threshold segmentation method is most applied in image segmentation. It is a method to separate the object from the background by threshold value, whose advantages are simple calculation, high efficiency and fast speed. It is widely used in real-time.

It can be seen that the key of threshold segmentation algorithm is to determine the threshold, so if a suitable threshold is determined, the image can be separated accurately. Because the actual target image shown in Fig. 3 has a single background, and the difference between foreground and background is obvious which can be seen from the gray histogram shown in Fig. 4, there is no need for complex threshold segmentation method, and the global threshold can be set for image segmentation.

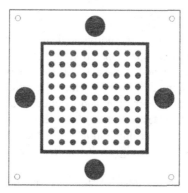

**Fig. 3.** Image of cooperative target

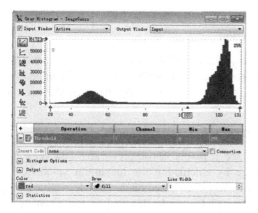

**Fig. 4.** Gray histogram

In order to improve reliability, the automatic global threshold method is adopted to image segmentation. Firstly, the histogram of image is counted. Then, the highest frequency of gray values is found. Finally, the grayscale value lower than a certain amount of gray value at the highest frequency is considered as the threshold. The corresponding instruction in HALCON is as follows.

```
gray_histo(ImageGauss,ImageGauss,AbsoluteHisto,RelativeHiso)
PeakGray:=sort_index(AbsoluteHisto)[255]
threshold(ImageGauss,Regions,0,PeakGray*0.85)
```

After selecting the foreground from the background in the target image, the marker circle region can be chosen mainly according to difference between the area of the marker circles and other objects, which can be realized by using the visual interface of the feature histogram in HALCON. Then, using the cutout technique [11], we extracted the marked circle region in the original image, so that the marked circle image was extracted.

(3)  Morphological Image Processing.

The methods constantly used in morphological image processing are dilation operation, corrosion operation, open operation and closed operation. The dilation operation makes object larger and is used to fill the holes and narrow gaps in the objects. In contrast, Corrosion can be used to eliminate small and meaningless objects. Open operation is dilation after corrosion, and its effect is similar to that of corrosion, which reduces the object. It has less effect on object reduction than corrosion operation. Closed operation is corrosion after dilation, and its effect is similar to that of dilation operation, which makes the object bigger. It has less effect on object expansion than dilation operation.

In order to better extract the edge of the circle in the target image, the closed operation is used to increase the circle area appropriately. The function named "closing_circle" in HALCON is used. Because the extracted objects are circles, the structure element is chosen as circle.

## 2.4  Feature Extraction

After image preprocessing and image segmentation, the image of marker circle is determined. Next, edge detection is needed to obtain the edge of marker circles. And then they are fitted into ellipse in order to determine the coordinates of the circles center.

(1)  Image Edge Detection

Edge of image is set of pixels whose grey value take on step change, so we can detect the image edge by calculating the maximum value of first-order derivative. Gradient operator commonly used based on the first derivative includes Sobel, Roberts, Prewitt. In addition, image edge can be detected by zero crossing of second derivative. Operator frequently used based on the second derivative is Laplacian of a Gaussian. These operators are on the basis of local window, so their algorithms are simple and easy to implement. However, if these algorithm are adopted, some information on the edge will be lost, the system is sensitive to noise, and edge detection effect is not very ideal. In 1986 John Canny proposed a new edge detection operator [12]. The operator is with the optimization idea, which has a large signal-to-noise ratio and a higher detection precision. It is currently considered as the most ideal edge detection method, and widely used [13, 14]. The process of Canny edge detection is as follows.

① The image is smoothed with a Gaussian filter to remove noise and interference. One-dimensional Gaussian function is used as filter. The original image $f(x,y)$ is convolved by row and column respectively. The image after smoothing is $I(x,y)$. The Gaussian filter and $I(x,y)$ are given by

$$G(x) = \frac{\exp(-x^2/2\sigma^2)}{2\pi\sigma^2} \tag{1}$$

$$I(x, y) = [G(x)G(y)] * f(x, y) \tag{2}$$

where $\sigma$ is the standard deviation of Gaussian, which is used to control the smoothness.

② For the smoothed image $I(x,y)$, the finite difference of first-order partial derivatives is calculated, and the amplitude M and the direction $\theta$ of the gradient are obtained, given by

$$\begin{cases} P_x[i,j] = (I[i+1,j] - I[i,j] + I[i+1,j+1] - I[i,j+1])/2 \\ P_y[i,j] = (I[i,j+1] - I[i,j] + I[i+1,j+1] - I[i+1,j])/2 \end{cases} \quad (3)$$

$$M[i,j] = \sqrt{P_x[i,j]^2 + P_y[i,j]^2} \quad (4)$$

$$\theta[i,j] = \arctan\left(\frac{P_y[i,j]}{P_x[i,j]}\right) \quad (5)$$

③ Non-maximum suppression is carried out. The image is scanned along the image gradient direction, and if pixels are not part of the local maxima they are set to zero.

④ Double threshold algorithm is used to detect and connect edges. High and low thresholds are set. The point where the gradient is greater than the high threshold is considered to be the real edge of the image, so it is retained. One where the gradient value is less than the low threshold is not edge and is removed. For the point between the two thresholds, those adjacent to the edge point are retained as edge, otherwise deleted.

Because Canny edge detection is better than other methods, it is used to detect the edge of the target circle. The instruction in HALCON is edges_sub_pix (ImageReduced, Edges, 'Canny', 4, 20, 40).

(2)  Elliptic Fitting

Edge detection is used to extract edges, which reduces the amount of data, eliminates irrelevant information, and retains the important structural attributes of the image. Because the result of edge detection is a collection of edge pixels, it is also necessary to select the contour as a whole. The function used in HALCON for contour selection is select_contours_xld.

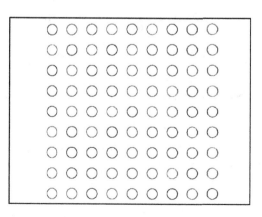

**Fig. 5.** Image after elliptic fitting

In order to calculate the centers of the target circles, their contours need to be fitted with an ellipse. Image after elliptic fitting is shown in Fig. 5. Compared with the edge detected, the data is reduced. The function of the elliptic fitting contour in HALCON is fit_ellipse_contour_xld.

## 2.5    Pose Calculation

Based on the mathematical model of the camera and the known camera interior parameters, pose calculation is to establish the relationship between 3D space coordinates of target features and 2D coordinates of image features, so as to determine the relative position and attitude between camera and target.

The commonly used features include point features, linear features, etc. Among them, the problem of pose calculation using point feature is called PnP (perspective-n-point) problem. When n of point number is less than or equal to 2, known condition is insufficient, so the pose parameters of the target cannot be determined. When n > 5, the problem can be solved linearly. When $3 \leq n \leq 5$, the PnP problem is usually non-linear, and there are possible multiple solutions. Due to the requirement of real time in practical application, points between 3 and 5 are often selected to calculate the position and attitude.

The P3P problem can be described as follows. The known conditions are that angles between two of three rays starting from a vertex O are α, β, γ, and three sides of a triangle are c, b, a, respectively. The distance between the vertex O of rays and three vertex of the triangle ABC, namely d1, d2, and d3, will be calculated. Description of P3P problem is shown as Fig. 6.

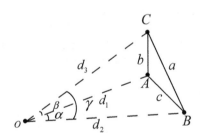

**Fig. 6.** Description of P3P problem

By using cosine theorem, we have

$$\begin{cases} d_1^2 + d_2^2 - 2d_1d_2\cos\alpha = c^2 \\ d_1^2 + d_3^2 - 2d_1d_3\cos\beta = b^2 \\ d_2^2 + d_3^2 - 2d_2d_3\cos\gamma = a^2 \end{cases} \tag{6}$$

Suppose

$$\begin{cases} d_2 = xd_1 \\ d_3 = yd_1 \end{cases} \tag{7}$$

Substitute (7) into (6), we can derive

$$\begin{cases} d_1^2(1 + x^2 - 2xc\cos\alpha) = c^2 \\ d_1^2(1 + y^2 - 2yc\cos\beta) = b^2 \\ d_1^2(x^2 + y^2 - 2xy\cos\gamma) = a^2 \end{cases} \tag{8}$$

If two of the ones in (8) are taken, and d1 is eliminated, we can obtain

$$\begin{cases} a^2(1 + y^2 - 2y\cos\beta) = b^2(x^2 + y^2 - 2xy\cos\gamma) \\ b^2(1 + x^2 - 2x\cos\alpha) = c^2(1 + y^2 - 2y\cos\beta) \end{cases} \tag{9}$$

Connecting two equations of (9) and removing $b^2x^2$, it is not difficult to derive that

$$a^2(1 + y^2 - 2y\cos\beta) - b^2(y^2 - 2xy\cos\gamma)$$
$$= c^2(1 + y^2 - 2y\cos\beta) - b^2(1 - 2x\cos\alpha) \tag{10}$$

Extracting $x$ from (10), we can obtain

$$x = \frac{a^2 + b^2 - c^2 + (a^2 - b^2 - c^2)y^2 - 2y(a^2 - c^2)\cos\beta}{2b^2(\cos\alpha - y\cos\gamma)} \tag{11}$$

Substituting (11) into the lower equation of (9), a quartic equation about y is obtained as

$$a_4 y^4 + a_3 y^3 + a_2 y^2 + a_1 y + a_0 = 0 \tag{12}$$

Four sets of solutions for y can be found. Substituting the solution of y into (11), the value of $x$ can be obtained. According the above equation of (8), we have

$$d_1 = \sqrt{c^2/(1 + x^2 - 2xc\cos\alpha)}$$

Thus $d_1$ can be calculated. According to (5), $d_2$ and $d_3$ can also be obtained. From the process derived above, it can be seen there are at most four solutions to the P3P problem. For P4P problem, when four control points are coplanar, there is only one solution to the problem. Any three points can be selected to determine the pose, and then the fourth point is used to verify it [15]. Since the actual target image is coplanar, four control points are used to determine the pose. The instruction of pose calculation in HALCON is vector_to_pose.

## 3  Performance Evaluation

AAR-2 above air bearing table, the MV-500SM CMOS camera, and the M1620-MPW2 optical lens produced by Computar company are used in the system to carry out the simulation experiment. The resolution is 2592 * 1944 pixels. Its pixel size is 2.2 μm. The AAR-2 experimental platform is shown in Fig. 7.

**Fig. 7.** AAR-2 experimental platform

In order to meet the real-time demand of the system, the operation efficiency of the system can be improved by simplifying the algorithm, reducing the complexity of image processing and transformation, and reducing the storage space. There are two main methods: one is to reduce the amount of data to be processed; the second is to adopt a simple or simplified algorithm. Reducing the amount of data to be processed plays a major role in real-time of image processing algorithms.

The measures to reduce the amount of data to be processed are taken to improve the operation efficiency of the program in the paper.

(1) The function of image zooming is applied to reduce the size of the image from 2592 * 1944 to 1024 * 768, which lifts the speed of the program. The corresponding instruction in HALCON is zoom_image_size (Image1, Image, 1024, 768, 'constant').

(2) By setting the interest area or selective region, the data volume of image processing is greatly reduced, and the operation efficiency is dramatically improved. The corresponding instruction in HALCON is reduce_domain (Image, ROI_Rectangle, ImageReduced1), where ROI_Rectangle is the area of interest, which include 9 circles and margin region with the size of the interval between circles, shown as Fig. 8.

**Fig. 8.** Region of interest

(3) The kernel size of the Gaussian filter is reduced from 5 to 3 which makes the running time reduced. The corresponding instruction in HALCON is gauss_image (Image, ImageGauss, 3).

By taking these main measures, the running time of the program is reduced from above 200 ms to less than 100 ms, which satisfies the system's requirement for real-time performance.

In addition, the accuracy of the system is verified by virtue of AAR-2 experimental platform, and the error is less than 0.6 mm, which meets the requirement of precision.

## 4   Conclusion

The pose measurement system of AAR-2 in space is effectively modeled and simulated. After images of the cooperation target are captured in real-time, image preprocessing, image segmentation, edge extraction and ellipse fitting process are carried out to extract the centers of target circles. Then, according to the 3D coordinates of cooperation targets feature points, their corresponding 2D image coordinates, and the interior camera parameters, the position and attitude of camera relative to the target are calculated using the PnP algorithm. Furthermore, the automatic global threshold method is adopted to improve the reliability. A series of measures to improve the real time are taken, such as reducing the image, setting the interest area and reducing the kernel size of the Gaussian filter. The system is simulated in the HALCON software, and the results demonstrate its satisfactory performance in terms of computational cost and precision.

**Acknowledgment.** The authors would like to acknowledge the support from the National Science Foundation of China under Grant No. 51775541, 51575412, 51575338 and 51575407, and the State Key Laboratory of Robotics Foundation under Grant No. Y5M3180301, Y5M3190301, and Y7M3250301, the EU Seventh Framework Programme (FP7)-ICT under Grant No. 611391, and the Research Project of State Key Lab of Digital Manufacturing Equipment & Technology of China under Grant No. DMETKF2017003.

# References

1. Gao, M., Zhao, G., Gu, Y.: Space science and application mission in China's space station. Bull. Chin. Acad. Sci. **30**(6), 721–732 (2015)
2. Eslinger, G., Saenz-Otero, A.: Electromagnetic formation flight control using dynamic programming. In: Proceedings of 36th Annual AAS Guidance and Control Conference, pp. 17–32 (2013)
3. Fong, T., Micire, M., Morse, T., Park, E., Provencher, C., To, V., et al.: Smart SPHERES: a telerobotic free-flyer for intravehicular activities in space. In: Proceedings of AIAA SPACE 2013 Conference and Exposition, pp. 1–15. AIAA, San Diego (2013)
4. Bualat, M., Barlow, J., Fong, T., Provencher, C., Smith, T., Zuniga, A.: Astrobee: developing a free-flying robot for the international space station. In: Proceedings of AIAA SPACE 2015 Conference and Exposition, pp. 1–10. AIAA, Pasadena (2015)
5. Coltin, B., Fusco, J., Moratto, Z., Alexandrov, O., Nakamura, R.: Localization from visual landmarks on a free-flying robot. In: Proceedings of International Conference on Intelligent Robots and Systems (IROS), pp. 4377–4382. IEEE/RSJ, Daejeon (2016)
6. Liu, J., Gao, Q., Liu, Z., Li, Y.: Attitude control for astronaut assisted robot in the space station. Int. J. Control Autom. Syst. **14**(4), 1082–1095 (2016)
7. Gao, Q., Liu, J., Tian, T., Li, Y.: Free-flying dynamics and control of an astronaut assistant robot based on fuzzy sliding mode algorithm. Acta Astronautica **138**, 462–474 (2017)
8. Zhang, Z.: A flexible new technique for camera calibration. IEEE Trans. Pattern Anal. Mach. Intell. **22**(11), 1330–1334 (2000)
9. Li, Z., Chen, X.: Research on image analysis of bank card based on HALCON software. Electron. Sci. Tech. **30**(9), 56–59 (2017)
10. Wang, H., Zhao, B., Sun, Z., Chen, X.: Defect detection of medical bags based on Halcon. Packag. Eng. **36**(7), 31–35 (2015)
11. Ai, X., Xing, J.: Computer Vision: Algorithms and Applications, p. 388. Tsinghua University Press, Beijing (2012)
12. Canny, J.: A computational approach to edge detection. IEEE Trans. Pattern Anal. Mach. Intell. **6**, 679–698 (1986)
13. ElAraby, W., Madian, A., Mahmoud, A., Farag, I., Nassef, M.: Fractional Canny edge detection for biomedical applications. In: Proceedings of IEEE International Conference on Microelectronics (ICM), pp. 265–268 (2016)
14. Nikolic, M., Tuba, E., Tuba, M.: Edge detection in medical ultrasound images using adjusted Canny edge detection algorithm. In: Proceedings of IEEE Telecommunications Forum (TELFOR), pp. 1–4 (2016)
15. Wang, T., Dong, W., Wang, Z.: Position and orientation measurement system based on monocular vision and fixed target. Infrared Laser Eng. **46**(4), 1–8 (2017)

# Control of a Mechanically Compliant Joint with Proportional-Integral-Retarded (PIR) Controller

Xixian Mo, Feng Jiang, Wenhui Wang, Bo Zhang$^{(\boxtimes)}$, and Ye Ding

State Key Laboratory of Mechanical System and Vibration,
School of Mechanical Engineering, Shanghai Jiao Tong University,
Shanghai 200240, China
{wanjushengfang,f.jiang,15026686939,b_zhang,y.ding}@sjtu.edu.cn

**Abstract.** In this paper, a mechanically compliant joint is designed and a novel proportional-integral-retarded (PIR) controller is adopted. Firstly, the structure of the invariant stiffness joint is introduced in details and linear springs are connected with the output link to obtain flexibility. Secondly, the control programs are built with Simulink and automatically compiled and downloaded into the controller. Thirdly, the physical mechanism and electromagnetic function of permanent magnet synchronous motor (PMSM) are combined to get the transfer function. Fourthly, a direct integration method (DIM) with Simpson method is proposed and used to analyze the stability of PIR controller with the joint. Finally, some compared experiments with conventional proportional-integral-derivative (PID) controller added a low-pass filter in the derivative term in speed loop and position loop are taken, and the results show that PIR controller is better in the speed of settlement and more stable in some condition.

**Keywords:** Mechanically compliant joint · PIR · DIM

## 1 Introduction

Robot compliance characteristic plays an important role in the interaction between robots and human or environment to ensure the safety of operation. The compliance or stiffness of robot joints can be obtained from the force control of rigid joints with torque sensors or compliant joints based on mechanical system like springs [5]. Because of the good performance of the bandwidth of force input and energy efficiency [4,13], the mechanically compliant joints have been widely studied in recent decades. Humanoid robots with mechanically compliant systems may have wide applications in fields like service, education and health care.

One type of mechanically compliant joints have invariant stiffness like Domo [2] and the Robonaut 2 (R2) designed by the National Aeronautics and Space

© Springer Nature Switzerland AG 2018
Z. Chen et al. (Eds.): ICIRA 2018, LNAI 10984, pp. 379–390, 2018.
https://doi.org/10.1007/978-3-319-97586-3_34

Administration (NASA) [1]. Another type of joints have variable stiffness like the Floating Spring Joint (FSJ) from Deutsches Zentrum für Luft- und Raumfahrt (DLR) (German Aerospace Center in English) [12], hybrid dual actuator unit (HDAU) [5] and The CompAct-VSA unit designed by Istituto Italiano di Tecnologia (IIT) (Italian Institute of Technology in English) [9]. There is only one motor in invariant stiffness joints to keep the output position that results in advantages in cost, weight and size while the stiffness cannot be adapted [8]. Variant stiffness joints have one motor to keep the position of the output and another motor to adjust the stiffness of the joint, so they can be used in different tasks because of the adaptable stiffness [6]. Considering the invariant stiffness joints are lighter, smaller and cost less, an invariant stiffness joint using linear springs with constant elastic coefficient as elastic elements was designed.

Recently, a novel type of controller called proportional-integral-retarded (PIR) controller is proposed [7,10]. For n-th order linear time-invariant (LTI) system with PIR controller

$$\begin{cases} x^{(n)}(t) + \sum_{k=0}^{n-1} a_k x^{(k)}(t) = \sum_{k=0}^{n-1} b_k u^{(k)}(t) \\ u(t) = k_p e(t) + k_i \int_0^t e(\tau)d\tau - k_r e(t - t_r) \end{cases} \tag{1}$$

where $u(t)$ is the input, $x(t)$ is the output, $e(t) = r(t) - x(t)$ is the error, $r(t)$ is the reference and $e(t < 0) = 0$. $k_p$ is the proportional gain, $k_i$ is the integral gain, $k_r$ is the retarded gain, and $t_r > 0$ is the delayed error. The difference between PIR controller and conventional PID controller is that the derivative term is replaced by a former error, thus noise amplification by derivation is avoided. In order to analyze the stability of PIR controller, direct integration method (DIM) is used [11]. Typically, time delay is approximated by Pade approximation to a rational function but losing accuracy when the delay is long and needs high order approximation. Frequency domain study on second order LTI systems has been taken [7], but not appropriate for higher order systems.

The rest of this paper is organized as follows. The design and model of the mechanically compliant joint are in Sect. 2. In Sect. 3, stability analysis with DIM on PIR controller is presented. Some experiments are taken and results are analyzed in Sect. 4, followed by the conclusion in Sect. 5.

## 2   Mechanically Compliant Joint

Figures 1 and 2 shows the structure of the joint. The joint consists a motor, a harmonic drive and a passive compliant module. The passive compliant module is fixedly connected with the output of the harmonic drive. An output link is connect to the base of the module with four linear springs. Given the physical mechanisms of the mechanically compliant joint, a linear torsion spring model approximately describes the flexibility of the joint. When the rotation angle of the output link is different with the output of harmonic drive, the passive

**Fig. 1.** Mechanically compliant joint.

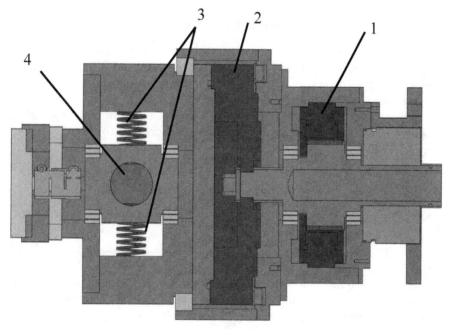

1. PMSM; 2. Harmonic drive; 3. Linear springs; 4. Extended output link.

**Fig. 2.** Section view of the joint.

compliant module products a torque which has linear relationship with the rotation angle difference, and the spring torque is

$$\tau_k = k(\frac{\theta_m}{r} - \theta_l) \tag{2}$$

where $k$ is the spring's torsion coefficient, $\theta$ is the rotation angle when subscript $m$ represents motor and subscript $l$ represents the link, and $r$ is the gearing reduction ratio.

At the motor side and the link side, assume there is no load and perform the torque balance

$$j_m\ddot{\theta}_m + b_m\dot{\theta}_m = \tau_m - \tau_k \tag{3}$$

$$j_l\ddot{\theta}_l + b_l\dot{\theta}_l = \tau_k \tag{4}$$

where $j$ is the moment of inertia, $b$ is the damping coefficient, and $\tau_m$ is the motor torque.

The model of the PMSM is described in direct ($d$) and quadrature ($q$) rotating coordinate system by

$$u_d = L_d\dot{i}_d + Ri_d - p\dot{\theta}_m L_q i_q \tag{5}$$

$$u_q = L_q\dot{i}_q + Ri_q + p\dot{\theta}_m(L_d i_d + \psi) \tag{6}$$

$$\tau_m = \frac{3}{2}pi_q((L_d - L_q)i_d + \psi) \tag{7}$$

where $u_d$ and $u_q$ are $d$ axis and $q$ axis voltage components, $i_d$ and $i_q$ are $d$ axis and $q$ axis current components, $L_d$ and $L_q$ are $d$ axis and $q$ axis inductance components, $R$ is the stator resistance, $p$ is the motor pole pairs, and $\psi$ is the permanent magnet flux linkage.

In order to simplify the system, PI controllers are used in current loops for controlling $i_d$ to zero, and $u_q$ is approximately proportional to $i_q$, so

$$u_q = k_q i_q \tag{8}$$

$$\tau_m = k_u u_q \tag{9}$$

After Laplace transformation, transfer function between input $u_q$ and output $\theta_l$ is

$$P(s) = \frac{\Theta_l(s)}{U_q(s)} = \frac{k_u k}{(rj_m s^2 + rb_m s + k)(j_l s^2 + b_l s + k) - k^2} \tag{10}$$

Notice that the model is a fourth order system with four poles and no zeros. For a large number of experiments and comparisons, the best identification can be given as follows:

$$\frac{7.148 \times 10^6}{s^4 + 130.2s^3 + 2.671 \times 10^4 s^2 + 6.030 \times 10^5 s + 4261} \tag{11}$$

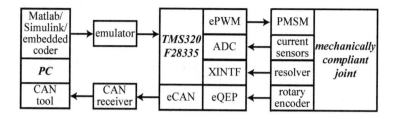

**Fig. 3.** Scheme of control and sensor system.

Figure 3 shows the control and sensor system. The main chip is a microcontroller TMS320F28335 from Texas Instruments. In general, programs are developed with C program language and debugged in the integrated development environment (IDE) named Code Composer Studio (CCS), but a more convenient way is developing with block diagram in the Matlab/Simulink environment and building by embedded coder toolbox. Applications are automatically compiled, downloaded through emulator and executed by only one click.

There are many integrated peripherals in F28335. With space vector pulse-width modulation (SVPWM), desired $d$ axis and $q$ axis voltage are encoded into pulse signals by enhanced pulse-width modulation (ePWM) modules to control six metal-oxide-semiconductor field-effect transistors (MOSFET). When the PMSM is actuated, the harmonic drive reduces speed and the output link of the joint waves. There are three types of sensors for data acquisition. Firstly, currents are detected by two current sensors Honeywell CSNE151-100 based on Hall effect and converted into digital signals by the analog-to-digital converter (ADC) module. Secondly, the position of motor from a resolver on the PMSM is read through external interface (XINTF). Finally, a rotary encoder ROD 486 from HEIDENHAIN gets the rotation angle of the output link and transferred to enhanced quadrature encoder (eQEP) module. When the program is running, the enhanced controller area network (eCAN) module sends data in 1 Mbps to PC through a CAN receiver and a CAN tool software.

## 3   DIM on PIR Controller

In the system Eq. 1 with PIR controller, assume $r(t) = 0$, then

$$x^{(n+1)}(t) + \sum_{i=0}^{n} c_i x^{(i)}(t) - \sum_{j=1}^{n} k_r b_{j-1} x^{(j)}(t - t_r) = 0 \qquad (12)$$

where $c_0 = k_i b_0$, $c_n = a_{n-1} + k_p b_{n-1}$, $c_k = a_{k-1} + k_p b_{k-1} + k_i b_k (k = 1, ..., n-1)$. Let $\boldsymbol{X}(t) = [x(t), \dot{x}(t), ..., x^{(n)}(t)]^T$, then

$$\dot{\boldsymbol{X}}(t) = \boldsymbol{B}\boldsymbol{X}(t - t_r) + \boldsymbol{C}\boldsymbol{X}(t) \qquad (13)$$

where

$$B = k_r \begin{pmatrix} & \mathbf{0}_{n\times(n+1)} & \\ 0\ b_0 & \cdots & b_{n-1} \end{pmatrix}, C = \begin{pmatrix} 0 & 1 & & \\ & & \ddots & \\ & & & 1 \\ -c_0 & -c_1 & \cdots & -c_n \end{pmatrix} \tag{14}$$

Assume the sample time is $\delta < t_r$ and the time period $[t - t_r, t]$ is separated into $t_k = t - t_r + \frac{k}{2}\delta(k = 0, 1, ..., N)$ ($N = \langle 2t_r/\delta \rangle$, $\langle x \rangle$ means the nearest even integer of $x$). So

$$\dot{X}(t) = BX(t - t_r) + C(X(t_k) + \int_{t_k}^{t} \dot{X}(\tau)d\tau) \tag{15}$$

If numerically integrated by the trapezoidal rule at the start and the Simpson's rule with the rest, then

$$\begin{cases} \dot{X}(t_0) = BX(t_0 - t_r) + CX(t_0) \\ \dot{X}(t_1) = BX(t_1 - t_r) + CX(t_0) + \frac{\delta}{2}C(\dot{X}(t_0) + \dot{X}(t_1)) \\ \dot{X}(t_i) = BX(t_i - t_r) + CX(t_{i-2}) + \frac{\delta}{6}C(\dot{X}(t_{i-2}) + 4\dot{X}(t_{i-1}) + \dot{X}(t_i)) \\ (i = 2, ..., N) \end{cases} \tag{16}$$

Let $\mathbb{Z}(t) = [X(t_0)^T, X(t_1)^T, ..., X(t_N)^T]^T$, so

$$D\dot{\mathbb{Z}}(t) = E\mathbb{Z}(t - t_r) + F\mathbb{Z}(t) \tag{17}$$

where

$$D = \begin{pmatrix} I_{n+1} & & & \\ -\frac{\delta}{2}C & I_{n+1} - \frac{\delta}{2}C & & \\ -\frac{\delta}{6}C & -\frac{4\delta}{6}C & I_{n+1} - \frac{\delta}{6}C & \\ & \ddots & \ddots & \ddots \\ & & -\frac{\delta}{6}C & -\frac{4\delta}{6}C\ I_{n+1} - \frac{\delta}{6}C \end{pmatrix} \tag{18}$$

$$E = \begin{pmatrix} B & & \\ & \ddots & \\ & & B \end{pmatrix} \tag{19}$$

$$F = \begin{pmatrix} C & & & \\ C & & & \\ C & & & \\ & \ddots & & \\ & & C\ \mathbf{0}_{n+1}\ \mathbf{0}_{n+1} \end{pmatrix} \tag{20}$$

$I_{n+1}$ is the identity matrix of size $n + 1$.

In the period $[t_0, t]$,

$$X(t) = X(t_k) + \int_{t_k}^{t} \dot{X}(\tau)d\tau \tag{21}$$

also using the numerical integration,

$$\begin{cases} X(t_1) = X(t_0) + \frac{\delta}{2}(\dot{X}(t_0) + \dot{X}(t_1)) \\ X(t_i) = X(t_{i-2}) + \frac{\delta}{6}C(\dot{X}(t_{i-2}) + 4\dot{X}(t_{i-1}) + \dot{X}(t_i))(i = 2, ..., N) \end{cases} \tag{22}$$

So

$$G\dot{Z}(t) = HZ(t) \tag{23}$$

where

$$G = \begin{pmatrix} \mathbf{0}_{n+1} & & & \\ \frac{\delta}{2}I_{n+1} & \frac{\delta}{2}I_{n+1} & & \\ \frac{\delta}{6}I_{n+1} & \frac{4\delta}{6}I_{n+1} & \frac{\delta}{6}I_{n+1} & \\ & \ddots & \ddots & \ddots & \\ & & \frac{\delta}{6}I_{n+1} & \frac{4\delta}{6}I_{n+1} & \frac{\delta}{6}I_{n+1} \end{pmatrix} \tag{24}$$

$$H = \begin{pmatrix} \mathbf{0}_{n+1} & & & \\ -I_{n+1} & I_{n+1} & & \\ -I_{n+1} & \mathbf{0}_{n+1} & I_{n+1} & \\ & \ddots & \ddots & \ddots \\ & & -I_{n+1} & \mathbf{0}_{n+1} & I_{n+1} \end{pmatrix} \tag{25}$$

With Eqs. 17 and 23,

$$(H - GD^{-1}F)Z(t) = (GD^{-1}E)Z(t - t_r) \tag{26}$$

Equation 26 cannot be solved because it is an indeterminate equation, thus adding the continuity (i.e. $X(t_0) = X(t_N - t_r)$)

$$PZ(t) = QZ(t - t_r) \tag{27}$$

where

$$P = \begin{pmatrix} I_{n+1} \\ & \mathbf{0} \end{pmatrix}, Q = \begin{pmatrix} & I_{n+1} \\ \mathbf{0} & \end{pmatrix} \tag{28}$$

then

$$Z(t) = \Phi Z(t - t_r) \tag{29}$$

where

$$\Phi = (H - GD^{-1}F + P)^{-1}(GD^{-1}E + Q) \tag{30}$$

$\Phi$ is the transition matrix. According to Floquet theory, the system is stable when all modules of eigenvalues of $\Phi$ is less than one [3]. In other words, if $\lambda$ presents eigenvalue function, then

$$\max(|\lambda(\Phi)|) < 1 \tag{31}$$

In order to demonstrate DIM is as reliable as frequency domain method, an analyzed system by Ramírez et al. [7] is taken as an example (Fig. 4).

*Example.* $n = 2$, $a_1 = 0.45056$, $a_0 = 309.76$, $b_1 = 0$, $b_0 = 31$, $k_p = 22.57$, $k_i = 0$, find stable region in the $(t_r, k_r)$ space.

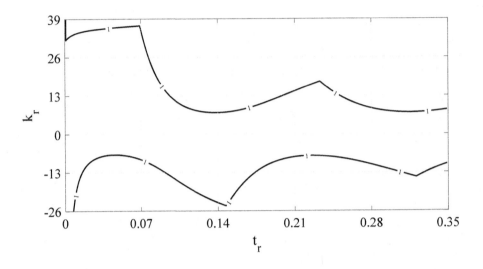

**Fig. 4.** Stable region of example.

## 4    Experiment

Figure 5 shows the block diagram of experiments. In order to control the position of the link, two controllers based on errors are used in position loop $F(s)$ and speed loop $H(s)$ and two types of controllers are compared: PIR and PIDN. PIDN controller is a conventional PID controller added a low-pass filter with coefficient $N$ in the derivative term to filter noise and sudden reference signal change, and the transfer function is

$$k_p + \frac{k_i}{s} + k_d s \frac{N}{s + N} \tag{32}$$

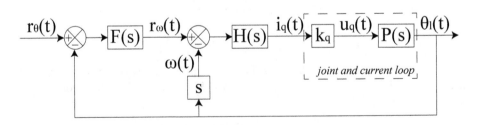

**Fig. 5.** Block diagram of experiments.

There are four parameters in both controllers, and parameters tuning is performed through minimizing the integral time absolute error (ITAE) function

$$ITAE = \int_0^T t|e(t)|dt \tag{33}$$

Because the ITAE function is hard to calculate derivative, a derivative-free Nelder-Mead method is used. In order to protect the joint, stability is analyzed by DIM before taking experiment.

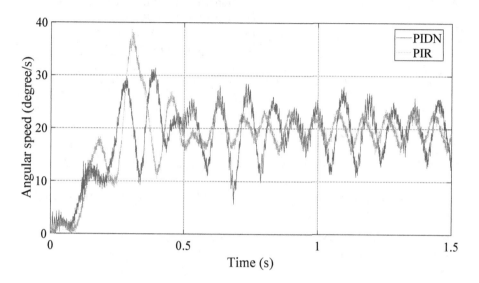

**Fig. 6.** Experiment results in speed loop.

**Table 1.** Experiment results in speed loop.

| Speed loop controller | ITAE (T = 4s) |
| --- | --- |
| PIDN | 14.32 |
| PIR | 9.51 |

Figure 6 and Table 1 show the optimal results in speed loop. Due to the amplification of noise and disturbance when using position difference to obtain speed, both angular speeds are oscillating around given speed 20 degree/s, but obviously the amplitude of oscillation and ITAE of PIR controller are smaller than those of PIDN controller.

The optimal results in position loop are shown in Fig. 7 and Table 2 when the parameters of controllers in speed loop are not tuned again. The results prove that using PIR controller in speed loop is much better than PIDN controller no

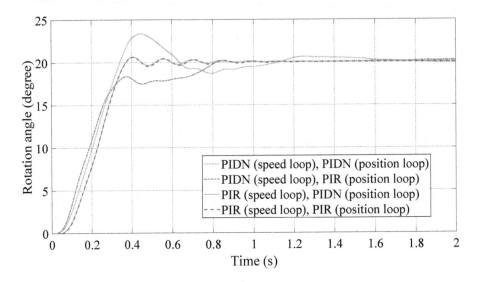

**Fig. 7.** Experiment results in position loop.

**Table 2.** Experiment results in position loop.

| Speed loop controller | Position loop controller | ITAE (T = 4s) | 2% settling time (s) |
|---|---|---|---|
| PIDN | PIDN | 1.518 | 1.548 |
| PIDN | PIR | 2.001 | 0.788 |
| PIR | PIDN | 1.028 | 0.563 |
| PIR | PIR | 0.799 | 0.571 |

matter which controller is used in position loop. When using PIR controller in speed loop, the performance of PIR controller and PIDN controller in position loop is almost the same. Even though using PIDN controller in speed loop, ITAE and settling time of PIR controller in position loop is smaller than those of PIDN controller. The conclusion is PIR controller is better in the speed of settlement and more stable than PIDN controller in some condition.

## 5    Conclusion

This paper presents a mechanically compliant joint. The joint is designed with linear springs to obtain flexibility. Physical model and the characteristic of PMSM are combined to get the transfer function. A PIR controller is proposed and DIM is used to analyze the stability. The compared experiment results show that PIR controller is better in the speed of settlement and more stable than PIDN controller in some condition. In the speed loop, the amplitude of oscillation and ITAE of PIR controller are smaller than those of PIDN controller. When using PIDN controller in speed loop, ITAE and settling time of PIR controller in position loop is smaller than those of PIDN controller. In the future work, some global optimization like generic algorithm and particle swarm optimization may be used to optimize the parameters of controllers in speed loop and position loop at the same time.

**Acknowledgments.** This work was supported by the National Natural Science Foundation of China under grant 91648112 for which the authors are grateful.

# References

1. Diftler, M.A., et al.: Robonaut 2 - the first humanoid robot in space. In: 2011 IEEE International Conference on Robotics and Automation, pp. 2178–2183, May 2011. https://doi.org/10.1109/ICRA.2011.5979830
2. Edsinger, A.: Robot manipulation in human environments (2007)
3. Farkas, M.: Periodic Motions, vol. 104. Springer, New York (2013)
4. Hurst, J.W., Chestnutt, J.E., Rizzi, A.A.: An actuator with physically variable stiffness for highly dynamic legged locomotion. In: Proceedings of the IEEE International Conference on Robotics and Automation, ICRA 2004, vol. 5, pp. 4662–4667, April 2004. https://doi.org/10.1109/ROBOT.2004.1302453
5. Kim, B.S., Song, J.B.: Hybrid dual actuator unit: a design of a variable stiffness actuator based on an adjustable moment arm mechanism. In: 2010 IEEE International Conference on Robotics and Automation, pp. 1655–1660, May 2010. https://doi.org/10.1109/ROBOT.2010.5509264
6. Park, J.J., Kim, H.S., Song, J.B.: Safe robot arm with safe joint mechanism using nonlinear spring system for collision safety. In: 2009 IEEE International Conference on Robotics and Automation, pp. 3371–3376, May 2009. https://doi.org/10.1109/ROBOT.2009.5152268
7. Ramrez, A., Mondi, S., Garrido, R., Sipahi, R.: Design of proportional-integral-retarded (PIR) controllers for second-order LTI systems. IEEE Trans. Autom. Control **61**(6), 1688–1693 (2016). https://doi.org/10.1109/TAC.2015.2478130
8. Tsagarakis, N.G., Laffranchi, M., Vanderborght, B., Caldwell, D.G.: A compact soft actuator unit for small scale human friendly robots. In: 2009 IEEE International Conference on Robotics and Automation, pp. 4356–4362, May 2009. https://doi.org/10.1109/ROBOT.2009.5152496
9. Tsagarakis, N.G., Sardellitti, I., Caldwell, D.G.: A new variable stiffness actuator (compact-VSA): Design and modelling. In: 2011 IEEE/RSJ International Conference on Intelligent Robots and Systems, pp. 378–383, September 2011. https://doi.org/10.1109/IROS.2011.6095006
10. Villafuerte, R., Mondi, S., Garrido, R.: Tuning of proportional retarded controllers: theory and experiments. IEEE Trans. Control Syst. Technol. **21**(3), 983–990 (2013). https://doi.org/10.1109/TCST.2012.2195664
11. Wen, Z., Ding, Y., Liu, P., Ding, H.: Direct integration method for time-delayed control of second-order dynamic systems. J. Dyn. Syst. Measur. Control **139**(6), 061001 (2017). https://doi.org/10.1115/1.4035359

12. Wolf, S., Eiberger, O., Hirzinger, G.: The DLR FSJ: Energy based design of a variable stiffness joint. In: 2011 IEEE International Conference on Robotics and Automation, pp. 5082–5089, May 2011. https://doi.org/10.1109/ICRA.2011. 5980303
13. Zinn, M., Khatib, O., Roth, B., Salisbury, J.K.: Playing it safe [human-friendly robots]. IEEE Robot. Autom. Mag. **11**(2), 12–21 (2004). https://doi.org/10.1109/ MRA.2004.1310938

# Sensors and Actuators

# Intelligent Control of Variable Ranging Sensor Array Using Multi-objective Behavior Coordination

Sasuga Kitai[1(✉)], Yuichiro Toda[2], Naoyuki Takesue[1], Kazuyoshi Wada[1], and Naoyuki Kubota[1]

[1] Graduate School of Systems Design, Tokyo Metropolitan University,
6-6 Asahigaoka, Hino, Tokyo, Japan
`kitai-sasuga@ed.tmu.ac.jp`
[2] Graduate School of Natural Science and Technology, Okayama University,
3-1-1 Tsushimanaka, Kita-ku, Okayama City, Okayama, Japan

**Abstract.** Recently, in order to reduce secondary disasters at the disaster site, the expectation of rescue robots has been increasing. In the disaster site, many kinds of environments are mixed. Therefore, in order to make robots perform the environmental sensing in the complicated environment, it is necessary to change the measurement area and density according to the environment information. Therefore, we developed a Variable Ranging Sensor Array as a 3D-distance measurement system for changing measurement area and density. In this paper, we propose an intelligent control method for the sensor array based on Multi-Objective Behavior Coordination, and show results of proposed method in a simulation experiment.

**Keywords:** Variable Ranging Sensor Array · Robot vision
3D point-cloud processing

## 1 Introduction

Recently, in order to reduce secondary disasters at a disaster site, development of a walking robot that can act robustly is desired. In order for robots to work instead of people in such a site, it is necessary for the remotely control or the robot to act autonomously. When robots act autonomously under unknown environments, it is necessary to construct three-dimensional map and environmental sensing. Environmental sensing means that extract environment information necessary for movement such as obstacles from the map. Research on three-dimensional distance measuring sensors such as LIDAR has been actively conducted in recent years in such three-dimensional map construction technology and environment recognition technology [1, 2].

At a disaster site, Robots need to act inside a mixture of flat areas and debris areas on the road, to ascend and descend stairs, and to avoid dynamic objects such as other robots and people. In this way, the robot must act in a very complicated environment mixed with various environments at the disaster site. Therefore, in environmental sensing under such a complicated environment, depending on the environment in which the robot is placed and the moving method, the required measurement area, range, and

© Springer Nature Switzerland AG 2018
Z. Chen et al. (Eds.): ICIRA 2018, LNAI 10984, pp. 393–403, 2018.
https://doi.org/10.1007/978-3-319-97586-3_35

density are different. Accordingly, it is difficult to construct a sufficient map for environmental sensing with only one measurement. Therefore, it is necessary to always update the map. However, it is necessary to update the whole map if the robot equips with a single LIDAR, so that it takes a long time to measure and calculate SLAM and environmental sensing because the amount of measurement data becomes enormous.

For these reasons, we have developed Variable Ranging Sensor Array using multiple Laser Range Finder (LRF), as a three-dimensional distance measurement system capable of changing the measurement region and density. And, we have proposed environmental recognition methodology [3–5], and control methodology of robots [6]. In this paper, we propose a methodology for controlling measurement range and density of the sensor array, in order to make measurement suitable for the environment where the robot is placed and the movement method, using Multi-Objective Behavior Coordination (MOBC) [7]. MOBC is a method of realizing smooth motion, by dynamically updating the weights of multiple basic behaviors from acquired environmental information. In addition, we show the effectiveness of the proposed method by comparing the measurement results in the case of controlling and the case of not controlling in the simulation environment.

## 2    Variable Ranging Sensor Array

### 2.1    Outline of Variable Ranging Sensor Array

In this chapter, we will explain the sensor we developed. Figure 1 shows the Variable Ranging Sensor Array developed in this research. It consists of 4 LRFs and 2 servo actuators. It has two LRFs on the top and two on the side. Also, one actuator that rotates the LRF is mounted on the upper side and the side. Tables 1 and 2 show the specifications of the LRF and the actuator.

**Table 1.**  Specification of LRF.

| Model | UST-20LX |
|---|---|
| Measurement range | 0.06–10[m] |
| Measurement accuracy | ±40[mm] |
| Scan angle | 270[°] |
| Scan time | 25[msec] |
| I/O | Ethernet 100BASE-TX |

**Table 2.**  Specification of actuator.

| Model | FHA-8C |
|---|---|
| Maximum torque | 3.3[N · m/A] |
| Maximum rotational speed | 120[r/min] |
| Positioning accuracy | 120[sec] |
| Reduction ratio | 50 |
| Encoder type | Absolute |

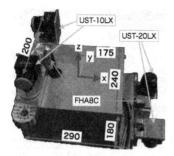

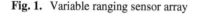

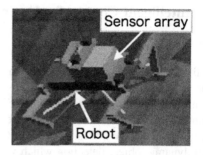

**Fig. 1.** Variable ranging sensor array

**Fig. 2.** Robot and sensor array model of simulation environment

The sensor array can be reconstructed three-dimensionally with minimum coordinate transformation by matching the rotation axis of the motor with the optical axis of the LRF. Moreover, by changing the mounting angle of the LRF installed on the upper side and the side, in the upper LRF, it is possible to evenly measure the whole environment, in the side LRF, it is possible to measure the robot forward direction more densely. The sensor array can control the measurement range by controlling the rotation speed of the servo actuator and change the measurement range by controlling the rotation range. In this paper, we aim to intelligently control the rotation range of the servo actuator according to the recognized environment.

## 2.2 Model of Robot and Sensor in Simulation Environment

In this paper, we created a model of a quadruped robot and the sensor array as shown in Fig. 2 using Open Dynamics Engine (ODE) which is an open source physical operation engine. We conducted a control experiment on the simulation environment. Moreover, distance data measured by the sensor array in the simulation environment can store data in the same format as the actual LRF, and it can be restored to a three-dimensional map as shown in Fig. 3.

**Fig. 3.** 3D-map of simulation environment.

# 3   Intelligent Control for Environmental Sensing Using MOBC

## 3.1   Outline of the Control Method

In this chapter, we propose a method to intelligently control the sensor array using the concept of MOBC. The control of sensor array means that the measurement range is controlled by changing the rotation speed and the rotation range of the servo actuator suitable for the surrounding environment of the robot perceived. A flowchart of the algorithm is shown in Fig. 4. We use 3D object recognition, feature point extraction, and dynamic object detection which we have proposed as a method of environment recognition [3–5]. The basic behavior of MOBC in this paper is defined as the following three measurement methods: 1. Flat ground walking measurement, 2. Dynamic object following measurement, 3. Ladder measurement. And each control output is calculated using fuzzy control [6].

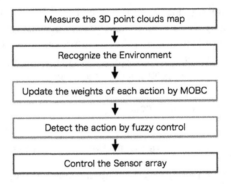

**Fig. 4.** Flowchart of control system.

In the conventional method using MOBC, we calculated the integrated output according to the weight of each basic behavior. However, in the case of the sensor array, the integrated output may not be available in some case. For example, it is the case that the weights of the two measurements methods take high values. If the positions of the two objects are far apart, there is a problem that it becomes impossible to measure either object by outputting to measure the intermediate position between two objects, in integrated output. Therefore, in this paper, we select the action that has become the highest weight among the updated weights as the final control output.

## 3.2   Determination of Output of Each Behavior by Fuzzy Control

The basic behavior in this paper is calculated by fuzzy control [6] using simplified fuzzy inference. Details and input information of the antecedent part of fuzzy control of each basic behavior are shown below.

1. Flat ground walking measurement:

It is a measurement method used when walking on a flat ground. Through this measurement method, intensive measurement for obstacles, localization, and dynamic object detection are performed. The two LRFs mounted on the side of the sensor array are fixed, and the two LRFs on the upper part of the sensor array are controlled so as to centrally measure the obstacle nearest to the robot. The input information of the antecedent part is the distance, direction, and size to the obstacle nearest to the robot.

2. Dynamic object following measurement:

It is a measurement method used when detecting dynamic objects. In order to confirm the behavior of dynamic objects, we control two LRFs mounted on the top of the sensor array so as to follow dynamic objects. The input information of the antecedent part is the position, direction, size, moving speed and moving direction of the dynamic object.

3. Ladder measurement:

It is a measurement method used when climbing a ladder. In order to acquire the detailed data of the ladder, all LRFs are controlled so as to measure intensively the area around the ladder. The input information of the antecedent part is the position and width of the ladder.

### 3.3 Measurement Method Determination by Multi-objective Behavior Coordination

In this chapter, we describe a method to calculate the weight of each action using MOBC based on environmental perception information obtained by environmental recognition. Assuming that the control output at action $k$ is $y_{k,j}$, j, the weight is $wgt_k$, and the total number of behaviors is r (=3), the weight $wgt_k$, is updated by (1) according to perceived time series sensor information. $\alpha$ is a forgetting factor. This factor has the purpose of preventing accumulated weight update width from returning to 0 when the sensor information is the same value as the initial value.

$$wgt \leftarrow \frac{\alpha \cdot wgt + \Delta wgt_k}{\sum_{k=1}^{r} (\alpha \cdot wgt + \Delta wgt_k)} \qquad (1)$$

$\Delta wgt_k$ is a weight updating rule and is set as follows.

$$\begin{bmatrix} \Delta wgt_1 \\ \Delta wgt_2 \\ \Delta wgt_3 \end{bmatrix} = \begin{bmatrix} dw_{1,1} & \cdots & dw_{1,5} \\ \vdots & \ddots & \vdots \\ dw_{3,1} & \cdots & dw_{3,5} \end{bmatrix} \begin{bmatrix} si_1 \\ \vdots \\ si_5 \end{bmatrix} \qquad (2)$$

$si_m$ represents environmental perception information and takes a value of 0 to 1 according to the input information. In this paper, we define the following five variables based on the information of the detected obstacle, dynamic object, ladder position information and position of the robot. $si_1$ represents the degree of danger to the obstacle. This variable is set to rise sharply when the distance $l_o$ between the robot and the obstacle closest to the robot is within 1.2 m. $\gamma_o, \eta_o$ are arbitrary constants which determine the gradient of the rise of risk, and in this paper, $\gamma_o = 1.0$, and $\eta_o = 2.0$. $si_2$ represents the degree of danger to dynamic objects. $l_m$ is the distance between the robot and the dynamic object, $L_m, \gamma_m, \eta_m$ are arbitrary constants, In this paper, $L_m = 1.5[m]$, $\gamma_m = 1.0[m]$, $\eta_m = 2.0[m]$. $si_3$ is also the degree of danger for dynamic object. Set $l_m$ as $l'_m$ acquired by previous measurement. If $l'_m > l_m$, that is when the dynamic object is approaching the robot, $si_3 = 1$ is returned. And if it is leaving the robot $si_3 = 0$ is returned. $si_4$ is set so that the ladder measurement weight increases sharply when the distance $L_l$ between the robot and the ladder is within 1 m.

$$si_1 = \begin{cases} \exp(\gamma_o \cdot (L_o - l_o)L_o)/\eta_o & (if\ L_o > l_o) \\ 1.0 & (otherwise) \end{cases} \tag{3}$$

$$si_2 = \begin{cases} \exp(\gamma_m \cdot (L_m - l_m)/L_m)/\eta_m & (if\ L_m > l_m) \\ 1.0 & (otherwise) \end{cases} \tag{4}$$

$$si_3 = \begin{cases} 1.0\ (if\ l'_m > l_m) \\ 0.0\ (otherwise) \end{cases} \tag{5}$$

$$si_4 = \begin{cases} 1.0\ (if\ L_l > l_l) \\ 0.0\ (otherwise) \end{cases} \tag{6}$$

$$si_5 = 1 - si_2 \tag{7}$$

In addition, $dw_{k,m}$ is a constant parameter that determines the influence of action $k$ on $si_m$. It was established as follows.

$$\begin{aligned} dw_1 &= \{0.6, 0.0, 0.0, 0.0, 0.4\} \\ dw_2 &= \{0.0, 0.7, 0.2, 0.0, 0.0\} \\ dw_3 &= \{0.0, 0.0, 0.0, 0.7, 0.2\} \end{aligned} \tag{8}$$

The weight of each action is calculated using these equations and the sensor array is controlled based on the output calculated by the fuzzy controller of the behavior with the largest weight.

## 4   Experiment

In this experiment, we confirm that sensor array is controlled by the proposed method so that the measurement suitable for the environment can be performed. For this reason, we conducted a control experiment of the sensor array on the simulation environment. Also, we compared the measurement results with and without control. An experimental environment with obstacles, dynamic objects and ladder shown in Fig. 5 was constructed on the above simulation environment.

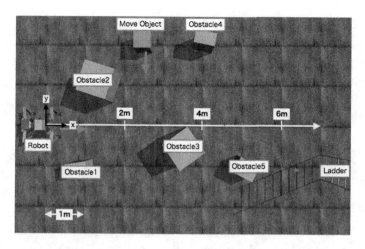

**Fig. 5.** Simulation environment of experiment.

By walking the robot towards the ladder placed at the position of 7 [m] in the x direction and −1 [m] in the y direction as seen from the robot, we confirm that the optimum measurement can be performed according to the surrounding environment of the robot by the proposed method. In this paper, the position of robot, obstacle, dynamic object, and ladder is treated as known information.

The walking behavior of the robot in the experiment is shown in Fig. 6. And, the transition of the weight of each behavior calculated by MOBC at that time is shown in Fig. 7. Experimental results show that at the start point there are only obstacles in the surroundings, so that the weight of Flat ground walking measurement is maximized. On the other hand, from the position $x = 1.4$ [m] before, we can see that the dynamic object measurement weight is rising as the dynamic object behind the obstacle can be seen. Thereafter, while waiting for the dynamic object to pass by at the position of $x = 1.6$ [m], as the dynamic object passes by, the weight of the dynamic object measurement decreases, and the weight of the Flat ground walking measurement rises again. The weight of the dynamic object measurement increases again around $x = 1.7$ [m], because the distance to the dynamic object approaches as the robot starts moving. Also, from the vicinity of $x = 6.0$ [m], since the distance to the ladder is within 1.0 [m], the weight of the ladder measurement rises and is found to be the maximum.

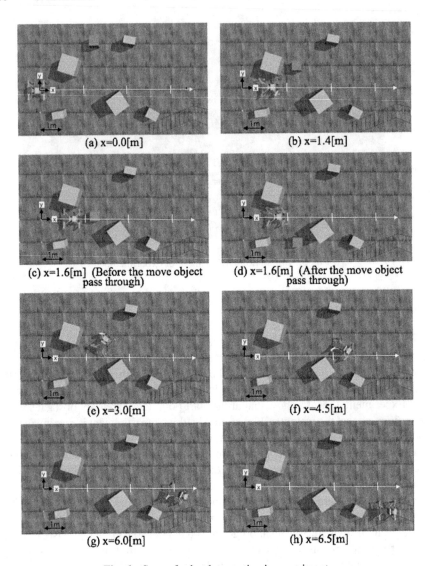

(a) x=0.0[m]

(b) x=1.4[m]

(c) x=1.6[m] (Before the move object pass through)

(d) x=1.6[m] (After the move object pass through)

(e) x=3.0[m]

(f) x=4.5[m]

(g) x=6.0[m]

(h) x=6.5[m]

**Fig. 6.** State of robot locomotion in experiment.

In addition, we compare the number of measured point clouds when control is performed by the proposed method and when the entire map is continuously updated without control. As shown in Table 3, looking at the number of point clouds measured while the robot moves from the position of 2 [m] to the position of 3 [m], there is not much difference in the number of the whole point clouds. However, the number of point cloud that measured Obstacle 2 and Obstacle 3 is more than doubled when control is performed. Also, looking at Table 4, in the point clouds measured while the robot moves from the position of 5 [m] to the position of 6 [m], the number of point clouds measuring Obstacle 5 and Ladder is nearly tripled when control is performed. Also, looking at the

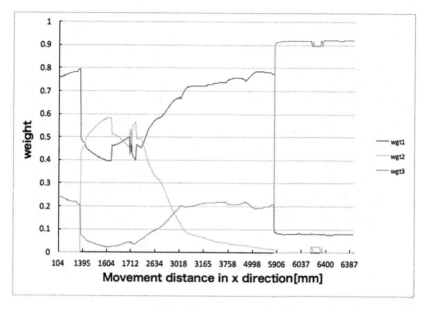

**Fig. 7.** Experimental result of transition of weight.

point clouds of the ladder measured while the robot moves from the position of 6 [m] to the position of 6.5 [m], as shown in Fig. 8, when control is not performed, a part of the ladder is missing, whereas when the control is performed t can be seen that the shape of the ladder can be measured completely.

**Table 3.** Number of point clouds while the robot has moved from 2[m] to 3[m].

|  | Number of point clouds | | | | | | | |
|---|---|---|---|---|---|---|---|---|
|  | Total num | Without ground | Obstacle 1 | Obstacle 2 | Obstacle 3 | Obstacle 4 | Obstacle 5 | Ladder |
| Controlled | 118075 | 51852 | 189 | 28235 | 19215 | 99 | 145 | 0 |
| Not controlled | 101972 | 29703 | 683 | 14152 | 8992 | 924 | 151 | 0 |

**Table 4.** Number of point clouds while the robot has moved from 5[m] to 6[m].

|  | Number of point clouds | | | | | | | |
|---|---|---|---|---|---|---|---|---|
|  | Total num | Without ground | Obstacle 1 | Obstacle 2 | Obstacle 3 | Obstacle 4 | Obstacle 5 | Ladder |
| Controlled | 108698 | 31634 | 1 | 1036 | 5641 | 154 | 14879 | 912 |
| Not controlled | 126871 | 19666 | 0 | 942 | 4256 | 495 | 5591 | 383 |

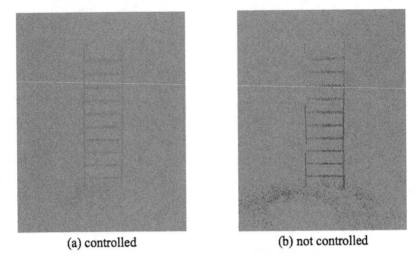

(a) controlled                          (b) not controlled

**Fig. 8.** Point clouds of ladder while the robot has moved from 6[m] to 6.5[m]

From these results, it was shown that by controlling the sensor array according to the surrounding environment of the robot by the proposed method, it is possible to measure information necessary for robot walking and work.

## 5    Conclusion

In this paper, we proposed a method to intelligently control the measurement range of the Variable Ranging Sensor Array in accordance with the environment around the robot, in order to perform environmental sensing that acquires information necessary for robot walking and work in a complicated environment such as a disaster site. In addition, we showed that we can perform environment measurement suitable for the surrounding environment of robot by measuring the environment while moving the robot on a simulation environment mixed with obstacles, dynamic objects, ladders. We also showed the usefulness of the proposed method by comparing the measurement results.

In this paper, we established empirically the various parameters in MOBC and the environmental perception information that decides what kind of actions will be selected given what kind of input information. However, in actual disaster sites, it is thought that more various environments exist, and the works of the robot is more diverse. Therefore, it becomes difficult for a person to design parameter setting and environment perception information. For this reason, future work is to enable learning of optimal control output even in unknown environments using reinforcement learning.

**Acknowledgment.** This work was funded by Impulsing Paradigm Change through Disruptive Technologies (ImPACT) Program of the Council for Science, Technology and Innovation.

# References

1. Cole, D.M., Newman, P.M.: Using laser range data for 3D SLAM in outdoor environments. In: Proceedings of 2006 IEEE International Conference on Robotics and Automation 2006 (ICRA 2006), pp. 1556–1563 (2006)
2. Kita, S., Jae, H.L., Okamoto, S.: 3D map building using mobile robot with scan device. Int. J. Mech. Eng. Robot. Res. (IJMERR) 5(3), 191–195 (2016)
3. Toda, Y., Narita, T., Kubota, N.: Information visualization based on 3D modeling for human-friendly teleoperation. In: Proceedings of 2012 IEEE World Congress on Computational Intelligence (WCCI 2012), pp. 3558–3564, 10–15 June 2012
4. Toda, Y., Yu, H., Ju, Z., Takesue, N., Wada, K., Kubota, N.: Real-time 3D point cloud segmentation using growing neural gas with utility. In: The 9th International Conference on Human System Interaction, pp. 418–422 (2016)
5. Miyake, S., Toda, Y., Kubota, N., Takesue, N., Wada, K.: Intensity histogram based segmentation of 3D point cloud using growing neural gas. In: International Conference of Intelligent Robotics and Applications, pp. 335–345 (2016)
6. Fukuda, T., Kubota, N.: An intelligent robotic system based on a fuzzy approach. Proc. IEEE 87(9), 1448–1470 (1999)
7. Nojima, Y., Kojima F., Kubota, N.: Behavior acquisition of mobile robots based on multi-objective behavior coordination. Trans. JSME(C) 68(671), 2067–2073 (2002)
8. Hokuyo Automatic Co., Ltd. http://www.hokuyo-aut.co.jp
9. Harmonic Drive Systems Inc. https://www.hds.co.jp

# Research on the Estimation of Sensor Bias and Parameters of Load Based on Force-Feedback

Nianfeng Wang, Jianbin Zhou[✉], and Xianmin Zhang

Guangdong Provincial Key Laboratory of Precision Equipment
and Manufacturing Technology, School of Mechanical
and Automotive Engineering, South China University of Technology,
Guangzhou 510640, China
menfwang@scut.edu.cn, katniss1037@163.com

**Abstract.** This paper proposes a method for estimating the force/torque sensor bias, and the parameters of the load(including the gravity component and the center of mass). We set the 6-axis force/torque sensor between robot and the end-effector, so that we can estimate the sensor bias and the parameters by reading data of sensor in 8 sets of robot orientations. These estimates can be subtracted from the sensor readings, in order to improve the accuracy of the force/torque measurements. In addition, this paper verifies that the installation angle bias of robot will increase measurement deviation. The experiments show that the error of resulting force compensation is not more than 3.1% of the gravity of the load, and the error of resulting torque compensation is not more than 6.1% of the gravitational torque. Moreover, considering the installation angle bias of robot, the measurements of sensor will be more accurate.

**Keywords:** Force sensing · Gravity compensation · Sensor bias
Installation angle bias

## 1 Instruction

In robotic applications where robot comes in contact with the environment, robots based solely on position servo loops often fail to meet expectations [1]. For example, when industrial robots are used in deburring and grinding [2, 3], the contact force and torque between robots and environment need to be monitored accurately. Almost always, the robot comes in contact with the environment only through its end-effector, the resulting contact forces can be monitored by using a force/torque sensor. At relatively low speed, the value of sensor can be divided into three parts: the gravity component, the bias of the sensor and the environmental contact force. And we usually expect to get the accurate measurements of contact force between the robot and the environment. Therefore, we need to compensate the influence of gravity and sensor bias when we are reading the force/torque sensor.

The gravity component and gravitational torque component should be compensated in real time with the orientations' change of the robot. When the gravity and center of mass of the load was known, the gravity component and gravitational torque component

Z. Chen et al. (Eds.): ICIRA 2018, LNAI 10984, pp. 404–413, 2018.
https://doi.org/10.1007/978-3-319-97586-3_36

can be eliminated [4]. However, in practical applications, the gravity and the center of mass of the load are all unknown quantities, which need to be determined by experiment or another way. In [5–8] it defaults that the Z-axis of the robot base frame is parallel to the direction of gravity acceleration *g*, and adjusts the robot to a series of orientations to calibrate the gravity of the load, then solved the center of mass of the load according to the relationship between force and torque, thus eliminating the gravitational torque component. But usually, the installation of robots will not ensure that the coordinate axis $Z_0$ of base frame is parallel to the direction of gravity acceleration *g*, so it will cause the measurement deviation due to the installation angle bias. In [9],the least square method is used to get the parameters of the load, but the compensation of sensor bias is not considered. In [10, 11], they discussed the compensation of sensor bias, they controlled the robot to a series of orientations in which the effect of gravity can be counteracted, so that they can eliminate the sensor bias without considering the effect of gravity.

In this paper, we assemble the sensor in a rigidly connected way to the end of robot, so the orientations of sensor will change with the change of robot's orientations, then the compensation for the gravity component needs to change with the change of the robot's orientations. As for the sensor bias, it is determined by the physical characteristics of the components of sensor (such as the internal strain gauge or the force beam of the sensor), it will not change with the change of the robot's orientations, but the gap between components of sensor will change with the load. So sensor bias should be compensated according to different loads. In addition, we need to transform the Z-axis direction of the robot base frame to the direction that parallel to the gravity acceleration, so the installation angle bias of robot should be taken into consideration.

## 2    The Discussion of the Establishment of Frame

In general, the definition of each frame will be defined as in Fig. 1(a): The world frame $O_0$-$X_0Y_0Z_0$ and the robot base frame $O_b$-$X_bY_bZ_b$ is coincident, and $Z_0$ and $Z_b$ is in the opposite direction as the gravity acceleration *g*. In the initial state, the axis of the terminal frame $O_t$-$X_tY_tZ_t$ is parallel to the axis of $O_b$-$X_bY_bZ_b$, and its direction of X-axis and Y-axis are the same, and its direction of Z-axis is opposite. However, due to the bias of installation angle of robot, the direction of the $Z_t$ is not parallel to the direction of the gravity acceleration when the robot is in the initial state. In other words, the direction of $Z_b$ and gravity acceleration *g* can not be considered parallel, that is, the robot has an installation angle bias, so the definition of each coordinate system should be shown in Fig. 1(b).

Since the $Z_0$ can be defined arbitrarily around its Z axis, it can be assumed that $O_b$-$X_bY_bZ_b$ can be obtained by $O_0$-$X_0Y_0Z_0$ rotating 0 degrees around the $Z_0$ first, then rotating $\alpha$ degrees around the $Y_0$, and finally rotating $\beta$ degrees around the $Z_0$ (this paper will represent the orientations conversion with RPY combination transformation). So the transformation matrix from $O_b$-$X_bY_bZ_b$ to $O_0$-$X_0Y_0Z_0$ will be:

$$_b^0R = R(Z_0, 0) \cdot R(Y_0, \alpha) \cdot R(X_0, \beta) \tag{1}$$

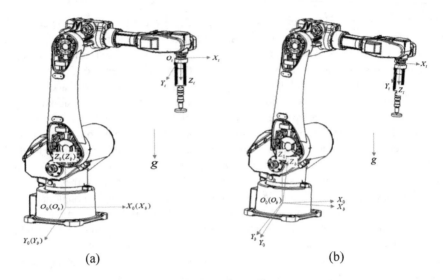

(a)                                                         (b)

**Fig. 1.** Definition of coordinate system

The transformation matrix of robot terminal frame $O_t$-$X_tY_tZ_t$ to base frame $O_b$-$X_bY_bZ_b$ can usually be directly acquired by robot control system. For the orientations conversion between the sensor frame and the terminal frame, the sensor and its connection elements in this paper have good mechanical positioning datum, so it is assumed that there is only a position transformation between the sensor frame and the terminal frame, and the orientations conversion between them have been ignored.

## 3   Methodology

### 3.1   The Installation Angle Bias of Robot

The gravity acceleration $g$ can be expressed in $O_0$-$X_0Y_0Z_0$ as:

$$g_0 = \begin{bmatrix} 0 \\ 0 \\ -g \end{bmatrix} \tag{2}$$

From the analysis of the Sect. 2, we can see that under the initial state of robot, the gravity of the load can be defined in sensor frame by:

$$G = \begin{bmatrix} G_x \\ G_y \\ G_z \end{bmatrix} = m \cdot g_t = {}_t^b R^{-1} \cdot \begin{bmatrix} m \cdot \cos \cdot \alpha \sin \beta \cdot g \\ -m \cdot \sin \alpha \cdot g \\ -m \cdot \cos \alpha \cdot \cos \beta \cdot g \end{bmatrix} \tag{3}$$

When the tool of robot comes in contact with the environment, the force of each channel of sensor can be expressed as:

$$
\begin{bmatrix} F_x \\ F_y \\ F_z \end{bmatrix} = \begin{bmatrix} F_{x0} \\ F_{y0} \\ F_{z0} \end{bmatrix} + G = \begin{bmatrix} F_{x0} \\ F_{y0} \\ F_{z0} \end{bmatrix} + {}^b_t R^{-1} \cdot \begin{bmatrix} m\cos\alpha\sin\beta \\ -m\sin\alpha \\ -m\cos\alpha\cos\beta \end{bmatrix} = [I|{}^b_t R^{-1}] \begin{bmatrix} F_{x0} \\ F_{y0} \\ F_{z0} \\ k_1 \\ k_2 \\ k_3 \end{bmatrix} \quad (4)
$$

where $I$ is three order identity matrix, and $k_1 = m \cdot \sin\alpha \cdot \sin\beta$, $k_2 = -m \cdot \sin\alpha$, $k_3 = -m \cdot \cos\alpha \cdot \cos\beta$.

So the installation angle bias of robot can be acquired:

$$
\begin{cases} \alpha = \arcsin\left(-\frac{k_2}{m}\right) \\ \beta = \arctan\left(-\frac{k_1}{k_3}\right) \end{cases} \quad (5)
$$

## 3.2    The Sensor Bias and the Parameters of the Load

We know that the value of sensor can be divided into the gravity of the load, the bias of the sensor and the environmental contact force. When the tool of robot are not contact with the environment, the torque on each channel can be expressed as:

$$
\begin{bmatrix} M_x \\ M_y \\ M_z \end{bmatrix} = \begin{bmatrix} M_{x0} \\ M_{y0} \\ M_{z0} \end{bmatrix} + \begin{bmatrix} M_{Gx} \\ M_{Gy} \\ M_{Gz} \end{bmatrix} \quad (6)
$$

As shown in Fig. 2, $x$, $y$, $z$ is the center of mass of the load in the sensor frame. $M_{Gx}$, $M_{Gy}$, $M_{Gz}$ is the component of torque cause by the gravity of the load, they can be acquired:

$$
\begin{cases} M_{Gx} = G_z \times y - G_y \times z \\ M_{Gy} = G_x \times z - G_z \times x \\ M_{Gz} = G_y \times x - G_x \times y \end{cases} \quad (7)
$$

Combine the Eqs. (4), (6), (7) we can obtain:

$$
\begin{bmatrix} M_x \\ M_y \\ M_z \end{bmatrix} = [F|I] \begin{bmatrix} x \\ y \\ z \\ a \\ b \\ c \end{bmatrix} \quad (8)
$$

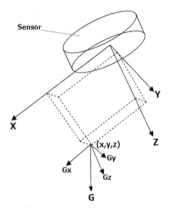

**Fig. 2.** The center of mass of the load in sensor frame

where $I$ is three order identity matrix, and $a = M_{x0} + F_{x0} \times z - F_{z0} \times y$, $b = M_{y0} + F_{z0} \times x - F_{x0} \times z$, $c = M_{z0} + F_{x0} \times y - F_{y0} \times x$, $F$ is expressed by:

$$\begin{bmatrix} 0 & F_z & -F_y \\ -F_z & 0 & F_x \\ F_y & -F_x & 0 \end{bmatrix} \tag{9}$$

In this paper, we control the robot to 8 sets of orientations to get 8 groups of sensor data, Eq. (8) will be converted to:

$$\begin{bmatrix} M_{x1} \\ M_{y1} \\ M_{z1} \\ \cdot \\ \cdot \\ \cdot \\ M_{x8} \\ M_{y8} \\ M_{z8} \end{bmatrix} = \begin{bmatrix} F_1|I \\ \cdot \\ \cdot \\ \cdot \\ F_8|I \end{bmatrix} \begin{bmatrix} x \\ y \\ z \\ a \\ b \\ c \end{bmatrix} \tag{10}$$

We reduced the Eq. (10) into:

$$m = F \cdot q \tag{11}$$

According to the matrix form of least square method, the solution of center of mass can be acquired:

$$q = \left(F^T F\right)^{-1} F^T m \tag{12}$$

Similarly, after obtaining 8 groups of different orientations, the Eq. (4) can be expressed as:

$$
\begin{bmatrix}
F_{x1} \\
F_{y1} \\
F_{z1} \\
\cdot \\
\cdot \\
\cdot \\
F_{x8} \\
F_{y8} \\
F_{z8}
\end{bmatrix}
=
\begin{bmatrix}
\left[ I\big|_t^b R_1^{-1} \right] \\
\cdot \\
\cdot \\
\cdot \\
\left[ I\big|_t^b R_8^{-1} \right]
\end{bmatrix}
\begin{bmatrix}
F_{x0} \\
F_{y0} \\
F_{z0} \\
k_1 \\
k_2 \\
k_3
\end{bmatrix}
\tag{13}
$$

reduced the Eq. (13) into:

$$
f = R \cdot p \tag{14}
$$

then we can obtain the solution of the sensor bias of force by:

$$
p = \left( R^T R \right)^{-1} R^T f \tag{15}
$$

At this point, we have obtained the constants $(a, b, c, k_1, k_2, k_3)$, the center of mass $(x, y, z)$, and the sensor force bias $(F_{x0}, F_{y0}, F_{z0})$. Then the sensor bias of torque can be acquired:

$$
\begin{cases}
M_{x0} = a - F_{y0} \times z + F_{z0} \times y \\
M_{y0} = b - F_{z0} \times x + F_{x0} \times z \\
M_{z0} = c - F_{x0} \times y + F_{y0} \times x
\end{cases}
\tag{16}
$$

The above analysis gives the method to obtain the sensor bias and the parameters of the load. It is necessary to note that it must get more than 3 groups of data of different robot orientations, and ensure that there are at least 3 groups of orientations that the direction of the end are non-coplanar. This is to guarantee the prevention of the occurrence of a morbid matrix.

## 4   Experiments

Several experiments using the proposed methods were conducted based on the 6-axis force/torque sensor of the company for SRI type M3714C. The technical parameters of M3714C are shown in Table 1.

*Non-linearity* represents the maximum deviation of the output of the sensor from the reference line (the line of the connecting zero point and the full range output point) by the percentage of the full range; *Hysteresis* represents the percentage of the maximum deviation of output when loading and unloading at the same point and the full scale. One part of the experimental system is shown in Fig. 3.

**Table 1.** Technical parameters of M3714C.

|  | $F_x$(N) | $F_y$(N) | $F_z$(N) | $M_x$(N) | $M_y$(N) | $M_z$(N) |
|---|---|---|---|---|---|---|
| Capacity (N/N) | ±1600 | ±1600 | ±3200 | ±88 | ±88 | ±88 |
| Non-linearity (%FS) | 0.03 | 0.02 | 0.08 | 0.04 | 0.08 | 0.05 |
| Hysteresis (%FS) | 0.02 | 0.08 | 0.04 | 0.27 | 0.10 | 0.14 |

**Fig. 3.** Experimental system

In order to verify the estimation method of this paper, the experiments conducted under the conditions of no-load and two different terminal tools (the grinding wheel make of fibre and stone) installed. The 8 groups of robot orientations taken in the experiment are shown in Table 2. As an example, the data of sensor when terminal tool was the fibre grinding wheel is given in Table 3. In order to reduce the influence of random error on the experimental results, the average of 4000 data that collected continuously under the same orientations were recorded. Table 4 is the results of the sensor bias, the parameters of load and the installation angle bias with three kinds of load. Note that the gravity $G$ calculated at the condition of no-load is actually the gravity of the sensitive side of sensor.

As we can see from Table 2, the final rotation of the robot is (Roll:0, Pitch:0, Yaw:90.66), so the direction of the end is approximately horizontal. When the tool are

**Table 2.** Rotation list of the robot

| No. | Roll (°) | Pitch (°) | Yaw (°) |
|-----|----------|-----------|---------|
| 1 | 0 | 0 | −180 |
| 2 | 51.49 | 37.46 | 97.56 |
| 3 | −50.99 | 36.48 | −97.63 |
| 4 | 117.22 | 32.48 | 140.23 |
| 5 | 126.56 | 37.30 | −115.21 |
| 6 | −137.72 | 37.30 | 179.96 |
| 7 | 126.48 | 37.21 | 115.57 |
| 8 | 0 | 0 | −90.66 |

**Table 3.** The data of sensor when load in fibre grinding wheel

| No. | $F_x$(N) | $F_y$(N) | $F_z$(N) | $M_x$(N·m) | $M_y$(N·m) | $M_z$(N·m) |
|-----|----------|----------|----------|------------|------------|------------|
| 1 | 22.25 | 86.69 | −3.59 | −1.29 | 1.31 | 0.19 |
| 2 | 39.35 | 111.08 | −27.96 | −3.88 | 3.16 | 0.23 |
| 3 | 38.77 | 65.74 | −25.33 | 1.16 | 3.46 | 0.06 |
| 4 | 38.16 | 72.54 | 1.59 | 0.39 | 3.03 | 0.04 |
| 5 | 40.27 | 108.83 | −12.82 | −3.59 | 3.10 | 0.18 |
| 6 | 39.76 | 88.32 | 2.30 | −1.32 | 3.17 | 0.11 |
| 7 | 40.34 | 67.47 | −6.66 | 0.97 | 3.34 | 0.01 |
| 8 | 24.53 | 115.49 | −26.89 | −4.44 | 1.16 | 0.26 |

**Table 4.** The results of the estimation

| Load | $F_{x0}$(N) | $F_{y0}$(N) | $F_{z0}$(N) | $M_{x0}$(N·m) | $M_{y0}$(N·m) | $M_{z0}$(N·m) |
|------|-------------|-------------|-------------|---------------|---------------|---------------|
| no-load | 19.28 | 84.57 | −28.21 | −1.36 | 1.08 | 0.11 |
| fibre | 21.81 | 88.9 | −26.10 | −1.31 | 1.22 | 0.13 |
| stone | 22.64 | 89.79 | −27.33 | −1.29 | 1.29 | 0.14 |
| Load | x(m) | y(m) | z(m) | G(N) | $\alpha$(m) | $\beta$(m) |
| no-load | 0.0026 | 0.0004 | 0.0141 | 1.27 | −0.58 | 0.86 |
| fibre | 0.0028 | 0.0007 | 0.1225 | 29.88 | −0.39 | 0.72 |
| stone | 0.0029 | 0.0007 | 0.1298 | 37.94 | −0.27 | 0.77 |

not in contact with the environment, the external force of the sensor is approximately the gravity of terminal load. The ratio of force which is compensated with gravity and the sensor bias to the gravity $G$ is taken as the force compensation error, and the ratio of torque (which is compensated as well) to gravitational torque $M_G$ (estimated by $G \times z$) is taken as the torque compensation error. The result are shown in Table 5.

$F_\Delta$ and $M_\Delta$ is the force and the torque which is compensated with gravity and the sensor bias (in the case of X-axis direction, $F_{\Delta x} = F_x - F_{x0} - G_x, M_{\Delta x} = M_x - M_{x0} - M_{Gx}$).

**Table 5.** The compensation error.

| Load | $F_\Delta$(N) | $G$(N) | $\delta_F$(%) | $M_\Delta$(N·m) | $M_G$(N·m) | $\delta_M$(%) |
|------|------|------|------|------|------|------|
| no-load | 0.025 | 1.27 | 2.3 | 0.0005 | 0.018 | 3.1 |
| fibre | 0.927 | 29.88 | 3.1 | 0.2086 | 3.66 | 5.7 |
| stone | 1.06 | 37.94 | 2.8 | 0.2991 | 4.92 | 6.1 |

We represented the estimated value of gravity of tool by $G_{fibre} - G_{no\text{-}load}$ or $G_{stone} - G_{no\text{-}load}$ (get from Table 4), and compared with the actual value obtained from an electronic scale. The result are shown in Table 6, note that $\Delta_G$ represented the measurement deviation.

**Table 6.** The measurement deviation (N)

| Load | Estimation value | Actual value | $\Delta_G$(N) |
|------|------|------|------|
| fibre | 28.61 | 28.13 | 0.48 |
| stone | 36.67 | 35.12 | 1.55 |

In addition, in order to investigate the influence of installation angle bias on the estimation results, the measurement deviation that installation angle bias is not considered is shown in Table 7. We compared the conditions of considering the installation angle bias (shown in Table 6) and ignoring the installation angle bias. As we can see, the measurement deviation will increase when we are estimating without considering the installation angle.

**Table 7.** The measurement deviation without considering the installation angle bias

| Load | Estimation value | Actual value | $\Delta_G$(N) |
|------|------|------|------|
| fibre | 31.78 | 28.13 | 3.65 |
| stone | 39.18 | 35.12 | 4.06 |

## 5   Conclusion

In this paper, we propose a method for estimating the sensor bias and the parameters of load in industrial robot. A 6-axis force/torque sensor was set between robot and the end-effector, and we analyze the relationship between the force and torque of the sensor to obtain two sets of linear equations. We solved them by least square method and compensated the gravity and the sensor bias in sensor readings. Besides, we verify that the installation angle bias of robot will lead to the measurement deviation. The experimental results show that by the proposed method, the error of force compensation is not more than 3.1% of the gravity of the load, and the error of torque compensation is not more than 6.1% of the gravitational torque. The compensation of gravity and sensor bias has improved the accuracy of the force/torque measurements, and the error of the force compensation is acceptable in most of the applications of sensor using in industrial robot.

**Acknowledgments.** The authors would like to gratefully acknowledge the reviewers' comment. This work is supported by National Natural Science Foundation of China (Grant Nos. 51575187, 91223201),Science and Technology Program of Guangzhou (Grant No. 2014Y2-00217), the Fundamental Research Funds for the Central University (Fund No. 2015ZZ007) and Natural Science Foundation of Guangdong Province (S2013030013355).

# References

1. Li, Z.Y.: Research and Application of Robot Force Position Control Methods for Robot-Environment Interaction. Huazhong University of Science and Technology (2011)
2. Domroes, F., Krewet, C., Kuhlenkoetter, B.: Application and analysis of force control strategies to deburring and grinding. Modern Mech. Eng. **03**(2), 11–18 (2013)
3. Liu, Z.: Research on Grinding Robot Control System based on Force Feedback. Harbin Institute of Technology (2017)
4. Zhang, Q.W., Han, L.L., Fang, X.U., et al.: Research on hybrid position/force control for grinding robots. Control Instrum. Chem. Ind. **39**, 884 (2012)
5. Liu, W.-B.: Research on Industrial Robot Grinding based on Force Control. South China University of Technology (2014)
6. Park, J.O., Kim, W.Y., Han, S.H., Park, S., Ko, S.Y.: Gravity compensation of a force/torque sensor for a bone fracture reduction system. In: Proceedings of the 13th International Conference on Control, Automation and Systems, pp. 1042–1045. IEEE, Gwangju (2013)
7. Du, H.P., Sun, Y.W., Feng, D.Y., Xu, J.T.: Automatic robotic polishing on titanium alloy parts with compliant force/position control. Proc. Inst. Mech. Eng. Part B J. Eng. Manuf. **229**(7), 1180–1192 (2015)
8. Tele-Teaching based on Force Sensing in Remote Welding. Harbin Institute of Technology (2006)
9. Wei, X.Q., Wu, L., Gao, H.-M., Li, H.C.: Research on gravity compensation algorithm for tool-assembling with force control in remote welding. Trans. China Weld. Inst. **30**(4), 109–112 (2009)
10. Lin, J.J.: Research in Active Compliant Assembly System for Industrial Robot with Force Sense. South China University of Technology (2013)
11. Zhang, X.H.: Research on the static calibration of six-axis force sensor of robot. Autom. Instrum. **3**, 86–89 (2004)

# A Novel Framework for Coverage Optimization of Sensor Network

Rui Xu[1], Li Chai[2], and Xi Chen[2(✉)]

[1] School of Information Science and Engineering,
Wuhan University of Science and Technology, Wuhan 430081, China
[2] Engineering Research Center of Metallurgical Automation and Measurement
Technology, Wuhan University of Science and Technology, Wuhan 430081, China
{chaili,chenxi_99}@wust.edu.cn

**Abstract.** In this paper, the coverage optimization of a circular sector sensor network in a region is considered. We propose a new framework to solve this problem, in which a so-called optimization matrix is constructed to addresses the coverage strength of every sensor to all sampling points, and the overlap coverage strength of every two sensors. Then, the coverage optimization of sensor network is equivalent to maximize the coverage strength of all sensors to all points in the target space subject to each overlap coverage strength of every two sensors is smaller than a threshold. This framework is general and can be applied to any networked sensors with different coverage models or geometries. Moreover, a simulation example is provided to illustrate the effectiveness of the proposed method.

**Keywords:** Sensor network · Coverage optimization

## 1 Introduction

In last decades, mobile sensor networks are widely used in practical engineering applications, such as inspection, sensing and collecting information from a variety of environments, monitoring targets via different integrated micro-sensors [1–3]. For the coverage problem of directional sensor networks, the objective is to propose an effective control/optimization algorithm to maximize the covered area [11]. The main challenge with the coverage problem for a directional sensor network is that the sensing geometry and the target space distribution, 2-D or 3-D, significantly increases the complexity of optimizing the sensing performance at both the individual sensor and the network level. Some relevant works have been stated in [3–13]. In [3], the authors make a *Voronoi* partition for a target area, and then pursue the optimization on a proposed coverage strength performance measure for the circular sector sensor network. The sensor networks

---

Supported by the National Natural Science Foundation of China under Grant 61703315, 61625305, 61471275.

Z. Chen et al. (Eds.): ICIRA 2018, LNAI 10984, pp. 414–422, 2018.
https://doi.org/10.1007/978-3-319-97586-3_37

optimization problem for multiple areas is presented in [4]. For heterogeneous sensors with *Voronoi* partition, this problem is solved by minimizing a cost function with no geometric property considered. A similar problem is also studied in [8], with some obstacles existing in the areas. A new distributed learning algorithm is proposed in [5] to solve the coverage problem of directional sensor networks. In [6], the authors consider the unknown areas or unknown obstacles existing in the monitored area. Moreover, a group of mobile sensors that can communicate intermittently is employed to solve the inhomogeneous coverage of the area in [7]. The asymptotical k-coverage of planar sensor networks is studied in [9] and the binary particle swarm optimization (BPSO) is introduced in [10] to solve the deployment of the sensor networks. However, the direction of sensors is fixed or can only be selected from a fixed discrete set. It is noted that these existing methods in developing the coverage optimization algorithm mainly rely on individual sensor or local information within the *Voronoi* partition.

In this paper, a novel framework is established to formulate the coverage optimization problem of sensor networks. This framework is general and can be applied to any different coverage models or geometries of sensors. To solve the coverage problem of sensor network, we firstly discretize the monitoring region into many sampling points, and construct a matrix to address the coverage strength information of every sensor to every single point. Then, we calculate the overlap coverage strength of every two sensors to analyze the inter-performance of the sensor network. And finally, based on the overlap coverage strength, a so-called *optimization matrix* is established, where the diagonal elements denote the coverage strength of each sensor to all points, and the off-diagonal elements represent the overlap coverage strength of every two sensors. Consequently, the coverage optimization problem of sensor network is equivalent to maximize the trace of the so-called *optimization matrix* subject to each off-diagonal element is smaller than a threshold. The final optimization is a nonlinear programming problem, of which various efficient algorithms exists, for instance, the particle swarm optimization algorithm [1], Distributed Primal-Dual Subgradient Method [12] and Regularized Primal-Dual Subgradient Method [13].

The paper is organized as follows: The coverage strength from [3] is presented for the circular sector directional sensor as a preliminary in Sect. 2. The framework to solve the coverage problem of sensor network is proposed in Sect. 3. Then we verify the effectiveness of the control strategy through simulation in Sect. 4. Finally, the paper is concluded in Sect. 5.

Notations. Let R and $R_+$ be the set of real and positive real numbers. Let $p_i = [x_i \, y_i \, \theta_i]^T \in R^3$ denotes the $i^{th}$ sensors center coordinate and the orientation angle in radian, and $P_s = \{p_1, p_2, \cdots, p_n\}$, with $n$ being the total number of sensors be the set of all the sensors. $d_E(,)$ stands for the Euclidean distance on $R^2$. $q \in Q$ is one sample point, with $k$ being the total number of grid point.

## 2    Model of Circular Sector Sensor

The considered 2D sensors have a circular sector sensing region, denoted by $S_i \subset R^2$ for sensor $i$, with the boundary $\partial S_i$. Let $r > 0$ be the radius of the sector, $\alpha$ the central angle of the circular sector in radians. In this subsection, we aim to construct a homeomorphism $\boldsymbol{H}_i\colon \mathrm{R}^2 \to \mathrm{R}^2$ to map the sector region $S_i$ into a unit circle. Without loss of generality, as shown in Fig. 1(a), we choose the middle point $o_i$ of the central axis within the sector as the 'sensing center'. $F(XoY)$ denotes the world coordinate system, $F_i(X_i o_i Y_i)$ is the $i^{th}$ sensor frame with its origin located at $o_i$, and its $Y_i$ axis is aligned with the central axis of the sector. $F_{ti}(X_{ti} o_{ti} Y_{ti})$ denotes the frame in the transformed space through the homeomorphism $\boldsymbol{H}_i$, as shown in Fig. 1(b) and (c) respectively.

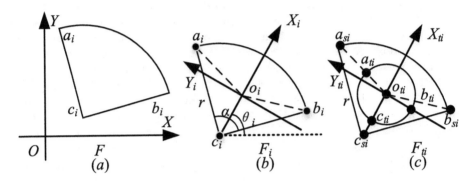

**Fig. 1.** The original sector region of sensor $s_i$ is mapped into a circle.

*(1)* Relationship between $F$ and $F_i$: in the frame $F$, the position of the sensor is represented by $c_i = [x_i\, y_i]^T \in \mathrm{R}^2$, while its orientation is $\theta_i$. $c_i$ and $\theta_i$ constitute the configuration of the $i^{th}$ sensor as $p_i = [c_i^T\, \theta_i]^T = [x_i\, y_i\, \theta_i]^T$. The relationship between $F$ and $F_i$ is expressed as

$$q_s = R^T(q - T), \tag{1}$$

where $q$ and $q_s$ are the coordinates of an arbitrary point expressed in $F$ and $F_i$, respectively, and

$$R = \begin{bmatrix} \cos\theta_i & -\sin\theta_i \\ \sin\theta_i & \cos\theta_i \end{bmatrix}, \quad T = \begin{bmatrix} x_i + \frac{r}{2}\cos\theta_i \\ y_i + \frac{r}{2}\cos\theta_i \end{bmatrix}. \tag{2}$$

*(2)* Homeomorphism between $F_i$ and $F_{ti}$: Let $q_s = [x_s\, y_s]^T$ denote the coordinates of an evaluated point in $F_i$. Through a homeomorphism, $q_s$ is mapped to $q_t = [x_t\, y_t]^T$. A function $G_i(q_s)$ is introduced to describe the homeomorphism between $F_i$ and $F_{ti}$ [3], as:

$$q_t = G_i(q_s), \tag{3}$$

with

$$G_i(q_s) = \begin{cases} \frac{q_s}{g(q_s)} & q_s \neq \mathbf{0}, \\ 0 & q_s = \mathbf{0}, \end{cases} \tag{4}$$

where $g(q_s)$ denote the distance of $o_i$ to the intersection point between the segment $o_i q_s$ and $\partial S_i$.

(3) *Homeomorphism between F and $F_{ti}$*: Based on the relationship in (1), together with the homeomorphism in (3), we can derive the homeomorphism $\boldsymbol{H}_i$: $\mathrm{R}^2 \rightarrow \mathrm{R}^2$ between $F$ and $F_{ti}$ as:

$$q_t = H_i(q) = G_i(R^T(q - T)). \tag{5}$$

Then, we have the following definition.

*Definition 1* [3]. *Coverage distance.* The coverage distance of sensor $p_i$ to the point $q$ in $F$, which is denoted by $d_c(p_i, q)$, is defined as the Euclidean distance from $o_{ti}$ to $q_t$ after the homeomorphism transformation, which expresses as

$$d_c(p_i, q) = d_c(o_i(p_i), q) = d_E(o_{ti}, q_t), \tag{6}$$

where $o_{ti}$ and $q_t$ are corresponding point of $o_i$ and point of $q$ after homeomorphism transformation, respectively.

*Coverage Strength.* Through the coverage distance, the coverage strength is defined as

$$C_s(p_i, q) = e^{-\rho d_c(p_i, q)}, \tag{7}$$

where $\rho > 0$ is the decaying rate. For a coverage model with better strength distribution, we choose $\rho = 0.5$ to ensure relatively large coverage strength on $\partial S_i$ in this paper.

Substituting (3) into (6) and (7), the coverage strength of every point in sensing region can be calculated, more details can be found in [3].

## 3   Framework for Coverage Optimization

In this section, we aim to establish a novel framework to solve the coverage optimization of the circular sector sensor network. We firstly discretize the whole region $Q$ into $k$ sampling points. Then, the coverage strength of one sensor to single point can be obtained by (7). For the region $Q$, we compose it by $k$ pieces, each piece $q_k$ is so small that each sensor either covers $q_k$ or not. The coordinates of the pieces $q_k$ is denoted by $q_k = (x_k, y_k)^T$. We construct a matrix $G = [g_{im}]$ to address these information as follows:

$$G(p_1, p_2, \cdots, p_n) = \begin{bmatrix} g_{1*} \\ g_{2*} \\ \vdots \\ g_{n*} \end{bmatrix}_{n \times k}, \tag{8}$$

let $g_{i*} = [g_{i1}, \ldots, g_{ik}], i = 1, \ldots, n$ be a row vector, of which

$$g_{im} = \begin{cases} C_s(p_i, q_m), & \text{if } p_i \text{ covers } q_m \\ 0, & \text{otherwise} \end{cases} \tag{9}$$

where $g_{im}, m = 1, \ldots, k$ is the coverage strength $C_s(p_i, q_m)$ by (7).

Based on the coverage strength of every sensor to every single point in region $Q$, we bring in a new variable which is called *overlap coverage strength* present the comprehensive coverage strength of the overlap area by two sensors and then to analyze the inter-performances of sensors in network (see Fig. 2). The overlap coverage strength of sensor $i$ and $j$ is denoted as $h_{ij}$ for $i, j = 1, \ldots, n, i \neq j$ as follows:

$$h_{ij}(p_i, p_j) = g_{i*} \cdot g_{j*}^\mathsf{T}$$
$$= \sum_{m=1}^{k} C_s(p_i, q_m) \cdot C_s(p_j, q_m). \tag{10}$$

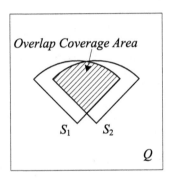

**Fig. 2.** Overlap coverage area between two sensors.

Then, the *optimization matrix* can be constructed as follows:

$$M(p_1, p_2, \cdots, p_n) = \begin{bmatrix} \|g_{1*}\|_1 & h_{12} & \cdots & h_{1n} \\ h_{21} & \|g_{2*}\|_1 & \cdots & h_{2n} \\ \vdots & \vdots & \ddots & \vdots \\ h_{n1} & h_{n2} & \cdots & \|g_{n*}\|_1 \end{bmatrix}_{n \times n}. \tag{11}$$

This optimization matrix addresses the coverage strength of single sensor to all sampling points in target space, and also the overlap coverage strength of any two sensors. A straightforward idea to optimize the coverage of sensor network is to maximize the coverage strength of every sensor to all sampling points, and simultaneously reduce the overlap coverage strength.

Now, it is ready to solve the coverage optimization of circular sector sensor network. The optimization objective and constraint can be concluded as follows:

(1) Maximize the trace of $M$ to ensure the sensors can have biggest coverage strength to all sampling points in region;
(2) All the off-diagonal elements of $M$ are requested to be smaller than a threshold $\varepsilon$, to reduce the overlap coverage redundances between the sensors of the network.

Based on the above two statements, the coverage optimization of the sensor network can be formulated as follows:

$$\max_{(p_1,p_2,\cdots,p_n)} \quad Trace(M) = \sum_{i=1}^{n} \|g_{i*}\|_1 \tag{12}$$
$$subject \ to \quad h_{ij} \leq \varepsilon, \quad for \ i,j = 1,\ldots,n, i \neq j.$$

The final coverage performance of sensor network depends on the constraint threshold $\varepsilon$. However, it is not a fact that setting a smaller threshold can reach a better coverage performance, due to the strictly constraint condition with a very small threshold will lead the sensors' coverage ranges being over the region. In addition, the selection of threshold $\varepsilon$ also depends on the total number of the sensors, the more sensors there are, the bigger $\varepsilon$ is selected. It should be noted however, that the total coverage area of the sensors is far greater than the target area, there is no great significance to solve sensor coverage, also no need to judge $\varepsilon$. Therefore, the total coverage area of sensor networks is close to the target area, threshold $\varepsilon$ is still suit.

*Remark 1.* The overlap coverage strength of more than two sensors can be calculated in a similar way. In this paper, however, we only concern with the overlap coverage strength of two sensors, due to it is absolutely greater than the overlap coverage strength of more than two sensors.

*Remark 2.* In this paper, the detailed process of the nonlinear optimization problem is omitted, which is similar to that in [1,12,13].

*Remark 3.* To simplify the optimization process and reduce the time, a distributed optimization algorithm is proposed as follows:

$$for \quad i = 1,2,\ldots,n :$$
$$\max_{(p_i)} \sum_{i=1}^{n} \|g_{i*}\|_1 - \alpha \cdot \sum_{j=1,j \neq i}^{n} h_{ij}, \tag{13}$$

where $\alpha$ is positive weight. In (13), the optimization is operated in $n$ steps, and only three variables (i.e., $x_i, y_i$ and $\theta_i$) are involved in step $i$ for $i = 1,\ldots,n$. The relationship of the optimization result and the selection of weight $\alpha$ will be studied in future.

## 4    Simulation

In this section, we consider that there are 15 circular sector sensors that will cover an area $Q$ of $200 \times 200$ (m), and the radius and central angle of the circular sector is set as $r = 60\,m$, $\alpha = \pi/2$. The decaying rate of the coverage strength model is set as $\rho = 0.5$, and the initial positions and orientations of sensors are given randomly, which is shown in Fig. 3.

Under the coverage model (7), the optimization matrix $M$ can be obtained. In this example, we set the coverage strength of a sensor to sampling points that out of the circular sector boundary as 0, to simplify the computing process and reduce the simulation time. In this case, the constraint threshold is set as $\varepsilon = 40$, and the final deployment of sensors are shown in Fig. 4 after optimization. Compare with the initial performance and the final performance of sensor network, it can be seen from Table 1 that the coverage ratio increases to 87.14% (final time) from 63.68% (initial time), and the overlap coverage ratio reduces to 5.62% (final time) from 16.70% (initial time).

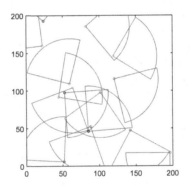

**Fig. 3.** Sensor networks distribution in initial conditions.

**Table 1.** Improvement of coverage rate.

|  | Initial state | Final state |
|---|---|---|
| Coverage rate | 63.68% | 87.14% |
| Overlap coverage rate | 16.70% | 5.62% |

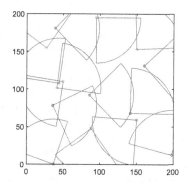

**Fig. 4.** Sensor networks distribution in final conditions.

## 5    Conclusion

In this paper, the coverage optimization of circular sector sensor network in a region is studied. A novel framework is proposed to solve this problem, and this framework is very general thus can be applied to other networked sensors with different coverage models or geometries. The coverage optimization of sensor network is analyzed based on the so-called optimization matrix which addresses some coverage information of sensors. Specifically, the diagonal of the optimization matrix denotes the coverage strength of sensors to all sampling points, and the off-diagonal of the matrix represents the overlap coverage strength of sensors. Then, the aforemention optimization problem is equivalent to maximize the trace of the optimization matrix subject to every off-diagonal element is smaller than a threshold.

## References

1. Mavrinac, A., Chen, X., Alarcon-Herrera, J.L.: Semiautomatic model-based view planning for active triangulation 3-D inspection systems. IEEE/ASME Trans. Mechatron. **20**(2), 799–811 (2015)
2. Zhang, X., Chen, X., Alarcon-Herrera, J.L., et al.: 3-D model-based multi-camera deployment: a recursive convex optimization approach. IEEE/ASME Trans. Mechatron. **20**(6), 3157–3169 (2015)
3. Zhang, X., Chen, X., Liang, X., et al.: Distributed coverage optimization for deployment of directional sensor networks. In: Proceedings of IEEE International Conference on Decision and Control, pp. 246–251 (2015)
4. Abbasi, F., Mesbahi, A., Velni, J. M.: Coverage control of moving sensor networks with multiple regions of interest. In: Proceedings of IEEE American Control Conference, pp. 3587–3592 (2017)
5. Varposhti, M., Saleh, P., Afzal, S., et al.: Distributed area coverage in mobile directional sensor networks. In: Proceedings of IEEE International Symposium on Telecommunications, pp. 18–23 (2017)

6. Parapari, H.F., Abdollahi, F., Menhaj, M.B.: Distributed coverage control for mobile robots with limited-range sector sensors. In: Proceedings of IEEE International Conference on Advanced Intelligent Mechatronics, pp. 1079–1084 (2016)

7. Miah, S., Bao, N., Bourque, A., et al.: Nonuniform coverage control with stochastic intermittent communication. IEEE Trans. Autom. Control **60**(7), 1981–1986 (2015)

8. Mahboubi, H., Sharifi, F., Aghdam, A.G., et al.: Distributed coordination of multi-agent systems for coverage problem in presence of obstacles. In: Proceedings of IEEE American Control Conference, pp. 5252–5257 (2012)

9. Shi, G., Hong, Y.: Region coverage for planar sensor network via sensing sectors. In: Proceedings of IFAC World Congress, vol. 41, no. 2, pp. 4156–4161 (2008)

10. Morsly, Y., Aouf, N., Djouadi, M.S., et al.: Particle swarm optimization inspired probability algorithm for optimal camera network placement. IEEE Sens. J. **12**(5), 1402–1412 (2012)

11. Zhang, J.W., Li, N., Wu, N., et al.: A coverage algorithm based on D-S theory for directional sensor networks. Int. J. Distrib. Sens. Netw. **12**(9), 1–13 (2016)

12. Yuan, D., Xu, S., Zhao, H.: Distributed primal-dual subgradient method for multiagent optimization via consensus algorithms. IEEE Trans. Cybern. **41**(6), 1715–1724 (2011)

13. Yuan, D., Ho, D.W., Xu, S.: Regularized primal-dual subgradient method for distributed constrained optimization. IEEE Trans. Cybern. **46**(9), 2109–2118 (2015)

# Interval Type-2 Fuzzy Control of Pneumatic Muscle Actuator

Xiang Huang[1], Hai-Tao Zhang[1(✉)], Dongrui Wu[1], and Lijun Zhu[2]

[1] School of Automation, State Key Laboratory of Digital Manufacturing Equipment and Technology, and the Key Laboratory of Imaging Processing and Intelligence Control, Huazhong University of Science and Technology, Wuhan 430074, China
{huangxiang92,zht,drwu}@hust.edu.cn
[2] Department of Electric and Electronic Engineering, The University of Hong Kong, Kowloon, Hong Kong, China
ljzhu@eee.hku.hk

**Abstract.** Pneumatic muscle actuator (PMA) is a highly nonlinear system and it is a challenging task to design the controller for it. In this paper, we aim to propose an interval type-2 fuzzy controller. Since the fuzzy sets of interval type-2 fuzzy controller are fuzzy themselves, it has better ability to deal with uncertainty than type-1 fuzzy controller. Both simulation and experiments are conducted to verify the effectiveness of the type-2 fuzzy control algorithm the results confirm that better performance can be achieved by the proposed controller.

**Keywords:** Pneumatic muscle actuator
Interval type-2 fuzzy control · Fuzzy membership function

## 1 Introduction

Pneumatic Muscle actuator (PMA), as one of modern actuation technologies, has promising applications in the field of robotics. The PMA consists of an inner nylon tube surrounded by an outer nylon braided mesh [1], and it expands and contracts as the supplying air flows in and out. In particular, when the pressure increases in PMA, it expands in radial direction and shorten in axial direction. As a result, the force is generated in the axial direction [2]. On the contrary, the PMA contracts when the internal pressure decreases and it restores to the initial state.

Compared with electric motor, PMA has advantages of high power/volume ratio, high power/mass ratio, lower price and being environment friendly. Since PMAs use the soft and flexible materials and have low stiffness, PMAs can be utilized in applications where the safety is the first priority such as in human-machine interaction scenario. Due to these advantages, PMA has been widely used in robotics [3–6].

Supported by National Natural Science Foundation of China with Grant No. U1713203.

Z. Chen et al. (Eds.): ICIRA 2018, LNAI 10984, pp. 423–431, 2018.
https://doi.org/10.1007/978-3-319-97586-3_38

The PMA system is a highly nonlinear system, therefore accurate modeling and control are very difficult. In recent years, efforts have been devoted to the modeling and control of PMAs. PID control is the most frequently used method [7], due to its simple structure. However a PMA is time-varying dynamical system, a simple PID control can not always guarantee a good performance. Various types of controllers were proposed to improve the performance, e.g., adaptive controller [8], neural network controller [9,10], sliding mode controller [11,12], mode predictive controller [13,14] etc.

Fuzzy controllers [15–17] are proposed to deal with system uncertainties. Most of these fuzzy controllers utilized Type-1 fuzzy controllers, but it is less capability to deal with uncertainties [18,19] than type-2 fuzzy controller. In this paper, an interval type-2 fuzzy controller is designed to control PMAs where the fuzzy membership functions are intervals. We carry out the simulation and experiment comparison between type-1 fuzzy controller and the proposed type-2 fuzzy controller, showing that the latter can achieve better performance.

The remainder of this paper is organized as follow. Section 2 presents the model of the PMA and introduces the control objective. In Sect. 3, the type-2 fuzzy controller is constructed. Both simulation and experiments on PMA systems are conducted in Sect. 4 to show the effectiveness of proposed controller. In Sect. 5, conclusions are finally drawn.

## 2    Modeling of PMA

Pneumatic muscles (PMs) are regarded as a combination of nonlinear elastic, viscous and contractile elements [2], and the dynamics can be represented as follows

$$M\ddot{X} + B\dot{X} + KX = F - Mg \tag{1}$$

where $M$ is the mass of the load and $X$ is the displacement of the PMA. The elastic coefficient $B$, viscous coefficient $K$ and contractile force $F$ are pressure-dependent and given as follows.

$$\begin{aligned} B &= b_1 P + b_2 \\ K &= k_1 P + k_2 \\ F &= f_1 P + f_2 \end{aligned} \tag{2}$$

where $b_1$, $b_2$, $k_1$, $k_2$, $f_1$ and $f_2$ are coefficients that might be time-varying. The schematic model is illustrated in Fig. 1.

In order to move the PM in two direction, an initial pressure $P_0$ is supplied such that the PM rests at $X = x_0$. According to (1), one has

$$(k_1 P_0 + k_2)x_0 = f_1 P_0 + f_2 - Mg \tag{3}$$

Let $P = P_0 + u$ and $X = x_0 + x_d$ where $u$ is pressure increment from initial pressure $P_0$ and $x_d$ is the corresponding displacement increment. With the change of variable and (3), dynamics (1) becomes

$$M\ddot{x}_d = -(b_1 P_0 + b_2)\dot{x}_d - (k_1 P_0 + k_2)x_d + (f_1 - k_1 x_0 - k_1 x_d - k_1 \dot{x}_d)u \tag{4}$$

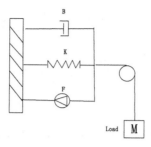

**Fig. 1.** Reynolds' model for a PMA.

Let $x = [x_d, \dot{x}_d]^T$, Eq. (4) can be rewritten as

$$\dot{x} = Ax + B(x)u$$
$$y = Cx \qquad\qquad (5)$$

with $A = \begin{bmatrix} 0 & 1 \\ -(k_1 P_0 + k_2)/M & -(b_1 P_0 + b_2)/M \end{bmatrix}$, $B(x) = [0, (f_1 - k_1 x_0 - k_1 x_d - b_1 \dot{x}_d)/M]^T$, $C = [1, 0]$. Then, the objective of the tracking problem is to design the controller $u$ for a given reference signal $r$ such that

$$\lim_{t \to \infty} e_1(t) = x_d(t) - r(t) = 0$$
$$\lim_{t \to \infty} e_2(t) = \dot{x}_d(t) - \dot{r}(t) = 0 \qquad\qquad (6)$$

where $\dot{e}_1(t) = e_2(t)$. Let $e = [e_1, e_2]^T$, the state-space for error is given as

$$\dot{e} = A_e e + B_e(r, e)u + G(r, \dot{r}, \ddot{r}) \qquad\qquad (7)$$

where $A_e = A$, $B_e(r, e) = [0, (f_1 - k_1 x_0 - k_1(r + e_1) - b_1(\dot{r} + e_2))/M]^T$, $G(r, \dot{r}, \ddot{r}) = [0, (-(k_1 P_0 + k_2)r - (b_1 P_0 + b_2)\dot{r} - \ddot{r})/M]^T$.

## 3   Controller Design

As illustrated in Fig. 2, the controller consists of two components

$$u = u_f + \overline{u} \qquad\qquad (8)$$

where $u_f$ is an interval type-2 fuzzy feedback controller and $\overline{u}$ is feedforward controller compensating for $G(r, \dot{r}, \ddot{r})$ at steady state, i.e., $B_z(r, 0)\overline{u} = -G(r, \dot{r}, \ddot{r})$. Note that at steady state ((6) is achieved), one has

$$\overline{u} = \frac{\ddot{r} + (b_1 P_0 + b_2)\dot{r} + (k_1 P_0 + k_2)r}{-b_1 \dot{r} - k_1 r + f_1 - k_1 x_0}. \qquad\qquad (9)$$

With Taylor series expansion, $\overline{u}_s$ can be simplified as

$$\overline{u} = \theta_1 r + \theta_2 \dot{r} + \theta_3 \ddot{r} \qquad\qquad (10)$$

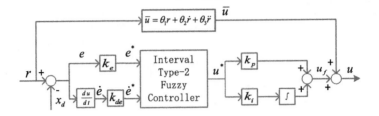

**Fig. 2.** The block diagram of the controller.

where $\theta_1, \theta_2, \theta_3$ are linear parameters.

For the the interval type-2 fuzzy controller, we regard $e^* = k_e e$ and $\dot{e}^* = k_{de} \dot{e}$ as the input to the fuzzy controller where $k_e$ and $k_{de}$ are scale factors.

Different from type-1 fuzzy membership functions (MF), every type-2 fuzzy membership function of interval type-2 controller is an interval. The membership functions (MFs) of $e^*$ and $\dot{e}^*$ are shown in Fig. 3 where the lower bound for each membership function is determined by the parameter $\mu_a$. Take $e^*$ as an example (Fig. 3(a)), it combines three type-2 fuzzy membership functions, i.e., $\tilde{x}_{11}, \tilde{x}_{12}, \tilde{x}_{13}$, with corresponding shadow areas in yellow, blue and green respectively. For the membership function $\tilde{x}_{11}$, $\overline{x}_1$ is the upper bound and $\underline{x}_1$ is the lower bound. The other fuzzy sets follow in the same way. The membership functions can be written as

$$
\begin{aligned}
[\mu_{\underline{x}_{i1}}(X), \mu_{\overline{x}_{i1}}(X)] &= [-\mu_a X, -X] \\
[\mu_{\underline{x}_{i2}}(X), \mu_{\overline{x}_{i2}}(X)] &= \begin{cases} [\mu_a X + \mu_a, X + 1] & (-1 \le X \le 0) \\ [-\mu_a X + \mu_a, -X + 1] & (0 \le X \le 1) \end{cases} \\
[\mu_{\underline{x}_{i3}}(X), \mu_{\overline{x}_{i3}}(X)] &= [\mu_a X, X]
\end{aligned}
\tag{11}
$$

with $i = 1$ for $X = e^*$ and $i = 2$ for $X = \dot{e}^*$.

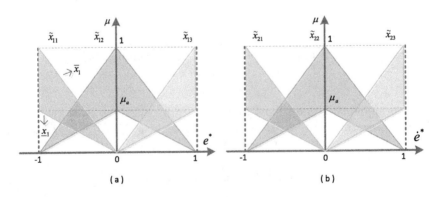

**Fig. 3.** (a) MFs for $e^*$, (b) MFs for $\dot{e}^*$. (Color figure online)

Rulebase of the proposed type-2 fuzzy controller is listed as follows:

$$R^1 : \text{IF } e^* \text{ is } \tilde{x}_{11} \text{ and } \dot{e}^* \text{ is } \tilde{x}_{21}, \text{ THEN } u^* \text{ is } U_1$$
$$R^2 : \text{IF } e^* \text{ is } \tilde{x}_{11} \text{ and } \dot{e}^* \text{ is } \tilde{x}_{22}, \text{ THEN } u^* \text{ is } U_2$$
$$R^3 : \text{IF } e^* \text{ is } \tilde{x}_{11} \text{ and } \dot{e}^* \text{ is } \tilde{x}_{23}, \text{ THEN } u^* \text{ is } U_3$$
$$R^4 : \text{IF } e^* \text{ is } \tilde{x}_{12} \text{ and } \dot{e}^* \text{ is } \tilde{x}_{21}, \text{ THEN } u^* \text{ is } U_4$$
$$R^5 : \text{IF } e^* \text{ is } \tilde{x}_{12} \text{ and } \dot{e}^* \text{ is } \tilde{x}_{22}, \text{ THEN } u^* \text{ is } U_5$$
$$R^6 : \text{IF } e^* \text{ is } \tilde{x}_{12} \text{ and } \dot{e}^* \text{ is } \tilde{x}_{23}, \text{ THEN } u^* \text{ is } U_6$$
$$R^7 : \text{IF } e^* \text{ is } \tilde{x}_{13} \text{ and } \dot{e}^* \text{ is } \tilde{x}_{21}, \text{ THEN } u^* \text{ is } U_7$$
$$R^8 : \text{IF } e^* \text{ is } \tilde{x}_{13} \text{ and } \dot{e}^* \text{ is } \tilde{x}_{22}, \text{ THEN } u^* \text{ is } U_8$$
$$R^9 : \text{IF } e^* \text{ is } \tilde{x}_{13} \text{ and } \dot{e}^* \text{ is } \tilde{x}_{23}, \text{ THEN } u^* \text{ is } U_9$$

where $U_n = [\underline{u}_n, \overline{u}_n]$. Note that we choose $\underline{u}_n = \overline{u}_n$ in this paper. $U_n$ are given in Table 1. The firing interval for each rule is

Table 1. Rulebase for interval type-2 controller.

|  | $\tilde{x}_{21}$ | $\tilde{x}_{22}$ | $\tilde{x}_{23}$ |
|---|---|---|---|
| $\tilde{x}_{11}$ | $U_1 = [-1, -1]$ | $U_2 = [-0.9, -0.9]$ | $U_3 = [0, 0]$ |
| $\tilde{x}_{12}$ | $U_4 = [-0.9, -0.9]$ | $U_5 = [0, 0]$ | $U_6 = [0.9, 0.9]$ |
| $\tilde{x}_{13}$ | $U_7 = [0, 0]$ | $U_8 = [0.9, 0.9]$ | $U_9 = [1, 1]$ |

$$F_n(e^*, \dot{e}^*) = [\underline{f}_n, \overline{f}_n] = [\mu_{\underline{x}_{1i}}(e^*) \times \mu_{\underline{x}_{2j}}(\dot{e}^*), \mu_{\overline{x}_{1i}}(e^*) \times \mu_{\overline{x}_{2j}}(\dot{e}^*)] \tag{12}$$

where $1 \le i \le 3$, $1 \le j \le 3$, $n = 3(i - 1) + j$. Compared with type-1 fuzzy controller, we also need a type-reducer. A center-of-sets type-reducer [18] is used and described as follows

$$u_l = \frac{\sum_{n=1}^{L} \overline{f}_n \underline{u}_n + \sum_{n=L+1}^{N} \underline{f}_n \underline{u}_n}{\sum_{n=1}^{L} \overline{f}_n + \sum_{n=L+1}^{N} \underline{f}_n}$$

$$u_r = \frac{\sum_{n=1}^{R} \underline{f}_n \overline{u}_n + \sum_{n=R+1}^{N} \overline{f}_n \overline{u}_n}{\sum_{n=1}^{R} \underline{f}_n + \sum_{n=R+1}^{N} \overline{f}_n}. \tag{13}$$

By means of Karnik-Mendel algorithms [18], $u_l$ and $u_r$ can be achieved. The crisp output is finally obtained by

$$u^* = \frac{u_l + u_r}{2} \tag{14}$$

and $u_f$ is

$$u_f = k_p u^* + k_i \sum u^* \tag{15}$$

## 4    Experiments

### 4.1    Experiment Setup

The platform for PMA experiments is shown in Fig. 4 where a Festo MAXM-20-AA type PMA is used. The air is supplied by air compressor. When the air flows in and out of PMA and causes the variation of the pressure, it results in the expansion and contraction of the PMA and thus the motion of the load attached at the end of the PMA. The displacement sensor (Firstmark NC 27622) is used to measure the displacement of moving end of the PMA. The sensor data is acquired by the A/D board of dSPACE (dSPACE 1103), and the proposed controller is executed in the dSPACE board, the voltage signal is sent to the proportional valve (Festo VPPM-6L-L-1-G18-0L6H-V1N) via the D/A board for the actuation. The proportional valve is employed to regulate the air pressure in the PMA.

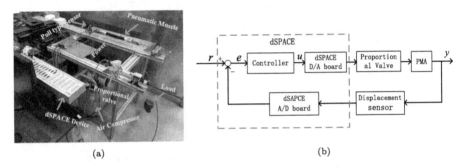

(a)                                                        (b)

**Fig. 4.** (a) The PMA experiment platform, (b) The block diagram of the PMA control system.

Parameters in terms $B, K, F$ are adapted from [8] as follows

$$B = \begin{cases} 2.2685 \times 10^{-4}P + 2435.3 & (inflation) \\ 0.0032P + 2522 & (deflation) \end{cases}$$
$$K = \begin{cases} -0.2132P + 9.0638 \times 10^4 & (P \geq 32542\,\text{Pa}) \\ 0.0105P + 1.8063 \times 10^4 & (P \leq 32542\,\text{Pa}) \end{cases} \tag{16}$$
$$F = 0.0022P - 202.32$$

The nominal pressure and displacement in Eq. (3) are $P_0 = 338536\,\text{Pa}$ and $x_0 = 0.2168\,\text{m}$ respectively.

### 4.2    Simulation

The simulation is conducted to verify the performance of proposed type-2 fuzzy controller. The tracking trajectory is a sine function

$$r(t) = 0.01\sin(2\pi f t) \tag{17}$$

with frequency $f = 0.5\,\text{Hz}$ and amplitude $0.01\,\text{m}$.

The parameters of the controller in (8) are $k_e = 200, k_{de} = 10, k_p = 2 \times 10^4$, $k_i = 2 \times 10^6$ and $\theta_1 = 2 \times 10^3$, $\theta_2 = 3 \times 10^3$, $\theta_3 = 200$. We will vary $\mu_a$ in the fuzzy controller for the purpose of comparison, i.e., $\mu_a = 0.5$ and $\mu_a = 1$. When $\mu_a = 1$, the intervals vanish (see Fig. 3) and the controller turns out to be a type-1 fuzzy controller.

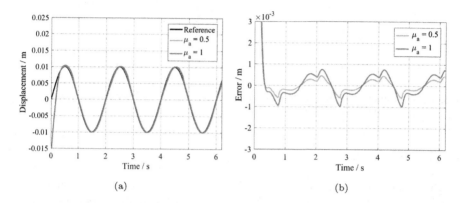

(a)                                    (b)

**Fig. 5.** Simulation results. Tracking performance comparison between the controllers with $\mu_a = 0.5$ and $\mu_a = 1$.

In Fig. 5, the performance comparison of the type-2 fuzzy controller ($\mu_a = 0.5$) and type-1 fuzzy controller ($\mu_a = 1$) are shown. Both simulations start with the same initial condition. As shown in Fig. 5, the approximated tracking can be achieved by both controllers. But it can be seen that the type-2 fuzzy controller ($\mu_a = 0.5$) can achieve better tracking performance than the type-1 fuzzy controller. The tracking error of type-2 fuzzy controller keeps at a lower level of $0.5\,\text{mm}$, while that of type-1 fuzzy controller nearly doubles.

### 4.3 Experiment Validation

In the experiment, the comparison between type-2 fuzzy controller ($\mu_a = 0.5$) and type-1 fuzzy controller ($\mu_a = 1$) is carried out. Moreover, type-2 fuzzy controller with $\mu_a = 0.4$ is added to test the influence of the parameter $\mu_a$. Parameters for experiment are $k_e = 200$, $k_{de} = 3$, $k_p = 5 \times 10^3$, $k_i = 8 \times 10^5$ and $\theta_1 = 2 \times 10^3$, $\theta_2 = 5 \times 10^3$, $\theta_3 = 10$.

Figure 6 illustrates the tracking performances of the three controllers. For the type-2 fuzzy controller with $\mu_a = 0.4$, the maximal error reaches 4 mm. The result of the type-2 fuzzy controller with $\mu_a = 0.5$ is slightly better than that of type-1 controller with $\mu_a = 1$. The corresponding Root Mean Square Error (RMSE) for each controller is presented in Fig. 6(b). It can be conclude that the performance of type-2 fuzzy controller relies on the parameter $\mu_a$, and the type-2 fuzzy controller with a proper $\mu_a$ achieves higher tracking accuracy than type-1 fuzzy controller.

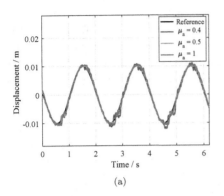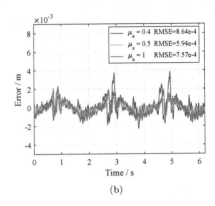

**Fig. 6.** Experiment results. Tracking performance comparison between controllers with $\mu_a = 0.4$, $\mu_a = 0.5$ and $\mu_a = 1$.

## 5    Conclusions

In this paper, an type-2 fuzzy controller is proposed. The fuzzy membership functions of the type-2 controller are intervals, which contain more expert experience and better for handling the uncertainties compared with type-1 fuzzy controller. Both simulation and experiment results illustrated the effectiveness of the proposed controller with the proper $\mu_a$. Our further work will be focused on enriching fuzzy rules of the type-2 fuzzy controller to achieve better performance.

## References

1. Caldwell, D.G., Medrano-Cerda, G.A., Goodwin, M.: Control of pneumatic muscle actuators. IEEE Control Syst. **15**(1), 40–48 (1995)
2. Reynolds, D.B., Repperger, D.W., Phillips, C.A., Bandry, G.: Modeling the dynamic characteristics of pneumatic muscle. Ann. Biomed. Eng. **31**(3), 310–317 (2003)
3. Tondu, B., Ippolito, S., Guiochet, J.: A seven-degrees-of-freedom robot-arm driven by pneumatic artificial muscles for humanoid robots. Int. J. Robot. Res. **24**(4), 257–274 (2005)
4. Ferris, D.P., Czerniecki, J.M., Hannaford, B.: An ankle-foot orthosis powered by artificial pneumatic muscles. J. Appl. Biomech. **21**(2), 189–197 (2005)
5. Hosoda, K., Sakaguchi, Y., Takayama, H., Takuma, T.: Pneumatic-driven jumping robot with anthropomorphic muscular skeleton structure. Auton. Robot. **28**(3), 307–316 (2010)
6. Rus, D., Tolley, M.T.: Design, fabrication and control of soft robots. Nature **521**(7553), 467 (2015)
7. Kawashima, K., Sasaki, T., Ohkubo, A., et al.: Application of robot arm using fiber knitted type pneumatic artificial rubber muscles. In: Proceedings of 2004 IEEE International Conference on Robotics and Automation, ICRA 2004, vol. 5, pp. 4937–4942. IEEE (2004)

8. Zhu, L., Shi, X., Chen, Z., Zhang, H.-T., Xiong, C.-H.: Adaptive servomechanism of pneumatic muscle actuators with uncertainties. IEEE Trans. Ind. Electron. **64**(4), 3329–3337 (2017)

9. Thanh, T.D.C., Ahn, K.K.: Nonlinear PID control to improve the control performance of 2 axes pneumatic artificial muscle manipulator using neural network. Mechatronics **16**(9), 577–587 (2006)

10. Jiang, X., Wang, Z., Zhang, C., Yang, L.: Fuzzy neural network control of the rehabilitation robotic arm driven by pneumatic muscles. Ind. Robot Int. J. **42**(1), 36–43 (2015)

11. Aschemann, H., Schindele, D.: Sliding-mode control of a high-speed linear axis driven by pneumatic muscle actuators. IEEE Trans. Ind. Electron. **55**(11), 3855–3864 (2008)

12. Jouppila, V.T., Gadsden, S., Bone, G.M., et al.: Sliding mode control of a pneumatic muscle actuator system with a pwm strategy. Int. J. Fluid Power **15**(1), 19–31 (2014)

13. Schindele, D., Aschemann, H.: Nonlinear model predictive control of a high-speed linear axis driven by pneumatic muscles. In: American Control Conference, 2008, pp. 3017–3022. IEEE (2008)

14. Bone, G.M., Xue, M., Flett, J.: Position control of hybrid pneumatic-electric actuators using discrete-valued model-predictive control. Mechatronics **25**, 1–10 (2015)

15. Chan, S.W., Lilly, J.H., Repperger, D.W., Berlin, J.E.: Fuzzy PD+I learning control for a pneumatic muscle. In: The 12th IEEE International Conference on Fuzzy Systems, FUZZ 2003, vol. 1, pp. 278–283. IEEE (2003)

16. Chang, M.K.: An adaptive self-organizing fuzzy sliding mode controller for a 2-DOF rehabilitation robot actuated by pneumatic muscle actuators. Control Eng. Pract. **18**(1), 13–22 (2010)

17. Xie, S.Q., Jamwal, P.J.: An iterative fuzzy controller for pneumatic muscle driven rehabilitation robot. Expert Syst. Appl. **38**(7), 8128–8137 (2011)

18. Mendel, J.M.: Uncertain Rule-Based Fuzzy Logic Systems: Introduction and New Directions. Prentice Hall PTR, Upper Saddle River (2001)

19. Wu, D.: A brief tutorial on interval type-2 fuzzy sets and systems. Fuzzy Sets Syst. (2010)

# Sliding-Mode Control of Soft Bending Actuator Based on Optical Waveguide Sensor

Chenglong Liu, Wenbin Chen$^{(\boxtimes)}$, and Caihua Xiong

School of Mechanical Science and Engineering, Huazhong University of Science and Technology,
Wuhan 430073, China
wbchen@hust.edu.cn

**Abstract.** Traditional robots were made of rigid metal materials. But in recent years, various robots based on soft materials and flexible parts have been developed. Compared to rigid actuators such as motors and hydraulic cylinders, soft actuators have the advantages of strong impact resistance and soft interaction, and are potentially attractive in the application related to the human-robot coexisted environment. However, the dynamic deformation of the soft actuator is complex, which brings the difficulty to integrate a proprioceptive sensor for precisely measuring its movement state. What's more, it is a challenge work to establish the mathematical models and analysis controllers for the soft actuator. To overcome these issues, this article develops a waveguide sensor with high sensitivity and low hysteresis, which is made of PMMA optical fiber. Based on the proprioceptive state information of the soft actuator from the optic waveguide sensor, we identify the system model of the soft bending actuator through step response experiments. Moreover, the system model is simplified and linearized, and the adaptive controller is designed by the sliding-mode control theory. Our experimental results show the efficiency of our proposed methods.

**Keywords:** Soft actuator · Optical waveguide sensor · System model · Sliding-mode controller

## 1 Introduction

Recently, soft robots have emerged as people's demands for exoskeleton robotics, human-computer interaction, and sports have increased [1]. As an important part of the soft robot, the soft actuator driven by fluid is made of a soft material or soft structure. Due to the material conformity, soft actuators adapt better to different environments when large continuous output deformations occur. Soft actuators are very adaptable but suffer from lots of challenges in sensing and control of continuously deformable bodies.

In order to measure the bending curvature of the soft actuator, the commercially available flexible bending sensor is conventionally used, but the size of the sensor is difficult to adapt to different requirements of the application. Rebecca K. Kramer et al. fabricate a curvature sensor composed of a thin, transparent elastomer film embedded with a microchannel of conductive liquid and a sensing element [2]. In order to integrate the actuator and sensor, Huichan Zhao et al. use a retractable optical waveguide in the

© Springer Nature Switzerland AG 2018
Z. Chen et al. (Eds.): ICIRA 2018, LNAI 10984, pp. 432–441, 2018.
https://doi.org/10.1007/978-3-319-97586-3_39

prosthetic hand for hand flexion, elongation, and pressure measurements [3]. The control of soft robots has always been a challenge because of structural uncertainty and large deformations. Giada Gerboni et al. design a feedback controller using a low-pass filter and a proportional-integral controller through a commercial sensor and ground electromagnetic tracking system [4]. Frederick Largilliere et al. achieve real-time control of soft robots through asynchronous finite element modeling [5]. Ming Luo et al. present a sliding-mode controller that directly drives the binary solenoid valve states to steadily hold the actuator without continuous pressurization and depressurization [6, 7]. In addition, considerable work has been done on the control of soft robots [8–11].

With respect to the proprioceptive curvature sensor, most of work in literature cannot guarantee the output linearity along with curvature variation, because the conductive ink based flexible sensor and elastomer based optic waveguide are both easily affected by the induced inner stress of the fabricated materials under inflated fluid pressure. Most of the modelling work related to the soft actuator do not consider nonlinear characteristics of the system in varied inflated pressure, thus the resulted model is too much simple and cannot reflect the dynamic performance well in wide range of inflated pressure. In this work, we develop a curvature sensor made of PMMA (Polymethylmethacrylate) optical fiber which is flexible and free of the effect from the inner stress under the inflated pressure. The system dynamic model is proposed based on the identified pressure dependent parameters by the step response experiment. By using the Taylor expansion around the input pressure, we give the state function corresponding to the system model with locally linearized the controlling variable. The sliding model controller is thus developed based on Lyapunov function. The trajectory tracking experiment verify the performance of the proprioceptive curvature sensor and validate the effect of the proposed system model.

## 2    Fabrication of Soft Actuator

The soft actuators are pneumatic driven to perform only bending motion. The soft actuator consists of two cast together chambers with a sinusoidal contour (see Fig. 1) for bidirectional bending deformation. Each chamber is reinforced with Kevlar fibers which are wound around the crest of the sinusoidal contour and crossly wound in the bottom to next crest. The non-stretchable glass fiber and the flexible optic waveguide are embedded in the middle layer between two chambers. Due to the strain limitation from non-stretchable glass fiber in the middle layer and the crest bulging limitation from the reinforced fiber, the soft actuator cannot elongate along the axis direction but only bend to one side when inflating one chamber.

Differing from the elastomeric waveguides which were used in strain sensing and position sensing [12, 13], the optic waveguide in this work is made of PMMA (Polymethylmethacrylate) fiber which is flexible to arbitrarily bend and hard to radial deformation. Thus, the cross sectional shape of the optic waveguide is not affected by the introduced inner stress of multi-materials under the inflated pressure. This feature eliminates the additionally nonlinear issues of light transmission due to the irregular deformation of the cross section. As the optic light quantity transmitted from one end to

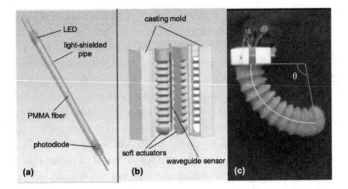

**Fig. 1.** (a) Structure of the waveguide sensor, (b) Structure of the soft actuator, (c) Image of the soft actuator and definition of bending angle.

another end is seriously dependent on the curved degree of the path when the light refraction is considered to be comparable with the light reflex. The more the path curves, the great the light amount losses. Thus, the amount of the light loss along its transmitted path directly reflects the degree of the path curvature. So, the optic waveguide should have the micro crude structure to let the light reflection for intentionally losing light. In this work, we uniformly scrape off the surface of the commercially available PMMA optic fiber. Then, the transmitter and the receiver which emits and receives the light are encapsulated respectively at the two ends of the waveguide, as show in Fig. 1(a). Finally, we integrate waveguide sensors into soft actuators, as show in Fig. 1(b).

We conduct series of simulation experiments on different structural parameters of soft actuator. The work of structural parameters optimization of soft actuators is not the focus of this paper, so we only give the final result. And the result of simulation experiments is shown in Fig. 2. We can observe that the soft actuator is bent about 270° when inflating one chamber 0.1 Mpa.

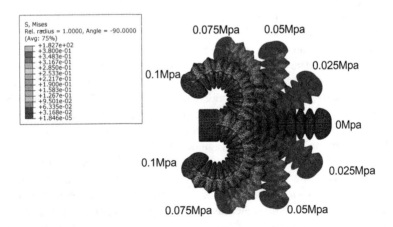

**Fig. 2.** The simulation results when separately inflating the two chambers under different pressures.

We define the output power of a waveguide without bending deformation as the baseline power $V_0$. With the output power as $V$, the output power loss in decibels is then defined as

$$L = 10 \log_{10} \left( V_0/V \right) \tag{1}$$

The central angle formed by the actuator centerline is defined as the bending angle shown in Fig. 1(c). Through the definition (1), we can calibrate the bending angle and light power loss. As shown in Fig. 3, we inflate one chamber under different pressures and record the subtended angle of the soft actuator and the corresponding value of output power loss by increasing 5° each time from 0 to 180°. From this result, we can get the particular relationship between bending angle and output power loss of sensor.

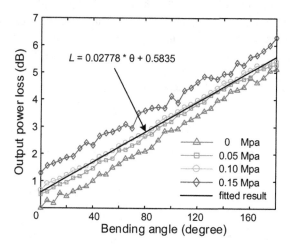

**Fig. 3.** Output power loss of sensor during bending artificially specified angles.

## 3    System Parameters Identification

The step response experiment is employed to identify the system parameters [14, 15]. The general procedure is to measure the response curve under each step input and further fit the model parameters of the system by least square.

The dynamic behavior of the soft actuators is governed by a higher-order nonlinear model. However, we simplify the system into a second-order nonlinear system which approximate the real dynamic behavior. The simplified system can be described as follows:

$$\ddot{x}(t) = a(u)\dot{x}(t) + b(u)x(t) + c(u)u \tag{2}$$

where $x(t)$ is the bending angle of soft actuator, $u$ is the input pressure. The parameters $a(u)$, $b(u)$, and $c(u)$ are nonlinearly related to the input pressure and reflect the characteristic of the flexible structure of the actuator. In order to obtain detailed formulas of these parameters, we perform step response experiments using different pressure inputs and fit them.

The time-domain representation of bending angle in step response can be generally formulated as:

$$x(t) = A_0 - A_1 e^{-\lambda_1 t} - A_2 e^{-\lambda_2 t} \tag{3}$$

which can be used to estimate $A_0$, $\lambda_1$ and $\lambda_2$ by fitting the response curve at a certain input pressure using least-squares techniques. For a specified input pressure, the least-square techniques is employed to obtain the particular value of the parameters $A_0$, $\lambda_1$ and $\lambda_2$. The fitting procedure is repeated for different step response with input pressure from 0.085 to 0.194 MPa in steps of 0.1 MPa. Then, a collection of $A_0$, $\lambda_1$ and $\lambda_2$ under different input pressure $u$ can be obtained (see Figs. 4 and 5). And the collection of $A_1$ and $A_2$ is not given because it is not used in later calculations. Thus, the corresponding relationship between the three parameters and the input pressure can be approximately established respectively. The fitted expressions of $A_0(u)$, $\lambda_1(u)$ and $\lambda_2(u)$ are estimated as follows:

$$\begin{aligned} A_0 &= \alpha_0 u - \alpha_1 u^2 \\ \lambda_1 &= \beta_0 - \beta_1 u \\ \lambda_2 &= \gamma_0 - \gamma_1 u \end{aligned} \tag{4}$$

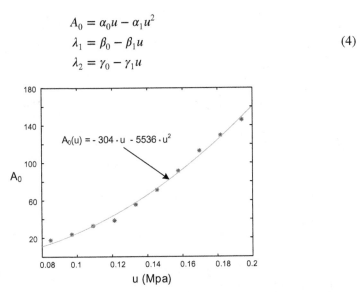

**Fig. 4.** The relationship between the parameter $A_0$ and input pressure $u$.

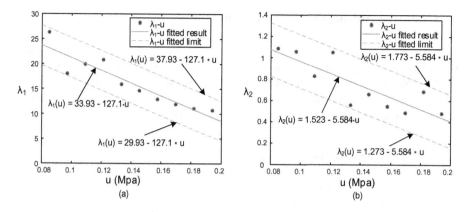

**Fig. 5.** The relationship between the parameter $\lambda_1$, $\lambda_2$ and input pressure $u$.

As shown in Figs. 4 and 5, the estimation of $A_0(u)$ is close to the experimental data compared to the estimation of $\lambda_1(u)$ and $\lambda_2(u)$. Two limit straight lines are drawn to determine the margin of error, and we can regard this error as the disturbance of the system.

Thus, the system parameter in (2) can be expressed as $a(u) = -(\lambda_1 + \lambda_2)$, $b(u) = -\lambda_1\lambda_2$, $c(u) = \lambda_1\lambda_2A_0/u$. Considering the system error, the differential equation of the system model (2) is reformulated as follows:

$$\ddot{x}(t) = -(\lambda_1 + \lambda_2)\dot{x}(t) - \lambda_1\lambda_2 x(t) + \lambda_1\lambda_2 A_0 + d(t) \tag{5}$$

Where $dt$ is the system model error. We define the error of $\lambda_1(u)$ and $\lambda_2(u)$ as $d\lambda_1$ and $d\lambda_2$ respectively. Thus, the value of $dt$ can be derived as:

$$d(t) = -(d\lambda_1 + d\lambda_2)\dot{x}(t) + d\lambda_1 \cdot d\lambda_2 (A_0 - x(t)) \tag{6}$$

From the Fig. 5, $d\lambda_1$ and $d\lambda_2$ are given:

$$\begin{aligned} |d\lambda_1| \le 4 \\ |d\lambda_2| \le 0.25 \end{aligned} \tag{7}$$

And from the general performance of the system, we suppose:

$$\begin{aligned} |\dot{x}(t)| \le 1000^\circ/s \\ |A_0 - x(t)| \le 250^\circ \end{aligned} \tag{8}$$

Then maximum fluctuation range of $dt$ can be determined by:

$$|d(t)| < D, D = 4500 \tag{9}$$

## 4  Design of Sliding-Mode Controller

Sliding mode control is also called variable structure control and is especially suitable for dealing with the system with parameter indeterminacy. However, it generally require that the system is linear about the control variable. As shown in (5), the system model is nonlinear about the control variable. Thus, we need to simplify it linearly. By using first order Taylor expansions, the parameters $a(u)$, $b(u)$ and $c(u)$ can be approximated as:

$$a(u) = \left( a(u_0) + \frac{da(u_0)}{du} \cdot \Delta u \right), \Delta u = u - u_0$$

$$b(u) = \left( b(u_0) + \frac{db(u_0)}{du} \cdot \Delta u \right), \Delta u = u - u_0 \qquad (10)$$

$$c(u) = \left( c(u_0) + \frac{dc(u_0)}{du} \cdot \Delta u \right), \Delta u = u - u_0$$

where $u_0$ is the reference input pressure used to simplify the system model near the input pressure. We limit $\Delta u$ to a small range in the controller so that small errors will occur by using the simplification formula.

We define functions $g(\dot{x}, x)$, $h(\dot{x}, x)$ as

$$g(\dot{x}, x) = a(u_0)\dot{x} + b(u_0)x + c(u_0)$$

$$h(\dot{x}, x) = \frac{da(u_0)}{du}\dot{x} + \frac{db(u_0)}{du}x + \frac{dc(u_0)}{du} \qquad (11)$$

Then the system model becomes

$$\ddot{x} = g(\dot{x}, x) + h(\dot{x}, x)\Delta u + d(t), \Delta u = u - u_0 \qquad (12)$$

For the trajectory tracking problem, if the desired trajectory is $x_d$, the sliding-mode function can defined as

$$s = ce + \dot{e} \qquad (13)$$

where $c > 0$ is positive constant and $e = x_d(t) - x(t)$ is the error in current position and desired position. The real-time bending angle $x(t)$ is measured by the optical waveguide sensor, the relationship between bending angle and output of sensor can be fitted by the data shown in Fig. 3. We define Lyapunov function as:

$$V = \frac{1}{2}s^2 \qquad (14)$$

Then

$$\dot{V} = s\dot{s} = s\left( c\dot{e} + \ddot{x}_t - g - h \cdot t - d(t) \right) \qquad (15)$$

In order to ensure the stability of the system, we can design the controlling input as

$$u_n = u_{n-1} + \frac{1}{h}\left(c\dot{e} + \ddot{x}_d - g + \eta s + D\,\mathrm{sgn}(s)\right) \tag{16}$$

where $u_n$ is the control input at the $n$th calculation, and $\eta > 0$ determines the convergence velocity of the controlled trajectory to the desired trajectory.

## 5  Trajectory Tracking Experiment

In order to verify the controlling effect of the designed controller for the fabricated soft actuator, we let the soft actuator track a reference sine trajectory. Although the fabricated soft actuator can bend in two opposite direction by giving different pressure to the two chambers, here, we do not consider the situation of simultaneously inflating two chambers, but only inflating just one chamber to verify the controlling effect. In the experiments, we use a sine signal to test the performance of the controller. The reference sine trajectory is with amplitude $a = 30°$ and frequency $\omega = 0.4\pi \ rad/s$. The parameters for the sliding-mode controller are $c = 10, \eta = 50$. In order to show the controlling performance, we compare the resulted trajectory of the proposed sliding-mode controller and PID controller. We conduct experiments in different parameters of PID controller and the best parameters are $P = 6.5 \times 10^{-3}$, and $D = 8.5 \times 10^{-2}$. The composition of the experimental platform is shown in Fig. 6.

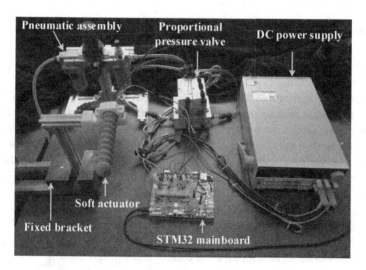

**Fig. 6.** The composition of the experimental platform.

As shown in Fig. 7, it is observed that the tracking error of the PID controller is significantly larger than sliding-mode controllers. The PID controller shows poor following performance due to the hysteresis of the pneumatic system. Comparing with the result of the PID controller, the tracking result of sliding-mode controller is closer to the reference trajectory. Because there are underdamped poles which are not considered in the system identification, the chattering of tracked trajectory cannot be avoided.

We will improve it by considering the stiffness control of the soft actuator, for example simultaneously inflating two chambers to regulate the system dampness, in future work.

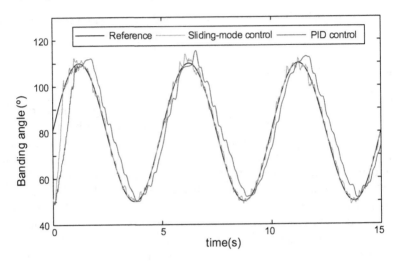

**Fig. 7.** Experiment results. Tracking performance comparison among the sliding-mode and PID controllers for a sinusoidal reference signal.

## 6    Conclusion and Future Work

In this paper, we propose a bidirectional bending soft actuator which has the proprioceptive curvature sensing capability by using the PMMA optic waveguide. The reliability of the optic waveguide about the linear output between light loss and bending angle under different inflated pressure is verified in this study. Aimed at the controlling issue of such soft actuator, we propose a simplified second order dynamic model of the soft actuator based on the step response experiment. By using the Taylor expansion, the linearized state function related to the control input is obtained to establish the sliding mode controller. The experimental results validate the effectiveness of the proposed controller.

For the soft actuators, accurate bending deformation control is a well-known issue in literature due to the high nonlinear characteristics. In the future work, we will consider precise control issues related to the developed soft actuator, including chatter depressing in the trajectory control, interaction force control and stiffness control.

**Acknowledgment.** This work was partially supported by the National Natural Science Foundation of China (Grant No. 91648203 and No. 51335004), the International Science & Technology Cooperation Program of China (Grant No. 2016YFE0113600) and the Natural Science Foundation of Hubei (Grant No. 2018CFB431).

# References

1. Cho, K.J., Koh, J.S., Kim, S., et al.: Review of manufacturing processes for soft biomimetic robots. Int. J. Precis. Eng. Manuf. **10**(3), 171–181 (2009)
2. Kramer, R.K., Majidi, C., Sahai, R., et al.: Soft curvature sensors for joint angle proprioception. In: IEEE International Conference on Intelligent Robots and Systems, pp. 1919–1926 (2011)
3. Zhao, H., O'Brien, K., Li, S., et al.: Optoelectronically innervated soft prosthetic hand via stretchable optical waveguides. Sci. Robot. **1**(1), eaai7529 (2016)
4. Gerboni, G., Diodato, A., Ciuti, G., et al.: Feedback control of soft robot actuators via commercial flex bend sensors. IEEE/ASME Trans. Mechatron. **22**(4), 1881–1888 (2017)
5. Largilliere, F., Verona, V., Coevoet, E., et al.: Real-time control of soft-robots using asynchronous finite element modeling. In: 2015 IEEE International Conference on Robotics and Automation (ICRA), pp. 2550–2555 (2015)
6. Luo, M., Skorina, E.H., Tao, W., et al.: Toward modular soft robotics: proprioceptive curvature sensing and sliding-mode control of soft bidirectional bending modules. Soft Robot. **4**(2), 117–125 (2017)
7. Skorina, E.H., Luo, M., Tao, W., et al.: Adapting to flexibility: model reference adaptive control of soft bending actuators. IEEE Robot. Autom. Lett. **2**(2), 964–970 (2017)
8. Niu, Y., Ho, D.W.C., Lam, J.: Robust integral sliding mode control for uncertain stochastic systems with time-varying delay. Automatica **41**(5), 873–880 (2005)
9. Polygerinos, P., Wang, Z., Overvelde, J.T.B., et al.: Modeling of soft fiber-reinforced bending actuators. IEEE Trans. Robot. **31**(3), 778–789 (2015)
10. Trivedi, D., Lotfi, A., Rahn, C.D.: Geometrically exact models for soft robotic manipulators. IEEE Trans. Robot. **24**(4), 773–780 (2008)
11. Majidi, C., Kramer, R., Wood, R.J.: A non-differential elastomer curvature sensor for softer-than-skin electronics. Smart Mater. Struct. **20**(10), 105017–105023 (2011)
12. Begej, S.: Planar and finger-shaped optical tactile sensors for robotic applications. IEEE J. Robot. Autom. **4**(5), 472–484 (1988)
13. Lagakos, N., Schnaus, E.U., Cole, J.H., et al.: Optimizing fiber coatings for interferometric acoustic sensors. IEEE J. Quantum Electron. **30**(4), 683–689 (1982)
14. Zhu, L., Shi, X., Chen, Z., et al.: Adaptive servomechanism of pneumatic muscle actuators with uncertainties. IEEE Trans. Ind. Electron. **64**(4), 3329–3337 (2017)
15. Polygerinos, P., Wang, Z., Overvelde, J.T.B., et al.: Modelling and control for soft finger manipulation and human-robot interaction. IEEE Trans. Robot. **31**(3), 778–789 (2010)

# Design and Experiment of a Fast-Soft Pneumatic Actuator with High Output Force

Amin Lotfiani[1,2], Xili Yi[1,2], Zhao Qinzhi[1,2(✉)], Zhufeng Shao[1,2(✉)], and Liping Wang[1,2]

[1] Beijing Key Lab of Precision/Ultra-precision Manufacturing Equipment and Control, Tsinghua University, Beijing 100084, China
zqz13@mails.tsinghua.edu.cn, shaozf@tsinghua.edu.cn
[2] State Key Laboratory of Tribology and Institute of Manufacturing Engineering, Tsinghua University, Beijing 100084, China

**Abstract.** This study proposes a novel high-speed and heavy-load soft pneumatic actuator, inspired by a snake that stores energy using its wavy body configuration. The snake-like soft pneumatic actuator consists of a wavy chamber attached to an inextensible layer. The finite element model is developed to analyze the static behavior. Since the wavy and continuous chamber is difficult to fabricate, a new fabrication method has been developed. Experiments conducted to evaluate the bending behavior, speed, output force, and payload capacity of the actuator in different air pressures. Experiment outcomes are in good agreement with the simulation results and illustrate that the proposed soft pneumatic actuator achieves a complete bending in only 65 ms with outstanding output force and load capacity.

**Keywords:** Soft pneumatic actuator · Design · Fabrication · High-speed
High output force

## 1 Introduction

Soft robotics is a branch of robotic science in which the robot is produced from soft materials. Compared with the traditional rigid robot, the soft structure introduces significant advantages in terms of adaptability, passive compliance, control simplicity, and fabrication cost. Additionally, soft robots provide a safe and bright solution for human–robot interaction. As a research hotspot of soft robotics, soft hands are comprised of soft actuators (SAs) as fingers operated in different ways depending on material composition [1]. Many researches have been developed recently to use different types of SAs in various areas from industrial hands for grasping to soft finger and gloves with rehabilitation applications [2–4]. Among all SAs, using compressed air to inflate the chamber in a silicone-based body is a promising way to produce cost-effective and fast actuators.

---

**Electronic supplementary material** The online version of this chapter (https://doi.org/10.1007/978-3-319-97586-3_40) contains supplementary material, which is available to authorized users.

Z. Chen et al. (Eds.): ICIRA 2018, LNAI 10984, pp. 442–452, 2018.
https://doi.org/10.1007/978-3-319-97586-3_40

Furthermore, owing to high biocompatibility of silicone, this kind of soft pneumatic actuator (SPA) has gradually become popular.

Many studies have been conducted to develop various SPAs to provide fast and stable grasping as soft fingers. The motion of SPAs results from elastomer elongation. Thus, mechanically programming the SPA pressurized configuration is possible by limiting strains in different directions. As a result, the SPA can generate axial motion [5, 6], twisting, or bending [7]. Bending capability is required for the SPA to be used as the finger of a soft hand. Bending can be produced by asymmetrical design with limited elongation on one side of the actuator. Asymmetrical cross section [8, 9] or a sleeve-like cover with several incisions can be adopted [10, 11]. Cross-sectional shape is also an important parameter that can affect the bending angle [9].

To meet grasping requirements, a soft finger should be able to bend very fast while having acceptable stiffness to perform a reliable grasping. Other than the amount of air pressure and silicone material properties (hardness, maximum elongation before break, and so on), the bending speed and stiffness of an SPA depend on the actuator design. Pneumatic network (PneuNet) [12] and PneuFlex [13] are designed to improve the grasping features of soft hands. However, the continuous chamber of these actuators demands a relatively large volume of gas for actuation, and consequently, the bending speed is not ideal [14]. One way to increase the speed of actuation is adopting separated chambers to make a fast pneumatic network (FPN) [14]. In this way, the required air volume for a complete bending decreases considerably, which, alongside the interaction force between two inflated adjacent units, leads to full bending in about 50 ms under the pressure of 345 kPa. Although FPN provides fast actuation, the input pressure is quite high, and the separated chambers decrease the stiffness, consequently leading to low stability and output force.

The weak structure of SPAs has also limited their application. One solution to address this issue is using passive or active particle/layer jamming to increase the stiffness of SPAs [15]. A vacuum is used to compress the particles in active particle jamming, whereas the particles are pressed to each other by the pressurized air in actuator air channel in passive jamming. In the latter method, the actuator becomes harder as the bending angle increases [16]. Both passive and active jamming solutions have been proven effective in increasing the stiffness of SPAs. However, the complex structure hinders the bending motion and slows down the action speed. The development of high-speed and heavy-load SPAs is the frontier of soft robotics and deserves further study.

A novel fast SPA with high output force is proposed in this article. The structure of this actuator is inspired by nature. Experiments are carried out to verify the performance of the proposed SPA. The remainder of this paper is organized as follows. The design concept of snake-like SPA (SSPA) is presented in the following sections. Unit design is discussed with the aim of choosing the best shape for the actuator. The behavior of the actuator is simulated with finite element modeling (FEM) method. Experiments are carried out to assess the actuator characteristics including bending angle, speed, output force, and payload capacity in real situation. Conclusion and discussion of this paper are listed in Sect. 5.

## 2   Design of Snake-Like Pneumatic Actuator

### 2.1   Conceptual Design

Like many other soft actuators, clues for the design of fast SPA with high output force are provided by nature. The radiated rat snake can quickly shoot its semi-soft body toward the bait. Before attacking, the snake curls up the front part of its body, as shown in Fig. 1. This snake benefits from a wavy configuration to store and release energy very rapidly. The SSPA inspired by the snake consists of a wavy silicone chamber as the curled-up body. When one side of the chamber is attached to an inextensible layer, the actuator can generate bending motion.

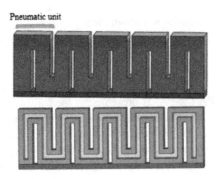

**Fig. 1.**   Curled-up body of radiated rat snake [17]

Figure 2 shows the SSPA is composed of a wavy chamber and an inextensible layer. Similar to the snake, the wavy chamber of SSPA can open very quickly with the pressurized air. When SSPA is inflated, the tendency of the wavy shape to unfold in company with strain in chamber walls results in a fast actuation, which is bending in the presence of the inextensible layer. From another point of view, the wavy chamber can be considered as the fusion of traditional continuous and separated chamber. Therefore, it is expected to generate high output force and fast actuation.

Pneumatic unit

**Fig. 2.**   Snake-like pneumatic actuator, wavy chamber (green), air channel (yellow), and inextensible layer (blue) (Color figure online)

## 2.2 Chamber Unit Design

Chamber is the kernel part of SSAP, which consists of several pneumatic units (Fig. 2). The deformation of the whole chamber is equal to the sum of all the units. The geometry optimization is carried out on the unit instead of the whole chamber to improve the simulation efficiency of FEM. The amount of deformations is only required in this stage. Therefore, attaching inextensible layer to the pneumatic unit is not needed in simulations. Circular, rectangular, and triangular units have been simulated and analyzed to find the best unit shape with the highest deformation in pressurized status. The height, length, width, and the section geometric parameters including wall thickness and air chamber height for all the three units are kept constant. Units are simulated in the ABAQUS software (details on simulation are illustrated in Sect. 2.3). Simulation results indicate that rectangular and circular units produce more elongation than the triangular unit (Fig. 3) because of their higher circumference. However, the circular unit tends to move up when opening, which is not desirable when bending is required. Finally, the rectangular unit is adopted to develop the final SSPA.

**Fig. 3.**  FEM results of circular, triangular, and rectangular pneumatic units.

## 2.3 FEM of the SSPA

A complete actuator with the rectangular unit shape is developed and simulated in this section. FEM approach is used to predict the performance of the actuator. Prior to simulation, material properties of the silicone must be identified. In this study Dragonskin 0030 (Smooth-on Inc.) shore hardness 30A was used as the silicone rubber. Samples of Dragonskin 0030 for both tensile and compression were tested according to ASTM D636 (Type IV) standard. Then, the average of tests' results was used to find the hyperelastic material model according to [18]. The Yeoh model with N = 2, C01 = 0.1122 MPa, and C02 = 0.025 MPa matches the test data. After identifying the silicone material model, FEM can be employed to simulate SSPA static behavior. Gravity was applied to the SSPA during simulations to match the model with real situation. The surface perpendicular to the actuator axes was constrained by encastre boundary conditions to fix the actuator at one end. The air inlet was not considered in the simulation, and the pressure was applied on the inner surface of the wavy air channel. Different unit lengths and heights were examined to find the best with highest bending angle versus pressure while keeping the actuator volume at the lowest possible change. The SSPA's configuration was selected according to the simulation results. The SSPA is composed of five rectangular units and an inextensible layer. The thickness of all the walls and the air channel height are 2 mm. The width, height, and length of the actuator are 20, 23, and 79 mm respectively. Simulation results of bending angle are presented in Figs. 7 and 8.

## 3　Fabrication

Novel wavy structure of the SSPA benefits the performance but challenges the fabrication process, and a new fabrication method must be developed. Generally, two ways are available in making a soft actuator: casting silicone with a mold or printing directly with silicone 3D printer. The silicone 3D printer is neither mature in technology nor popular. Due to the lack of support for the walls over the air channel, the existing silicone 3D printer cannot print the chamber of SSPA. Casting seems to be the only way. However, the wavy chamber is impossible to be fabricated with the ordinary casting method. A four-part mold was designed to solve this problem (Fig. 4). The idea is casting the liquid silicone in the assembled mold. After the silicone was cured, the wavy part (part 4), which was printed with thermoplastic polyurethane (TPU), was easily extracted from the chamber. The other parts (parts 1 to 3) were printed with polylactic acid (PLA) material. Two holes, which were covered by the same half cured silicone or a silicone glue, were left at both ends. Finally, the inextensible layer was attached to the chamber. This layer was made of silicon with cotton bandage embedded inside. Two major problems were observed when adopting papers as the extensible layer: 1. The paper does not stick to silicone very well. 2. After several bending actions, the paper cracks and makes the bottom layer extensible. A simple cotton bandage was selected as the inextensible layer to address these issues. The bandage can join with silicone very well and survive for a longer time compared with paper. Moreover, it is lighter than paper and has less resistance to bending. The whole manufacturing process of the SSPA is illustrated in Fig. 5. Attaching the open chamber to the inextensible layer in FPN fabrication increases the probability of leakage or even breakage from the connection surface. The closed air chamber of SSPA is a promising solution for such problems. Moreover, the closed air chamber eliminates air channel blockage which may happen when connecting the FPN chamber to the inextensible layer.

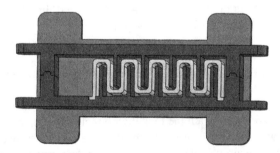

**Fig. 4.** SSPA mold: part 1: mold wall 1 (red); part 2: mold wall 2 (blue); part 3: mold bottom cap (green); and part 4: wavy part (yellow). (Color figure online)

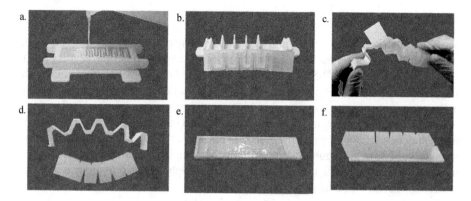

**Fig. 5.** Fabrication process of the SSPA: (a) silicone casting, (b) cured chamber including wavy part, (c) extracting wavy part from silicone chamber, (d) SSPA chamber and wavy part, (e) inextensible layer with embedded bandage, and (f) chamber connected to the inextensible layer

## 4   Experiments

FEM is a helpful method to analyze the static behavior of the actuator such as the relationship between bending angle and input pressure, but performing dynamic analysis of SPA with FEM is difficult. The experiment on bending angle was carried out first to verify the simulation results with FEM. Then, typical experiments were developed to determine the SSPA characteristics of speed and load. One of the significant dynamic characteristics of SPAs is speed of actuation (bending speed). Several experiments were carried out to find the bending speed of SSPA and verify the FEM results in Sect. 4. SSPA was connected to a robotic arm via a 3D printed connector to fix one end of the actuator and set the actuator axel in any required direction. A big air tank was used to store pressurized air. A pressure sensor connects SSPA to the air tank, and an Arduino micro controller was employed to read the data from the sensor (Fig. 6). Experiments were performed to find bending angle, speed of actuation, output force, and maximum payload capacity of the actuator in different amounts of inflation pressure.

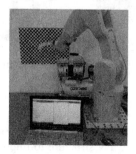

**Fig. 6.**  Experimental set up

### 4.1 Bending Angle

The actuator is fixed at one end to find the bending angle of the SSPA in practice. In this way, the displacement of the other end of the actuator (the tip) shows the bending angle. The simulations results of SSPA inflation are presented in Fig. 7 with blue points. The SSPA was inflated with corresponding pressure as the simulation, and the actual bending angles were measured and presented in Fig. 7 with red squares. The high coincidence of simulation and experiment results indicates good accuracy of FEM model. Furthermore, the complete bending of the SSPA is reached at only 60 kPa, which is a low pressure in SPA field. Figure 8 demonstrates the complete bending of the SSPA. An initial bending angle (angle before inflation) in the graph is related to the gravity effect which depends on the actuator weight and stiffness.

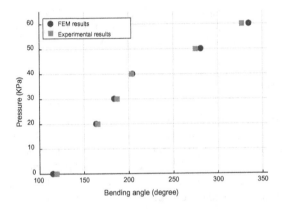

**Fig. 7.** Bending angle of the actuator in different input pressures (Color figure online)

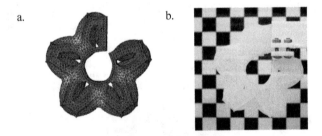

**Fig. 8.** Complete bending with 60 kPa pressure. (a) FEM simulation result, (b) Experiment result

### 4.2 Speed of the SSPA

As mentioned above, speed of actuation is a key index to evaluate SPAs' performance. The SSPA is designed with the wavy structure to improve the action speed, and it is expected to bend very quickly after inflation. In the speed experiment, the bending action was recorded with a high-speed camera (1,000 fps) in different input pressures to determine the actual speed of the SSPA. Complete bending times and the related input pressures are presented in Table 1. Results indicate that the actuator can bend very quickly even at a low pressure of 60 kPa. In addition to the wavy shape of the chamber, the consistent large cross-section area of the air channel helps the air spread in the chamber rapidly and smoothly. The fastest actuation achieved in maximum allowable pressure of 200 kPa took 65 ms. Experiments in higher pressure were limited by material properties (300% maximum elongation of Dragonskin 0030). Another set of experiments was carried out when the actuator was horizontally installed to show the effect of gravity on the speed of actuation. The actuator bending time decreased to 68 ms under the pressure of only 140 kPa. Three frames of the fastest recorded actuations of vertical and horizontal installations are given in Fig. 9 to illustrate the actuation configuration during inflation.

**Table 1.**  Speed of actuation under vertical installation

| Input pressure (KPa) | Time (ms) |
|---|---|
| 60 | 80 |
| 100 | 75 |
| 200 | 65 |

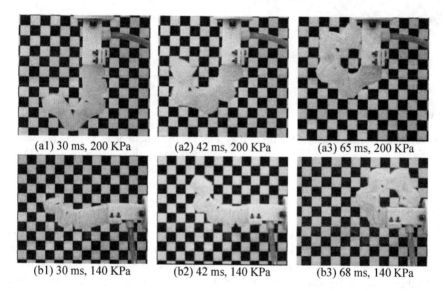

|  |  |  |
|---|---|---|
| (a1) 30 ms, 200 KPa | (a2) 42 ms, 200 KPa | (a3) 65 ms, 200 KPa |
| (b1) 30 ms, 140 KPa | (b2) 42 ms, 140 KPa | (b3) 68 ms, 140 KPa |

**Fig. 9.**  SSPA actuation modes: above; vertical installation; below: horizontal installation

Experiments show that almost the same circular shapes (bending curvature) are obtained under different pressures, whereas final states of FPN alter with pressure as reported in [11].

## 4.3   Output Force

Another important parameter for evaluating the actuator capability is the maximum output force. This is an effective way to measure the actuator body strength. The same approach as in [6] was adopted to test such capability. The force at the distal tip of the actuator is measured while the vertical motion of the actuator units is constrained by a hard plate. In this study, a high accuracy scale was used to measure the force, as shown in Fig. 10. Force was measured at the tip of the actuator. As the pressure increased, the scale showed higher force at the tip of the actuator. Measured forces in Fig. 11 demonstrate higher output force of SSPA than the force–pressure data in [8]. The maximum force is 22.3 N under 100 kPa input pressure, whereas the maximum force reported in [6] is less than 10 N under about 240 kPa input pressure. Although the finger structure and dimension in [6] are different from the SSPA, comparing their performances is possible. With the recorded force-pressure data, curve fitting was used to find the best fitting function for the data. The cubic polynomial can perfectly fit the data, and it can be used to predict the actuator output force under higher or lower pressures.

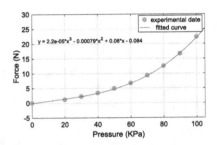

**Fig. 10.** Force measurement at the tip of the actuator.

**Fig. 11.** Actuator Forces vs input pressure. A cubic polynomial is perfectly fitted to the experimental data.

## 4.4   Payload Capacity

Payload capacity is also used as an index to evaluate the performance of the soft actuator. This index is important especially when the actuator is used as a finger of a grasping hand. The actuator was pressurized to an initial pressure to hold the weight to find the payload capacity of the SSPA. Then, the pressure increased gradually until the actuator reached complete bending. Four weights, 100, 200, 400, and 595 g, were tested, and the results are presented in Fig. 12 demonstrate that the SSPA can handle above loads under 80, 85, 100, and 110 kPa, respectively, which are less than half working pressure for SPAs [8]. This experiment not only shows the high-load capacity of the SSPA at low pressure but also indicates the stability of the actuator under load (Movie S1).

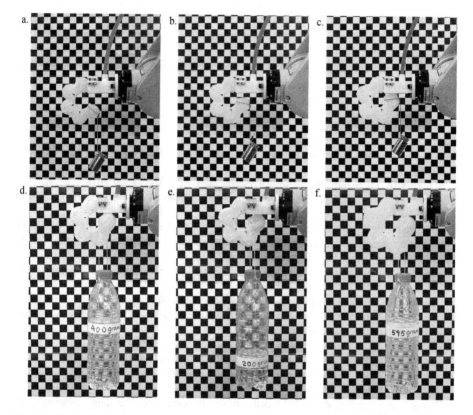

**Fig. 12.** Payload capacity experiments: (a) weight: 100 g, pressure: 60 kPa; (b) weight: 100 g, pressure: 70 kPa; (c) weight: 100 g, pressure: 80 kPa (complete bending); (d) weight: 200 g, pressure: 85 kPa; (e) weight: 400 g, pressure: 100 kPa; and (f) weight: 595 g (full bottle), pressure: 110 kPa.

## 5    Conclusions

In this study, a new SPA with advantages of high speed and large ratio of load to pressure was proposed, inspired by the energy storage of radiated rat snake. The SSPA consists of a wavy chamber attached to an inextensible layer. The wavy chamber challenges the ordinary manufacturing process of SPAs. Therefore, a novel way was implemented to fabricate the SSPA chamber. Compared with the FPN, the closed wavy chamber improves the performance of the actuator by eliminating the leakage and breakage on the connection surface of the chamber and inextensible layer.

A finite element model of the actuator was developed in ABAQUS software to investigate static behavior of the actuator. An experimental platform was prepared to evaluate the accuracy of the model and other key performances of the actuator. Typical experiments were carried out to measure the bending angle, speed of actuation, output force, and payload capacity under different input pressures. Experiment results of bending angle verify good accuracy of the FEM simulation. Moreover, the proposed

SSPA can bend completely in 65 ms under 200 kPa pressure, and it possesses higher output force with higher payload capacity compared with the reported results of similar experiments of other SPAs [6, 16]. However, the actuator presented in this paper is only a preliminary version. Optimization design and further experiments will be conducted to analyze and improve the performance of the SSPA.

# References

1. Rus, D., Tolley, M.T.: Design, fabrication and control of soft robots. Nature **521**, 467–475 (2015)
2. SOFT ROBOTICS. https://www.softroboticsinc.com. Accessed 17 Apr 2018
3. Salvietti, G., Hussain, I., et al.: Design of the passive joints of underactuated modular soft hands for fingertip trajectory tracking. IEEE Robot. Autom. Lett. **2**(4), 2008–2015 (2017)
4. Polygerinos, P., et al.: Towards a soft pneumatic glove for hand rehabilitation. In: IEEE/RSJ International Conference on Intelligent Robots and Systems (IROS), Japan, pp. 1512–1517 (2013)
5. Klute, G.K., Czemiecki, J.M., Hannaford, B.: McKibben artificial muscles: pneumatic actuators with biomechanical intelligence. In: Proceedings of the IEEE/ASME International Conference on Advanced Intelligent Mechatronics, USA, pp. 221–226 (1999)
6. Daerden, F., Lefeber, D., Verrelst, B., Ham, R.V.: Pleated pneumatic artificial muscles: actuators for automation and robotics. In: IEE/ASME International Conference on Advanced Intelligent Mechatronics, Italy, pp. 738–743 (2001)
7. Polygerinos, P., Wang, Z., et al.: Soft robotic glove for combined assistance and at-home rehabilitation. Robotics and Autonomous Systems **73**, 135–143 (2015)
8. Shapiro, Y., Wolf, A., Gabor, K.: Bi-bellows: pneumatic bending actuator. Sens. Actuators, A **167**, 484–494 (2011)
9. Polygerinos, P., Wang, Z., et al.: Modeling of soft fiber-reinforced bending actuators. IEEE Trans. Robot. **31**(3), 778–789 (2015)
10. Lazeroms, M., Haye, A.L., et al.: A Hydraulic forceps with force-feedback for use in minimally invasive surgery. Mechatronics **6**(4), 437–446 (1996)
11. Galloway, K.C., Polygerinos, P., Wood, R.J., WalshK, C.J.: Mechanically programmable bend radius for fiber-reinforced soft actuators. In: IEEE 16th International Conference on Advanced Robotics (ICAR) (2014)
12. Ilievski, F., Mazzeo, A.D., et al.: Soft robotics for chemists. Angew. Chem **123**, 1930–1935 (2011)
13. Deimel, R., Brock, O.: A novel type of compliant and underactuated robotic hand for dexterous grasping. Int. J. Robot. Res. **35**, 161–185 (2016)
14. Mosadegh, B., Polygerinos, P., et al.: Pneumatic networks for soft robotics that actuate rapidly. Adv. Funct. Mater. **24**(15), 2163–2170 (2015)
15. Wall, V., Deimel, R., Brock, O.: Selective stiffening of soft actuators based on jamming. In: IEEE International Conference on Robotics and Automation (ICRA), Washington, pp. 252–257 (2015)
16. Li, Y., Chen, Y., Yang, Y., et al.: Passive particle jamming and its stiffening of soft robotic grippers. IEEE Trans. Robot. **33**(2), 446–455 (2017)
17. Photo by C.K. Oon, Amphibia. http://amphibia.my/page.php?pageid=s_foundr&s_foundr&s_id=149&search1=Coelognathus%20radiatus&species=Coelognathus%20raditus&type=reptiles. Accessed 10 Apr 2018
18. Yeoh, O.H.: Some forms of the strain energy function for rubber. Rubber Chem. Technol. **66**(5), 754–771 (1993)

# A Pulse Condition Reproduction Apparatus for Remote Traditional Chinese Medicine

Sensen Liu, Lei Hua, Pengyu Lv, Yang Yu, Yiqing Gao, and Xinjun Sheng[✉]

State Key Laboratory of Mechanical System and Vibration,
Shanghai Jiao Tong University,
800 Dongchuan Road, Minhang District, Shanghai, China
xjsheng@sjtu.edu.cn
http://bbl.sjtu.edu.cn/

**Abstract.** Pulse diagnosis is an integral part in Traditional Chinese Medicine (TCM). However, the feeling of pulse condition is subjective and mysterious based on lots of practice and experience which brings many obstacles to pass on the skill. Additionally, in order to take full advantage of the limited resource of skillful TCM experts, remote TCM by Internet was proposed in recent years. Obviously, reproduction apparatus of pulse condition is of great importance in remote TCM system as well as pulse diagnosis education. In this paper, a pulse condition reproduction apparatus, which is used to reproduce corresponding pulse vibration according to the pulse condition data received from universal serial bus, was developed. To evaluate the performance of the apparatus, the reproduced pulse wave was measured and compared with the input pulse wave. Consequently, the correlation coefficients between the average output pulse waveforms and the input pulse waveforms is 0.9911 for normal pulse. The flowing sense of the pulse under fingers is indirectly demonstrated by the phase difference between adjacent fingers. Results indicated that the developed apparatus is available for pulse diagnosis education and remote palpation.

**Keywords:** Pulse diagnosis
Remote Traditional Chinese Medicine (RTCM)
Reproduction apparatus · Flowing sense

## 1 Introduction

Pulse diagnosis, belonging to palpation, is one of the four pillars (interrogation, visual inspection, auscultation and olfaction observation, and palpation) in Traditional Chinese Medicine(TCM) [1–9]. In pulse diagnosis, TCM practitioners put their three fingers of middle, index and ring fingers on the terminal region of the radial artery to feel the nuances of pulse wave characteristics, which is used to judge the health condition of patients [1,3,4,8,10–14]. In general, the region

© Springer Nature Switzerland AG 2018
Z. Chen et al. (Eds.): ICIRA 2018, LNAI 10984, pp. 453–464, 2018.
https://doi.org/10.1007/978-3-319-97586-3_41

of the pulse-taking is divided into three adjacent intervals called Cun, Guan and Chi, as shown in Fig. 1. TCM clinicians use their three fingers simultaneously and individually by different pressure on respective points (shown in Fig. 1) to feel the change of the pulse in frequency, strength, shape, width and size, etc. It seems that Chinese sphygmology is suspect without direct physiological basement. Nevertheless, the formation of pulse condition is fairly complex than people might think. From the view of anatomy, the pulse waveform contains incident wave from the ventricular output of the blood and the reflect wave caused by the peripheral blood vessels [4]. There is a long way to go for the blood, from heart to radial arteries driven by the heart. Thus, the pulse condition may reflect the conditions of the heart, nerves, skin, arterial walls, muscle, blood parameters, etc. [12]. Hence, pulse diagnosis is still important in modern medicine and TCM.

However, the pulse diagnosis is very difficult to master because of the subjectivity and the strong dependence on experience. The TCM practitioners usually use the metaphorical expression to vaguely describe the pulse condition. The inheritance difficulty of this ancient skill causes the shortage of the quality TCM resource. In order to relieve the shortage of the TCM resource, the concept of remote TCM system has been proposed in recent years, through which the patients can communicate with the TCM experts by Internet. In the remote TCM system, the acquisition and reproduction of the pulse condition is the challengeable part to accomplish the remote pulse diagnosis. With the development of the sensor technology, a lot of devices to acquire the pulse condition have been studied [1,2,14]. Nevertheless, the researches on the reproduction apparatus of pulse condition are relatively deficient. Therefore, it is urgent to develop a reliable reproduction apparatus of the pulse condition through which the experts can pass on the skill to students and the patients can be treated by the remote TCM doctors.

Previous studies on the reproduction apparatus of pulse condition mainly contains three groups which are bionic circulatory system based on hydraulic system, linear motor simulator based on voice coil motor (VCM) and the electric signal simulator. In bionic hydraulic system, the formation of the pulse is constructed by the bionic circulatory system comprised of the hydraulic bump, elastic tube and working fluid [6,15–18]. This system can simulate some classic pulse condition very well by controlling the action of the bump and valves. However, the pulse types these systems can reproduce are always limited because of the inaccuracy controlling of the hydraulic system and the immutability of working fluid for a specific system. Another simulator is a kind of electrical generator based on electrical theory and pulse mechanism [19]. Obviously, the apparatus cannot be felt by fingers. The third group used the linear motors, voice coil motors mostly, as the actuators to reproduce the vibration of the pulse wave [8,11,20]. In existing research, although these linear motor simulator systems can be controlled flexibly to reproduce arbitrary pulse vibration, the flowing sense of pulse condition except for simple reciprocating vibration cannot be reproduced. Hence, we designed a new pulse condition reproduction apparatus with artificial vessel used to generate the flowing sense based on VCMs. Moreover, the developed apparatus was designed with serial communication port through which the apparatus can be driven by arbitrary pulse condition data in real-time way which is necessary in remote TCM application.

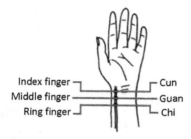

**Fig. 1.** Three regions of the pulse-taking (left hand).

## 2   Design and Methods

### 2.1   Apparatus Structure

The overall hardware framework is shown in Fig. 2. The system is mainly comprised of three parts: control board with USB communication port, motor driver board and VCM actuator. The device can receive pulse condition data from the upper computer and then the pulse data can be converted into control signal by the control board. The driver board can amplify the control signal to drive the VCM (VCM, TMEC0005-005-000, Kunshan, China) to reproduce the pulse condition. In this system, the configuration of the VCM is of crucial importance.

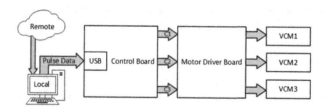

**Fig. 2.** The framework of the reproduction apparatus.

The main structures of the actuators are shown in Fig. 3 below. Figure 3(a) is the assembly diagram of the actuators which contains three VCMs with three pins, springs, an artificial vessel mounted on the pins and the framework to support other components. Figure 3(b) view shows the details of the structures. The sliding bearing can reduce the resistance to motion and the springs can offset the gravity of the VCMs and pins as well as enlarge the system stiffness to increase the frequency response range of the VCMs. Figure 3(c) demonstrates the shell, artificial skin and other accessories. Figure 3(d) shows the real product appearance.

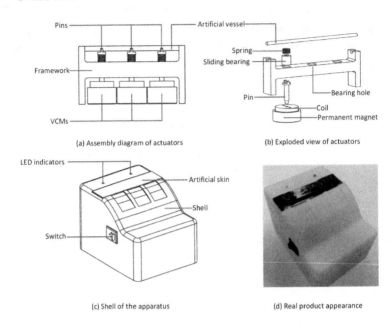

(a) Assembly diagram of actuators        (b) Exploded view of actuators

(c) Shell of the apparatus        (d) Real product appearance

**Fig. 3.** The main structures of the actuators based on VCMs.

## 2.2  Device Test

Two experiments were performed to evaluate the availability of the device. They were comparison between the output waveform and the input waveform as well as the flowing sense verification test. To reduce the artifact influence during acquiring the reappearance pulse, we designed the measurement platform based on flexible hose shown in Fig. 4. All of the test were performed on this platform.

**Output Pulse Characteristics Measurement.** Pulse condition is mainly described by the pulse wave. Pulse wave is the pressure collected from the surface of the radial artery terminal which contains at least 28 classic variations of pulse patterns in TCM [11,12]. In various patterns, normal pulse is the common one for most healthy people. Therefore, we selected the classic normal pulse as the input pulse condition data to the apparatus. The classic normal pulse is shown in Fig. 5. The normal pulse contains 4 feature points which are described by 7 parameters ($t_1$ to $t_3$, $h_1$ to $h_3$ and $t_{end}$). After the output of the developed apparatus is measured by the pulse collection equipment (HK-2010/3, Hefei, China), these 7 feature parameters would be compared with the input waveform to evaluate the performance of the reproduction. Moreover, the correlation coefficient between reproduced and input waveforms would be calculated. In fact, it is not easy to acquire the classic normal pulse from the human. However, some of the studies on the pulse characteristics extraction claim that the pulse wave can be decomposed into several Gaussian functions [10,20]. That indicates some pulse wave can be

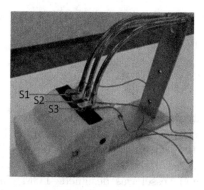

**Fig. 4.** The platform for test. S1 to S3 mean the number and location of the pulse sensors.

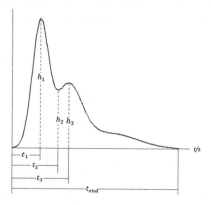

**Fig. 5.** Feature points of the normal pulse.

the summation of the several Gaussian functions. According to the literature [20], the pulse function $x(t)$ can be simulated by the formula below:

$$x(t) = \sum_{i}^{3} V_i e^{\frac{-(t-T_i)^2}{U_i}} \tag{1}$$

In the formula, $V_i$ is the amplitude of the Gaussian. $T_i$ is the center time point and $U_i$ is the width of the Gaussian. Theoretically, various pulse wave can be simulated by selecting the proper parameters in formula 1. Table 1 shows the corresponding parameters of the normal pulse in which the absolute value of the parameters have no real physical meaning just denoting the shape and position of simulated wave [20].

**Flowing Sense Verification Measurement.** Flowing sense is an abstract quantity which is caused by the pulse wave propagating along radial artery vessel. In traditional linear motor reproduction apparatus design, just the linear

**Table 1.** Normal pulse conditions fitting parameters

| $V_0$ | $V_1$ | $V_2$ | $T_0$ | $T_1$ | $T_2$ | $U_0$ | $U_1$ | $U_2$ |
|----|----|----|----|----|----|----|----|-----|
| 40 | 20 | 5 | 10 | 20 | 35 | 16 | 49 | 196 |

reciprocating movement of the motors was considered without involving the propagating of the waveforms. Hence, doctors may only feel the pulse strength and rhythm, although they are important, which are only parts of the pulse condition. The flowing sense under the finger cannot be ignored. To deal with the problem, the artificial vessel was designed. TCM clinicians can put their index and ring fingers on the artificial vessel between two pins mounted on respective VCMs when the apparatus is employed. And the position of the pulse sensors used for collecting the output pulse is consistent with the position of the fingers. Figure 6 shows the specific position of the fingers and the sensors when the apparatus is used.

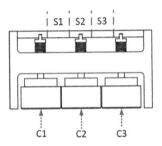

**Fig. 6.** The position of fingers and pulse wave sensors on the artificial vessel for the test. S1 to S3 mean the position of the three pulse sensors. When the apparatus is employed, the S1 to S3 mean the position of index, middle and ring finger respectively. C1 to C3 mean the three channels of stimulation data to driven the VCMs respectively.

In order to verify the propagation of the pulse wave along the artificial vessel, we utilized three channels of normal pulse data which were derived from formula 1 with a specific phase difference (5 data points) between two adjacent channels (see Fig. 7). The pressure on the middle pin and adjacent both sides artificial vessel segments was measured simultaneously. At last, we analyzed the acquired pulse phase difference between two adjacent sensors to verify the flowing sense caused by the waveform propagation.

### 2.3   Data Processing

**Obtain the Average Output Pulse Waveform And Compare with Input Pulse Waveform.** We used the pulse sensors to acquire a serial periods of the reproduced pulse waves and then the pulse wave signals were divided into

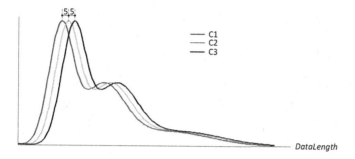

**Fig. 7.** Three channels normal pulse wave with phase difference between adjacent channels.

a serial single periods based on the lowest feature points. In order to get the location of highest point relative to two adjacent lowest points, we attained the location of the highest points and the location difference in data length formation between highest points and the adjacent lowest points. After the calculating, we knew that the average difference between highest point and front lowest point is 34 data points and the average difference between highest point and back lowest point is 162 data points. Hence, the average length of the pulse are 196 data points. According to this preprocessing, we can align these divided pulse waveforms by the highest points and then selected 34 points before the highest point as well as 162 points after the highest points as a new pulse waveform to ensure every new pulse waveform has same data length. Then we could superimpose these new pulse waveforms and then got the average pulse waveforms. The process is shown in Fig. 8.

After we obtained the average waveforms of the pulse wave, the signal should be normalized before the comparison of input waveforms and the average output waveforms. In this study, we normalized the two signals by subtracting the respective minimum from the whole signal to ensure the minimum of the two signals both are 0. Then we could calculate the correlation coefficient $R_{xy}$ between the average waveform and the input waveform by the formula 2 below:

$$R_{xy} = \frac{\sum_{i=1}^{N} (Y_i - \bar{Y})(X_i - \bar{X})}{\sqrt{\sum_{i=1}^{N}(X_i - \bar{X})^2}\sqrt{\sum_{i=1}^{N}(Y_i - \bar{Y})^2}} \tag{2}$$

where:

$R_{xy}$ is the correlation coefficient varying from $-1$ to 1.
$X_i$ is the output average pulse waveform.
$Y_i$ is the input normal pulse waveform.

And then the characteristics error between the average output waveform and the input waveforms could be attained from following formula 3:

$$Error = \frac{Input pulse feature value - reproduced pulse feature value}{Input pulse feature value} \tag{3}$$

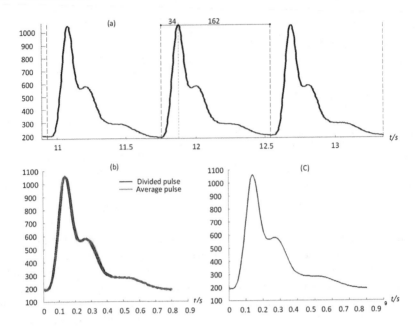

**Fig. 8.** The method to get the average pulse waveforms. (a) The segmentation of the pulse data. (b) The process of superimposing the segmented pulse wave. (c) The average pulse waveforms of the segmented pulse wave.

**Verify the Phase Difference Between Adjacent Channels.** After recording the three channels of reproduced pulse wave simultaneously, the phase difference between two adjacent channels was analyzed. The parts of acquired three channels of pulse waves are shown in Fig. 9. The raw data had been normalized based on the maximum amplitude of the pulses so that the highest point can be found easily and then we could get the phase difference by calculating the horizontal axis difference between the adjacent highest points.

## 3    Results

### 3.1    The Average Pulse Wave Characteristics

The average reproduced pulse waveform and the input normal pulse waveform are shown as Fig. 10. The correlation coefficient between these two waveforms is 0.9911 and the reproduced feature error is shown in the Table 2 below.

### 3.2    The Phase Difference Between Two Adjacent Channels

Table 3 demonstrates the phase difference between two adjacent channels in the form of data point number with ten continuous pulse waveforms. The first row is the difference between second channel and first channel, and the second row is the difference between third channel and the second channel.

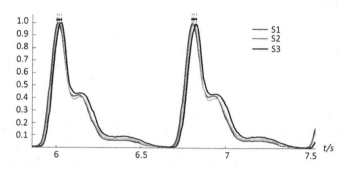

**Fig. 9.** The phase difference between adjacent channels. S1 to S3 mean the channels corresponding to the sensors in Fig. 4 or Fig. 6.

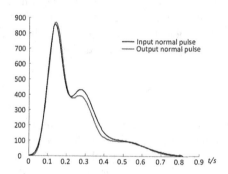

**Fig. 10.** The comparison of input and average output normal pulse.

**Table 2.** Main feature value comparison of reproduced pulse and input pulse

| Feature coordinate | $t_1/s$ | $h_1$ | $t_2/s$ | $h_2$ | $t_3/s$ | $h_3$ | $t_{end}/s$ |
|---|---|---|---|---|---|---|---|
| Input pulse | 0.1388 | 857 | 0.2245 | 384 | 0.2735 | 431 | 0.8163 |
| Reproduced pulse | 0.1429 | 870 | 0.2327 | 375 | 0.2694 | 390 | 0.8041 |
| Input pulse | | −2.95% | −1.52% | −3.65% | 2.34% | 1.50% | 9.51% | 1.49% |

**Table 3.** The phase difference between two adjacent channels (unit in data length)

| S2-S1 | 2 | 2 | 1 | 1 | 1 | 1 | 1 | 1 | 2 | 1 |
|---|---|---|---|---|---|---|---|---|---|---|
| S3-S2 | 3 | 3 | 2 | 3 | 4 | 3 | 3 | 3 | 3 | 4 |

## 4   Discussion

The comparison of the reproduced pulse wave and the input pulse wave indicates that the feature errors are generally small (mostly less than 3.65%) and the apparatus can reflect the main characteristics of input pulse. The correlation coefficient is 0.9911 which indicates the reproduced pulse condition and the input pulse condition have significant correlation. The structure based on VCM and spring may contribute to the performance. However, the biggest error occurred at the second peak of the pulse (9.51%, see Fig. 10 and Table 2). That maybe caused by the open-loop control lack of instant feedback. The vibration of the test platform and the measurement equipment can also cause some error.

The results of phase difference show that different test location has different phase along the artificial vessel and the difference always has same sign and similar value (2, 1 and 3, 4, see Table 3) which means the phase difference has consistency in direction and quantity. Meanwhile, we can find that the phase difference changed subtly. This may be caused by the sampling frequency instability. This reflects the pulse wave propagation along the artificial vessel and the flowing sense can be felt under the finger tips. This phenomenon seems owe to the design of the artificial vessel and artificial skin.

## 5   Conclusion

We developed an improved pulse condition reproduction apparatus which can reappear the pulse condition in real-time used for remote TCM or pulse diagnosis education. The dynamic characteristic of the apparatus can cover the main frequency range of human pulse. The reproduced pulse has significant correlation with input pulse and the characteristic error are mostly less than 3.65%. The most significant error 9.51% exists in second peak which seems caused by the open-loop control strategy. Differently, the new design with artificial vessel mounted on the pins can generate the pulse flowing sense which is more natural and real than simple reciprocating of the VCM. Further research is necessary for more accurate reproduction by closed-loop control of VCM. The property of the artificial vessel and the influence on the performance of the apparatus also need more investigation.

**Acknowledgments.** This research was supported by the Science and Technology Commission of Shanghai Municipality under Grant 17JC1402700. The authors thank the whole participants for their devotion in this research.

# References

1. Chu, Y.-W., Luo, C.-H., Chung, Y.-F., Chung-Shing, H., Yeh, C.-C.: Using an array sensor to determine differences in pulse diagnosisthree positions and nine indicators. Eur. J. Integr. Med. **6**(5), 516–523 (2014)
2. Hu, C.-S., Chung, Y.-F., Yeh, C.-C., Luo, C.-H.: Temporal and spatial properties of arterial pulsation measurement using pressure sensor array. Evid. Based Complement. Altern. Med. **2012**, 9 (2012)
3. Wang, H. Cheng, Y.: A quantitative system for pulse diagnosis in traditional Chinese medicine. In: 2005 IEEE Engineering in Medicine and Biology 27th Annual Conference, pp. 5676–5679
4. Jeon, Y.J., et al.: A clinical study of the pulse wave characteristics at the three pulse diagnosis positions of chon, gwan and cheok. Evid. Based Complement. Altern. Med. **2011** (2011)
5. King, E., Cobbin, D., Ryan, D.: The reliable measurement of radial pulse: gender differences in pulse profiles. Acupunct. Med. **20**(4), 160–167 (2002)
6. Lee, J.-Y., Jang, M., Shin, S.-H.: Study on the depth, rate, shape, and strength of pulse with cardiovascular simulator. Evid. Based Complement. Altern. Med. **2017**, 11 (2017)
7. Ma, L., Dong, M., Xin, W., Wei, Z., Liu, X.: Innovative design of human pulse reappearance system. In 2011 4th International Conference on Biomedical Engineering and Informatics (BMEI), vol. 2, pp. 1118–1121
8. Shin, K.Y., et al.: A pulse wave simulator for palpation in the oriental medicine. In: 2011 Annual International Conference of the IEEE Engineering in Medicine and Biology Society, pp. 4163–4166
9. Shin, K.Y., et al.: Characteristics of the pulse wave in patients with chronic gastritis and the healthy in Korean medicine. In: 2012 Annual International Conference of the IEEE Engineering in Medicine and Biology Society, pp. 992–995
10. Chen, Y., Zhang, L., Zhang, D., Zhang, D.: Wrist pulse signal diagnosis using modified gaussian models and fuzzy c-means classification. Med. Eng. Phys. **31**(10), 1283–1289 (2009)
11. Heo, H., Kim, E.G., Nam, K.C., Huh, Y.: Radial artery pulse wave simulator using a linear motor. In: 2008 30th Annual International Conference of the IEEE Engineering in Medicine and Biology Society, pp. 4895–4898
12. Lee, C.T., Wei, L.Y.: Spectrum analysis of human pulse. IEEE Trans. Biomed. Eng. BME **30**(6), 348–352 (1983)
13. Ma, C., Xia, C., Wang, Y., Yan, H., Li, F.: An improved approach to the classification of seven common TCM pulse conditions. In: 2011 4th International Conference on Biomedical Engineering and Informatics (BMEI), vol. 2, pp. 621–624
14. Wang, D., Zhang, D., Lu, G.: A robust signal preprocessing framework for wrist pulse analysis. Biomed. Signal Process. Control **23**, 62–75 (2016)
15. Di, J.-j., Chen, S., Wang, X.-m.: Study on the traditional Chinese medicine pulse-taking machine. BME Clin. Med. **12**(06), 503–506 (2008)
16. Lee, J.-Y., Jang, M., Lee, S., Kang, H., Shin, S.-H.: A cardiovascular simulator with elastic arterial tree for pulse wave studies. J. Mech. Med. Biol. **15**(06), 1540045 (2015)
17. Lee, J.-Y., Jang, M., Shin, S.-H.: Mock circulatory system with a silicon tube for the study of pulse waves in an arterial system. J. Korean Phys. Soc. **65**(7), 1134–1141 (2014)

18. Wang, X., Song, P., Zhou, P., Lu, X., Wang, Y., Cao, H.: System of traditional chinese medicine pulse replay system based on multi-sensor technology. Chin. J. Sens. Actuators **26**(11), 1604–1609 (2013)
19. Hassania, K., Navidbakhshb, M., Rostami, M.: Simulation of the cardiovascular system using equivalent electronic system. Biomed. Pap. Med. Fac. Univ. Palacky Olomouc Czech Repub. **2006**, 7 (2006)
20. Zhang, P.: Research and design of pulse wave simulation and reproduction system. Master thesis (2013)

# A Feasibility Study on an Intuitive Teleoperation System Combining IMU with sEMG Sensors

Heng Zhang, Zeming Zhao, Yang Yu, Kai Gui, Xinjun Sheng$^{(\boxtimes)}$, and Xiangyang Zhu

State Key Laboratory of Mechanical System and Vibration,
Shanghai Jiao Tong University,
800 Dongchuan Road, Minhang District, Shanghai, China
xjsheng@sjtu.edu.cn
http://bbl.sjtu.edu.cn/

**Abstract.** In this paper, we proposed an intuitive teleoperation system combining surface-electromyogram (sEMG) with inertia measurement unit (IMU) sensors. Two IMU sensors were worn in the upper arm and forearm to capture arm motion. The results were sent as control command to the remote robotic manipulator in task space. Pattern recognition algorithm was employed to decode sEMG signals collected from the forearm to control a humanoid mechanical hand. As a validation of the proposed system, we evaluated the performance of the system which included a Universal Robot 10 (UR10), a homemade humanoid pro-prosthetic hand (SJT-6) and two homemade armbands. Final experiments showed that the success rate on moving spherical object can be up to 86.7%.

**Keywords:** Teleoperation · Surface-electromyogram (sEMG)
Inertia Measurement Unit (IMU) · Manipulator

## 1 Introduction

Thanks to the great progress in the past decades, robot automation has successfully applied in various fields like intelligent warehousing, automotive mount and domestic service. Better robot intelligence is higher required by more and more complicated tasks and unknown environments. However, current artificial intelligence (AI) technology in robotics cannot support robot to autonomously accomplish complex tasks without the help of human being. So robot teaching is necessary in a great many scenarios. But there are still some harsh and inaccessible environments where we human cannot enter, so teleoperation system will be a better choice in those environments like deep sea scientific exploration [1], radiation detection [2] and surgical robot [3].

To fulfill the teleoperation, local motion should be transferred to remote robotic system. Motion capture systems are often divided into two categories,

© Springer Nature Switzerland AG 2018
Z. Chen et al. (Eds.): ICIRA 2018, LNAI 10984, pp. 465–474, 2018.
https://doi.org/10.1007/978-3-319-97586-3_42

vision-based sensing interface and wearable device interface/joystick. In [4], a Kinect (RGB-D camera) was employed to capture and identify hand motion of subjects when picking object. The manipulator copied the motion pattern to complete the same task. A Leap Motion sensor-based motion capture system was developed in [5]. The movement of subjects was recorded to teach a welding robot to complete weld tasks. However, vision-based system is still not self-contained because it relies on external camera. Furthermore, it would be unsteady in poor light environment [6]. There were some attempts focusing on wearable device. Two IMU sensors were employed in [7] to capture arm motion. The manipulator would follow the arm based on the neural learning network. Considering ease of use, intuition and stability, IMU based interface was superior to others in motion capture [8].

Although arm motion can be successfully detected, dynamic property of hand motion like grasping and opening has not achieved satisfying status by only IMU sensors. In recent years, sEMG has been widely used for human machine interface (HMI) because of its human-friendly interaction and short time delay. In [9], an anthropomorphic robot arm was controlled in real time by kinematic variables extracted from only sEMG signals of human arm. There were also some other works focused on EMG-based pattern recognition [10,11]. sEMG can detect dynamic property of human motion and make up for the lack of IMU sensor. Therefore, it will be a much more intuitive human-machine interface combining IMU and sEMG sensors. Some related work was introduced in [12]. In their system, arm motion and hand motion were captured by four IMU sensors and four sEMG electrodes respectively. Seven joints value of upper limb were decoded to control a seven degrees of freedom (DOF) redundant manipulator. However, the conversion between human arm joints and redundant manipulator joints was not intuitive but with some coupling, so it will still bring a heavy burden to control the robot in joint space.

In this paper, we proposed an intuitive teleoperation system combining the advantages of IMU and sEMG sensors. Two IMU sensors were worn in the upper arm and forearm to capture arm motion. The results were sent as control commands to the remote robotic manipulator in task space instead of config-uration space. Because in teleoperation system, we paid much more attention to the motion of the end effector. Meanwhile, a pattern recognition algorithm was employed to decode two kinds of hand motions (open and grasp) based on sEMG signals collected from forearm to control a humanoid mechanical hand. Finally we validated the performance of the proposed system by using a UR10, a SJT-6 hand and two armbands.

The paper is organised as follows. In Sect. 2, we introduce the hardware com-ponents of the system. IMU decoding algorithm and sEMG pattern recognition algorithm are illustrated in Sect. 3. We also show the system architecture and the way to control UR10 and SJT-6 hand in this part. Experiments and results are shown in Sect. 4. Section 5 shows the conclusion of this paper.

## 2   System Architecture

This section gives a description of the components of the system. As shown in Fig. 1, the system consists of two armbands and a UR10 manipulator with a SJT-6 hand.

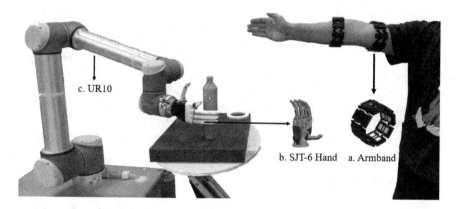

**Fig. 1.** The teleoperation system including Armband, UR10 and SJT-6 hand. The motion of human arm was captured by two IMU sensors in armbands. sEMG signals in forearm were used to get status of human hand.

### 2.1   Integrated Armband

In our previous study [13], an integrated armband consists of eight sEMG electrodes and an IMU sensor was designed to measure human motion and sEMG signals. In what follows, we will give a short description of the system.

**IMU Sensors.** IMU sensors are physical device which can measure attitude angle and accelerated speed. IMU sensor often consists of three-axis gyroscope, triaxial accelerometer and triaxial magnetometer. The output data format is Euler angle with a frequency about 50 Hz in the system. In order to make this device more portable and ease of use, we use bluetooth 2.0, a wireless way, to communicate with principle computer. In this system, operator should wear the armband on the upper arm and forearm. The accurate position is not specific because in IMU calibration process (Sect. 3.2), we will calculate the relative transformation between IMU frame and arm frame.

**sEMG Sensors.** Each armband consists of eight electrodes made by gold-plated copper. It can acquire ample signals to distinguish two kinds of hand motion via pattern recognition algorithm. Like IMU sensor, the communication interface is also bluetooth 2.0 with frequency up to 1 KHz. sEMG signals and IMU data processings are both in the architecture of Robot Operating System (ROS) on principle computer. The result can be easily used to control UR10 and SJT-6.

## 2.2    UR10 Manipulator with SJT-6 Hand

UR10 is a collaborative robot with a payload about 10 Kg. It has six degrees of freedom (DOF) and can reach any position with any orientation in workspace. Urscript is a set of control commands interacting with real robot via TCP/IP. The maximum communication frequency between robot and computer is 125 Hz. According to urscript manual, we can only control UR10 in speed loop and position loop as current loop is invisible to developer.

SJT-6 is a homemade humanoid pro-prosthetic hand. Due to the feature of underactuation, it has only six DOF, two in thumb and each in other four fingers, but can make a lot of gestures. On the other hand, humanoid hand shape feature gives it the capacity to grasp different shapes of object like spherical, cylindrical and cubic because of its high fault tolerance. SJT-6 is actuated by six brushless DC motors, with a maximum payload 2 Kg. The control interface can be wire and wireless. In this experiment, we use bluetooth 2.0 between SJT-6 and computer.

## 3    Methodology

This part describes algorithms we employed in the system. First in Sect. 3.1, we show the overall framework of the system. The way to capture arm motion by IMU sensors is shown in Sect. 3.2. In Sect. 3.3, we give an account of pattern recognition algorithm in sEMG signals processing. Robot manipulator and SJT-6 hand control strategies are presented in Sect. 3.4.

### 3.1    System Framework

The system framework can be seen in Fig. 2. All the subsystems are working in the architecture of ROS, the communication interface between UR10 manipulator and computer is TCP/IP, other communication interfaces like SJT6 and armband to computer are bluetooth.

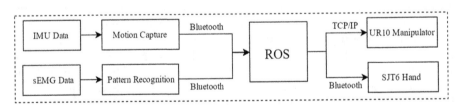

**Fig. 2.** Overall framework of teleoperation system.

### 3.2    Arm Motion Capture by IMU Sensors

Arm motion capture system can be divided into calibration stage and work stage. In calibration stage, we calibrate IMU sensors because as we said above, IMU sensor can be arbitrary in the upper arm and forearm. The model of human arm can be equivalent to two linked spherical hinges. So in calibration stage, operator should keep arm and IMU sensors horizontal for three seconds and the values of the two IMU sensors are recorded as $\mathbf{R}_{1init}(r_{1init},p_{1init},y_{1init})$, $\mathbf{R}_{2init}(r_{2init},p_{2init},y_{2init})$.

After preparation part, according to IMU values $\mathbf{R}_1(r_1,p_1,y_1)$, $\mathbf{R}_2(r_2,p_2,y_2)$, we can calculate the position increment of the end point of our arm.

$$\begin{cases} \Delta x = l_1 cos\Delta p_1 sin\Delta y_1 + l_2 cos\Delta p_2 sin\Delta y_2 \\ \Delta y = l_1 cos\Delta p_1 cos\Delta y_1 + l_2 cos\Delta p_2 cos\Delta y_2 \\ \Delta z = l_1 sin\Delta p_1 + l_2 sin\Delta p_2 \end{cases} \tag{1}$$

$l_1, l_2$ above are the length of human arm. $\Delta r$, $\Delta p$ and $\Delta y$ are the angle increment of roll, pitch and yaw. $\Delta x$, $\Delta y$ and $\Delta z$ are translation increment of end point of the arm.

### 3.3 Pattern Recognition by sEMG Sensors

According to the physiological knowledge, the contraction of skeletal muscle is controlled by the bioelectrical impulses, i.e. motor unit action potential trains (MUAPts), propagating from central and peripheral nervous systems. Therefore, sEMG signals could reflect the activities of human muscles with the fact that it is generated from MUAPts with volume conduction. Pattern recognition is a mainstream method for the recognition of different gestures in the past decades [14], especially in which multiple degrees of freedom are needed, e.g. dexterous humanoid prostheses. In general, sEMG pattern recognition mainly contains four steps: data acquisition, data segmentation, feature extraction and classification. This scheme has nearly become a standard paradigm since Englehart et al. [15] reported an novel method for sEMG pattern recognition. Besides, constrained by supervised learning, the whole process were divided into two sessions: training and testing session. The main destination of training session was to determine the parameters of the classifier. In our study, we tried to construct the mappings between human gestures and the states of SJT-6 hand.

As mentioned above, the acquisition of EMG signals was accomplished through a wireless armband. For each channel EMG signals, we employed a 100 ms overlapped sliding window with an 50 ms increment to segment the data into pieces. And for data in each window, a feature set in time domain [16] was extracted, including mean absolute value (MAV), waveform length (WL), zero crossings (ZC) and slope sign changes (SSC). These four kinds of features are defined in Eqs. (2)–(5).

Supposing a piece of N-length EMG data, $x_i$ and $\varepsilon$ are the magnitude of EMG data at epoch and a controlled threshold respectively.

$$MAV = \frac{1}{N} \sum_{i=1}^{N} |x_i| \tag{2}$$

$$WL = \sum_{i=1}^{N} |x_{i+1} - x_i| \tag{3}$$

$$\begin{cases} \{x_i > 0 \ and \ x_{i+1} < 0\} \ or \ \{x_i < 0 \ and \ x_{i+1} > 0\} \\ |x_i - x_{i+1}| > \varepsilon \end{cases} \tag{4}$$

$$\begin{cases} \{x_i > x_{i+1} \ and \ x_i > x_{i+1}\} \ or \ \{x_i < x_{i-1} \ and \ x_i < x_{i+1}\} \\ \{|x_i - x_{i+1}| < \varepsilon \ or \ |x_i - x_{i-1}| \ge \varepsilon\} \end{cases} \tag{5}$$

Concatenating all of the features in every channel (8 channels), we finally received an 32 dimensional feature vector. And then the vector was fed into trained classifier in which test data would be classified into different classes corresponding to different gestures. Linear discriminant analysis (LDA) was adopted in this study with the advantages that it needs less calculation and easy to be implemented. The detailed information on implementing LDA classifier into sEMG classification can be obtained in [17].

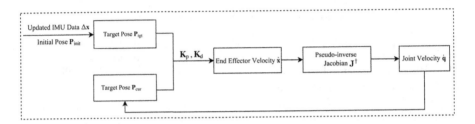

**Fig. 3.** The control strategy of robot manipulator.

### 3.4   UR10 and SJT6 Hand Control Strategy

In this experiment, we fix the orientation of the end effector and treat the result from arm motion capture module as incremental translation of UR10. The control strategy of UR10 can be seen in Fig. 3. The input of the subsystem $\Delta\mathbf{x}$ is added to the initial pose $\mathbf{P}_{init}$ to get target pose $\mathbf{P}_{tgt}$. So we can easily use a PD control law to trace the target pose in task space shown below.

$$\dot{\mathbf{x}} = \mathbf{K}_p(\mathbf{P}_{tgt} - \mathbf{P}_{cur}) + \mathbf{K}_d\Delta(\mathbf{P}_{tgt} - \mathbf{P}_{cur}) \tag{6}$$

$\mathbf{P}_{cur}$ in the equation is current pose. And all the poses are $6 \times 1$ row vector with last three elements constant. Because we only consider translation in Cartesian space. So the first three elements in $\dot{\mathbf{x}}$ represents translation velocity and last three are all zero. In order to control the manipulator and avoid singularity, we should use pseudo-inverse Jacobian matrix $\mathbf{J}^\dagger$ to map the Cartesian velocity into joint velocity.

$$\dot{\mathbf{q}} = \mathbf{J}^\dagger\dot{\mathbf{x}} \tag{7}$$

And according to [18], we can get the relationship between (pseudo-)inverse Jacobian $\mathbf{J}^\dagger$ and Jacobian matrix $\mathbf{J}$.

$$\mathbf{J}^\dagger = \mathbf{J}^T(\mathbf{J}\mathbf{J}^T + \lambda\mathbf{I})^{-1} \tag{8}$$

$\lambda$ can be calculated as

$$\lambda = \begin{cases} 0 & \|\mathbf{J}\| \geq \delta \\ \lambda^* & \|\mathbf{J}\| < \delta \end{cases} \quad (9)$$

$\delta$ is the threshold and $\lambda^*$ is a specific value that can be used in Eq. (8).

The strategy to control SJT6 hand is very easy. When it grasps object, current $I$ in the control system will increase until reaching threshold $I_{thres}$, then motors will stop and keep the state until receiving an open command. Meanwhile, we will not send control command to SJT6 when operating UR10 since variation in arm position leads to a poor classification results for sEMG pattern recognition [19,20].

## 4 Experiments and Results

Experiments and results on validating the proposed system are shown in this part. In the system, we mainly care about the tracking error and time delay of robotic manipulator. In the meantime, success rate of pattern recognition is another important index we concern. Finally the combination performance of the system is evaluated with grasping task.

### 4.1 Arm Motion Tracking

Arm motion tracking performance is the most important index of this system. A good tracking performance should be less time delay and high tracking accuracy. In this part, subjects should keep their upper limb and IMU sensors horizontal for three seconds to fulfill calibration stage. In order to control the robot manipulator, we recommend them placing their arms in front of the body and maintaining a semi-collapsed state. Because they can move the arm freely in all directions in this position. Finally we start to read IMU data to control robot manipulator.

Figure 4 shows the result of UR10 following desired position from IMU sensors. Tracking performance is good with average tracking error about 1.51 cm. It can be also calculated from Fig. 4 that average time delay is 0.2 s. The most possible reasons are bluetooth communication from armband to computer, TCP/IP communication between UR10 and computer and inherent delay attribute of PD control law.

### 4.2 Gesture Recognition for SJT-6 Hand Control

The grasping performance and efficiency of the whole system rely largely on the classification accuracy of gesture recognition during manipulation. The variation of arm position during grasping contributes to misclassification owing to changes of muscles states in the forearm, especially when the elbow moves. Consequently, it is of necessity to validate the gesture recognition accuracy while manipulating.

To enhance the stability of the system, the prosthetic hand has no response to control commands delivered by principle computer before the arm motion is

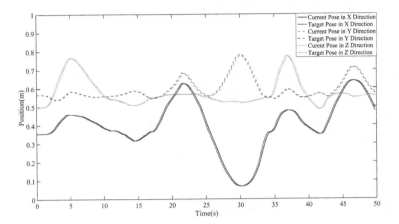

**Fig. 4.** Arm motion tracking results.

**Table 1.** Hand motion control results (%)

| Position | Front | Left | Right | Up | Down | Average |
|---|---|---|---|---|---|---|
| Subject1 | 100 | 93.3 | 96.7 | 86.7 | 93.3 | 94 |
| Subject2 | 96.7 | 93.3 | 93.3 | 90 | 90 | 92.7 |
| Subject3 | 100 | 90 | 93.3 | 83.3 | 86.7 | 90.7 |
| Subject4 | 100 | 96.7 | 96.7 | 93.3 | 96.7 | 96.7 |
| Subject5 | 93.3 | 93.3 | 86.7 | 90 | 96.7 | 92 |

approximately still. Based on this control strategy, we designed our experiment protocol. We selected five classic limb positions in the whole motion space, i.e. forearm parallel to the ground in the middle, arm pointing to left/right, arm obliquely upwards/downwards. Data acquired in the middle position was used to train the classifier. Classification performance was tested in all five positions. The training process contained 5 s of rest and grasping. Thirty trials were performed in each position for validation. Five subjects participated in this experiment and were informed with contents of the experiments. The classification accuracy in different positions was presented in Table 1. The average classification accuracy is up to 93.22% which is considered an available level in prosthetic hand control. For arm position in the up and down, it is reasonable for that result as in which elbow joint flexes more obviously than that in other positions.

### 4.3   Real-Time Grasping Experiments

Real-time grasping experiments consider comprehensively arm motion tracking performance, gesture recognition accuracy and system stability. In this part, we tested the combination performance of the proposed system by grasping bottle-like object. As can be shown in Fig. 1, subjects should first keep their arm and

IMU sensors horizontal for 3 s to accomplish IMU sensor calibration session. Then subjects need keep their arm in front of the body with a comfortable pose to fulfill sEMG training session. Finally, bottle should be moved from initial place to final place by teleoperation. We repeated the task 15 times, the totally success rate can be up to 86.7%.

## 5    Conclusions

This paper proposed an intuitive teleoperation system combining IMU motion data and sEMG signals. Linear discriminant analysis (LDA) algorithm was capable to distinguish two kinds of hand gestures. IMU sensor can capture arm motion to control remote manipulator in task space. The advantages of the system are as follows. First, we control the manipulator in task space which is more intuitive than joint space. Second, armband-based human machine interface is more portable and does not rely on external devices. Experiments and results showed that the integrated system can fulfill activities-of-daily-living like grasping drinks well.

**Acknowledgement.** This research was supported by National Key Technology Research and Development Program of China under Grant 2015BAF01B02.

## References

1. Saltaren, R., Aracil, R., Alvarez, C., Yime, E., Sabater, J.M.: Field and service applications-exploring deep sea by teleoperated robot-an underwater parallel robot with high navigation capabilities. IEEE Robot. Autom. Mag. **14**(3), 65–75 (2007)
2. Wei, W., Yuan, K.: Teleoperated manipulator for leak detection of sealed radioactive sources. In: Proceedings of the IEEE International Conference on Robotics and Automation, ICRA 2004, vol. 2, pp. 1682–1687 (2004)
3. Funda, J., Taylor, R.H., Eldridge, B., Gomory, S., Gruben, K.G.: Constrained Cartesian motion control for teleoperated surgical robots. IEEE Trans. Robot. Autom. **12**(3), 453–465 (1996)
4. Levine, S.J., Schaffert, S., Checka, N.: Natural user interface for robot task assignment (2010)
5. Liu, Y., Zhang, Y., Fu, B., Yang, R.: Predictive control for robot arm teleoperation. In: Proceedings of 39th Annual Conference of the IEEE Industrial Electronics Society, IECON 2013, pp. 3693–3698. IEEE (2013)
6. Vlasic, D., Adelsberger, R., Vannucci, G., Barnwell, J., Gross, M., Matusik, W.: Practical motion capture in everyday surroundings. ACM Trans. Graph. **26**(3), 35 (2007)
7. Yang, C., Chen, J., Chen, F.: Neural learning enhanced teleoperation control of Baxter robot using IMU based motion capture. In: Proceedings of 22nd International Conference on Automation and Computing, ICAC 2016, pp. 389–394. IEEE (2016)
8. Brigante, C.M., Abbate, N., Basile, A., Faulisi, A.C., Sessa, S.: Towards miniaturization of a MEMS-based wearable motion capture system. IEEE Trans. Ind. Electron. **58**(8), 3234–3241 (2011)

9. Artemiadis, P.K., Kyriakopoulos, K.J.: EMG-based control of a robot arm using low-dimensional embeddings. IEEE Trans. Robot. **26**(2), 393–398 (2010)
10. Liu, H.J., Young, K.Y.: Upper-limb EMG-based robot motion governing using empirical mode decomposition and adaptive neural fuzzy inference system. J. Intell. Robot. Syst. **68**(3–4), 275–291 (2012)
11. Naik, G.R., Kumar, D.K., Singh, V.P., Palaniswami, M.: Hand gestures for HCI using ICA of EMG. In: Proceedings of the HCSNet Workshop on Use of Vision in Human-Computer Interaction, vol. 56, pp. 67–72. Australian Computer Society, Inc. (2006)
12. Kim, M.K., Ryu, K., Oh, Y., Oh, S.R., Kim, K.: Implementation of real-time motion and force capturing system for tele-manipulation based on sEMG signals and IMU motion data. In: Proceedings of the IEEE International Conference on Robotics and Automation, ICRA 2014, pp. 5658–5664. IEEE (2014)
13. Guo, W., Sheng, X., Liu, J., Hua, L., Zhang, D., Zhu, X.: Towards zero training for myoelectric control based on a wearable wireless sEMG armband. In: IEEE International Conference on Advanced Intelligent Mechatronics, pp. 196–201 (2015)
14. Farina, D., et al.: The extraction of neural information from the surface EMG for the control of upper-limb prostheses: emerging avenues and challenges. IEEE Trans. Neural Syst. Rehabil. Eng. **22**(4), 797–809 (2014)
15. Englehart, K., Hudgins, B.: A robust, real-time control scheme for multifunction myoelectric control. IEEE Trans. Biomed. Eng. **50**(7), 848–854 (2003)
16. Hudgins, B., Parker, P., Scott, R.N.: A new strategy for multifunction myoelectric control. IEEE Trans. Biomed. Eng. **40**(1), 82–94 (1993)
17. Rechy-Ramirez, E.J., Hu, H.: Stages for developing control systems using EMG and EEG signals: a survey (2011)
18. Chiaverini, S., Egeland, O., Kanestrom, R.: Achieving user-defined accuracy with damped least-squares inverse kinematics. In: Proceedings of the Fifth International Conference on Advanced Robotics, ICAR 1991. 'Robots in Unstructured Environments', pp. 672–677. IEEE (1991)
19. Geng, Y., Samuel, O.W., Wei, Y., Li, G.: Improving the robustness of real-time myoelectric pattern recognition against arm position changes in transradial amputees. BioMed Res. Int. 2017(April) 1–11 (2017)
20. Yu, Y., Sheng, X., Guo, W., Zhu, X.: Attenuating the impact of limb position on surface EMG pattern recognition using a mixed-LDA classifier. In: Proceedings of the IEEE International Conference on Robotics and Biomimetics, ROBIO 2017, pp. 1497–1502. IEEE (2017)

# Data-Driven Modeling of a Coupled Electric Drives System Using Regularized Basis Function Volterra Kernels

Jeremy G. Stoddard[(✉)] and James S. Welsh

School of Electrical Engineering and Computing, University of Newcastle,
Callaghan, Australia
jeremy.stoddard@uon.edu.au

**Abstract.** In this paper, we consider the problem of data-driven modeling for systems containing nonlinear sensors. The issue is explored via an established nonlinear benchmark in the system identification community, referred to as the "coupled electric drives." In the benchmark system, nonlinearity emerges in the pulse transducer used to measure the angular velocity of a pulley, which is invariant to the direction of rotation. In order to model the nonlinear dynamics without the use of extensive prior knowledge, we estimate a nonparametric Volterra series model using a regularized basis function approach. While the Volterra series is typically an impractical modeling tool due to the large number of parameters required, we obtain accurate models using only a short estimation dataset, by directly regularizing the basis function expansions of each Volterra kernel in a Bayesian framework.

**Keywords:** System identification · Volterra series · Regularization

## 1 Introduction

Data-driven modeling of dynamic systems is often complicated by the presence of nonlinear sensors. If the nonlinearity is sufficiently weak, then it may be acceptable to estimate an approximate linear model of the system, for which there is an abundance of established methods and literature available, see e.g. [4, 7]. There are many circumstances, however, where a linear model is not sufficient, and we must move to the nonlinear setting to make more accurate predictions about the system behaviour. The "coupled electric drives" benchmark system [14] is one such example, where the output of the system is a filtered pulse count produced by a pulse transducer used to measure the angular velocity of a pulley. The sensor's inability to differentiate direction of rotation amounts to an 'absolute value' nonlinearity, and the estimation of a nonlinear model for prediction and control purposes will be the focus of this paper.

One of the biggest challenges in nonlinear system identification is the selection of a model class from the vast array of possible choices available in the

© Springer Nature Switzerland AG 2018
Z. Chen et al. (Eds.): ICIRA 2018, LNAI 10984, pp. 475–485, 2018.
https://doi.org/10.1007/978-3-319-97586-3_43

literature, particularly since most choices require significant prior knowledge of the system under study to provide a good estimate. In this paper, we will all but avoid the problem of prior knowledge, by considering a nonparametric Volterra series approach to modeling. The Volterra series [9] can be seen as a nonlinear and multidimensional generalization of the linear Finite Impulse Response (FIR), and has the ability to approximate any time-invariant and fading-memory nonlinear system to an abritrary accuracy level [2]. Its main shortcoming is the large number of parameters required for an accurate representation, which grows rapidly with the memory length and truncation order of the series. In an identification context, the estimated parameters are known to suffer high variance when using short data records in the presence of measurement noise.

The variance issue has recently been addressed, by extending the Bayesian regularization techniques developed for FIR estimation in [5]. In [1], each Volterra series term or 'kernel' was modeled as a Gaussian process, and prior covariances were constructed to impose smoothness and exponential decay across the entire hypersurface of each kernel. In this paper, we instead consider an Orthonormal Basis Function (OBF) formulation of the Volterra series estimation problem, and apply regularization to multi-dimensional basis function expansions as opposed to the original time-domain kernels. Careful design of the bases allows these expansions to be far more compact than their time-domain counterparts, reducing computation time and variance of the estimates. The OBF-formulated method was proposed in [10], and the optimization of hyperparameters (a requirement in many such empirical Bayes methods) is performed using a recently introduced expectation-maximization (EM) method presented in [11].

When applied to the coupled electric drives benchmark data, the estimated OBF-expanded Volterra kernels provide good prediction accuracy on a validation dataset, despite using a short estimation record and imposing very little prior knowledge. When compared to models obtained using block-oriented and nonlinear autoregressive exogenous (NARX) model classes, the Volterra series approach is seen to be very competitive even though the number of estimated parameters is much larger. This competitiveness can be attributed to the compact OBF formulation and efficient regularization scheme.

The remainder of the paper is organised as follows. Section 2 provides some background and details on the coupled electric drives benchmark system, and introduces the identification problem. Section 3 summarises the theory of Volterra series, basis functions, and the regularized estimation method being proposed. The proposed method is applied to the benchmark system in Sect. 4, and compared to other established model classes and methods. Some concluding remarks are given in Sect. 5.

## 2   The Coupled Electric Drives System

### 2.1   Physical Description

The coupled electric drives [14] is one of a set of nonlinear benchmark systems used within the system identification community (the full set can be found at www.nonlinearbenchmark.org). This particular system is inspired by machines which transfer a belt of material from one spool to another using two electric drives and a pulley [13]. The benchmark system is configured as shown in Fig. 1, where the input is applied to two motors driving a flexible belt, which in turn rotates a pulley being held by a spring. The angular speed of the pulley is measured by a pulse transducer, whose measurement is sent through an analogue low-pass filter and a digital anti-aliasing filter to produce the final speed output which can be used for control purposes.

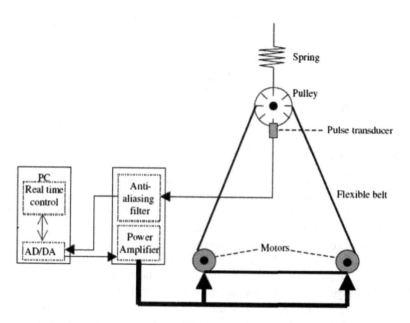

**Fig. 1.** Schematic of the coupled electric drives system. Source: [14]

The spring and flexible belt add lightly damped dynamic modes to the system, and there will also be time constants associated with the motors, however these are all linear effects. The modeling complication is brought about by the pulse transducer, which is insensitive to the *direction* of rotation of the pulley, since it simply counts pulses to approximate angular velocity. As the motors can be driven in either direction, the transducer can be considered as having an 'absolute value' nonlinearity, such that the total system is well approximated by a Wiener-Hammerstein block structure, as depicted in Fig. 2.

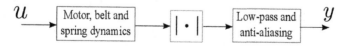

**Fig. 2.** Wiener-Hammerstein block structure for the coupled electric drives

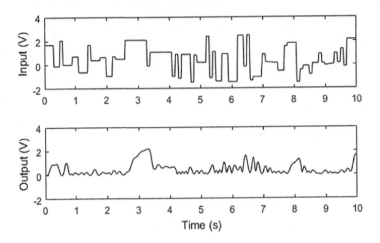

**Fig. 3.** Input (top) and output (bottom) for the estimation dataset

## 2.2 Identification Problem

While some physical modeling insights can be obtained by analysing the system set-up, it is the experimental input-output data which will be used to identify and assess candidate models. For the coupled electric drives, two sets of data have been recorded, an estimation set and a validaton set, both containing 500 input-output samples recorded at 50 Hz. The inputs are contstructed from a Pseudo-Random Binary Sequence (PRBS) which is multiplied by a noise sequence with uniform distribution. The estimation data set is shown in Fig. 3.

The identification problem is to estimate a nonlinear model for the drives system using the estimation data, and evaluate the accuracy of this model by using the validation input to predict the validation output. The physical insights discussed in Sect. 2.1 suggest that a Wiener-Hammerstein structure would be an appropriate model class, and this idea was explored in the original technical report for the benchmark, found in [14]. In this paper, however, we wish to consider the broader problem of modeling systems with nonlinear sensors, and thus we will approach the identification from a position of no prior knowledge.

# 3    Regularized Basis Function Estimation of Volterra Kernels

## 3.1    Volterra Series and the Basis Function Formulation

We assume that the output, $y$, of the system can be approximated by a discrete Volterra series structure [9],

$$y(k) = h_0 + \sum_{m=1}^{M} \left[ \sum_{\tau_1=0}^{n_m-1} \cdots \sum_{\tau_m=0}^{n_m-1} h_m(\tau_1, \ldots, \tau_m) \prod_{\tau=\tau_1}^{\tau_m} u(k-\tau) \right] + e(k), \qquad (1)$$

where $u$ is the input signal, $e$ is white Gaussian measurement noise and $h_m$ is the $m$'th order Volterra kernel (in the time domain) truncated to memory length $n_m$. The maximum considered order is denoted by $M$.

If for each kernel $h_m$, we define a set of orthonormal basis functions $\{f_{m,l}\}_{l=1}^{B_m}$, then the model structure in (1) can be reformulated using basis function expansions [8], i.e.

$$y(k) = \alpha_0 + \sum_{m=1}^{M} \left[ \sum_{i_1=1}^{B_m} \cdots \sum_{i_m=1}^{B_m} \alpha_m(i_1, \ldots, i_m) \prod_{j=1}^{m} u_{m,i_j}^f(k) \right] + e(k), \quad (2)$$

where $u_{m,l}^f$ denotes the input signal convolved with $f_{m,l}$, and $\alpha_m$ is the $m$'th order multidimensional expansion coefficient set, which we will refer to as a 'basis function kernel' in the sequel. If the basis is well chosen, then the basis function kernel can be much more compact than the equivalent time domain kernel, as demonstrated in Fig. 4 for a 2nd order example.

In this paper, we will consider (for their simplicity) the set of OBFs known as the Laguerre Basis Functions (LBFs), which are parameterized by a single real pole [12]. Thus, $f_{m,l}$ can be expressed in the z-domain as,

$$F_{m,l}(z) = \frac{\sqrt{1 - |a_m|^2}}{z - a_m} \left( \frac{1 - a_m z}{z - a_m} \right)^{l-1}, \quad a_m \in (-1, 1). \quad (3)$$

Having defined the basis function filters, the model description in (2) can now be placed in a least squares framework,

$$Y = \phi\alpha + E, \quad (4)$$

where $Y$ is the output measurement vector, $\phi$ is a regressor containing products of filtered inputs, $E$ is a Gaussian white noise vector, and $\alpha$ is a vector containing the set of all expansion coefficients from every order, $m$. Note that to ensure a unique solution, $\phi$ is designed to enforce symmetry in the $m$-dimensional basis function kernels.

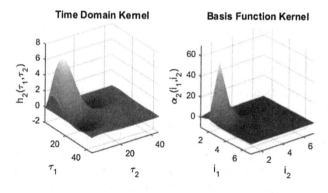

**Fig. 4.** A 2nd order time domain kernel (left) and corresponding well-parameterized basis function kernel (right)

## 3.2  Bayesian Regularization

Following the method proposed in [10], we place a Gaussian assumption on the parameter vector,

$$\alpha \sim \mathcal{N}(0, P),$$

where the covariance $P$ becomes the (inverse) regularization penalty in the cost function,

$$\hat{\alpha} = \arg \min_{\alpha} \|Y - \phi\alpha\|_2^2 + \sigma^2 \alpha^T P^{-1} \alpha, \tag{5}$$

and $\sigma^2$ is the variance of the measurement noise, $e$.

The covariance, $P$, has a block diagonal form,

$$P = \begin{bmatrix} P_0 & & \mathbf{0} \\ & \ddots & \\ \mathbf{0} & & P_M \end{bmatrix}, \tag{6}$$

where $P_m$ is the prior covariance of the $m$'th (vectorised) basis function kernel. As in the time domain case of [1], $P_m$ is designed to impose smoothness and stable decay across the hypersurface of each basis function kernel $\alpha_m$, by applying a standard Diagonal/Correlated (DC) covariance structure [6] along $m$ perpendicular regularizing directions.

If we denote the set of orthogonal directions as $(\mathcal{V}_m^1, \ldots, \mathcal{V}_m^m)$, then the set can be used to define a rotated coordinate system. Applying the DC covariance structure along a direction $\mathcal{V}_m^j$ will produce a 'partial covariance', whose $x, y$'th element is given by,

$$P_m^j(x, y) = (\lambda_m^j)^{|x' - y'|}(\beta_m^j)^{(x' + y')/2}, \tag{7}$$

where $\lambda_m^j$ and $\beta_m^j$ are decay hyperparameters, and $x'$ denotes the coordinate, on the $\mathcal{V}_m^j$ axis, of the kernel coefficient corresponding to the $x$'th index of $P_m$.

The full covariance matrix is obtained by an Hadamard product of all $m$ partial covariances formed by applying the DC structure along each regularizing direction, i.e.

$$P_m = c_m \cdot P_m^1 \circ \ldots \circ P_m^m, \tag{8}$$

where $c_m$ is a normalization hyperparameter.

## 3.3  LBF Pole Selection

The set of basis function poles, $\eta_a = [a_1 \ldots a_M]$, should be chosen to reflect the dynamics at each nonlinear order, but this choice is difficult to make without prior knowledge of the system under study. If we consider that in the case where a system's Volterra series expansion is exactly known, optimal LBF poles can be computed analytically [3], then it is possible to form a recursive least squares method which iteratively updates the pole selection from an initial guess. Such an algorithm is detailed in [10], and has the advantage of separating pole selection from the computationally intensive regularization process. This is the method which will be used in this paper for the coupled electric drives system.

## 3.4  Hyperparameter Tuning

After the basis function poles have been chosen, there are a large number of hyperparameters remaining in the regularization problem; the decay and normalization hyperparameters for each $\alpha_m$ and the noise variance $\sigma^2$. Traditionally, and in the linear case [5], this hyperparameter tuning would be done via a marginal likelihood maximization on the output, i.e.

$$\hat{\eta} = \arg \min_{\eta} Y^T \Sigma_Y^{-1} Y + \log \det \Sigma_Y, \tag{9}$$

where $\Sigma_Y$ is the covariance matrix of $Y$ obtained from the joint distribution of $[\alpha^T \ Y^T]^T$ as

$$\Sigma_Y = \phi(\eta_a) P(\eta) \phi(\eta_a)^T + \sigma^2 I, \tag{10}$$

and $I$ is an identity matrix of appropriate dimension.

For the nonlinear Volterra series case, the large hyperparameter space and non-convexity in (9) result in a high computational burden when optimizing hyperparameters [1]. An alternate approach to tuning was developed in [11], which uses expectation-maximization to break the large global optimization problem into several small problems in an iterative scheme. Robustness to poor hyperparameter intialization in the optimization is also typically improved. Interested readers are directed to [11] for the details and derivations, while the final iterative scheme is summarised below.

First, we define some required notation,

$$\bar{\eta}_m = [\lambda_m^1, \ldots, \lambda_m^m, \beta_m^1, \ldots, \beta_m^m], \tag{11}$$

$$P_m = c_m \cdot \bar{P}_m(\bar{\eta}_m), \tag{12}$$

$$\mathbf{E}_\alpha = P[\phi^T \phi P + \sigma^2 I]^{-1} \phi^T Y, \tag{13}$$

$$\mathbf{E}_{\alpha\alpha^T} = \sigma^2 P[\phi^T \phi P + \sigma^2 I]^{-1} + \mathbf{E}_\alpha \cdot \mathbf{E}_\alpha^T, \tag{14}$$

$$W = \begin{bmatrix} W_1 & & * \\ & \ddots & \\ * & & W_M \end{bmatrix} = \mathbf{E}_{\alpha\alpha^T}, \tag{15}$$

where $W_m$ is a square matrix with equal dimension to $P_m$. Now, the hyperparameter updates at each iteration are given by,

$$\hat{\bar{\eta}}_m^{(k+1)} = \arg \min_{\bar{\eta}_m} \left( \log \mathrm{Tr}[\bar{P}_m^{-1}(\bar{\eta}_m) W_m] + \log \det \bar{P}_m(\bar{\eta}_m) \right), \tag{16}$$

$$\hat{c}_m^{(k+1)} = \frac{\mathrm{Tr}[\bar{P}_m^{-1}(\hat{\bar{\eta}}_m^{(k+1)}) W_m]}{d_m}, \tag{17}$$

$$[\sigma^2]^{(k+1)} = \frac{Y^T Y - 2Y^T \phi \mathbf{E}_\alpha + \mathrm{Tr}[\phi^T \phi \mathbf{E}_{\alpha\alpha^T}]}{N}, \tag{18}$$

where $d_m$ is the number of unique parameters in $\alpha_m$, and $N$ is the length of $Y$. Due to the properties of EM sequences [15], iterating until convergence provides hyperparameters, $\hat{\eta}$, which correspond to a stationary point of (9) as desired.

Following hyperparameter optimization, the final regularized basis function kernels can be analytically computed from (5) as

$$\hat{\alpha} = (P(\hat{\eta})\phi(\hat{\eta}_a)^T\phi(\hat{\eta}_a) + \hat{\sigma}^2 I)^{-1}P(\hat{\eta})\phi(\hat{\eta}_a)^T Y. \tag{19}$$

## 4    Estimation Results

### 4.1    Regularized LBF Volterra Kernels

Laguerre basis function expanded Volterra kernels were estimated for the benchmark system using the methods discussed in Sect. 3. Due to the limited number of estimation samples available, the order of the Volterra series was limited to $M = 2$, and using no prior knowledge of the system, the number of basis functions for each order were chosen to be $B_1 = B_2 = 20$, such that the total number of unique parameters requiring estimation (231) was roughly half the number of data samples available. All non-convex optimization problems were solved using the MATLAB GlobalSearch function and fmincon to find local minima.

The estimated first order kernel ($\hat{\alpha}_1$) was negligible, which is expected since the nonlinearity is an even function. The second order LBF kernel estimate ($\hat{\alpha}_2$) is given in Fig. 5 alongside its time domain equivalent ($\hat{h}_2$). The constant term, $\alpha_0$, was estimated as 0.102.

Applying the validation input to the estimated model produced the output prediction shown in Fig. 6, which is plotted alongside the given validation output. It is clear that although the model contains many parameters, and even though we did not make use of prior knowledge in the estimation, the regularized basis function approach provides a very accurate model which is suitable for prediction or control purposes.

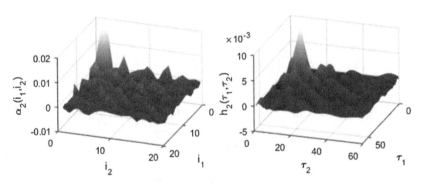

**Fig. 5.** Estimated 2nd order LBF kernel (left) and its time domain equivalent (right)

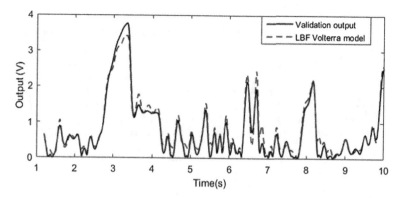

**Fig. 6.** True and model-predicted validation outputs for the LBF Volterra kernels

## 4.2   Comparison with Alternate Methods

In order to provide some context when assessing the performance of the method proposed here, several alternate model classes and methods were also applied to the coupled electric drives benchmark. The other models considered were:

1. Linear state space
2. Linear transfer function
3. NARX
4. Wiener block structure

For each model class, the estimation dataset was used to estimate a model using an appropriate method provided by MATLAB's System Identification toolbox. As in the proposed method, no prior knowledge was used in the estimation, and the model order for each case was chosen based on a cross-validation approach. High order polynomials were used for nonlinear function estimates. Note that Wiener-Hammerstein models cannot be estimated using the toolbox, but this case has already been treated in [14].

The predicted validation output from each model is plotted in Fig. 7, along with the true experimental output (in black). It is clear that for such a short estimation dataset, and without the use of prior knowledge, none of the estimated models provide accurate predictions for the coupled electric drives system. Of the four models, the Wiener structure produced the best validation results, which can be attributed to the Wiener-Hammerstein nature of the true system, where the final linear filter block contains dynamics which could be considered negligible.

The performance of each model class on the validation data was also assessed numerically, using a normalised root mean square error (NRMSE) metric defined by

$$\text{NRMSE} = 1 - \frac{||Y_{val} - Y_{mod}||}{||Y_{val} - \text{mean}(Y_{val})||}, \tag{20}$$

where $Y_{val}$ is the true validation output vector, $Y_{mod}$ is the model-predicted output vector, and $|| \cdot ||$ indicates the 2-norm of a vector. The error metrics for

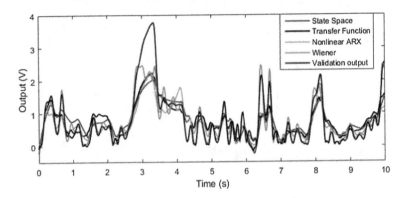

**Fig. 7.** True and model-predicted validation outputs for SysID toolbox methods

**Table 1.** NRMSE in validation for proposed and toolbox methods

| Model | State space | Transfer function | NARX | Wiener | *LBF-Volterra* |
|---|---|---|---|---|---|
| **NRMSE** | 0.402 | 0.457 | 0.446 | 0.639 | 0.811 |

each toolbox method are compared in Table 1 against the method proposed in this paper (italicized), highlighting the superior performance of the regularized nonparametric approach.

## 5    Conclusion

The Volterra series has historically been regarded as an impractical tool for data-driven modeling, due to its nonparametric nature and thus the large numbers of parameters required to be estimated. Using a more compact basis function formulation, however, and regularizing the estimation to impose known properties of smoothness and decay on the estimated basis function kernels, we have shown that the Volterra series *can* be a broad and powerful tool for nonlinear system identification.

In this paper, we considered the modeling of a coupled electric drives system, as a representative of dynamic systems containing nonlinear sensors. Despite using only a short estimation dataset of 500 input-output samples, and without incorporating prior knowledge of the drives system into the modeling process, the estimated basis function Volterra kernels provided accurate predictions of the system under study. Furthermore, the proposed method outperformed linear and nonlinear models provided by the established System Identification toolbox in MATLAB.

# References

1. Birpoutsoukis, G., Marconato, A., Lataire, J., Schoukens, J.: Regularized nonpara-metric Volterra kernel estimation. Automatica **82**, 324–327 (2017)
2. Boyd, S., Chua, L.O.: Fading memory and the problem of approximating nonlinear operators with Volterra series. IEEE Trans. Circ. Syst. **32**(11), 1150–1161 (1985)
3. Campello, R., Favier, G., do Amaral, W.: Optimal expansions of discrete-time Volterra models using Laguerre functions. Automatica **40**, 815–822 (2004)
4. Ljung, L.: System Identification: Theory for the User. Prentice Hall Information and System Sciences Series, Prentice Hall PTR (1999)
5. Pillonetto, G., De Nicolao, G.: A new kernel-based approach for linear system identification. Automatica **46**(1), 81–93 (2010)
6. Pillonetto, G., Dinuzzo, F., Chen, T., De Nicolao, G., Ljung, L.: Kernel meth-ods in system identification, machine learning and function estimation: a survey. Automatica **50**(3), 657–682 (2014)
7. Pintelon, R., Schoukens, J.: System Identification: A Frequency Domain Approach. Wiley, New York (2012)
8. Rugh, W.J.: Nonlinear System Theory: The Volterra-Wiener Approach. Johns Hopkins University Press, Baltimore (1980)
9. Schetzen, M.: The Volterra and Wiener Theories of Nonlinear Systems. Wiley, New York (1980)
10. Stoddard, J.G., Welsh, J.S.: Volterra kernel identification using regularized orthonormal basis functions (2018). https://arxiv.org/abs/1804.07429
11. Stoddard, J.G., Welsh, J.S., Hjalmarsson, H.: EM-based hyperparameter optimiza-tion for regularized Volterra kernel estimation. IEEE Control Syst. Lett. **1**(2), 388–393 (2017)
12. Wahlberg, B.: System identification using Laguerre models. IEEE Trans. Autom. Control **AC–36**(5), 551–562 (1991)
13. Wellstead, P.E.: Introduction to Physical System Modelling. Academic Press, London (1979)
14. Wigren, T., Schoukens, M.: Coupled electric drives data set and reference models. Technical report, Department of Information Technology, Uppsala University (2017). http://www.it.uu.se/research/publications/reports/2017-024/2017-024-nc.pdf
15. Wu, C.F.J.: On the convergence properties of the EM algorithm. Ann. Stat. **11**(1), 95–103 (1983)

# A Novel On-Machine Measurement Method Based on the Force Controlled Touch Probe

Hao Li, Huan Zhao[✉], and Han Ding

State Key Laboratory of Digital Manufacturing Equipment and Technology,
Huazhong University of Science and Technology, Wuhan 430074,
Hubei, People's Republic of China
huanzhao@hust.edu.cn

**Abstract.** For free form surfaces such as blades, their measurement accuracy and efficiency depend on the planning strategy of discrete sampling points and the compensation precision of probe tip error. However, the force information is not considered in the traditional adaptive sampling and error compensation methods, which may affect the measurement efficiency and accuracy. In this paper, a novel force controlled touch probe for on-machine measurement system with a 5-DOF force/moment sensor is proposed. According to the projection of the force on cross section of blade, the proposed adaptive sampling algorithm can automatically select measuring points. Meanwhile, the direction of measuring force is used to estimate the compensation direction for probe tip error. The experiments are conducted on the on-machine measurement system, and the results demonstrate that the proposed methods are effective, which can provide important reference value for free-form surface measurement.

**Keywords:** Force controlled touch probe · On-machine measurement
Adaptive sampling algorithm · Measurement error compensation

## 1 Introduction

Blades with complex surface are the most numerous and the most critical components of aero engines. The freeform surface of blades not only make it difficult to manufacture, but also bring the surfaces measurement complicated. Traditional coordinate measuring machine (CMM) shows high precision in the measurement of complex surface, but it is undeniable that CMM cannot carry out on-machine measurement of the blades, and the reposition error caused by the transformation of coordinate system and the second fixture of the blades is unavoidable. In addition, the measuring path of CMM requires off-line programming of the blade in advance, which would reduce the measurement efficiency especially for different sizes blades. Therefore, it is necessary to develop the on-machine measurement system, which is beneficial to improve the efficiency and accuracy of measurement.

On-machine measurement system and manufacturing equipment constitute a closed loop, which could feedback the measurement results of the workpiece to the processing system and provide allowance compensation for further processing [1]. There are lots

© Springer Nature Switzerland AG 2018
Z. Chen et al. (Eds.): ICRA 2018, LNAI 10984, pp. 486–497, 2018.
https://doi.org/10.1007/978-3-319-97586-3_44

of researches focused on the measurement accuracy of the on-machine measurement system with touch probe which mainly includes three types, based on CNC machine [2, 3], five-axis machine [4, 5], and three-axis machine [6]. The measurement accuracy of complex geometric features on blades depend on not only the kinematic accuracy of measuring equipment, but also the distribution of sampling points and the compensation precision of the probe tip error.

The optimal principle for distribution of sampling points is to obtain the minimum measuring error with the least sampling points. However, the high precision measurement with a CMM requires small scanning steps, which means that many sampling points will be collected. Mansour proposed a method that reduces the number of points and achieves the goal of a certain allowable deviation with the 3rd degree curves [7]. Suleiman M. developed three algorithms for sampling points of freeform surfaces at a patch scale, and collected the most critical points that depend on maximum and minimum Gaussian curvature of each patch, which effectively reduce the number of sampling points [8]. He et al. built the machining error model for freeform surface measuring on CMM and proposed an improved algorithm for determining the distribution of sampling points adaptively [9]. Compared with the four traditional strategies, He's method is more reliable and efficient. The freeform surface of blade is measured section-by-section in the engineering, so the distribution of the sampling points consists of determining the distribution of the cross section and determining the sampling points on each cross section. Jiang proposed a practical sampling method for determining leading edge curve and trailing edge curve of blade based on maximum chordal deviation [10]. For each sectional curves, the chordal deviation criterion was used to determine the sampling points.

The coordinates which were recorded by CMM represent the position of probe tip center. The actual contact point between probe and workpiece is not easy to calculate. In this aspect, there are many schemes are known that can be used to estimate the position of actual contact point. They can be mainly divided into two parts, the main method for compensating the probe tip error is to calculate the geometric relation of the measured points near the target point [11, 12], and the other is to analyze the CAD model of the workpiece [13–15]. However, there are few studies determine the error compensation direction considering the force information. Park et al. first proposed a new touch probe with a 3-DOF force sensor [16]. Liang et al. [17] and Lee et al. [18] then developed new probes and compensation method using 5-DOF force/moment sensor for CMM. The direction of the measuring force and elastic deformation of probe were considered in these papers, but they focused on developing force sensor.

In this study, in order to improve the measurement efficiency and accuracy of blade, a novel on-machine measurement method based on the force controlled touch probe equipped with 5-DOF force/moment sensor is proposed. The measuring force between probe and blade is controlled as a constant force by force controlled touch probe. The adaptive sampling algorithm is proposed to select the measuring points on cross section of blade based on force information. For the compensation of measurement error, the direction of measuring force is regarded as compensation vector direction of probe tip error. Finally, the experimental verifications are conducted on the on-machine measurement system.

The remaining sections of this paper is organized as follows: Sect. 2 proposes a method of measurement planning based on force information. The compensation algorithm for probe tip error is expressed in Sect. 3. Section 4 carries out experimental verification for the proposed method. Finally, some conclusions are given in Sect. 5.

## 2    Measurement Path Planning

### 2.1    Force Controlled Touch Probe

Contact trigger probes are widely used in CMM, when the probe contact with the workpiece, the circuit of probe will send a trigger signal to the CMM for sampling. However, contact trigger probe cannot provide effective information for the position of actual contact point, which is extremely important for the probe tip's error compensation. In this study, the proposed force controlled touch probe is used to collect the measuring force between the probe and the workpiece, as shown in Fig. 1(a).

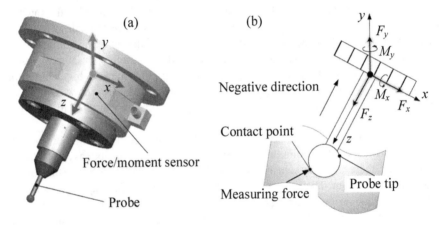

**Fig. 1.**  (a) Proposed force controlled touch probe; (b) Model of measuring force for probe

The force controlled touch probe can be simplified as a rigid cantilever beam with an ideal sphere tip, as shown in Fig. 1(b). We assume that the probe contact with the workpiece at a point. Thus the measuring force points to the ideal sphere center from the contact point. The direction of measuring force is same as the surface normal direction, and the force is equivalent to the orthogonal force on the three axis, which can be measured by a high precision force/moment sensor. Thus the measuring force can be calculated by

$$\vec{F} = \vec{F}_x + \vec{F}_y + \vec{F}_z \tag{1}$$

$$|F| = \sqrt{\left|F_x\right|^2 + \left|F_y\right|^2 + \left|F_z\right|^2} \tag{2}$$

where $\vec{F}$ represents the measuring force. $\vec{F}_x$, $\vec{F}_y$ and $\vec{F}_z$ represents the force vector of each axis of force/moment sensor, respectively.

The force controlled touch probe is mounted on the Z-axis of CMM. Therefore, the measuring force can be kept constant such as $-0.5$ N by controlling the movement of Z-axis. When the measuring force is less than $-0.5$ N, the force controlled touch probe moves away from the surface of workpiece, and otherwise the probe closes to the surface until the measuring force is approximate to $-0.5$ N. As shown in Fig. 2, the PID control algorithm is used to control the measuring force and the movement of Z-axis.

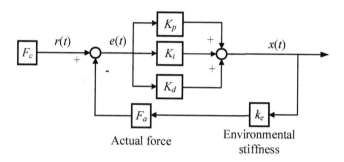

**Fig. 2.** PID force control algorithm

## 2.2 Adaptive Sampling Algorithm

The contour of blade is constructed by a group of 2D curves. As shown in Fig. 3, the leading edge point and the trailing edge point divide each 2D curve into two parts, back curve and abdominal curve. The back curve is the convex curve near the suction side, and the abdominal curve is the concave curve near the pressure side.

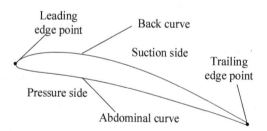

**Fig. 3.** Structure of 2D curve for blade

The design process for blade is to construct different 2D curves into freeform surface [9], as shown in Fig. 4. In this study, the reverse process of design is used to measure blade. First, a series of equidistant planes perpendicular to the Z-axis are used to divide the blade, the intersection curves of planes and blades are the cross sections for measuring. Then, the adaptive sampling algorithm based on force information is proposed to distribute the measuring points on each cross section.

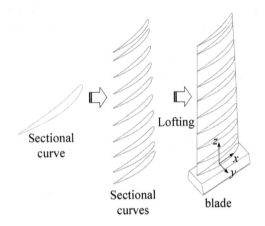

**Fig. 4.** Design process of blade

Traditional sampling method of measurement is shown in Fig. 5(a). The measurement path can be expressed as "S1-S2-S3-S4-S5-S6". The probe tip must move to the retract points S3 and S6. The acceleration and deceleration of probe all the time will reduce the measuring efficiency of CMM. In this paper, the probe always contacts with the workpiece and scans the trajectory slowly, thus the point positon can be measured at any position, as shown in Fig. 5(b).

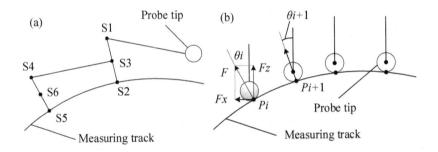

**Fig. 5.** Sampling points distribution: (a) traditional method; (b) constant force contact

The measuring force is used to select feature points on each curve. The probe is always in the same direction as the Z axis, thus the measuring force projected onto the cross section can be decomposed into $F_x$ and $F_z$.

Let $\theta_i(i = 1, 2, \cdots)$ be the angle between probe and the projection of measuring force on the cross section, and then

$$\theta_i = \arctan(F_x / F_z) \tag{3}$$

The curvature of the measuring track on the blade surface is reflected by the force information. When the probe moves on a measuring trajectory with curvature, the direction of the measuring force will change continuously.

Set an angle threshold $\bar{\theta}$, thus if the changed angle $\Delta\theta_i$ between the point $P_i$ and the point $P_{i+j}$ is less than $\bar{\theta}$, $\Delta\theta_i$ can be calculated by

$$\Delta\theta_i = \left|\theta_i - \theta_{i+j}\right| (j = 0, 1, 2, \cdots) \tag{4}$$

If $\Delta\theta_i$ is equal to $\bar{\theta}$, it means a new sampling point is found, and then $i, j$ and $\Delta\theta_i$, need to be recalculated.

$$i = i + j; j = 0; \Delta\theta_i = 0 \tag{5}$$

In this method, the smaller curvature of measuring curves with less sampling points. As the curvature increases, the changed angle between measuring points will also increase, and more sampling points will be found.

The obvious advantage of the proposed method is that the sampling points don't need to be planned in advance. The sampling points are adaptively collected according to the force information by force/moment sensor in the measurement process.

## 3 Measurement Error Compensation

The system error of CMM is composed of the probe tip error and the elastic deformation. When the probe contacts with the workpiece, the position recorded by CMM represents the location of the probe tip center. The actual contact point between probe and workpiece is difficult to estimate. In addition, the elastic deformation of probe caused by measuring force cannot be ignored.

The probe can be simplified into a cantilever beam with an ideal spherical tip, as shown in Fig. 6. The deviation between actual position $P_a$ and ideal position $P_d$ of probe tip center is expressed as $e_1$, and the probe tip error between actual position $P_a$ and actual contact point $P_i$ is expressed as $e_2$.

In this study, the direction of measuring force is used to estimate the compensation vector direction. The ruby probe with high stiffness can be considered as an ideal sphere. The measuring force points to the actual position $P_a$ from actual contact point $P_i$. Thus the measuring force direction can be regarded as the direction of the error compensation vector, and the error compensation vector can be calculated by

$$\vec{R} = |R| \cdot \vec{n} \tag{6}$$

where $|R|$ is the effective radius of probe tip, and $\vec{n}$ is the unit vector of measuring force which can be calculated by

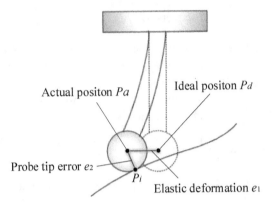

**Fig. 6.** Elastic deformation and probe tip error

$$\vec{n} = \frac{\vec{F}}{|F|} = \frac{\vec{F}_x + \vec{F}_y + \vec{F}_z}{\sqrt{F_x^2 + F_y^2 + F_z^2}} \tag{7}$$

Projecting the compensation vector to the three axes, the compensation value on each axis can be calculated by

$$\begin{cases} M_x = F_z \cdot y_b - F_y \cdot (L + z_b) \\ M_y = F_x \cdot (L + z_b) + F_z \cdot x_b \\ x_b^2 + y_b^2 + z_b^2 = |R|^2 \end{cases} \tag{8}$$

where $(x_b, y_b, z_b)$ represent the compensation value on each axis, and $L$ is the distance from probe tip center to force/moment sensor, $F_x$, $F_y$, $F_z$, $M_x$, and $M_y$ are the force information of force/moment sensor.

In addition, the elastic deformation $e_1$ is caused by measuring force, thus the value can be calculated by the three orthogonal forces from force/moment sensor, as follows

$$e_1 = [e_x, e_y, e_z]^T = KF \tag{9}$$

where $e_x$, $e_y$ and $e_z$ represent the deformation of each axis respectively. $K \in R^{3\times3}$ is the stiffness matrix of probe, and $F = [F_x, F_y, F_z]$.

## 4    Experiments

With the development of industrial automation technology, robot machining has received much attention. In this study, the blade is polished by a robot system. When the polishing is completed, the blade moves to the detection station for measurement. The robot end-effector is replaced by a rotary axis in this paper.

As shown in Fig. 7(a), the on-machine measurement equipment consists of three translational axes and a rotary axis. The three FAGOR grating rulers with 2500 pulses/mm are used to measure the displacement of each translational axis, while the displacement of the rotary axis is obtained by the encoder at the back of servo motor with a reduction ratio of 72:1. The force controlled touch probe contains a Renishaw probe with a $\phi 3$ mm ruby sphere and an ATI Gamma force/moment sensor. The process of blade measuring is detailed as shown in Fig. 7(b).

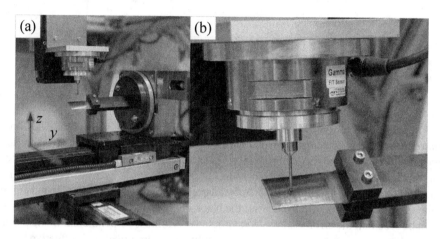

**Fig. 7.** Experimental equipment and measurement process. (a) On-machine measurement equipment (b) Blade measurement process

The measuring force between blade and the probe is measured by force/moment sensor with the sampling frequency of 1000 Hz. The force signal is used to not only select the sampling points, but also compensate the probe tip error and the elastic deformation. There are various kinds of noise in the original force signal, such as electromagnetic interference and motor vibration. Therefore, the collected force signal need to be pretreated at first.

In this study, the moving average filter is used to smooth the original force signal. The continuous sampling data is regarded as a queue with length of $N$ in the moving average filter algorithm which is based on statistical principle. When the new data is measured, the first data of the above queue is removed, and the rest $N - 1$ data move forward in turn, and the new sampling data is inserted as the tail of the new queue. The mean value of the new sequence is the measurement result at the point. The filtered force signals can be calculated by

$$\bar{f}_t = \frac{1}{n} \sum_{i=1}^{i=n} f_{t-i} \tag{10}$$

where $\bar{f}_t$ indicates the force signal after filtering, and $f_{t-i}$ is the original force signal, $n$ represents the width of the filter. And the contrast before and after the filtering is shown in Fig. 8.

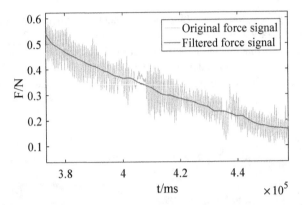

**Fig. 8.** Original force signal and filtered force signal

First, the back curve of one cross section on the blade is measured. The force controlled touch probe keeps a steady contact with the blade, and the adaptive sampling algorithm is used to find the feature points. As shown in Fig. 9, the compensation direction represents the projection of measuring force on the cross section. The sampling points mean the feature points extracted by adaptive sampling algorithm. The red curve is the trajectory of probe tip center and the black curve is the contour of blade surface. It can be seen that the error compensation direction is almost perpendicular to the probe tip center trajectory, which means that the adaptive sampling method is effective.

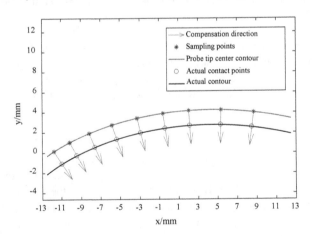

**Fig. 9.** Measurement error compensation (Color figure online)

In addition, the repeatability precision is an important criterion for evaluating the stability of mechanical system. To guarantee the accuracy of the proposed on-machine measurement method and the effectiveness of the force controlled touch probe, the repeatability precision of the on-machine measurement system is tested first. The back curve on the same cross section of blade is measured three times, and the contour profile coordinates are plotted in Fig. 10. The three curves represent the results of three measurements respectively, and the local magnification shows the maximum deviation of the three curves. It can be seen that the repeated precision of the on-machine measurement equipment is less than 0.05 mm, which meets the measurement requirements for blade. Thus, the on-machine measurement system is stable.

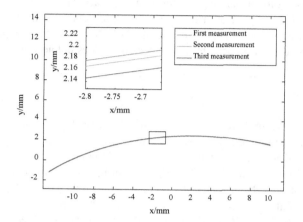

**Fig. 10.** Experimental results of repeated precision

Finally, the experimental results of on-machine measurement are compared with the 3D model of the blade and the measurement results of the three coordinate measuring machine. As shown in Fig. 11, the model sampling points are planned by CMM system software on the 3D model of blade, where the CMM measuring points and the OMM measuring points are the measurement result obtained by three coordinate measuring machine and on-machine measurement system, respectively. The three curves are obtained by spline fitting of sampling points and measuring points.

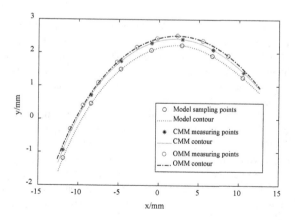

**Fig. 11.** Comparison of measurement results with CMM and 3D model

By comparison, we can see that the number of sampling points planned by CMM methods is similar to the proposed adaptive method. For CMM, since the blade is not processed into finished products, and there is still allowance on its surface, the results of CMM are deviate from the desired model contour; for the proposed on-machine measurement system, since there exists systematic error of coordinate transformation, the motion error of the measuring platform and the calibration error of the probe, the results of the proposed measurement system are also deviate from the desired model contour, and may even be worse than the results of CMM. However, the proposed adaptive sampling algorithm is easier to realize because it does not need to plan in advance.

## 5  Conclusions

A novel on-machine measurement method using force controlled touch probe with a 5-DOF force/moment sensor is proposed. The measuring force between probe and blade can be controlled as a constant force by using PID control algorithm. The adaptive sampling algorithm can automatically select measuring points when the changed angle of measuring force on the cross section is equal to the angle threshold, and the number of measuring points selected by the proposed method is similar to the CMM system software. The actual contact point of probe and blade can be estimated by regarding the direction of measuring force as the compensation direction of probe tip error. Finally, the experiments are conducted on the on-machine measurement system, and the results show that the proposed adaptive sampling method and the error compensation principle are effective.

**Acknowledgement.** This work was supported by the National Key Research and Development Program of China under Grant No. 2017YFB1303401.

## References

1. Eldessouky, H., Newman, S., Nassehi, A.: Closed loop CNC machining and inspection of interlinked manufacturing features for "Right First Time" production. In: 25th International Conference of Flexible Automation Integrated FAIM 2015. University of Bath (2015)
2. Kim, K.D., Chung, S.C.: Accuracy improvement of the On-Machine inspection system by correction of geometric and transient thermal errors. Technical Papers-Society of Manufacturing Engineers-All Series- (2003)
3. Jung, J.H., Choi, J.P., Lee, S.J.: Machining accuracy enhancement by compensating for volumetric errors of a machine tool and on-machine measurement. J. Mater. Process. Technol. **174**(1–3), 56–66 (2006)
4. Huang, N., Bi, Q., Wang, Y., Sun, C.: 5-axis adaptive flank milling of flexible thin-walled parts based on the on-machine measurement. Int. J. Mach. Tools Manuf. **84**, 1–8 (2014)
5. Ibaraki, S., Iritani, T., Matsushita, T.: Calibration of location errors of rotary axes on five-axis machine tools by on-the-machine measurement using a touch-trigger probe. Int. J. Mach. Tools Manuf. **58**, 44–53 (2012)

6. Choi, J.P., Min, B.K., Lee, S.J.: Reduction of machining errors of a three-axis machine tool by on-machine measurement and error compensation system. J. Mater. Process. Technol. **155**, 2056–2064 (2004)

7. Mansour, G.: A developed algorithm for simulation of blades to reduce the measurement points and time on coordinate measuring machine (CMM). Measurement **54**, 51–57 (2014)

8. Obeidat, S.M., Raman, S.: An intelligent sampling method for inspecting free-form surfaces. Int. J. Adv. Manuf. Technol. **40**(11–12), 1125–1136 (2009)

9. He, G., Sang, Y., Pang, K., Sun, G.: An improved adaptive sampling strategy for freeform surface inspection on CMM. Int. J. Adv. Manuf. Technol. **96**, 1521–1535 (2018)

10. Jiang, R.S., Wang, W.H., Zhang, D.H., Wang, Z.Q.: A practical sampling method for profile measurement of complex blades. Measurement **81**, 57–65 (2016)

11. Wozniak, A., Mayer, J.R.R.: A robust method for probe tip radius correction in coordinate metrology. Meas. Sci. Technol. **23**(2), 025001 (2011)

12. Lai, J., Fu, J., Wang, Y., Shen, H., Xu, Y., Chen, Z.: A novel method of efficient machining error compensation based on NURBS surface control points reconstruction. Mach. Sci. Technol. **19**(3), 499–513 (2015)

13. Cho, M.W., Seo, T.I.: Inspection planning strategy for the on-machine measurement process based on CAD/CAM/CAI integration. Int. J. Adv. Manuf. Technol. **19**(8), 607–617 (2002)

14. He, G., Huang, X., Ma, W., Sang, Y., Yu, G.: CAD-based measurement planning strategy of complex surface for five axes on machine verification. Int. J. Adv. Manuf. Technol. **91**(5–8), 2101–2111 (2017)

15. Han, D.Y.L.D.Z., Juan, W.W.W.: CAD model-based intelligent inspection planning for coordinate measuring machines. Chin. J. Mech. Eng. **24**(4), 1 (2011)

16. Park, J.J., Kwon, K., Cho, N.: Development of a coordinate measuring machine (CMM) touch probe using a multi-axis force sensor. Meas. Sci. Technol. **17**(9), 2380 (2006)

17. Liang, Q., Zhang, D., Wang, Y., Ge, Y.: Development of a touch probe based on five-dimensional force/torque transducer for coordinate measuring machine (CMM). Robot. Comput. Integr. Manuf. **28**(2), 238–244 (2012)

18. Lee, M., Cho, N.G.: Probing-error compensation using 5 degree of freedom force/moment sensor for coordinate measuring machine. Meas. Sci. Technol. **24**(9), 095001 (2013)

# Author Index

Printed in the United States
By Bookmasters